AF509021

Performance-Related Material Properties of Asphalt Mixture Components

Performance-Related Material Properties of Asphalt Mixture Components

Editors

Meng Ling
Yao Zhang
Haibo Ding
Yu Chen

MDPI • Basel • Beijing • Wuhan • Barcelona • Belgrade • Manchester • Tokyo • Cluj • Tianjin

Editors

Meng Ling
School of Civil and
Transportation Engineering
Beijing University of Civil
Engineering and Architecture
Beijing
China

Yao Zhang
College of Civil Science and
Engineering
Yangzhou University
Yangzhou
China

Haibo Ding
School of Civil Engineering
Southwest Jiaotong
University
Chengdu
China

Yu Chen
School of Transportation and
Logistics Engineering
Wuhan University of
Technology
Wuhan
China

Editorial Office
MDPI
St. Alban-Anlage 66
4052 Basel, Switzerland

This is a reprint of articles from the Special Issue published online in the open access journal *Materials* (ISSN 1996-1944) (available at: www.mdpi.com/journal/materials/special_issues/asphalt_mixtures_components).

For citation purposes, cite each article independently as indicated on the article page online and as indicated below:

LastName, A.A.; LastName, B.B.; LastName, C.C. Article Title. *Journal Name* **Year**, *Volume Number*, Page Range.

ISBN 978-3-0365-7427-1 (Hbk)
ISBN 978-3-0365-7426-4 (PDF)

Contents

Preface to "Performance-Related Material Properties of Asphalt Mixture Components" . . . . **vii**

Sudi Wang, Weixiao Yu, Yinghao Miao and Linbing Wang
Review on Load Transfer Mechanisms of Asphalt Mixture Meso-Structure
Reprinted from: *Materials* **2023**, *16*, 1280, doi:10.3390/ma16031280 . . . . **1**

Yao Zhang, Ye Wang, Aihong Kang, Zhengguang Wu, Bo Li and Chen Zhang et al.
Analysis of Rejuvenating Fiber Asphalt Mixtures' Performance and Economic Aspects in
High-Temperature Moisture Susceptibility
Reprinted from: *Materials* **2022**, *15*, 7728, doi:10.3390/ma15217728 . . . . **21**

Yiming Li and Simon A. M. Hesp
Enhanced Acceptance Specification of Asphalt Binder to Drive Sustainability in the Paving
Industry
Reprinted from: *Materials* **2021**, *14*, 6828, doi:10.3390/ma14226828 . . . . **47**

Zeyu Zhang, Julian Kohlmeier, Christian Schulze and Markus Oeser
Concept and Development of an Accelerated Repeated Rolling Wheel Load Simulator
(ARROWS) for Fatigue Performance Characterization of Asphalt Mixture
Reprinted from: *Materials* **2021**, *14*, 7838, doi:10.3390/ma14247838 . . . . **61**

Wenqi Wang, Ali Rahman, Haibo Ding and Yanjun Qiu
Assessment of Aging Impact on Wax Crystallization in Selected Asphalt Binders
Reprinted from: *Materials* **2022**, *15*, 8248, doi:10.3390/ma15228248 . . . . **87**

Gang Zhou, Chuanqiang Li, Haobo Wang, Wei Zeng, Tianqing Ling and Lin Jiang et al.
Preparation of Wax-Based Warm Mixture Additives from Waste Polypropylene (PP) Plastic and
Their Effects on the Properties of Modified Asphalt
Reprinted from: *Materials* **2022**, *15*, 4346, doi:10.3390/ma15124346 . . . . **101**

Xueyang Jiu, Peng Xiao, Bo Li, Yu Wang and Aihong Kang
Pavement Performance Investigation of Asphalt Mixtures with Plastic and Basalt Fiber
Composite (PB) Modifier and Their Applications in Urban Bus Lanes Using Statics Analysis
Reprinted from: *Materials* **2023**, *16*, 770, doi:10.3390/ma16020770 . . . . **115**

Yan Li, Weian Xuan, Ali Rahman, Haibo Ding and Bekhzad Yusupov
Effects of Geometry and Loading Mode on the Stress State in Asphalt Mixture Cracking Tests
Reprinted from: *Materials* **2022**, *15*, 1559, doi:10.3390/ma15041559 . . . . **137**

Zhimin Ma, Xudong Wang, Yanzhu Wang, Xingye Zhou and Yang Wu
Long-Term Performance Evolution of RIOHTrack Pavement Surface Layer Based on DMA
Method
Reprinted from: *Materials* **2022**, *15*, 6461, doi:10.3390/ma15186461 . . . . **151**

Xiaolong Li, Junan Shen, Tianqing Ling and Qingbin Mei
Pyrolysis Combustion Characteristics of Epoxy Asphalt Based on TG-MS and Cone Calorimeter
Test
Reprinted from: *Materials* **2022**, *15*, 4973, doi:10.3390/ma15144973 . . . . **169**

Yanzhu Wang, Xudong Wang, Zhimin Ma, Lingyan Shan and Chao Zhang
Evaluation of the High- and Low-Temperature Performance of Asphalt Mortar Based on the
DMA Method
Reprinted from: *Materials* **2022**, *15*, 3341, doi:10.3390/ma15093341 . . . . **183**

Xu Liu, Mo Zhang and Wanqiu Liu
The Relationship between Poisson's Ratio Index and Deformation Behavior of Asphalt Mixtures
Tested through an Optical Fiber Bragg Grating Strain Sensor
Reprinted from: *Materials* **2022**, *15*, 1882, doi:10.3390/ma15051882 . . . . **195**

Wenqi Wang, Azuo Nili, Ali Rahman and Xu Chen
Effects of Wax Molecular Weight Distribution and Branching on Moisture Sensitivity of Asphalt
Binders
Reprinted from: *Materials* **2022**, *15*, 4206, doi:10.3390/ma15124206 . . . . **217**

**Rongyan Tian, Haoyuan Luo, Xiaoming Huang, Yangzezhi Zheng, Leyi Zhu and Fengyang
Liu**
Correlation Analysis between Mechanical Properties and Fractions Composition of
Oil-Rejuvenated Asphalt
Reprinted from: *Materials* **2022**, *15*, 1889, doi:10.3390/ma15051889 . . . . **233**

Runmin Zhao, Jinzhi Gong, Yangzezhi Zheng and Xiaoming Huang
Research on the Combination of Firefly Intelligent Algorithm and Asphalt Material Modulus
Back Calculation
Reprinted from: *Materials* **2022**, *15*, 3361, doi:10.3390/ma15093361 . . . . **263**

Binshuang Zheng, Junyao Tang, Jiaying Chen, Runmin Zhao and Xiaoming Huang
Investigation of Adhesion Properties of Tire—Asphalt Pavement Interface Considering
Hydrodynamic Lubrication Action of Water Film on Road Surface
Reprinted from: *Materials* **2022**, *15*, 4173, doi:10.3390/ma15124173 . . . . **285**

Preface to "Performance-Related Material Properties of Asphalt Mixture Components"

This reprint contains 1 review paper and 15 scientific papers which focus on the most recent research in pavement engineering, which mainly addresses the performance-related properties of asphalt mixtures and asphalt pavements. We hope that this reprint is helpful for pavement researchers and pavement industries to generate a better and deeper understanding of pavement materials, pavement performance, and pavement evaluation. The authors would like to express the acknowledgment of the assistance from the publisher and editors for their support.

Meng Ling, Yao Zhang, Haibo Ding, and Yu Chen
Editors

Review

Review on Load Transfer Mechanisms of Asphalt Mixture Meso-Structure

Sudi Wang [1], Weixiao Yu [1], Yinghao Miao [1,*] and Linbing Wang [1,2,*]

1 National Center for Materials Service Safety, University of Science and Technology Beijing, Beijing 100083, China
2 School of Environmental, Civil, Agricultural and Mechanical Engineering, University of Georgia, Athens, GA 30602, USA
* Correspondence: miaoyinghao@ustb.edu.cn (Y.M.); linbing.wang@uga.edu (L.W.)

Abstract: Asphalt mixture is a skeleton filling system consisting of aggregate and asphalt binder. Its performance is directly affected by the internal load transfer mechanism of the skeleton filling system. It is significant to understand the load transfer mechanisms for asphalt mixture design and performance evaluation. The objective of this paper is to review the research progress of the asphalt mixture load transfer mechanism. Firstly, this paper summarizes the test methods used to investigate the load transfer mechanism of asphalt mixtures. Then, an overview of the characterization of load transfer mechanism from three aspects was provided. Next, the indicators capturing contact characteristics, contact force characteristics, and force chain characteristics were compared. Finally, the load transfer mechanism of asphalt mixtures under different loading conditions was discussed. Some recommendations and conclusions in terms of load transfer mechanism characterization and evaluation were given. The related work can provide valuable references for the study of the load transfer mechanism of asphalt mixtures.

Keywords: asphalt mixture; load transfer mechanism; testing method; contact force; force chain

1. Introduction

Asphalt mixture is a typical composite material comprising aggregates, asphalt binders, fillers, and voids. The extrinsic uncertainty, irregularity, vagueness, and nonlinearity of asphalt mixtures' properties (such as mechanical properties) are the reflection of its microstructural complexity. The skeleton structure is regarded as the main body to transfer external loads in asphalt mixture and the aggregate skeleton contributes the most to the high temperature stability of asphalt mixture. Therefore, the performance of asphalt mixtures can be evaluated from the load transfer mechanism of the materials. The study of the load transfer mechanism of the meso-structure of asphalt mixture provides a new perspective for the study of classical problems, and can also induce new understanding. In the past decades, a variety of methods have been introduced to study the load transfer mechanisms, such as the photoelastic experiment method, digital speckle method, digital image processing (DIP) technology, discrete element method (DEM) simulation, finite element method (FEM) simulation, and the combination of multiple methods [1–12].

The load transfer mechanism mainly refers to the characteristics of load transfer in an asphalt mixture under external load [5,13]. Aggregates in contact constitute the skeleton of asphalt mixtures and affect load transmission in the mixture, which determines its deformation resistance [14,15]. Meanwhile, the load transfer characteristics can be used to predict the load-bearing capacity of the aggregates and asphalt mixtures. Recent studies of the load transfer mechanism in asphalt mixture focus on three aspects: contact mechanical behavior, contact force characteristics, and force chain characteristics. The research on contact mechanical behavior of asphalt mixture is mainly focused on analyzing the aggregate contact, such as aggregate contact zone [16], aggregate contact chain

Citation: Wang, S.; Yu, W.; Miao, Y.; Wang, L. Review on Load Transfer Mechanisms of Asphalt Mixture Meso-Structure. *Materials* **2023**, *16*, 1280. https://doi.org/10.3390/ma16031280

Academic Editor: Simon Hesp

Received: 27 November 2022
Revised: 29 January 2023
Accepted: 1 February 2023
Published: 2 February 2023

structure [2,17–19], and aggregate contact number characteristics [20,21]. For contact force characteristics, the most common studies are statistical analysis of contact force, including contact force distribution characteristics [9,22,23] and contact force evolution characteristics in the compaction process [5,24]. Studies of force chain fall into two categories, force chains identification criteria, including force chains length [25,26] and force chains number [27,28], and force chains structural characteristics [6,29]. From the existing studies, it is found that the contact skeleton structure of asphalt mixture is the load transfer foundation, and the contact force and force chain characteristics can reflect load transfer mechanism of skeleton structure of asphalt mixtures. Meanwhile, massive research attention has been given to develop the evaluation parameters of load transfer mechanisms according to load transfer characteristics, such as contact number, contact chain structure, load-bearing contributions (based on average contact force), and force chains morphologies (force chains length, force chains angle, and force chains number) [5,16,21,28].

There are different load transfer mechanisms of asphalt mixtures under different test states, from compaction to service to destruction. The common compaction methods are the Marshall impact compaction (MIC) method [30], rotary compaction experimental method [31], and field compaction [32,33]. Exploring the load transfer mechanisms in different compaction processes provides references for evaluating the stability of the load-bearing structure of asphalt mixtures after compaction. Meanwhile, the load transfer mechanism of asphalt mixtures under different external load conditions, such as tension–compression conditions [34], shear conditions [35], and bending conditions [36], is the mechanical response of asphalt mixture to different field service conditions, which can provide the mechanical explanation of the macro-scale damage of asphalt mixture from the meso-scale mechanical mechanism.

The objectives of this review are the following, and the flowchart of this paper is shown in Figure 1.

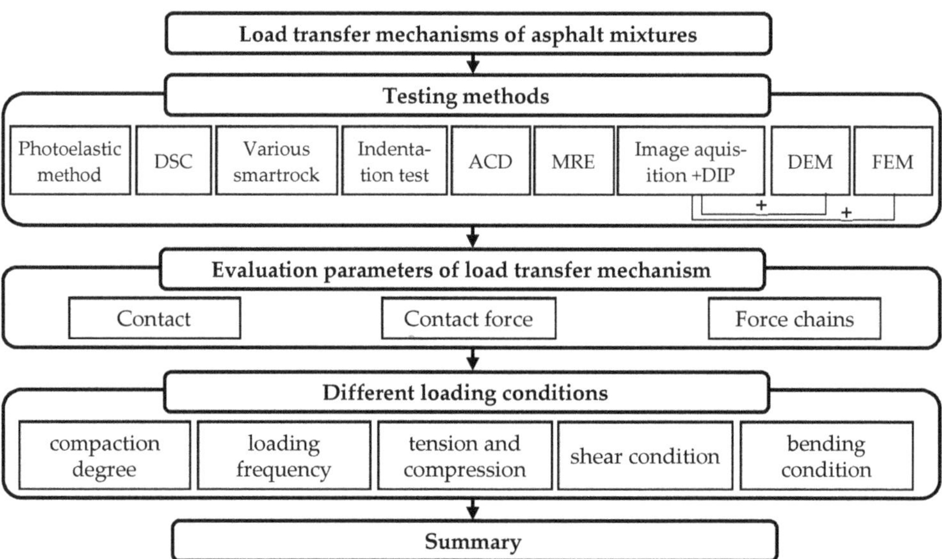

Figure 1. Flowchart of this paper. Note: DEM—Discrete element method; FEM—Finite element method; DIP—Digital image processing; DSC—Digital speckle correlation; ACD—Aggregate contact device; MRE—Magnetic Resonance Elastography; Various smartrock—includes Intelligent Aggregate, Intelligent Attitude Aggregate, and SmartRock.

(1) Present the comprehensive review of the test methods and techniques for the load transfer mechanism of asphalt mixtures.

(2) Collect and discuss the characteristic parameters of the load transfer mechanism.

(3) Compare the load transfer mechanism of asphalt mixtures under different loading conditions including compaction degree, different loading frequency, tension–compression, shear condition, and bending condition.

(4) Provide recommendations to select and improve the load transfer characteristics and evaluation parameters of the load transfer mechanism in asphalt mixtures.

This paper mainly reviews and summarizes the meso-scale load transfer (mainly referring to contact structure, contact force, and force chain) and related evaluation indicators to provide a theoretical reference for the investigation of the macro-scale properties of asphalt mixture at the level of meso-scale mechanical mechanisms.

For this review, the search papers across the three databases, Web of Science (WoS), Scopus, and China national knowledge infrastructure (CNKI), were used. The keywords searched mainly include asphalt mixture, meso-scale structure, load transfer, contact, contact force, force chain, pavement, aggregate blend, DEM, and FEM, etc. Most of the selected articles in this paper are published as journal articles.

2. Definitions and Test Methods for Load Transfer Mechanism

2.1. Definition

For the load transfer mechanism, the most common studies fall into three categories: contact [3], contact force [5], and force chain [13]. A schematic diagram is shown in Figure 2. The corresponding concepts are defined as follows.

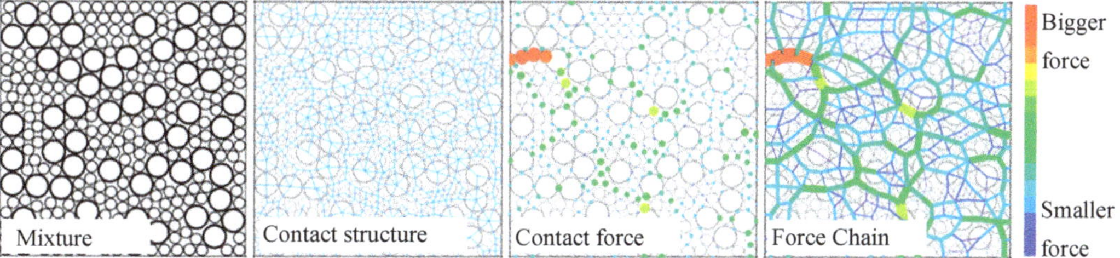

Figure 2. Load transfer characterizations of asphalt mixture: contact (blue line), contact force (colorful sphere), force chain (colorful line).

(1) Contact: This reflects the contact information between particles in asphalt mixtures. The contact skeleton of load transfer is formed between particles. Typically, the total contact number in asphalt mixtures can be defined by Equation (1) [5].

$$
\begin{bmatrix}
C_{m\sim m} & C_{m\sim(m+\Delta s_1)} & \cdots & C_{m\sim M} \\
 & C_{(m+\Delta s_1)\sim(m+\Delta s_1)} & \cdots & C_{(m+\Delta s_1)\sim M} \\
 & & \ddots & \vdots \\
 & & & C_{M\sim M}
\end{bmatrix} = C
\tag{1}
$$

where $C_{i\sim j}$ represents the contact number between particles of A_i and A_j, m is the smallest particle size in the aggregate blend, M is the largest particle size in the aggregate blend, and ΔS_i is the gap between adjacent sieve sizes; i represents the particle size, $i = 1, 2, \ldots, M$.

(2) Contact force: This represents the force transferred between particles through each contact point.

(3) Force chain: This presents the path of transferring external loading in asphalt mixtures and consists of the contact between particles and contact force.

2.2. Test Methods and Comparative Evaluation

There is no standardized test method to assess the load transfer mechanism of asphalt mixtures. According to the previous study, the test methods of load transfer mechanism characterizations can be divided into simulation methods and laboratory test methods. The basic information regarding the corresponding methods to investigate the load transfer mechanism is presented in Table 1. The relationship between the various test methods and the types of load transfer mechanisms is shown in Figure 3.

Table 1. Test methods for characterizing load transfer mechanisms.

No.	Method	Main Features	Obtain	Studies
		Laboratory test method		
1	Charge coupled device camera (CCD) or Computed tomography (CT) + DIP	[8]	Contact number; contact chain.	[8,17,37]
2	Photoelastic experiment		Contact number; contact chain; force chain.	[38–41]
3	Digital speckle correlation (DSC)		Strain state.	[42,43]
4	Intelligent aggregate (20 mm)	[44]	Stress at one point.	[44]
5	SmartRock	[45]	Change of contact structure; stress at one point.	[33,45]
6	Intelligent Attitude Aggregate (IAA).	[46]	Change of contact structure.	[46]
7	Indentation test	–	Contact force.	[9,47]
8	Aggregate contact device (ACD)	–	Contact structure.	[48]
9	Magnetic Resonance Elastography (MRE)	–	Contact structure.	[49]

Table 1. *Cont.*

No.	Method	Main Features	Obtain	Studies
		Simulation test method		
9	Discrete element method (DEM)	[5]	Contact number; contact chain; contact force; force chain.	[5,7,28]
10	Finite element method (FEM)	[50]	Contact chain; force chain.	[2,3,50,51]
11	DIP and numerical simulation	[52]	Contact number; contact chain; contact force; force chain.	[52,53]

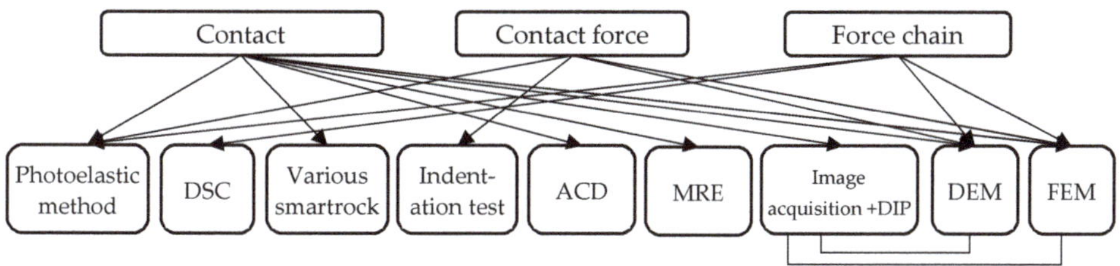

Figure 3. The relationship between test methods and load transfer mechanism characteristics: Various smartrock includes Intelligent Aggregate, Intelligent Attitude Aggregate, and SmartRock.

For the laboratory test methods, the use of image acquisition and DIP to analyze the contact characteristics of asphalt mixtures is more accurate [8,17,37], but the test workload and expenses are large. Furthermore, the load transfer mechanism cannot be well characterized. The photoelastic test can be used to analyze both contact characteristics and contact forces as well as the force chain [38–41]. However, there are high material property requirements and cost limitations for the photoelastic particles, and the photoelastic test limits the study to 2D. The digital speckle correlation test is applicable to the analysis of strain variation of the specimen cross-section [42,43]. Various smartrocks, including Intelligent Aggregate [44], Intelligent Attitude Aggregate [46], and SmartRock [33,45], are only applicable to a point in the asphalt mixture, and the test results have limitations due to the relatively few collected data. The indentation test can be used to characterize the magnitude and distribution of contact forces on a single contact surface, but it cannot characterize the load transfer in the whole blend structure. The Aggregate contact device (ACD) equipment can be used to interpret the relationship between contact structure changes and energy dissipation during compaction, while it cannot be used to characterize load transfer [48]. For the Magnetic Resonance Elastography (MRE) method, the contact structure of the mixture under external loading can be imaged using the MRE pulse sequence [49]. The application of the MRE method requires high technical requirements, and the contact structure obtained has certain errors because the data processing accuracy cannot be reached. For the numerical simulation test methods, such as the finite element method [2,3,50,51] and discrete element method [5,7,28], it is more convenient to analyze the influence of a single factor and to exclude the cross influence of many factors on the load transfer mechanisms. It can also reduce time consumption and cost, using the numerical simulation test method. The model parameters or material characteristics must be determined, but this can be challenging. Generally, the model parameters or material properties used in the aforementioned studies were obtained through reverse modeling rather than directly from experimental data. From the above analysis, there are some limitations when laboratory test methods are used to characterize certain load transfer mechanisms. Numerical simulation tests are more suitable for characterizing the load transfer mechanism of asphalt mixtures. Furthermore, the combination of DIP and a numerical simulation method, which is the recommended method to be used in current and future research, can provide a better way to investigate the load transfer mechanism of an asphalt mixture [52,53]. Based on the meso-scale structure of asphalt mixture obtained by the DIP method, the asphalt mixture load transfer characteristics can be studied by a numerical simulation. It should be noted that for the study of load transfer characteristics, there is no feasible method to obtain the load transfer (contact structure, contact force, or force chain) in real three-dimensional mixtures.

3. Characterization of Load Transfer Mechanism

This section provides a generalized overview of the characterization of contact, contact force, and force chain. The typical evaluation indicators are also illustrated and compared as follows.

3.1. Contact Characteristics

Aggregate contact characteristics, including the contact number, contact distance distribution, contact length distribution, and contact orientation, etc., are descriptions of the adjacent aggregate particle interaction, which play an important role in the load transmission in asphalt mixtures. The studies of contact in asphalt mixtures fall into two categories: (1) parameterization of aggregate contact [20,30,54–57] and (2) contact chain characterization [58–61]. The parameters such as aggregate contact point, average contact length, total contact length, and contact orientation can be obtained from a section image of the asphalt mixture. The contact chain network in an asphalt mixture is a complex topological structure, which changes with the change of external load action [30,62–65].

Initially, contact points are usually used to describe contact characteristics and evaluate the meso-structure performance of asphalt mixtures [56]. However, several researchers found that the contact points cannot fully judge the quality of the mixture structure [20,66]. With the development of computer technology, some other contact indicators are developed to distinguish the contact structure of different mixtures using an image processing method. Jiang et al. [21] established a series of contact structure indicators in respect of contact distance distribution, contact length distribution, and contact orientation by using a two-dimensional image acquisition and processing program. They found that the optimal contact distance threshold for contact line analysis is 0.5 mm. The contact length distributions are varying in different mixtures. Kutay et al. [61] proposed a calculation method for the aggregate contact point and direction in hot mixture asphalt (HMA) using image processing technology based on CT images, and analyzed the change of contact characteristics with increasing compaction degree. The results show that the number of contact points increases significantly with the increase of compaction degree. Cai et al. [67,68] characterized the contact characteristics of an asphalt mixture under different compaction cycles by introducing indicators such as coarse aggregate contact point and aggregate inclination based on digital image processing technology. The results show that when the compaction force exceeds the limit of the skeleton bearing capacity, the contact point decreases and the inclination of the aggregate fluctuates. With the increase of compaction repetitions, the average contact number increases first and then decreases. Xing et al. [69] proposed a calculation method for aggregate classification and contact performance, discussed the impact of failure on meso-structure and aggregate contact, and analyzed the relationship between disruption factor and contact characteristics based on X-CT and digital image processing technology. It was found that a higher disruption factor could reduce the number of aggregate contacts in the main skeleton and increase the number of contacts of broken aggregates. Shi et al. [17,19,70,71] characterized the contact chains in asphalt mixture using DIP. The parameter of modularity according to the spectral clustering method was used to evaluate the contact chains [72]. The skeleton mechanical behavior can be improved by obtaining a maximum amount of coarse aggregate in the contact chains. In conclusion, the indexes such as contact length and aggregate inclination obtained by digital image technology are able to well describe the contact characteristics. However, image processing technology can only analyze the acquired images and it is time-consuming and laborious.

DEM is a useful method for examining the meso-mechanical characteristics of granular materials and is crucial for understanding the meso-structures of asphalt mixtures. The contact properties of asphalt mixtures based on DEM have been the subject of numerous studies, including those on the impact of coarse aggregate morphology on the mechanical properties of the skeleton [34,73–75], contact meso-structure evaluation indices [76], and the impact of the contact skeleton on impairing the movement of coarse aggregates [77,78]. The volume indices, rutting resistance, durability, and road performance of asphalt mixtures are all positively correlated with the meso-scale properties of aggregate contact [79]. The contacted coarse aggregate is what makes up the contact chains in asphalt mixture, which together form a complex network that affects the macro-mechanical characteristics of the asphalt mixture. These contact chains operate as a bridge between the micro- and macro-scale properties of materials. Qian et al. [30] analyzed the influence of different compaction methods (Marshall impact compaction and static compaction) on the distribution characteristics of contact number with depth of specimen using DEM simulation test method. Tan et al. [16] established 3D FEM models based on CT scanning images by incorporating AC cores into the numerical model and quantified the impact of aggregate contact zone ratio on the visual properties of matrix phase. It was discovered that although the contact zone only makes up a minuscule volume proportion of AC, due to its substantially greater modulus than asphalt matrix, it can significantly raise the modulus of AC within the low-frequency region. After conducting a number of studies, Jin et al. [2–4,18,80] proposed a novel method based on graph theory for the prediction and evaluation of mixture stability. This method characterizes and assesses the initial and evolutionary morphologies of 3D aggregate contact chains during simulations and offers a significant new direction for the study of asphalt mixture contact chains.

To sum up, the contact skeleton structure formed by the particles in contact with each other is the load transfer path. The internal force transfer is the key to reveal the effect of the contact structure on mechanical properties. Then, it is essential to obtain the internal contact force response of asphalt mixtures to investigate the load transfer mechanism.

3.2. Contact Force Characteristics

In many studies, the contact force between particles refers to the normal component [5, 7,9,22]. Generally, contact forces are classified as strong and weak, and the strong constitutes the main load-bearing system in asphalt mixtures. Initially, it is considered that the contact forces between coarse particles form the strong and the contact forces between fine particles constitute the weak contact forces [81]. With the development of numerical simulation techniques, the values of contact forces can be extracted, which promotes the quantitative studies. It is common to define contact forces greater than the average as strong and those less than the average as weak [6,22].

Due to the anisotropy of the contact force distribution, there is a certain deficiency to characterize the contact force distribution by the strong or weak alone. Therefore, a series of studies are carried out to characterize the contact force distribution, mainly including two categories: contact force probability distribution [82,83] and contact force statistical characteristics [23,28,84,85]. Shashidhar and Shenoy [41] studied the contact force distribution in asphalt mixtures by means of a photoelastic experiment. The results show that different gradations exhibit different contact force distribution characteristics. Jiang et al. [83,86] explored the contact force distribution in the tight arrangement of single size particles by indentation experiments. The contact force of each layer particles is detected by the indentation of the impress board. It was found that the contact force probability distribution is approximately parabolic. Chang et al. [9] studied the contact force probability distribution characteristics for different grain size compositions using the indentation test. It was found that the probability of contact force distribution decayed exponentially for all different two-size compositions. To further explore the contact force distribution within different asphalt mixtures, Chang et al. [22,23,87] compared the contact force probability distribution of three type asphalt mixtures using DEM, Stone Matrix Asphalt (SMA) gradation, dense asphalt concrete (AC) gradation, and open-graded asphalt friction course (OGFC) gradation with a nominal maximum aggregate size (NMAS) of 13.2 mm. It was found that for the three type asphalt mixtures, the probability distributions of the normal contact forces show no significant difference. The probability of contact forces ($P(f)$) decreases with the increase of f_n (the normal contact force to the mean normal contact force) when $f_n \leq 0.75$. When $0.75 < f_n \leq 1.65$, $P(f)$ is directly proportional to f_n, and when $f_n > 1.65$, $P(f)$ is inversely proportional to f_n; $P(f)$ remains essentially unchanged at $f_n \geq 4$.

Some researchers also used various statistical parameters to characterize the features of contact forces. Zhu et al. [62] defined the vertical contact unbalanced force, which is calculated as the sum of the contact force vectors and the gravity of aggregate. It was found that the larger the particle size, the more the contact number, the greater the contact unbalance force in the Marshall impaction process. Generally, certain size particles have different load transfer characteristics in different grain size compositions. Zhang et al. [79] studied the contribution of each size aggregate to forming a skeleton structure by contact force analysis. The force occupation contributing to the formation of the aggregate skeleton is defined as the ratio of the contact force bigger than the total average force in one sieve size to the contact force larger than the total average force in all sieve sizes. The force extraction analysis demonstrated that, regardless of the total number of particles in the various sieve sizes, the bigger size included more contact force in each particle. It was found that 2.36 mm and 4.75 mm, which together contribute more than 50% of the main load carrying capacity, are the key sieve sizes in the primary structure. While 0.3 mm to 1.18 mm, which also contributes more than 50%, is a crucial sieve size range for stabilizing the basic structure. A series of DEM tests were conducted by Miao et al. [7] to examine the contact force characteristics of various sized particles in aggregate blends. To characterize the load-bearing contribution of each size particle in aggregate blends, an indicator was suggested. Wang et al. [5] also used DEM to examine the load-bearing contributions of various aggregate blends while taking into account the morphology of the aggregates. The critical load-bearing contribution particle for the SMA16 gradation was discovered to be 2.36 mm, while for the AC25 gradation, it was found to be 4.75 mm.

From the above studies, a lot of research has been carried out on the characterization of contact force distribution. The contact force distribution can only explain the overall force state of the asphalt mixture, but it cannot fully reveal how the load is transferred in the asphalt mixture and whether the load is transferred uniformly. The composition of asphalt mixtures has an influence on the contact force distribution, and the load-bearing capacity of each size aggregates also has effects on the load transfer mechanism. Based on current studies, the contact force characteristics of asphalt mixture need to be further analyzed.

3.3. Force Chain Characteristics

Numerous studies have found that the discontinuous and non-uniform arrangement of particles forms complex contact networks, which is the load transfer path [64,88]. However, the force chain is conceptually different from that of the contact network. The force chain is a selective force transfer path along the contact network, while the contact network is a geometric structure with granular particles in arrangement. The force chain is extremely sensitive to the loading method and the geometric characteristics of the system. Even in the same contact network, a slight change in the external loading can make the force chain very different. Dantu [89] specified the non-uniformity of force distribution inside the particles in the photoelastic experiments. It was found that the force chain has a tree-like structure. Edwards and Oakeshott [90] found the force chains arching in granular blends in 1989. Then, Bouchaud and Cates [91] further studied the force chain and explicitly introduced the concept of force chains.

Initially, laboratory experimental methods were used to study force chain characteristics in granular materials. He et al. [92] utilized the photoelastic method to study the load transfer in the asphalt mixture. Wang et al. [93] obtained the evidence of structural transformation of force chains under shear vibrations using mechanical spectroscopy. Sanfratello et al. [49] used magnetic resonance elastography (MRE) to observe and describe the three-dimensional force chain in granular materials. Generally, the main drawback of the experimental method is the inability to detect weak contact forces and the inability to detect the contact forces inside the blend without interference. Using the DEM, the force chain can be elaborately characterized. Sun et al. [64] studied the load transfer characteristics of granular blends under uniaxial compression by means of 2D DEM, proposed the angular criterion of the force chain, and found that the length of the force chain is distributed by the power rate. Zhang et al. [94,95] quantified the force chain characteristics during the high-speed compression of granular blends. It was found that the higher the initial impact velocity, the more the number of force chains, and the shorter their length. Additionally, the force chain direction showed anisotropy and formed an irregular distribution.

The mechanical characteristics of granular materials are influenced by force chains, which are a key component of the granular material mechanics theory [96]. Several studies furthering skeletal contact force statistical analysis try to assess the force chains in asphalt mixes. By using DEM, Chen et al. [81] qualitatively investigated the force chains in crumb rubber asphalt mixtures and categorized them only according to contact force magnitude. Based on a CT scanning picture, Wang et al. [97] created a 3D FEM model of an asphalt mixture that showed how internal load transmission develops in asphalt mixes. It was discovered that the aggregates bear the highest stress and that force chains build practically along their skeleton. Shi et al. [98] presented aggregate contact point efficacy parameters and evaluated force chains in SMA13 and AC13 asphalt mixes. These are all qualitative evaluations, which have limitations in terms of disclosing the properties of force chains. Chang et al. [23] developed force chain direction angles to evaluate asphalt mixture force chains morphological characteristics, and discovered asphalt mixture internal loading transfer law in order to quantitatively assess asphalt mixture force chains. In addition, Liu et al. [85] constructed asphalt mixed force chains by concurrently taking into account granular quantity and contact angle. According to the findings, the performance of various aggregate mixes can be reflected in considerable force chain differences. Liu et al. [82] evaluated the force chains number of dense-suspended and dense-skeleton asphalt blends based on the aforementioned force chains identification criteria. Additionally, systematic analysis of the asphalt mixture force chains identification criteria of Liu et al. [84] revealed that the suggested threshold values for contact force and angle are the average contact force and 45°, respectively. Based on known force chains identification criteria, Liu et al. [13] looked into the length distribution of force chains. The findings indicate that raising NMAS can contribute in the formation of force chains in asphalt mixes that are longer in length.

The study of force chains can systematically reflect the load transfer mechanism of the asphalt mixture, which effectively avoids the limitation of evaluating the overall load-bearing capacity due to the unilateral analysis of contact or contact force. However, the evaluation characterization of force chains has not formed a completed system, and the characterization of force chains is only at the stage of basic statistical analysis.

3.4. Comparison of Different Indicators

According to previous studies, a series of indicators have been proposed to characterize the load transfer mechanisms of asphalt mixtures. The typical indicators for load transfer quantitative characterization are summarized in Table 2. The indicators of contact characteristics can better characterize the contact skeleton structure and evaluate the goodness of the contact geometry structure. However, it cannot explain the force state in asphalt mixtures. The contact force characteristics can be used to characterize the overall force state of asphalt mixtures, to determine the overall load-bearing structure, and to evaluate the load-bearing capacity of each aggregate size. However, the contact force characteristics are only a statistical analysis of the contact forces at all contact points and do not provide an assessment of the load-bearing capacity of the load transfer structure. For the indicators of force chain characteristics, they can characterize the load transfer paths in asphalt mixtures and the force state of the contact structure.

Table 2. Typical indicators characterizing the load transfer mechanism.

Classification		Schematic Diagram	Indicators	References
Contact characteristics	Contact number	Contact point [37]	$n_s = \dfrac{\sum\limits_{i=1}^{g} n_{s,i}}{g}$ 1	[21,37]
	Contact orientation		$\theta = \dfrac{\sum_{k=1}^{N} |\theta_k|}{N}$ 2 $\Delta = \dfrac{100}{N}\sqrt{\left(\sum\limits_{k=1}^{N} \sin 2\theta_k\right)^2 + \left(\sum\limits_{k=1}^{N} \cos 2\theta_k\right)^2}$ 3	
	Contact angle	–	$\theta_c = \dfrac{180}{\langle Z \rangle'}$ 4	[64]
	Contact chain	Contact chains Aggregate	$E_{chain_i} = \sum\limits_{path} \dfrac{h_j}{\sum_{path} h_j} E_{path_j}$ 5 $E_{skeleton} = \sum\limits_{chain} \dfrac{v_{agg\_chain_k}}{c_V V_{specimen}} E_{chain_k}$ 6	[4]
	Contact distance	–	The minimum distance between one aggregate and its neighboring aggregates	[69]
	Contact length	–	$\overline{L_c} = \dfrac{1}{N_c} \sum\limits_{i,j \neq i}^{N_c} l_{ij}$ 7	[17]
Contact force characteristics	Contact force	[5]	$\overline{F_s} = \dfrac{\sum\limits_{k=1}^{C_s} f_k}{C_s}$ 8 $P_s = \dfrac{\overline{F_s}}{\sum\limits_{i=1}^{l} \overline{F_{s_i}}}$ 9	[5,7]
	Unbalanced contact force	Color line: contact force	$\begin{cases} F_{tavg} = \dfrac{\sum_{i=1}^{m}\left(\sum_{j=1}^{n_j} F_{ij}\right)}{\sum_{i=1}^{m} n_i} \\ f_{ti>tavg} = \dfrac{\sum_{j=1}^{n_i}(F_{ij}>F_{tavg})}{\sum_{i=1}^{m}\left(\sum_{j=1}^{n_i} F_{ij}>F_{tavg}\right)} \\ f_{ti\leq tavg} = \dfrac{\sum_{j=1}^{n_i}(F_{ij}\leq F_{tavg})}{\sum_{i=1}^{m}\left(\sum_{j=1}^{n_i} F_{ij}\leq F_{tavg}\right)} \end{cases}$ 10	[79]
	Contact force probability distribution	[23]	$P(f) = a(1 - b\exp^{-cf^2})\exp^{-df}$ 11	[22,23]

Table 2. *Cont.*

	Classification	Schematic Diagram	Indicators	References
Force chain	Force chain number		$R_{MFC} = \frac{S_{MEC}}{S_{FC}} \times 100\%$ [12]	[13,28]
	Force chain length	–	$\bar{l} = \frac{\sum_{i=1}^{N_1} l_{1i} + \sum_{j=1}^{N_3} l_{3i}}{N_1 + N_3}$ [13]	
	Force chain alignment coefficient		$\delta_i = 1 - \frac{\sum_{p=1}^{M} \alpha_p}{90^\circ \times M}$ [14]	
	Force chain direction angle	–	$\theta = \arccos\left(\frac{x_B - x_A}{\sqrt{(x_B - x_A)^2 + (y_B - y_A)^2}} \right)$ [15]	

[1] Where n_s is the contact number; g represents the total particle numbers of A_s; $n_{s, I}$ is the contact number of the i_{th} particle of A_s. [2,3] Where θ is the average angle of inclination; Δ is the vector magnitude; θ_k is the angle between the horizontal axis and the major axis of an individual particle or the orientation of an individual contact line in one section; N is the total number of aggregates or contact lines in one section. [4] Where θ_c is the contact angle; $<Z>$ represents the properties of materials, such as elastic modulus, Poisson's ratio, and gradation et al. [5,6] Where E_{chaini} is the morphology of the contact chain; h_j is the height of the bounding box of $path_j$; $E_{skeleton}$ is the morphology of the contact skeleton; $v_{agg\_chain_k}$ is the volume sum of aggregates involved in $chain_k$; $V_{specimen}$ is the volume of the specimen; c_V is the coefficient. [7] Where $\overline{L_c}$ is the average contact length; l_{ij} is the length of the contact chain between coarse aggregates i and j; Nc is the quantity of coarse aggregates in the contact chain. [8,9] Where $\overline{F_s}$ is the average contact force of size s particles; f_k is the contact force at contact point k; C_s is the total contact number of size s particles; l is the size l particles; P_s is the load-bearing contributions of size s particles. [10] Where F_{tavg} is the total average force; F_{ij} is the contact force; $f_{ti} > t_{avg}$ is the proportion of contact force larger than the total average force in one sieve size to the contact force larger than the total average force in whole sieve sizes; $f_{ti} \leq t_{avg}$ is the proportion of contact force smaller than the total average force in one sieve size to the contact force smaller than the total average force in whole sieve sizes. [11] Where $P(f)$ is the probability distribution of force chains; f is the ratio of the normal or shear contact force to the mean normal contact force; $a, b, c,$ and d are the fitted parameters. [12] Where S_{MFC} the asphalt mixture total number; S_{FC} is the total number of force chains; S_{MFC} is the total number of main force chains. [13] Where $\bar{l}$ is the average length of main force chain; l_{1i} represents the length of ith I type force chains; l_{3i} represents the length of ith III type force chains; N_1 and N_3 are the I and III type of force chain number, respectively. [14] Where δ_i is the main force chains alignment coefficient; α_p represents the pth angle between adjacent normal directions of the ith MFC; M is the total number of adjacent contacts. [15] Where θ is the force chain direction angle; (x_A, y_A) and (x_B, y_B) are location coordinates of aggregates A and B, respectively.

Based on the review reported above, the existing parameters are insufficient for investigating the characteristics of the load transfer mechanism of each particle size. The contact structure is a complex topological structure where contact forces are transferred at various nodes to form force chains. Through the above research summary, no definite quantitative index is given to clearly define the load transfer characteristics. Meanwhile, it does not reveal the essential issues of grade design, material selection, and service performance quality of asphalt mixtures from the level of load transfer mechanism. Further studies are still needed to be carried out to combine the contact, contact force, and force chains, and thus, to reveal the relevant mechanical mechanisms in depth.

4. Load Transfer Mechanism of Asphalt Mixture under Different Loading Conditions

There are different load transfer mechanisms in asphalt mixtures from compaction to service to destruction. Many researchers have conducted studies to analyze the load transfer mechanism of asphalt mixtures under different test states. Firstly, the load transfer mechanisms of asphalt mixture during the compaction process are summarized. Then, the internal load transfer mechanism of asphalt mixture under different test conditions is introduced. A comparative analysis of the load transfer mechanism under different test conditions is also carried out.

4.1. Load Transfer Mechanism in the Compaction Process

In the process of asphalt mixture compaction, with the application of external load and different load frequency, the spatial position of aggregate changes [78], then the load transfer characteristics in the contact structure, are changed. Exploring the changes of external load transfer characteristics in different compaction processes provides an important reference basis for evaluating the goodness of the load-bearing structure of an asphalt mixture after compaction.

The load transfer characteristics in compaction are usually investigated by simulating laboratory experiments. The Marshall impact compaction (MIC) method is most commonly used for fabricating asphalt mixture specimens. However, it was found that the load transfer characteristics inside the specimen under double-sided Marshall impact compaction and single-sided Marshall impact compaction are different [30]. The middle part has a large number of contacts, and the two sides have a small number of contacts. The contact number during double-sided compaction is more uniform than it is under single-sided compaction. In order to assess the variance in load transfer during the Marshall impact compaction process of asphalt mixes, Zhu et al. [62] employed the vertical contact unbalanced force. Using DEM modeling, it was possible to measure the vertical contact imbalanced force and the contact number for compaction numbers of 2, 4, 6, 8, 10, 15, 20, 25, 30, 35, 40, 45, 50, 55, 60, 70, 80, 90, and 100 [62]. It was discovered that when the number of strikes increased, the imbalanced force of each particle size progressively reduced. With each increase in strikes, the force of the contact imbalance decreased. The contact imbalance force in the MIC process increases with aggregate particle size, coordinate number, and contact unbalance. As the NMAS of the aggregate is larger, the poorer the Marshall compaction effect. In order to make the compaction process closer to the roller compaction in field, more attention is paid to fabricate the specimen using the rotary compaction method. Gong et al. [74,78] took both gyration angle and rotation action into account and investigated the displacement variation of aggregates in the compaction process. Miao et al. [7] investigated the contact force distribution and transfer characteristics of asphalt mixtures in rotary compaction using DEM. Different gradation asphalt mixtures have different contact force distribution characteristics, under the same external load, different size aggregates have different average contact force. The contact force evolution of different asphalt mixtures during compaction is also different [5]. Liu et al. [31] studied the mechanisms of aggregate movement and contact force changes within asphalt mixtures during a simulated compaction test. The results showed that the contact forces are mainly generated between aggregates.

The load transfer in the compaction process of laboratory specimen cannot completely reflect that of actual pavement in the field. Therefore, it is necessary to establish the relationship between field compaction and laboratory compaction for understanding the load transfer mechanism of asphalt mixtures. Dan et al. [32,33] designed a field test program and used SmartRocks to measure the load transfer response during vibrating compaction. Meanwhile, the load transfer of asphalt mixture during different gyratory compaction degree was also analyzed. It was found that the gyratory compaction degree and the peak acceleration of the vibration drum exhibit a strong linear correlation. By controlling the gyratory compaction degree of asphalt mixtures, the load transfer mechanism in the actual pavement compaction process can be better simulated.

4.2. Load Transfer Mechanism under Different Loading Conditions

According to previous studies, there are different test conditions for the study of the load transfer mechanism of asphalt mixtures. Under different external loads and dynamic loading frequencies, the load transfer response in the asphalt mixture is different. Furthermore, the contact forces in the top part of the sample are always higher than those in the lower part [99]. Chang et al. [23] investigated the contact force probability distribution of asphalt mixtures under haversine loading. It was found that the probability distributions of smaller contact forces are greater than that of larger contact forces, and the probability distribution of larger contact forces is the largest when the ratio of contact force to mean contact force is 1.75. Considering the actual vehicle loads, some researchers have studied the load transfer mechanism of asphalt mixtures under uniaxial loading and biaxial loading. Liu et al. [28,82] investigated the load transfer characteristics of AC, SMA, and OGFC under the single-wheel pressure surface load. There are different load transfer characteristics of different gradations of asphalt mixtures. The external loads were mainly transferred along the vertical direction, although a small amount of loads tended to extend horizontally. Meanwhile, the load transfer characteristics between the different structural layers of the pavement have an important influence on the whole structure's load-bearing capacity. The contact force distribution in high-modulus asphalt concrete (HMAC) pavement structure after double circular static loading was studied [100]. It was found that the application of HMAC decreased the vertical force in all structural layers except the upper surface layer, and the HMAC decreased the horizontal force in the subbase layer.

The tension–compression conditions, shear conditions, and bending conditions are the typical loading conditions of asphalt pavement [101]. Ma et al. [34] constructed a virtual tracking test model for asphalt mixtures and analyzed the micro-mechanical response of load transfer in the asphalt mixture under compression conditions. The results showed that contact forces primarily exist underneath the loading pressure area. Under the virtual wheel tracking test, Xue et al. [102] investigated the load transfer characteristics for different gradations of asphalt mixtures, including AC13 and SMA13. The average contact force increases continuously during the loading time from 5 min to 60 min. It was also found that the average contact force between coarse aggregates was the largest, followed by the average contact force between aggregate–mastic and the average contact force between mastics. Peng and Sun [103] used image analysis and DEM to simulate the indirect tensile (IDT) test of asphalt mixtures under tension–compression conditions (shown in Figure 4). The contact force distribution at microcracks was analyzed. Under the vertical loads, the contact forces exhibit compression and tension along the vertical and horizontal directions, respectively.

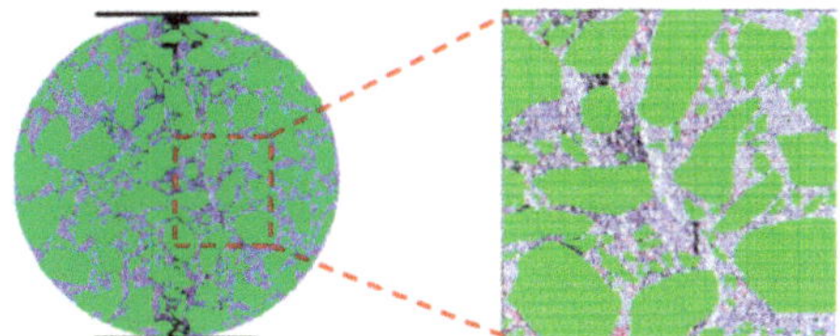

Figure 4. The indirect tensile (IDT) test of asphalt mixtures [52]: the green balls represent aggregates; the blue balls represent mastic; the white balls represent air voids; the black line represents the force chain.

Chen et al. [35] utilized the DEM numerical simulation penetration test to explore the contact force characteristics of aggregate particles under shear conditions. It was found that under the same penetration depth, the average contact force of the larger size aggregates is greater than that of the smaller size aggregates, and the contact force proportion taken by different aggregates depends on aggregate sizes. The larger the aggregate size, the more proportion of the contact force. Peng and Sun [104] simulated the uniaxial penetration test of asphalt mixtures using DEM. Ding et al. [75] analyzed the load transfer response of AC13 and SMA13 asphalt mixtures in virtual penetration tests. It is known that the contact forces of AC13 and SMA13 were mainly distributed in 0–5 N, accounting for 75–85%. The bigger the contact force, the smaller the corresponding probability distribution proportion.

A random heterogeneous DEM model was employed by Xue et al. [36] to simulate the semi-circular bending (SCB) test of asphalt mixes. It showed that, prior to cracking, tension contact forces were primarily focused in the notch tip and compression contact forces were primarily concentrated in the specimen's top and bottom. After the specimen cracked, the tension contact force concentration zone traveled from the fracture tip to the top of the specimen over time, but it was always there. It was believed that the primary cause behind crack propagation was the tension force.

From above review, the researchers have studied the load transfer characteristics of asphalt mixtures under different test conditions from various loading aspects. It is indicated that the load transfer mechanism of asphalt mixtures under different test conditions are not the same. Although the analysis of the load transfer mechanism was carried out in the mentioned studies, only a preliminary statistical analysis of contact force distribution characteristics under the corresponding test conditions were carried out. There is no relevant evaluation system of the load transfer mechanism under different loading conditions established. The meso-scale load transfer mechanism under different loading conditions is the mechanical response of the macro-scale properties of the asphalt mixtures. It is necessary to carry out further relevant studies to investigate load transfer mechanisms of asphalt mixtures and establish the corresponding evaluation system of the load transfer mechanism, which will provide a theoretical basis for explaining the macro-mechanical properties from the meso-scale mechanical mechanism.

5. Conclusions

Quantitatively capturing the relationship between the load transfer mechanism and the mechanical response of asphalt mixtures can provide a meso-mechanical basis for optimizing asphalt mixture design to improve the performance of asphalt pavement. This paper reviews the research progress of the load transfer mechanism of asphalt mixtures. Some conclusions are drawn as follows.

(1) The study of the load transfer mechanism consists of three main aspects: contact characteristics, contact force characteristics, and force chain characteristics. Various technical methods for studying the load transfer mechanism are summarized and comparatively analyzed. With the comprehensive analysis of different methods used in characterizing the load transfer mechanism, the X-ray CT and DIP and numerical simulation is highly recommended to be used for investigating the load transfer mechanism of asphalt mixtures.

(2) A systematic summary analysis of load transfer mechanism evaluation indexes revealed that the application of several evaluation indicators in combination could be better for characterizing load transfer mechanisms, and the statistical methods can obtain better typical quantitative indicators.

(3) The meso-scale load transfer mechanism under different loading conditions is the mechanical response of the macro-scale properties of the asphalt mixtures. So, it is important to carry out further relevant studies to investigate load transfer mechanisms of asphalt mixtures and establish the corresponding evaluation system of the load transfer mechanism, which can provide a theoretical basis for explaining the macro-mechanical properties from the meso-scale mechanical mechanism.

(4) To date, systematic evaluation methods on the load transfer mechanism of asphalt mixtures are not well developed. These should be considered to efficiently obtain how the load transfer mechanism functions during the actual service of asphalt mixture through the analysis algorithm. Further, a reasonable evaluation system of the load transfer mechanism should be established in order to realize the effective evaluation of the actual road structure's load-bearing capacity.

6. Recommendations

Extensive research has been carried out in the past to study the load transfer mechanisms of asphalt mixtures. The following points enlist the recommendations for future studies.

(1) The X-ray CT and DIP and numerical simulation is highly recommended to be used for investigating load transfer mechanism of asphalt mixtures.

(2) The contact structure is a complex topological structure where contact forces are transferred at various nodes to form force chains. The load transfer (contact, contact force, and force chain) mathematical model can be established according to the statistics method and graph method. Meanwhile, the optimal load-bearing structure of asphalt mixture can be quantified and analyzed by using the topology theory. It can provide the theoretical guidance for the mixture design.

(3) It is strongly recommended that several evaluation indicators (corresponding to contact, contact force, and force chain) are used in combination to characterize load transfer mechanisms.

(4) Based on the quantitative definition of the load transfer structure and characteristics, a series of studies should be conducted to explore the mechanism of the relationship between load transfer characteristics and performance. Numerical simulation experiments are carried out to explore the load transfer mechanism, and corresponding laboratory experiments are carried out to explore the macro-scale performance, so as to establish the mathematical model between the load transfer mechanism characterization indicator and macro-scale performance.

Author Contributions: Conceptualization, Y.M. and S.W.; methodology, Y.M. and S.W.; validation, Y.M. and S.W.; investigation, S.W. and W.Y.; writing—original draft preparation, S.W., W.Y., and Y.M.; writing—review and editing, Y.M., S.W., and L.W.; supervision, L.W.; project administration, L.W.; funding acquisition, Y.M. All authors have read and agreed to the published version of the manuscript.

Funding: This research was funded by the National Natural Science Foundation of China, grant number 51978048.

Institutional Review Board Statement: Not applicable.

Informed Consent Statement: Not applicable.

Data Availability Statement: Not applicable.

Conflicts of Interest: The authors declare no conflict of interest.

References

1. Yu, H.; Yao, D.; Qian, G.; Zhu, X.; Shi, Z.; Zhang, C.; Li, P. Review on Digital Twin Model of Asphalt Mixture Performance Based on Mesostructure Characteristics. *China J. Highw. Transp.* **2022**, 1–38. Available online: http://kns.cnki.net/kcms/detail/61.1313.U.20220909.1520.004.html (accessed on 1 January 2023).
2. Jin, C.; Zhang, W.; Liu, P.; Yang, X.; Oeser, M. Morphological Simplification of Asphaltic Mixture Components for Micromechanical Simulation Using Finite Element Method. *Comput. Aided Civ. Inf.* **2021**, *36*, 1435–1452. [CrossRef]
3. Jin, C.; Wan, X.; Yang, X.; Liu, P.; Oeser, M. Three-Dimensional Characterization and Evaluation of Aggregate Skeleton of Asphalt Mixture Based on Force-Chain Analysis. *J. Eng. Mech.* **2021**, *147*, 04020147. [CrossRef]
4. Jin, C.; Wan, X.; Liu, P.; Yang, X.; Oeser, M. Stability Prediction for Asphalt Mixture Based on Evolutional Characterization of Aggregate Skeleton. *Comput. Aided Civ. Inf.* **2021**, *36*, 1453–1466. [CrossRef]
5. Wang, S.; Miao, Y.; Wang, L. Investigation of the Force Evolution in Aggregate Blend Compaction Process and the Effect of Elongated and Flat Particles Using Dem. *Constr. Build. Mater.* **2020**, *258*, 119674. [CrossRef]
6. Wang, J.; Yang, L.; Li, F.; Wang, C. Force Chains in Top Coal Caving Mining. *Int. J. Rock Mech. Min.* **2020**, *127*, 104218. [CrossRef]
7. Miao, Y.; Yu, W.; Hou, Y.; Guo, L.; Wang, L. Investigating the Functions of Particles in Packed Aggregate Blend Using a Discrete Element Method. *Materials* **2019**, *12*, 556. [CrossRef]
8. Shi, L.; Yang, Z.; Wang, D.; Qin, X.; Xiao, X.; Julius, M.K. Gradual Meso-Structural Response Behaviour of Characteristics of Asphalt Mixture Main Skeleton Subjected to Load. *Appl. Sci.* **2019**, *9*, 2425. [CrossRef]
9. Chang, M.; Pei, J.; Huang, P.; Xiong, R. Analysis on the Distribution Probability of Force Chain of Contact Force Among Granular Matter Considering Gradation. *Mater. Rep.* **2018**, *32*, 3618–3622. [CrossRef]
10. Abbas, A.; Masad, E.; Papagiannakis, T.; Harman, T. Micromechanical Modeling of the Viscoelastic Behavior of Asphalt Mixtures Using the Discrete-Element Method. *Int. J. Geomech.* **2007**, *7*, 131–139. [CrossRef]
11. Apuzzo, M.D.; Evangelisti, A.; Nicolosi, V. Preliminary Investigation on a Numerical Approach for the Evaluation of Road Macrotexture. In Proceedings of the 17th International Conference on Computational Science and Applications (ICCSA 2017), Trieste, Italy, 3–6 July 2017.
12. D'Apuzzo, M.; Evangelisti, A.; Santilli, D.; Nicolosi, V. *3D Simulations of Two-Component Mixes for the Prediction of Multi-Component Mixtures' Macrotexture: Intermediate Outcomes*; Springer International Publishing: Cham, Switzerland, 2021; pp. 495–511.
13. Liu, G.; Han, D.; Jia, Y.; Zhao, Y. Asphalt Mixture Skeleton Main Force Chains Composition Criteria and Characteristics Evaluation Based on Discrete Element Methods. *Constr. Build. Mater.* **2022**, *323*, 126313. [CrossRef]
14. Zhang, H.; Anupam, K.; Scarpas, A.; Kasbergen, C.; Erkens, S. Effect of Stone-On-Stone Contact on Porous Asphalt Mixes: Micromechanical Analysis. *Int. J. Pavement Eng.* **2020**, *21*, 990–1001. [CrossRef]
15. Wang, X.; Gu, X.; Jiang, J.; Deng, H. Experimental Analysis of Skeleton Strength of Porous Asphalt Mixtures. *Constr. Build. Mater.* **2018**, *171*, 13–21. [CrossRef]

16. Tan, Z.; Leng, Z.; Jiang, J.; Cao, P.; Jelagin, D.; Li, G.; Sreeram, A. Numerical Study of the Aggregate Contact Effect on the Complex Modulus of Asphalt Concrete. *Mater. Des.* **2022**, *213*, 110342. [CrossRef]
17. Shi, L.; Xiao, X.; Wang, X.; Liang, H.; Wang, D. Mesostructural Characteristics and Evaluation of Asphalt Mixture Contact Chain Complex Networks. *Constr. Build. Mater.* **2022**, *340*, 127753. [CrossRef]
18. Jin, C.; Zou, F.; Yang, X.; Liu, K.; Liu, P.; Oeser, M. Three-Dimensional Quantification and Classification Approach for Angularity and Surface Texture Based on Surface Triangulation of Reconstructed Aggregates. *Constr. Build. Mater.* **2020**, *246*, 118120. [CrossRef]
19. Shi, L.; Wang, D.; Xiao, X.; Qin, X. Meso-Structural Characteristics of Asphalt Mixture Main Skeleton Based on Meso-Scale Analysis. *Constr. Build. Mater.* **2020**, *232*, 117263. [CrossRef]
20. Li, P.; Su, J.; Ma, S.; Dong, H. Effect of Aggregate Contact Condition on Skeleton Stability in Asphalt Mixture. *Int. J. Pavement Eng.* **2020**, *21*, 196–202. [CrossRef]
21. Jiang, J.; Ni, F.; Gao, L.; Yao, L. Effect of the Contact Structure Characteristics on Rutting Performance in Asphalt Mixtures Using 2D Imaging Analysis. *Constr. Build. Mater.* **2017**, *136*, 426–435. [CrossRef]
22. Chang, M.; Pei, J.; Zhang, J.; Xing, X.; Xu, S.; Xiong, R.; Sun, J. Quantitative Distribution Characteristics of Force Chains for Asphalt Mixtures with Three Skeleton Structures Using Discrete Element Method. *Granul. Matter* **2020**, *22*, 87. [CrossRef]
23. Chang, M.; Huang, P.; Pei, J.; Zhang, J.; Zheng, B.; Hamzah, M.O. Quantitative Analysis on Force Chain of Asphalt Mixture Under Haversine Loading. *Adv. Mater. Sci. Eng.* **2017**, *2017*, 7128602. [CrossRef]
24. Hurley, R.; Zhai, C. Challenges and Opportunities in Measuring Time-Resolved Force Chain Evolution in 3D Granular Materials. *Pap. Phys.* **2022**, *14*, 140003. [CrossRef]
25. Fu, L.; Zhou, S.; Guo, P.; Wang, S.; Luo, Z. Induced Force Chain Anisotropy of Cohesionless Granular Materials During Biaxial Compression. *Granul. Matter.* **2019**, *21*, 52. [CrossRef]
26. Peters, J.F.; Muthuswamy, M.; Wibowo, J.; Tordesillas, A. Characterization of Force Chains in Granular Material. *Phys. Rev. E* **2005**, *72*, 041307. [CrossRef]
27. Zhang, W.; Zhang, S.; Tan, J.; Zhang, N.; Chen, B. Relation Between Force Chain Quantitative Characteristics and Side Wall Friction Behaviour During Ferrous Powder Compaction. *Granul. Matter* **2022**, *24*, 86. [CrossRef]
28. Liu, G.; Han, D.; Zhao, Y.; Zhang, J. Effects of Asphalt Mixture Structure Types on Force Chains Characteristics Based on Computational Granular Mechanics. *Int. J. Pavement Eng.* **2022**, *23*, 1008–1024. [CrossRef]
29. Tordesillas, A.; Walker, D.M.; Lin, Q. Force Cycles and Force Chains. *Phys. Rev. E Stat. Nonlinear Soft Matter Phys.* **2010**, *81 Pt 1*, 011302. [CrossRef]
30. Qian, G.; Hu, K.; Li, J.; Bai, X.; Li, N. Compaction Process Tracking for Asphalt Mixture Using Discrete Element Method. *Constr. Build. Mater.* **2020**, *235*, 117478. [CrossRef]
31. Liu, W.; Gong, X.; Gao, Y.; Li, L. Microscopic Characteristics of Field Compaction of Asphalt Mixture Using Discrete Element Method. *J. Test. Eval.* **2019**, *47*, 20180633. [CrossRef]
32. Dan, H.; Yang, D.; Liu, X.; Peng, A.; Zhang, Z. Experimental Investigation on Dynamic Response of Asphalt Pavement Using Smartrock Sensor Under Vibrating Compaction Loading. *Constr. Build. Mater.* **2020**, *247*, 118592. [CrossRef]
33. Dan, H.; Yang, D.; Zhao, L.; Wang, S.; Zhang, Z. Meso-Scale Study on Compaction Characteristics of Asphalt Mixtures in Superpave Gyratory Compaction Using Smartrock Sensors. *Constr. Build. Mater.* **2020**, *262*, 120874. [CrossRef]
34. Ma, T.; Zhang, D.; Zhang, Y.; Hong, J. Micromechanical Response of Aggregate Skeleton within Asphalt Mixture Based on Virtual Simulation of Wheel Tracking Test. *Constr. Build. Mater.* **2016**, *111*, 153–163. [CrossRef]
35. Chen, J.; Li, H.; Wang, L.; Wu, J.; Huang, X. Micromechanical Characteristics of Aggregate Particles in Asphalt Mixtures. *Constr. Build. Mater.* **2015**, *91*, 80–85. [CrossRef]
36. Xue, B.; Pei, J.; Zhou, B.; Zhang, J.; Li, R.; Guo, F. Using Random Heterogeneous Dem Model to Simulate the Scb Fracture Behavior of Asphalt Concrete. *Constr. Build. Mater.* **2020**, *236*, 117580. [CrossRef]
37. Shi, L.; Wang, D.; Wang, J.; Jiang, Z.; Liang, H.; Qin, X.; Hao, W.; Wang, H. A New Method for Designing Dense Skeleton Asphalt Mixture Based on Meso Parameter. *Adv. Civ. Eng.* **2020**, *2020*, 3841291. [CrossRef]
38. Wang, S.; Miao, Y.; Wang, L. Effect of Grain Size Composition on Mechanical Performance Requirement for Particles in Aggregate Blend Based on Photoelastic Method. *Constr. Build. Mater.* **2022**, *363*, 129808. [CrossRef]
39. Li, F.; Yang, L.; Wang, J.; Wang, C. A Quantitative Extraction Method of Force Chains for Composite Particles in a Photoelastic Experiment. *Chin. J. Eng.* **2018**, *40*, 302–312. [CrossRef]
40. Daniels, K.E.; Kollmer, J.E.; Puckett, J.G. Photoelastic Force Measurements in Granular Materials. *Rev. Sci. Instrum.* **2017**, *88*, 051808. [CrossRef]
41. Naga, S.; Aroon, S. Investigating the Role of Aggregate Structure in Asphalt Pavements. In Proceedings of the International Center for Aggregates Research 8th Annual Symposium: Aggregates-Asphalt Concrete, Bases and Fines, Denver, CO, USA, 12–14 April 2000.
42. Pei, Z.; Lou, K.; Kong, H.; Wu, B.; Wu, X.; Xiao, P.; Qi, Y. Effects of Fiber Diameter on Crack Resistance of Asphalt Mixtures Reinforced by Basalt Fibers Based on Digital Image Correlation Technology. *Materials* **2021**, *14*, 7426. [CrossRef]
43. Tan, Y.; Zhang, K.; Hou, M.; Zhang, L. Studying the Strain Field Distribution of Asphalt Mixture with the Digital Speckle Correlation Method. *Road Mater. Pavement Des.* **2014**, *15*, 90–101. [CrossRef]

44. Zhang, C.; Wang, H. A New Method for Compaction Quality Evaluation of Asphalt Mixtures with the Intelligent Aggregate (Ia). *Materials* **2021**, *14*, 2422. [CrossRef]
45. Zhang, D.; Cheng, Z.; Geng, D.; Xie, S.; Wang, T. Experimental and Numerical Analysis on Mesoscale Mechanical Behavior of Coarse Aggregates in the Asphalt Mixture During Gyratory Compaction. *Processes* **2022**, *10*, 47. [CrossRef]
46. Zhang, C.; Zhang, Z. Study on Migratory Behavior of Aggregate in Asphalt Mixture Based on the Intelligent Acquisition System of Aggregate Attitude Data. *Sustainability* **2021**, *13*, 3053. [CrossRef]
47. Mueth, D.M.; Jaeger, H.M.; Nagel, S.R. Force Distribution in a Granular Medium. *Phys. Rev. E* **1998**, *57*, 3164–3169. [CrossRef]
48. Li, J.; Li, P.; Su, J.; Xue, Y.; Rao, W. Effect of Aggregate Contact Characteristics on Densification Properties of Asphalt Mixture. *Constr. Build. Mater.* **2019**, *204*, 691–702. [CrossRef]
49. Sanfratello, L.; Fukushima, E.; Behringer, R.P. Using Mr Elastography to Image the 3D Force Chain Structure of a Quasi-Static Granular Assembly. *Granul. Matter* **2008**, *11*, 1–6. [CrossRef]
50. Huang, P.; Zhao, Y.; Niu, Y.; Ren, X.; Chang, M.; Sun, Y. Mesoscopic Finite Element Method of the Effective Thermal Conductivity of Concrete with Arbitrary Gradation. *Adv. Mater. Sci. Eng.* **2018**, *2018*, 2352864. [CrossRef]
51. Han, D.; Liu, G.; Xi, Y.; Zhao, Y.; Tang, D. Performance Prediction of Asphalt Mixture Based on Dynamic Reconstruction of Heterogeneous Microstructure. *Powder Technol.* **2021**, *392*, 356–366. [CrossRef]
52. Yao, H.; Xu, M.; Liu, J.; Liu, Y.; Ji, J.; You, Z. Literature Review on the Discrete Element Method in Asphalt Mixtures. *Front. Mater.* **2022**, *9*, 236. [CrossRef]
53. Peng, Y.; Bao, J. Comparative Study of 2D and 3D Micromechanical Discrete Element Modeling of Indirect Tensile Tests for Asphalt Mixtures. *Int. J. Geomech.* **2018**, *18*, 04018046. [CrossRef]
54. Wang, F.; Xiao, Y.; Cui, P.; Ma, T.; Kuang, D. Effect of Aggregate Morphologies and Compaction Methods on the Skeleton Structures in Asphalt Mixtures. *Constr. Build. Mater.* **2020**, *263*, 120220. [CrossRef]
55. Gao, J.; Wang, H.; Bu, Y.; You, Z.; Hasan, M.R.M.; Irfan, M. Effects of Coarse Aggregate Angularity on the Microstructure of Asphalt Mixture. *Constr. Build. Mater.* **2018**, *183*, 472–484. [CrossRef]
56. Coenen, A.R.; Kutay, M.E.; Sefidmazgi, N.R.; Bahia, H.U. Aggregate Structure Characterisation of Asphalt Mixtures Using Two-Dimensional Image Analysis. *Road Mater. Pavement Des.* **2012**, *13*, 433–454. [CrossRef]
57. Sefidmazgi, N.R.; Tashman, L.; Bahia, H. Internal Structure Characterization of Asphalt Mixtures for Rutting Performance Using Imaging Analysis. *Road Mater. Pavement Des.* **2012**, *13* (Suppl. S1), 21–37. [CrossRef]
58. Jiang, J.; Ni, F.; Dong, Q.; Yao, L.; Ma, X. Investigation of the Internal Structure Change of Two-Layer Asphalt Mixtures During the Wheel Tracking Test Based on 2D Image Analysis. *Constr. Build. Mater.* **2019**, *209*, 66–76. [CrossRef]
59. Khalilitehrani, M.; Sasic, S.; Rasmuson, A. Characterization of Force Networks in a Dense High-Shear System. *Particuology* **2018**, *38*, 215–221. [CrossRef]
60. Wang, D.; Zhou, Y. Statistics of Contact Force Network in Dense Granular Matter. *Particuology* **2010**, *8*, 133–140. [CrossRef]
61. Kutay, M.E.; Arambula, E.; Gibson, N.; Youtcheff, J. Three-Dimensional Image Processing Methods to Identify and Characterise Aggregates in Compacted Asphalt Mixtures. *Int. J. Pavement Eng.* **2010**, *11*, 511–528. [CrossRef]
62. Zhu, X.; Qian, G.; Yu, H.; Yao, D.; Shi, C.; Zhang, C. Evaluation of Coarse Aggregate Movement and Contact Unbalanced Force During Asphalt Mixture Compaction Process Based on Discrete Element Method. *Constr. Build. Mater.* **2022**, *328*, 127004. [CrossRef]
63. Giusti, C.; Papadopoulos, L.; Owens, E.T.; Daniels, K.E.; Bassett, D.S. Topological and Geometric Measurements of Force-Chain Structure. *Phys. Rev. E* **2016**, *94*, 032909. [CrossRef]
64. Sun, Q.; Jin, F.; Wang, G.; Zhang, G. Force Chains in a Uniaxially Compressed Static Granular Matter in 2D. *Acta Phydica Sin.* **2010**, *59*, 30–37.
65. Sun, Q.; Xin, H.; Liu, J.; Jin, F. Skeleton and Force Chain Network in Static Granular Material. *Rock Soil Mech.* **2009**, *30* (Suppl. S1), 83–87. [CrossRef]
66. Cai, X.; Zhu, F.; Wu, K.; Wan, C. Steady-State Parameters and Model for Asphalt Mixture Skeletons. *China J. Highw. Transp.* **2019**, *32*, 39–46+96. [CrossRef]
67. Cai, X.; Wu, K.; Huang, W. Study on the Optimal Compaction Effort of Asphalt Mixture Based on the Distribution of Contact Points of Coarse Aggregates. *Road Mater. Pavement Des.* **2021**, *22*, 1594–1615. [CrossRef]
68. Cai, X.; Wu, K.H.; Huang, W.K.; Wan, C. Study on the Correlation Between Aggregate Skeleton Characteristics and Rutting Performance of Asphalt Mixture. *Constr. Build. Mater.* **2018**, *179*, 294–301. [CrossRef]
69. Xing, C.; Xu, H.; Tan, Y.; Liu, X.; Ye, Q. Mesostructured Property of Aggregate Disruption in Asphalt Mixture Based on Digital Image Processing Method. *Constr. Build. Mater.* **2019**, *200*, 781–789. [CrossRef]
70. Shi, L.; Wang, D. Evaluation Indexes of Asphalt Mixture Main Skeleton Based on Digital Image Processing. *China J. Highw. Transp.* **2017**, *30*, 52–58+73.
71. Shi, L.; Wang, D.; Cai, X.; Wu, Z. Distribution Characteristics of Coarse Aggregate Contacts Based on Digital Image Processing Technique. *China J. Highw. Transp.* **2014**, *27*, 23–31. [CrossRef]
72. Newman, M.E.J. The Structure and Function of Networks. *Comput. Phys. Commun.* **2002**, *147*, 40–45. [CrossRef]
73. Liu, P.; Hu, J.; Canon Falla, G.; Wang, D.; Leischner, S.; Oeser, M. Primary Investigation on the Relationship Between Microstructural Characteristics and the Mechanical Performance of Asphalt Mixtures with Different Compaction Degrees. *Constr. Build. Mater.* **2019**, *223*, 784–793. [CrossRef]

74. Gong, F.; Liu, Y.; Zhou, X.; You, Z. Lab Assessment and Discrete Element Modeling of Asphalt Mixture During Compaction with Elongated and Flat Coarse Aggregates. *Constr. Build. Mater.* **2018**, *182*, 573–579. [CrossRef]
75. Ding, X.; Ma, T.; Gao, W. Morphological Characterization and Mechanical Analysis for Coarse Aggregate Skeleton of Asphalt Mixture Based on Discrete-Element Modeling. *Constr. Build. Mater.* **2017**, *154*, 1048–1061. [CrossRef]
76. Yuan, G.; Li, X.; Hao, P.; Li, D.; Pan, J.; Li, A. Application of Flat-Joint Contact Model for Uniaxial Compression Simulation of Large Stone Porous Asphalt Mixes. *Constr. Build. Mater.* **2020**, *238*, 117695. [CrossRef]
77. Liu, W.; Gao, Y.; Huang, X.; Li, L. Investigation of Motion of Coarse Aggregates in Asphalt Mixture Based on Virtual Simulation of Compaction Test. *Int. J. Pavement Eng.* **2020**, *21*, 144–156. [CrossRef]
78. Gong, F.; Zhou, X.; You, Z.; Liu, Y.; Chen, S. Using Discrete Element Models to Track Movement of Coarse Aggregates During Compaction of Asphalt Mixture. *Constr. Build. Mater.* **2018**, *189*, 338–351. [CrossRef]
79. Zhang, Y.; Luo, X.; Onifade, I.; Huang, X.; Lytton, R.L.; Birgisson, B. Mechanical Evaluation of Aggregate Gradation to Characterize Load Carrying Capacity and Rutting Resistance of Asphalt Mixtures. *Constr. Build. Mater.* **2019**, *205*, 499–510. [CrossRef]
80. Jin, C.; Cheng, Y.; Yang, X.; Li, S.; Hu, J.; Lan, G. Adaptive Classification of Aggregate Morphologies Using Clustering for Investigation of Correlation with Contact Characteristics of Aggregates. *Constr. Build. Mater.* **2022**, *349*, 128802. [CrossRef]
81. Chen, Y.; Zhenxia, L.I. Meso-Structure of Crumb Rubber Asphalt Mixture Based on Discrete Element Method. *J. Harbin Inst. Technol.* **2013**, *45*, 116–121.
82. Liu, G.; Pan, Y.; Zhao, Y.; Zhou, J.; Li, J.; Han, D. Research on Asphalt Mixture Force Chain Identification Criteria Based on Computational Granular Mechanics. *Can. J. Civil Eng.* **2021**, *48*, 763–775. [CrossRef]
83. Jiang, H.; Lu, J.; Miao, T. Force Distribution in Three-Dimensional Granular Piles. *J. Lanzhou Univ. (Nat. Sci.)* **2007**, *43*, 134–139. [CrossRef]
84. Liu, G.; Han, D.; Zhu, C.; Wang, F.; Zhao, Y. Asphalt-Mixture Force Chains Length Distribution and Skeleton Composition Investigation Based on Computational Granular Mechanics. *J. Mater. Civil Eng.* **2021**, *33*, 04021033. [CrossRef]
85. Liu, G.; Han, D.; Zhao, Y. Quantitative Investigation of Aggregate Skeleton Force Chains of Asphalt Mixtures Based on Computational Granular Mechanics. *Adv. Civ. Eng.* **2020**, *2020*, 2196503. [CrossRef]
86. Jiang, H.; Lu, J. Experiment on and Analysis of Force Transfer of Axial Load in Granular Packs. *J. Lanzhou Univ. Technol.* **2006**, *32*, 117–121. [CrossRef]
87. Chang, M.; Huang, P.; Pei, J.; Zhang, J. Quantitative Analysis on Evolution and Distribution of Force Chain for Asphalt Mixture Using Discrete Element Method. *Mater. Rep.* **2017**, *31*, 155–159.
88. Sun, Q.; Wang, G.; Hu, K. Some Open Problems in Granular Matter Mechanics. *Prog. Nat. Sci.* **2009**, *19*, 523–529. [CrossRef]
89. Dantu, P. A Contribution to the Mechanical and Geometrical Study of Non-Conhesive Masses. In Proceedings of the International Conference on Soil Mechanics and Foundation Engineering, London, UK, 12–24 August 1957; pp. 144–148.
90. Edwards, S.F.; Oakeshott, R.B.S. The Transmission of Stress in an Aggregate. *Phys. D Nonlinear Phenom.* **1989**, *38*, 88–92. [CrossRef]
91. Bouchaud, J.P.; Cates, M.E.; Claudin, P. Stress Distribution in Granular Media and Nonlinear Wave Equation. *J. Phys. I* **1995**, *5*, 639–656. [CrossRef]
92. He, L.; Pan, G.; He, Z.; Luo, Y.; Pei, Y. A Photo_Elastic Study of Crack Caused by Thermal Stress in Asphalt Concrete Pavament. *J. South China Univ. Technol. (Nat. Sci. Ed.)* **2001**, *6*, 68–71. [CrossRef]
93. Wang, W.J.; Kong, X.Z.; Zhu, Z.G. Friction and Relative Energy Dissipation in Sheared Granular Materials. *Phys Rev E Stat. Nonlin Soft Matter Phys.* **2007**, *75 Pt 2*, 041302. [CrossRef]
94. Zhang, W.; Zhou, J.; Yu, S.; Zhang, X.; Liu, K. Investigation of the Stress Transmission Characterization in High Velocity Powder Compaction Based on Mechanics of Granular Materials. *Chin. J. Appl. Mech.* **2018**, *35*, 154–160+234.
95. Zhang, W.; Zhou, J.; Yu, S.; Zhang, X.; Liu, K. Quantitative Investigation on Force Chains of Metal Powder in High Velocity Compaction by Using Discrete Element Method. *J. Mech. Eng.* **2018**, *54*, 85–92. [CrossRef]
96. Sun, Q.; Jin, F.; Liu, J.; Zhang, G. Understanding Force Chains in Dense Granular Materials. *Int. J. Mod. Phys. B* **2010**, *24*, 1005578. [CrossRef]
97. Wang, H.; Huang, Z.; Li, L.; You, Z.; Chen, Y. Three-Dimensional Modeling and Simulation of Asphalt Concrete Mixtures Based on X-Ray Ct Microstructure Images. *J. Traffic Trans. Eng.(Engl. Ed.)* **2014**, *1*, 55–61. [CrossRef]
98. Shi, L.; Wang, D.; Xu, C.; Liang, H. Investigation Into Meso Performance of Asphalt Mixture Skeleton Based on Discrete Element Method. *J. South China Univ. Technol. (Nat. Sci. Ed.)* **2015**, *43*, 50–56.
99. Feng, H.; Pettinari, M.; Hofko, B.; Stang, H. Study of the Internal Mechanical Response of an Asphalt Mixture by 3-D Discrete Element Modeling. *Constr. Build. Mater.* **2015**, *77*, 187–196. [CrossRef]
100. Si, C.; Zhou, X.; You, Z.; He, Y.; Chen, E.; Zhang, R. Micro-Mechanical Analysis of High Modulus Asphalt Concrete Pavement. *Constr. Build. Mater.* **2019**, *220*, 128–141. [CrossRef]
101. Peng, Y.; Ying, L.; Kamel, M.M.A.; Wang, Y. Mesoscale Fracture Analysis of Recycled Aggregate Concrete Based on Digital Image Processing Technique. *Struct. Concr. J. FIB* **2021**, *22* (Suppl. S1), E33–E47. [CrossRef]
102. Xue, B.; Xu, J.; Pei, J.; Zhang, J.; Li, R. Investigation on the Micromechanical Response of Asphalt Mixture During Permanent Deformation Based on 3D Virtual Wheel Tracking Test. *Constr. Build. Mater.* **2021**, *267*, 121031. [CrossRef]

103. Peng, Y.; Sun, L. Aggregate Distribution Influence on the Indirect Tensile Test of Asphalt Mixtures Using the Discrete Element Method. *Int. J. Pavement Eng.* **2017**, *18*, 668–681. [CrossRef]

104. Peng, Y.; Sun, L.J. Micromechanics-Based Analysis of the Effect of Aggregate Homogeneity on the Uniaxial Penetration Test of Asphalt Mixtures. *J. Mater. Civil Eng.* **2016**, *28*, 04016119. [CrossRef]

Article

Analysis of Rejuvenating Fiber Asphalt Mixtures' Performance and Economic Aspects in High-Temperature Moisture Susceptibility

Yao Zhang [1],*, Ye Wang [1], Aihong Kang [1], Zhengguang Wu [1], Bo Li [1], Chen Zhang [1] and Zhe Wu [2]

[1] Department of Transportation Engineering, College of Architectural Science and Engineering, Yangzhou University, Yangzhou 225127, China
[2] Jiangsu Aoxin Science & Technology Company, Yangzhou 225800, China
* Correspondence: yaozhang@yzu.edu.cn

Abstract: Non-renewable resources such as natural stone and asphalt are in short supply. Recycling technology, with its lower cost, has been used as the primary approach to asphalt pavement maintenance engineering. The inclusion of reclaimed asphalt pavement materials in producing new asphalt pavements may increase the risk of cracking. The strength and toughness of the asphalt mixture can be reduced. In this study, Hamburg wheel tracking tests (HWTT) were performed on rejuvenated asphalt mixtures with distinct maintenance processes. Different kinds of fibers have been used as additives to reinforce the rejuvenated asphalt mixtures. The HWTT rutting curve was identified as having three stages, including the post-compaction stage, the creep stage, and the stripping stage. The three-stage rutting curve model was used to determine the intersection point between the creep stage and stripping stage. The other two feature points (i.e., the post-compaction point and the stripping inflection point) were redefined with a new calculation method. Then, the rutting effect and stripping effect were separated with these feature points. The performance and economic benefits of fiber-reinforced rejuvenated asphalt mixtures were investigated through grey correlation analysis under the three maintenance processes. The feature points of the HWTT curve and the cost of the corresponding maintenance process were selected as the impact factors. Finally, the optimal scheme was developed by analyzing the influence of each factor on both performance and economic benefits.

Keywords: hot rejuvenated asphalt mixtures; Hamburg wheel tracking test; moisture damage; economic benefit; fiber

Citation: Zhang, Y.; Wang, Y.; Kang, A.; Wu, Z.; Li, B.; Zhang, C.; Wu, Z. Analysis of Rejuvenating Fiber Asphalt Mixtures' Performance and Economic Aspects in High-Temperature Moisture Susceptibility. *Materials* **2022**, *15*, 7728. https://doi.org/10.3390/ma15217728

Academic Editor: Gilda Ferrotti

Received: 1 October 2022
Accepted: 29 October 2022
Published: 2 November 2022

Publisher's Note: MDPI stays neutral with regard to jurisdictional claims in published maps and institutional affiliations.

1. Introduction

The use of recycled materials such as reclaimed asphalt pavement (RAP) in hot mix asphalt (HMA) has increased due to the rising cost of petroleum-based products and the negative environmental impact of carbon emissions associated with asphalt binder production [1]. However, the main challenge in incorporating the RAP into the asphalt mixture is the aged binder, which makes the mixture more susceptible to cracking [2–4]. As a result, the aged binder in the RAP tends to limit the high content of recycled material in the mixture. The need to reverse the negative impact of recycled materials in the HMA has prompted researchers to identify and implement innovative and feasible approaches.

Nahar et al. detected the fusion interface between the old and new asphalt by using atomic force microscopy (AFM) [5]. It was found that the old and new asphalt were not entirely miscible. The RAP was portrayed more like a "black stone", which was directly encapsulated by the new asphalt. Therefore, it is unable to truly address the performance degradation brought on by the old asphalt. There are weak links in the recycled asphalt mixture. Zhou et al. proposed that the performance metrics of the rutting and fatigue cracking resistance of the SBS-RAP blended binders were opposed to each other [6]. New experimental methods need to be introduced to equilibrate two performances to choose the optimal RAP content. Cheng

P et al. found that the freeze–thaw splitting strength ratio of hot rejuvenated asphalt mixtures (HRAM) presented a decreasing trend with the increase in RAP content [7]. The decreasing rate was slow when the RAP content was within the range of 15% to 40%, while the decreasing rate became faster beyond this range. Since the aromatic phenol and saturated hydrocarbons in the aged binder transformed into resins and asphaltenes, various rejuvenators have been developed to restore the ratio of aromatic phenol and saturated hydrocarbons in rejuvenated asphalt binder [8,9]. The rejuvenators can reverse the aging process and restore part of the binder properties [10,11]. However, the properties of the cracking resistance and moisture susceptibility of the HRAM are still poor.

The Hamburg wheel tracking test (HWTT) was introduced for the evaluation of cracking resistance and moisture susceptibility with a more accurate reflection of the properties of asphalt mixtures [12]. The test involved the real-time monitoring of the correlation between the number of load cycles and the rutting depth (RD). Test curves were divided into three main phases: (a) the post-compaction phase, (b) the creep phase, and (c) the stripping phase. The stripping inflection point (SIP) was determined by counting the number of load cycles on the HWTT curve at which a sharp increase in rut depth occurs [13]. The SIP was represented at the intersection of the fitted lines that characterize the creep phase and the stripping phase. However, different fitting point selections could lead to an inaccurate finding of SIP. This may cause a misjudgment of the cracking resistance and moisture susceptibility of asphalt mixtures. In investigations on the rutting of hot recycled asphalt mixtures in the HWTT, researchers have found that the rutting depth increased more quickly in the third stage but developed more slowly in the first two stages when increasing the RAP. The findings give an indication that the properties of the cracking resistance and moisture susceptibility of hot recycled asphalt mixtures need to be enhanced, especially at increasing recycled binder ratios and RAP contents [14].

Fiber is a kind of natural or synthetic material, which acts as a reinforcement of strength and stability in the HMA. The fiber network structure enhances the binder properties of the asphalt mixtures [15]. It can withstand a portion of the force, preventing the cracking of the asphalt mixtures [16]. The moisture susceptibility can be also improved with the addition of fibers, and different types of fibers have varied functions in asphalt mixtures [17,18]. Chen Y Z et al. found that fibers could significantly improve the residual stability and freeze–thaw splitting strength of asphalt mixtures [19]. In comparison to polyester and wood algal fibers, basalt fibers are more effective at increasing the water stability of asphalt mixtures. Zhang K et al. found that asphalt mixture containing basalt fiber achieved better results in resisting salinity and moisture environments [20]. Although adding fibers and recycling agents as well as new aggregates can perform good results in asphalt mixtures, doing so will certainly drive up the price of the HRAM. On the other side, the excessive use of RAP can reduce the cost of the HRAM, but it degrades the serviceability of asphalt mixtures.

Hence, how to select a hot rejuvenated asphalt mixture that is both affordable and high-performing among available possibilities has been the main motivation for study. This is also the actual need of the pavement maintenance industry. Meanwhile, more innovative methods are needed to gain the feature points of the HWTT curve. Moreover, the properties of the cracking resistance and moisture susceptibility of hot recycled asphalt mixtures can be improved by controlling these feature points. Therefore, the objective of this paper is to investigate the mix design of hot rejuvenated asphalt mixtures with different fiber types under three pavement maintenance processes. The analysis approach for obtaining the feature points under three stages is modified after conducting the HWTT to obtain rutting curves. Subsequently, the costs of asphalt mixtures are obtained through an economic benefit analysis. Finally, the performance and economic benefits evaluation is conducted using grey correlation analysis to provide a rational solution and guidance for asphalt pavement maintenance projects. The flowchart of this study is shown below (Figure 1).

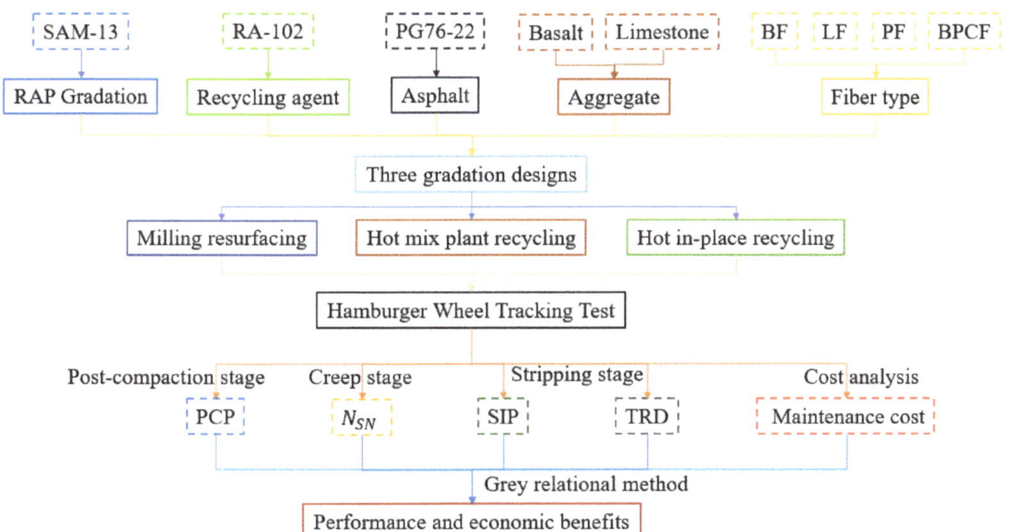

Figure 1. Flowchart of the research.

2. Construction Technology and Raw Materials

Three hot regeneration maintenance techniques were employed to create hot rejuvenated asphalt mixtures to investigate the high-temperature performance and moisture susceptibility of each asphalt mixture under various pavement maintenance operations in the HWTT.

2.1. Construction Technology

There are three typical maintenance technologies in current pavement construction, including the milling resurfacing (MR) process, the hot mix plant recycling (HMPR) process, and the hot in-place recycling (HIPR) process. The RAP dosage in the asphalt mixture differs greatly for the three maintenance processes. The RAP dosage in HIPR is generally more than 70%, while it is generally less than 50% in the HMPR process. There is no RAP used in the MR process; in other words, the material is a 100% new asphalt mixture. The basic principles of the three pavement maintenance processes are shown in Figures 2–4.

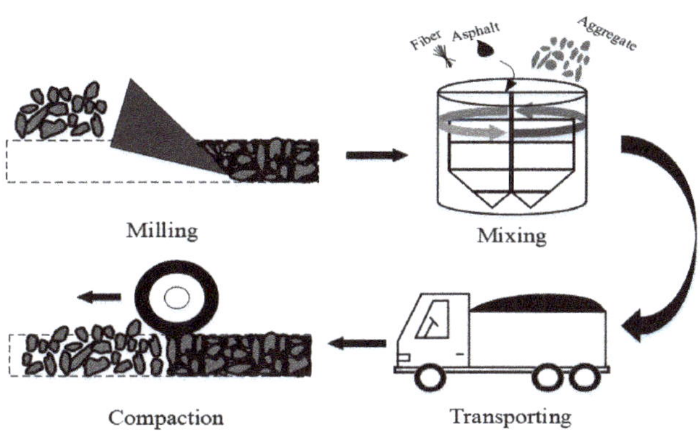

Figure 2. Milling resurfacing process.

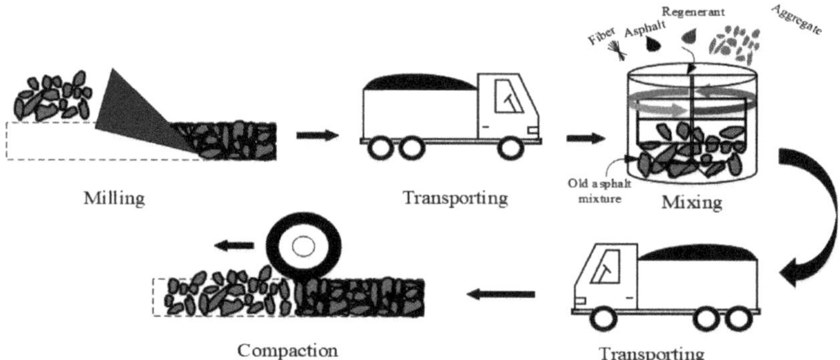

Figure 3. Hot mix plant recycling process.

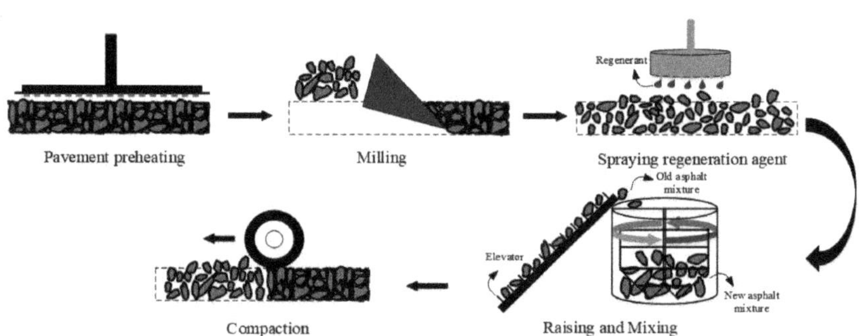

Figure 4. Hot in-place recycling process.

In the MR process, the old asphalt pavement is milled, and then 100% of the new asphalt mixture from the asphalt mixing plant is spread and compacted on the milled pavement. During the HMPR process, the RAP is transported to the asphalt mixing plant as aggregates, where it is mixed with the new asphalt mixture after adding some of the recycling agents. In the HIPR process, 10% to 30% of the new materials are transported to the construction site. The old pavement is loosened by hot raking with the HIPR equipment. Then, the recycling agent and new asphalt mixtures are stirred with the old asphalt mixtures to create rejuvenated asphalt mixtures. Finally, the rejuvenated asphalt mixtures are paved in situ and compacted to form new pavement.

By comparison of the three maintenance technologies, it can be seen that the milling resurfacing process uses the highest quality raw materials to ensure excellent pavement performance. However, the cost of this process is rather high, and the old materials cannot be reclaimed, which can pollute the environment. The hot mix plant recycling process can solve part of the RAP recycling problem, and more than 70% of the raw materials are added in this process, which can also ensure the performance of asphalt pavements. The hot in-place recycling process allows for the complete use of old materials, significantly reducing material costs. It can also increase the efficiency of construction and reduce traffic. However, the incorporation of excessive RAP materials may affect the crack resistance of asphalt pavements. As seen from these pavement maintenance technologies, it is found that each of them has its characteristics. When choosing the appropriate technology for pavement maintenance, the construction process should take both the technical and financial benefits into account.

2.2. Raw Materials

2.2.1. RAP

The recycled asphalt pavement (RAP) used in the study was from the large maintenance engineering project of the Wuxi section in the Huning Expressway. The RAP gradation was stone matrix asphalt with a nominal aggregate size of 13.2 mm (SMA-13) located at the upper layer of the pavement structure. The original pavement was reclaimed using a hot milling technique to reduce the impact of the milling operation on the RAP gradation. Figures 5–7 depict the RAP reclaiming and the on-site milling construction.

Figure 5. Pavement heating car.

Figure 6. Manual milling hot pavement.

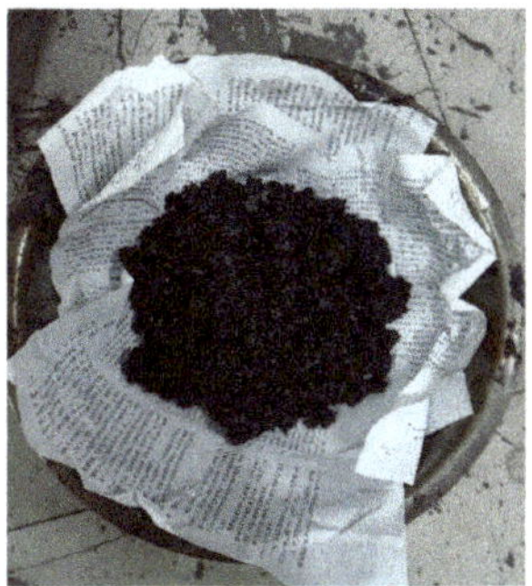

Figure 7. Reclaimed asphalt mixture.

2.2.2. Recycling Agent

The recycling agent of RA-102 was produced by the Subbo7t Company. The physical and mechanical properties, including the viscosity, flash point, saturated fraction content, aromatic content, viscosity ratio, and quality changes before and after the rolling thin film oven test (RTFOT), were measured with corresponding test specifications. The test results are shown in Table 1.

Table 1. Technical index of Subtherm RA-102 Recycling agent.

Technical Index	RA-102	National Normative Criterion	Test Method
90 °C Viscosity, cP	4000	-	T0619
Flash point/°C	248	$\geq$220	T0633
Saturated fraction content/%	25.6	$\leq$30	T0618
Aromatic content/%	53	$\geq$30	T0618
Viscosity ratio before and after RTFOT	1.34	$\leq$3	T0610
Quality changes before and after RTFOT/%	1.02	$\leq$4%	T0603

2.2.3. Asphalt

The performance graded 76-22 SBS-modified asphalt was used in this study. The asphalt properties were tested following the Chinese specification of the "Test Procedure for Asphalt and Asphalt Mixture for Highway Engineering" (JTG E20-2011). The test results are shown in Table 2.

Table 2. SBS-modified asphalt (PG76-22) technical index.

Test Items		Specification Requirement	Test Result	Test Method
Penetration degree (25 °C)/0.1 mm		60~80	71	T0604
Softening point/°C		$\geq$55	64	T0606
Ductility (5 cm/min, 5 °C)/cm		$\geq$30	48	T0605
Penetration index (PI)		-0.4~1.0	0.5	T0604
Segregation (softening point difference)/°C		$\leq$2.5	1.4	T0661
Elasticity recovery (25 °C)/%		$\geq$65	76	T0662
Residue after RTFOT	Quality change/%	$\pm$1.0	-0.08	T0610
	Penetration ratio/%	$\geq$60	86	T0604
	15 °C Residual ductility/cm	$\geq$20	37	T0605

2.2.4. Aggregate

The coarse aggregate was made of basalt with hard, clean, rough, and tough properties. The fine aggregate and mineral filler were made of limestone with properties of dry, clean, non-clumpy, and free from impurities. The technical indexes are shown in Tables 3 and 4.

Table 3. Technical indexes of coarse and fine aggregates.

Aggregate Type		Apparent Specific Gravity	Gross Volume Relative Density
Basalt aggregate	1#	2.953	2.908
	2#	2.962	2.902
	3#	2.902	2.853
Limestone aggregate	4#	2.704	/

Table 4. Technical indexes of mineral fillers.

Targets of Test		Test Result	Specification Requirement	Test Method
Water content/%		0.3	≤1	Drying method
Relative density		2.654	≥2.5	T0352
Appearance		no agglomerates	no agglomerates	/
Hydrophilic coefficient		0.63	<1	T0353
Particle size range	<0.6 mm	100	100	
	<0.15 mm	92.6	90~100	T0351
	<0.075 mm	92.2	75~100	

2.2.5. Fiber

Three types of fibers, including basalt fiber (BF), lignin fiber (LF), and polyester fiber (PF), were considered to create rejuvenated fiber asphalt mixtures. The basalt fiber and polyester fiber were short-cut fibers produced by Jiangsu Tianlong Basalt Continuous Fibers Co Ltd. The BF was golden brown in color, flat, and free from impurities, whereas the PF is pure white, flat, with no impurities. The lignin fiber used was the white flocculent fiber type ZZ8/1 from Riedenmei and Sons. Table 5 presents the results of the technical performance tests conducted on the three fibers, and Figure 8 displays the fiber morphology.

Table 5. Technical performances of fibers.

Technical Performances	Basalt Fiber	Lignin Fiber	Polyester Fiber
Diameter (μm)	16.23	0.045	16.58
Density (g/cm^2)	2.71	0.82	0.95
Tensile strength (MPa)	2785	265	1010
Breaking elongation (%)	5.72	17.26	34.72

(**a**) Basalt fiber   (**b**) Lignin fiber   (**c**) Polyester fiber

Figure 8. Morphology of used fibers.

2.3. Testing of Reclaimed Asphalt Mixture

2.3.1. RAP Gradation and Asphalt Content

The old asphalt and old aggregates were recycled using the centrifugal separation process in accordance with T0722-1993 in the Chinese specification of JTG E20-2011. Trichloroethylene was used as the extraction solvent. The asphalt mixture mineral gradation test method (T0725-2000) was conducted to determine the asphalt content and aggregate gradation of the reclaimed asphalt mixture. The results show that the asphalt content in the RAP is 5.8%, and the gradation of the RAP is shown in Figure 9.

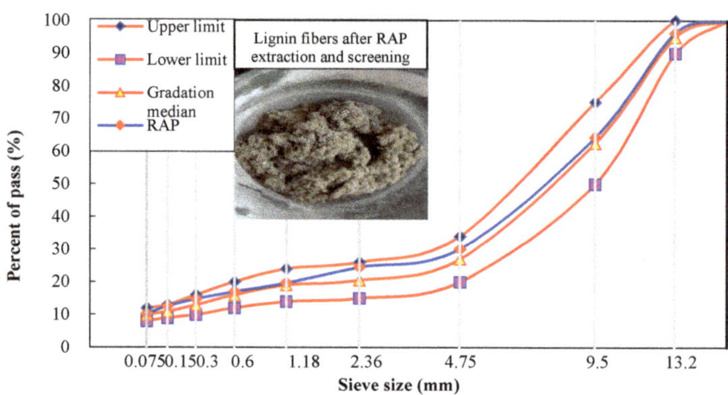

Figure 9. Gradation of RAP.

As can be seen in Figure 2, the RAP gradation is more closely aligned with the median gradation curve following the extraction and screening processes used in the layered hot milling technology. The old pavement used lignin fiber as a stabilizer, which was also extracted and shown in Figure 2. It was clear to see the quality loss and the aging phenomenon of the LF. Hence, the appropriate number of new fibers should be added to revitalize the rejuvenated SMA-13 mixture in the regeneration mix design.

2.3.2. Properties of Old Asphalt

The properties of old asphalt including the penetration degree, softening point, ductility, and viscosity were tested. The test results are shown in Table 6.

Table 6. Old asphalt properties.

Technical Index	Old Asphalt	SBS-Modified Asphalt Requirements	Test Method
Penetration degree (25 °C)/0.1 mm	39	50–80	T0604
Softening point/°C	69	>60	T0606
Ductility/cm	7.8	>30	T0605
Viscosity (135 °C)/Pa·s	2.33	≤3	T0613

It can be seen from this table that the technical indexes of the softening point and viscosity at 135 °C for the old asphalt can meet the SBS-modified asphalt specification requirements, while the values of the penetration degree and ductility are lower than the requirements. The old asphalt belongs to grade II aging, and the degree of aging is light [21].

2.3.3. Dosage of Recycling Agent

Since the aged asphalt might cause negative impacts on the rejuvenated asphalt mixture, the recycling agent was thought to be able to restore part of the aged asphalt properties. The performance design method was used to determine the proper dosage of the recycling agent in the total asphalt. The dosage of the recycling agent was selected at four levels, including 4%, 6%, 8%, and 10%, respectively. General performance tests of rejuvenated asphalt were conducted. The test results are shown in Table 7.

Table 7. Test results of rejuvenated asphalt.

Technical Index	Recycling Agent Dosage (%)					New SBS-Modified Asphalt	Test Method
	0	4	6	8	10		
Penetration degree (25 °C)/0.1 mm	39	60	68	74	78	71	T0604
Softening point/°C	69	65	63	61	56	64	T0606
Ductility/cm	7.8	22.4	28.6	31.4	34.6	48	T0605

Table 7 demonstrates that as the dosage of the recycling agent is increased, the penetration degree and ductility of rejuvenated asphalt increase, while the softening point declines. When the dosage of the recycling agent reaches 6%, the penetration degree and softening point of rejuvenated asphalt return to the performance level of the original SBS asphalt. Hence, the recycling agent dosage of 6% was chosen to formulate the hot rejuvenated asphalt mixtures.

2.4. Mix Proportion Design of HRAM

The HRAM was supplemented with basalt fibers (BF), lignin fibers (LF), polyester fibers (PF), and basalt–polyester compound fibers (BPCF) to evaluate the impact of various fibers on high-temperature performance and moisture susceptibility. The RAP content in the hot rejuvenated asphalt mixture for HIPR was set at 80%, while the RAP content for HMPR was set at 30%, taking into account the distinct maintenance procedures between the HIPR and the HMPR technologies.

Gradation Design

Marshall test was carried out for new gradation design based on the requirements of SMA-13 asphalt mixture in the Chinese specification of JTG F40-2004. The designed gradation curves under the two maintenance technologies (HIPR and HMPR) were shown in Figure 10. The gradation curve of the milling resurfacing (MR) technology was also shown in this figure as a control group, and the RAP content is 0% in this group. The optimal asphalt content of original and rejuvenated fiber asphalt mixtures is shown in Table 8, and the volumetric parameters of these mixtures are listed in Table 9.

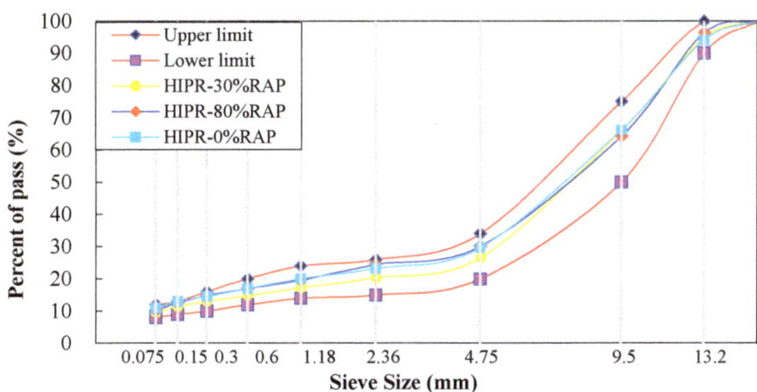

Figure 10. Gradation curves with three maintenance technologies.

Table 8. Fiber dosage and asphalt content under three maintenance processes.

Number	Grading Type	Fiber Types	Fiber Dosage/‰	Optimal Asphalt Content/%
1	MR SMA-13	LF	3	6.0
2	MR SMA-13	BF	3	5.8
3	MR SMA-13	PF	3	5.8
4	MR SMA-13	BPCF	1.5/1.5	5.8
5	HMPR SMA-13	LF	3	6.1
6	HMPR SMA-13	BF	3	5.9
7	HMPR SMA-13	PF	3	5.9
8	HMPR SMA-13	BPCF	1.5/1.5	5.9
9	HIPR SMA-13	LF	1	6.0
10	HIPR SMA-13	BF	3	6.0
11	HIPR SMA-13	PF	3	6.0
12	HIPR SMA-13	BPCF	1.5/1.5	6.0

Table 9. Volumetric parameters of aggregate gradations.

Volume Parameters	MR-BF	MR-LF	HMPR-BF	HMPR-LF	HIPR-BF	HIPR-LF	Specification Requirements
VV/%	4.3	3.8	4.1	4.0	4.1	4.0	3~4.5
VMA/%	16.8	17.2	17.1	17.3	17.5	17.5	≥16.5
VFA/%	74.4	78.0	76.0.	76.9	76.9	78.3	70~85
Stability/kN	11.50	11.93	12.05	11.5	12.45	11.28	≥6
Gross volume relative density	2.514	2.507	2.520	2.507	2.518	2.521	-
Maximum theoretical relative density	2.627	2.605	2.628	2.611	2.623	2.621	-

3. Experiments and Results

3.1. Hamburg Wheel Tracking Test

According to AASHTO T 324, the Superpave gyratory compactor (SGC) was selected to create cylinder specimens with a diameter of 150 mm, a thickness of 60 mm, and a target void content of $7.0 \pm 0.5\%$. After the SGC specimen had been cooled, a wet saw was used to cut along the equidistant line so that the specimen could be embedded in the mold. The water bath temperature of the HWTT apparatus was set at 60 °C, and the specimen was submerged in water for more than 20 mm. The reciprocating motion of the vehicle load on the specimen was simulated by a steel wheel with a diameter of 203.2 ± 2.0 mm and a width of 47 ± 0.5 mm under a load of 703 ± 4.5 N. The position of the steel wheel changed sinusoidally over time, traversing the specimen 52 times per minute, and reaching a top speed of 0.305 ± 0.02 m/s at the midpoint of the specimen. The linear variable differential displacement transducer (LVDT) was used to record the rut depth at 11 points along the direction of wheel crush in real time. The HWTT apparatus and specimen are shown in Figures 11 and 12. After the test was completed, the morphology of specimens such as HIPR-BF and HMPR-BF is shown in Figures 13 and 14. As can be seen from these figures, there exists a big difference in the rut depths at the end of the test with distinct maintenance technologies.

Figure 11. SGC specimen.

Figure 12. HWTT apparatus.

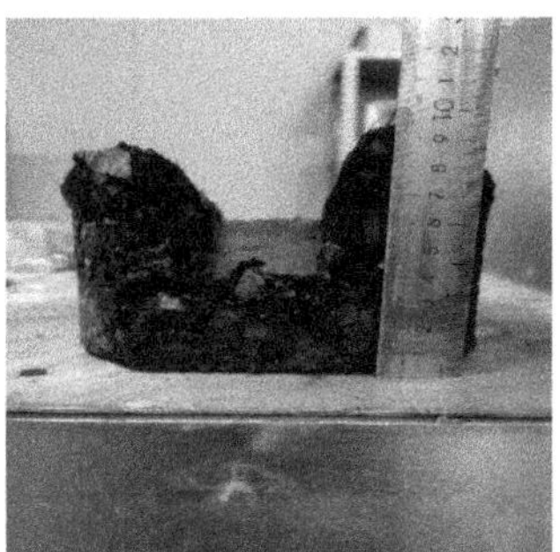

Figure 13. Specimen of HIPR-BF.

Figure 14. Specimen of HMPR-BF.

3.2. Test Results

The rut depth of eight rejuvenated asphalt mixtures and four new asphalt mixtures was recorded during the HWTT test, and the total rut depth of each mixture is shown in Figure 15. It can be seen from this figure that the HMPR asphalt mixture (i.e., 30% RAP) has the lowest rut depth when compared to the other two types of combinations, while the HIPR asphalt mixture (i.e., 80% RAP) has the maximum rut depth at the end of the test. The hydrothermal coupling performance of HWTT tends to improve and then decline with the increase in RAP materials. This is primarily caused by the modification to the aged asphalt content in the RAP materials, which increases the early-stage rutting resistance of rejuvenated asphalt mixtures. However, the ability of rejuvenated asphalt mixtures to resist moisture damage is insufficient to support cyclic heavy loads when the RAP content goes above a certain point. Based on three maintenance techniques, the BF combinations show better rutting resistance than the other three types of rejuvenated fiber asphalt mixtures.

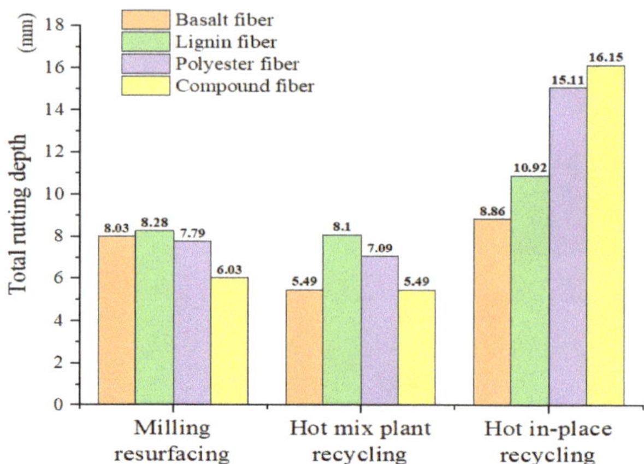

Figure 15. Total rutting depth at the end of HWTT.

Since the HWTT can simultaneously characterize the rutting resistance and water stability of the asphalt mixture, many researchers have tried to find the feature points of the HWTT curve to study its performance at different periods. There are three main stages for the specimen under wheel load, including the post-compaction stage (PCS), the creep stage (CS), and the stripping stage (SS) [22], as shown in Figure 15. The specimen in the PCS is subjected to a high compression force given by the HWTT wheels, and the air voids are post-compacted. Hence, the rut depth in the PCS changes at a faster rate and for a shorter period as the number of wheels increases. At the CS, the rut depth increases steadily, and materials in the specimen tend to have shear flow and continue to produce rutting damage under the wheel load. At the SS, the asphalt mixture is subjected to the double effects of wheel load as well as high-temperature moisture damage; part of the asphalt loses its adhesive bonds, the asphalt mixture begins to peel off, and the rutting and stripping of the specimen are accelerated.

Currently, the total rut depth (TRD) and stripping inflection point (SIP) are widely recognized as HWTT performance indicators. The SIP is the inflection point from CS to SS, which is commonly recognized as the intersection of the two straight lines fitted by CS points and SS points, as seen in Figure 16. In a sense, the value of SIP is greatly affected by the slope of two fitting lines. However, the slope of two fitting lines in the CS and SS seems not to be unified. Hence, a new methodology or mathematical model is needed to be developed to give a higher accuracy of SIP determination.

The HWTT curve is made up of two parts, the first of which has a negative curvature and is followed by the other, which has a positive curvature, according to the findings of earlier studies on rutting prediction models conducted by our research group [14] and Fan Y et al. [22]. This curve can be completely represented by Equation (1).

$$RD = \rho * \left[ln\left(\frac{N_{ult}}{N - N_0} \right) \right]^{-\frac{1}{\beta}} \tag{1}$$

where N = the number of load cycles; RD = the rut depth at a certain number of load cycles (mm); N_0 = the number of load cycles where rut depth occurs; ρ, N_{ult}, and β = model coefficients.

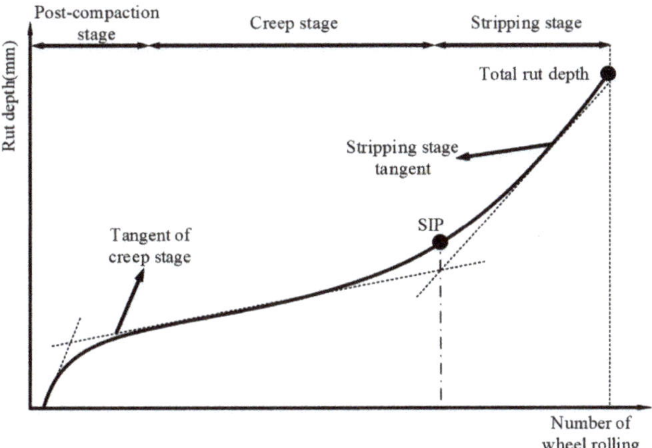

Figure 16. Schematic diagram of a conventional HWTT curve.

Using the above formula to fit the HWTT data under the three maintenance technologies, the fitting curves are shown in Figures 17–20, and the relevant fitting parameters are shown in Table 10. From this figure, it can be seen that all four fiber asphalt mixtures have been wheeled 20,000 times. The HIPR asphalt mixtures show good resistance to high-temperature rutting in the post-compaction and creep stages, but they present poor water stability in the stripping stage.

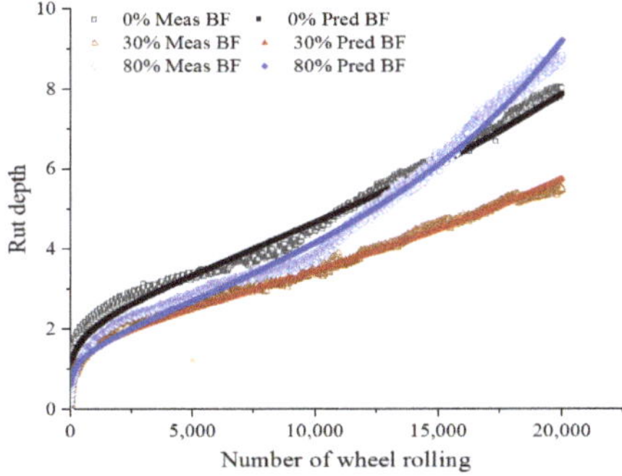

Figure 17. HWTT curve of BF asphalt mixture.

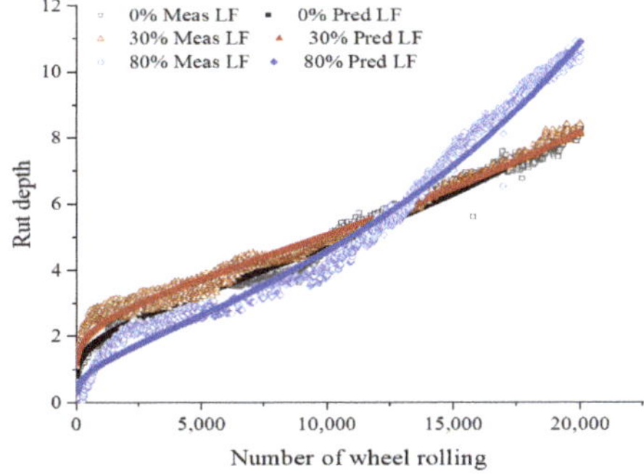

Figure 18. HWTT curve of LF asphalt mixture.

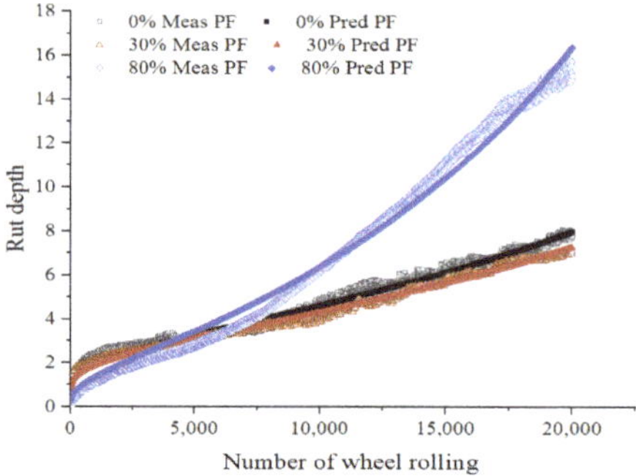

Figure 19. HWTT Curve of PF Asphalt Mixture.

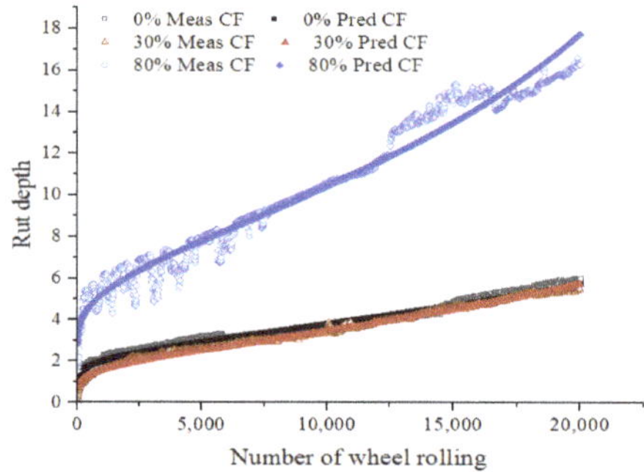

Figure 20. HWTT curve of BPCF asphalt mixture.

Table 10. HWTT fitting parameters of each asphalt mixture.

Mixture Type	HWTT in Wet Condition			R-Squared	Root Mean Square Error (RMSE)
	ρ	β	N_∞		
MR-BF	7.199	1.084	4.948×10^4	0.996	0.039
MR-LF	7.565	1.005	4.948×10^4	0.996	0.063
MR-PF	7.275	1.041	4.948×10^4	0.996	0.068
MR-BPCF	5.407	1.382	5.000×10^4	0.972	0.039
HMPR-BF	5.263	1.122	4.948×10^4	0.996	0.030
HMPR-LF	7.518	1.182	4.948×10^4	0.996	0.068
HMPR-PF	6.639	1.081	4.948×10^4	0.996	0.041
HMPR-BPCF	5.263	1.119	4.948×10^4	0.980	0.032
HIPR-BF	5.078	1.044	3.419×10^4	0.977	0.049
HIPR-LF	9.394	0.655	4.948×10^4	0.981	0.151
HIPR-PF	14.180	0.593	5.000×10^4	0.998	0.277
HIPR-BPCF	16.131	1.133	5.000×10^4	0.973	0.643

There are three feature points in the HWTT rutting curve, which can be used to separate the post-compaction stage, the creep stage, and the stripping stage. Finding the feature points in the curve fitting model can better characterize the HWTT results. The first derivative of Equation (1) is taken to obtain the slope of the HWTT curve, as given in Equation (2).

$$RD' = \frac{\rho}{N * \beta}\left(ln\frac{N_{ult}}{N}\right)^{-\frac{1}{\beta}-1} \tag{2}$$

When the curvature of the HWTT curve turns from negative to positive, the second inflection point can be determined. The second derivative of Equation (1) is given as,

$$RD'' = -\frac{\rho}{\beta * N^2}\left[\left(ln\frac{N_{ult}}{N}\right)^{-\frac{1}{\beta}-1} + \left(-\frac{1}{\beta}-1\right)\left(ln\frac{N_{ult}}{N}\right)^{-\frac{1}{\beta}-2}\right] \tag{3}$$

When the second derivative is set to zero, two solutions exist in Equation (3).

$$x_1 = N_{ult}, \quad x_2 = \frac{N_{ult}}{e^{1+\frac{1}{\beta}}}$$

Since N_{ult} is much larger than 20,000 loads, the unique solution of x_2 is taken. This solution is the intersection (i.e., N_{SN}) from the positive curvature to the negative curvature, as shown in Figure 21. The value of N_{SN} for each asphalt mixture is shown in Figure 22. In front of the point of N_{SN}, the adhesion between the aggregate and asphalt in the specimen is good, and the specimen is in the post-compaction stage with a positive curvature. After the point of N_{SN}, the rutting curve enters the negative curvature phase. The asphalt adhesion performance decreases when the asphalt mixture starts to produce rapid damage, and the rut depth of the specimen is accelerated by the moisture effect.

The intersection point (i.e., N_{SN}) manifests itself earlier as the RAP content increases, indicating that the RAP is more moisture-sensitive than new asphalt mixtures. The RAP in the mixture has the potential to accelerate moisture damage, causing the mixture to reach the stripping stage earlier.

To distinguish the damage to the asphalt mixture under the effects of high temperature and water immersion, the HWTT curve before the point of N_{SN} is fitted by Equation (4), and the schematic diagram is shown in Figure 18. The curve fitting parameters were obtained, as shown in Table 11.

$$RD = A \times NP^B \tag{4}$$

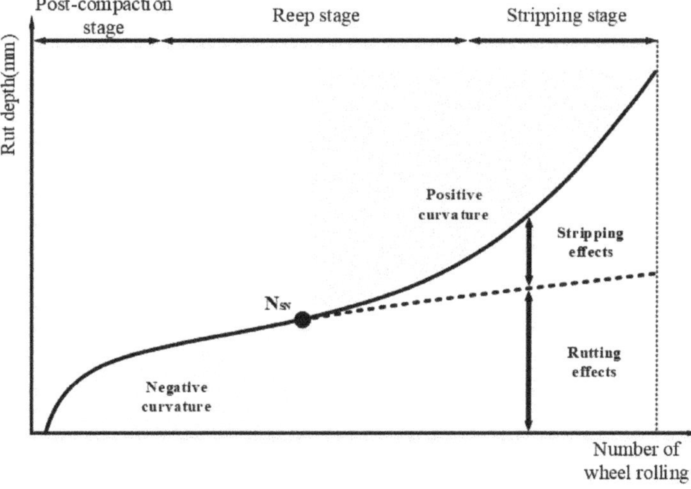

Figure 21. Intersection from positive curvature to negative curvature.

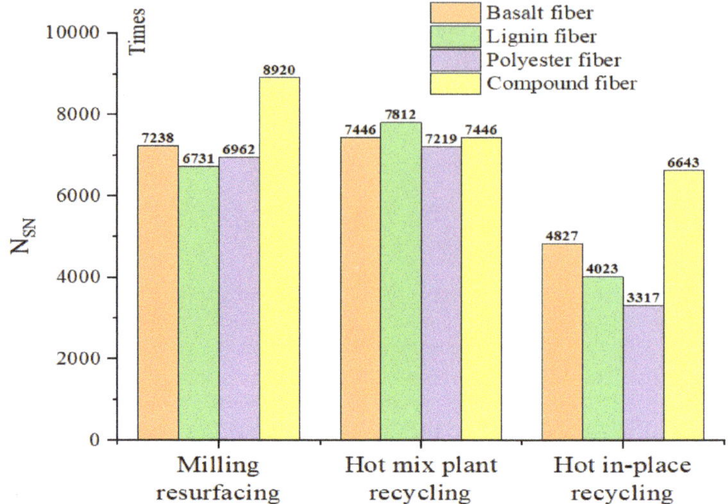

Figure 22. N_{SN} of each fiber asphalt mixture.

Table 11. Fitting parameters of the negative curvature phase.

Mixture Type	A	B	RMSE
MR-BF	0.303	0.283	0.015
MR-LF	0.429	0.238	0.027
MR-PF	0.570	0.199	0.054
MR-BPCF	0.616	0.187	0.021
HMPR-BF	0.486	0.203	0.062
HMPR-LF	0.691	0.196	0.052
HMPR-PF	0.585	0.190	0.047
HMPR-BPCF	0.488	0.204	0.070
HIPR-BF	0.507	0.182	0.030
HIPR-LF	0.423	0.182	0.056
HIPR-PF	0.41	0.193	0.074
HIPR-BPCF	0.928	0.248	0.066

The fitting curves can be extended to the end of the test to obtain the high-temperature rutting effects in the HWTT. The rut depth at 20,000 loads calculated by Equation (4) was obtained, as shown in Figure 23. The stripping effects can be obtained by using the total rut depth minus the high-temperature rutting effects, which are expressed in Figure 18. The results of the rut depth related to the stripping effects at 20,000 loads are shown in Figure 24. Figure 23 shows that the high-temperature rutting performance decreases with the increase in the RAP content in terms of single-doped fiber asphalt mixtures, and the LF asphalt mixture in the HIPR process performed the best. Figure 24 shows that the stripping resistance displays a contrary tendency compared with high-temperature rutting resistance. The asphalt mixture with basalt-polyester compound fibers in the MR process has the best performance.

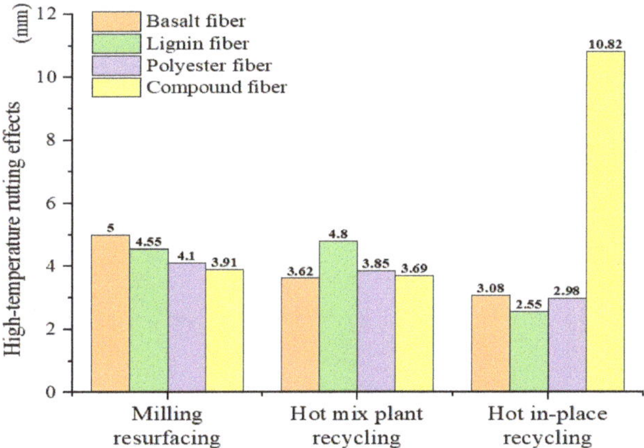

Figure 23. High-temperature rutting effects of fiber asphalt mixtures.

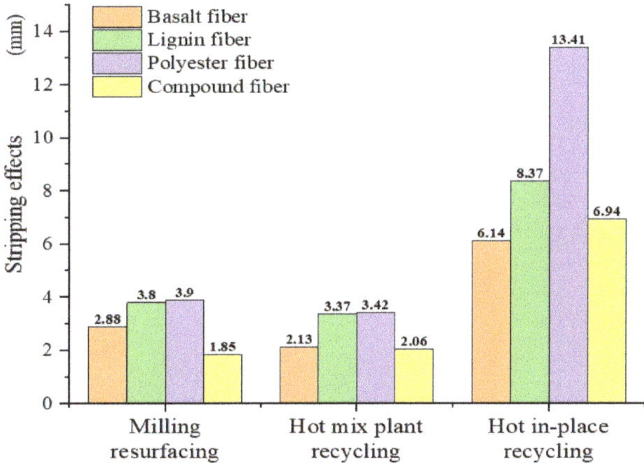

Figure 24. Stripping effects of fiber asphalt mixtures.

From the above study, it is found that the inflection point of N_{SN} can be obtained by setting the second derivative to zero in the three-stage curve fitting model. However, the other two critical points separate the three stages of the development of the HWTT curve, which cannot be determined directly from this model. Hence, the concept of a stationary point in the HWTT curve is introduced to mathematically redefine the post-compaction critical point and the stripping critical point. Take the BF asphalt mixture in the MR process as an example, connect the N_{SN} with the curve starting point to find the first line segment: $y_1 = 5.44 \times 10^{-4}x$, $(0 \leq x \leq 7238.3)$. Then, connect the N_{SN} with the point at end of the test to obtain the second line segment: $y_2 = 3.09 \times 10^{-4}x + 1.71$, $(7238.3 \leq x \leq 20000)$. Form the vertical line between the HWTT curve and the two line segments to obtain the maximum value of Δh_1 and Δh_2. The loading cycle at the maximum value of Δh_1 is defined as the first stationary point (i.e., PCP) between the post-compaction stage and the creep stage. Likewise, the loading cycle at the maximum value of Δh_2 is defined as the second stationary point (i.e., SIP) between the creep stage and the stripping stage. A schematic diagram of the stationary point is shown in Figure 25. The corresponding line segments and three feature points (i.e., first stationary point, inflection point, and second stationary point) are obtained in this method for all HWTT curves of asphalt mixtures, as shown in Table 12.

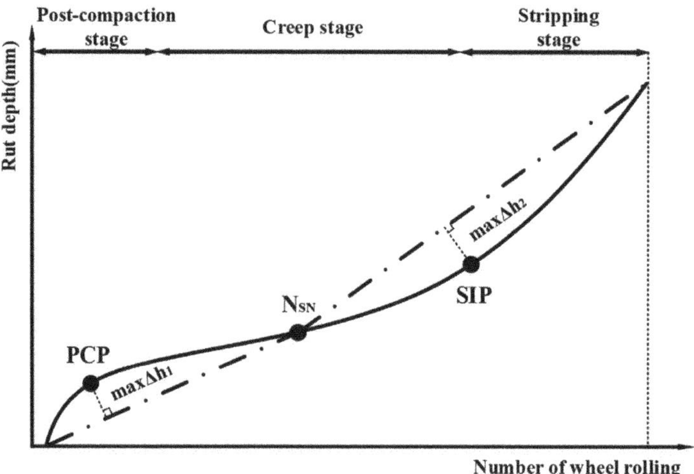

Figure 25. Schematic diagram of PCP and SIP redefinition.

Table 12. Line segments and feature points under the new method.

Mixture Type	First Line Segment	(PCP, RD)	Second Line Segment	(SIP, RD)
MR-BF	$y_1 = 5.44 \times 10^{-4}x$	(810, 1.95)	$y_2 = 3.09 \times 10^{-4}x + 1.71$	(14,500, 5.96)
MR-LF	$y_1 = 5.65 \times 10^{-4}x$	(780, 1.40)	$y_2 = 3.42 \times 10^{-4}x + 1.50$	(14,325, 6.11)
MR-PF	$y_1 = 5.47 \times 10^{-4}x$	(790, 1.43)	$y_2 = 3.22 \times 10^{-4}x + 1.57$	(14,410, 5.95)
MR-BPCF	$y_1 = 3.84 \times 10^{-4}x$	(950, 1.63)	$y_1 = 2.11 \times 10^{-4}x + 1.56$	(18,120, 5.35)
HMPR-BF	$y_1 = 3.99 \times 10^{-4}x$	(810, 1.16)	$y_2 = 2.21 \times 10^{-4}x + 1.33$	(14,615, 4.41)
HMPR-LF	$y_1 = 5.73 \times 10^{-4}x$	(820, 1.81)	$y_2 = 3.03 \times 10^{-4}x + 2.11$	(14,780, 6.41)
HMPR-PF	$y_1 = 5.02 \times 10^{-4}x$	(800, 1.39)	$y_2 = 2.86 \times 10^{-4}x + 1.56$	(14,515, 5.50)
HMPR-BPCF	$y_1 = 4.00 \times 10^{-4}x$	(810, 1.16)	$y_2 = 2.21 \times 10^{-4}x + 1.33$	(14,610, 4.41)
HIPR-BF	$y_1 = 5.53 \times 10^{-4}x$	(550, 1.00)	$y_2 = 4.32 \times 10^{-4}x + 0.59$	(13,960, 5.64)
HIPR-LF	$y_1 = 5.73 \times 10^{-4}x$	(570, 0.63)	$y_2 = 5.40 \times 10^{-4}x + 0.14$	(13,260, 6.17)
HIPR-PF	$y_1 = 7.79 \times 10^{-4}x$	(505, 0.63)	$y_2 = 8.27 \times 10^{-4}x - 0.16$	(12,930, 8.56)
HIPR-BPCF	$y_1 = 1.30 \times 10^{-3}x$	(600, 4.34)	$y_2 = 6.82 \times 10^{-4}x + 4.11$	(14,530, 13.4)

As can be seen from Table 12, the HIPR asphalt mixtures pass in the creep stage at around 500 to 600 cycles of the wheel rolling with a low value of rut depth. The MR asphalt mixtures and HMPR asphalt mixtures entered the creep stage at around 800 to 950 cycles of wheel millings. The results show that the PCP of the HIPR asphalt mixture is earlier than that of the MR asphalt mixture and the HMPR asphalt mixture. However, the SIP seems to have a reverse trend in these mixtures. Compared with the new definition of SIP and the conventional calculation of SIP, it is found that this new method can provide more scientific theoretical support for the HWTT results for different asphalt mixtures. The PCP and SIP calculation methods can provide a theoretical basis for decision-making in maintenance projects.

4. Comprehensive Analysis of Cost and Performance

4.1. Economic Cost Analysis

The economic benefit is one of the key factors which should be also considered. The costs incurred under different maintenance methods are different, and the cost is an unavoidable problem in the practical application of the project. By constructing an economic model to analyze the costs under the three maintenance methods, it is possible to guide the concrete implementation of maintenance works.

The cost of the asphalt pavement maintenance process can be divided into four parts, namely, the milling cost, mixing cost, paving and rolling cost, and transportation cost [23].

To intuitively reflect the economic effects, the unit of price is standardized in this paper as Yuan (CNY) per ton and the expression of economic cost is shown in Equation (5).

$$O = \sum_{i=1}^{n} M_i + \sum_{i=1}^{m} C_i + \sum_{i=1}^{l} P_i + \sum_{i=1}^{g} T_i \tag{5}$$

where M is the milling cost, C is the asphalt mixture and mixing cost, P is the paving and rolling cost, T is the transportation cost.

The asphalt mix and mixing costs are modeled as Equation (6).

$$C = \left(C_{as} P_{as} + C_{ag} P_{ag} + C_{fi} P_{fi} \right) \times \left(1 - R_p \right) + R_p \times C_{ra} + C_{mi} \tag{6}$$

where C_{as} is the cost per ton of asphalt, C_{ag} is the cost per ton of aggregates, C_{fi} is the cost per ton of fiber, C_{ra} is the cost per ton of regeneration agent, C_{mi} is the mixing cost per ton if asphalt mixture; P_{as} is the asphalt ratio, P_{ag} is the aggregate content, P_{fi} is the fiber dosage, and R_p is the RAP content. The calculation models for milling, paving and rolling, and transportation are simply cumulative and will not be repeated here.

In the MR maintenance process, the RAP content is zero percent, and there is a regeneration agent cost. The transportation cost only includes one-way transportation of the new asphalt mixture, so the cost function is given in Equations (7) and (8).

$$O_{re} = M + \sum_{i=1}^{3} C_i + P + T \tag{7}$$

$$C_{re} = C_{as} P_{as} + C_{ag} P_{ag} + C_{fi} P_{fi} + C_{mi} \tag{8}$$

In the HMPR maintenance process, the RAP content is 30%. The transportation cost includes the round-trip cost of transporting the old asphalt mixture to the mixing plant and the recycled asphalt mixture back to the construction site. The cost calculation model is Equations (9) and (10).

$$O_{pl} = M + \sum_{i=1}^{4} C_i + P + \sum_{i=1}^{2} T_i \tag{9}$$

$$C_{pl} = \left(C_{as} P_{as} + C_{ag} P_{ag} + C_{fi} P_{fi} \right) \times 70\% + 30\% \times C_{ra} + C_{mi} \tag{10}$$

In the HIPR maintenance process, the RAP content is 80%. Since the HIPR equipment is used, the fuel consumption and vehicle rental cost of the HIPR equipment are used to replace the separate costs of milling and paving, and the model is shown in Equations (11) and (12).

$$O_{ip} = \sum_{i=1}^{4} C_i + \sum_{i=1}^{3} T_i \tag{11}$$

$$C_{ip} = \left(C_{as} P_{as} + C_{ag} P_{ag} + C_{fi} P_{fi} \right) \times 20\% + 80\% \times C_{ra} + C_{mi} \tag{12}$$

Through the on-site investigation of the actual project, it is assumed that the asphalt mixing plant is 30 km away from the construction site. The cost list under the three maintenance processes and the cost details of each maintenance stage are obtained, as shown in Tables 13 and 14. The final costs for the three maintenance processes are shown in Figure 26.

Table 13. Price list of each maintenance process.

Raw Materials	Cost (¥/t)	Construction Technics	Cost (¥/t)
Asphalt	4500	Milling	25
Aggregate	250	Mixing	80
Regeneration agent	16,000	Transportation	5
BF	22,000	Vehicle group rental	160
LF	4200	Vehicle group fuel consumption	58
PF	5500	Paving and rolling	40

Table 14. Cost details of each maintenance process.

Mixture Type	Milling	Asphalt Mixture	Mixing	Paving and Rolling	Transportation
MR-BF	25	562.5	80	40	5
MR-LF	25	517.6	80	40	5
MR-PF	25	513.0	80	40	5
MR-BPCF	25	537.8	80	40	5
HMPR-BF	25	413.7	80	40	10
HMPR-LF	25	382.9	80	40	10
HMPR-PF	25	379.1	80	40	10
HMPR-BPCF	25	396.4	80	40	10
HIPR-BF	0	163.5	0	0	223
HIPR-LF	0	149.6	0	0	223
HIPR-PF	0	150.4	0	0	223
HIPR-BPCF	0	155.3	0	0	223

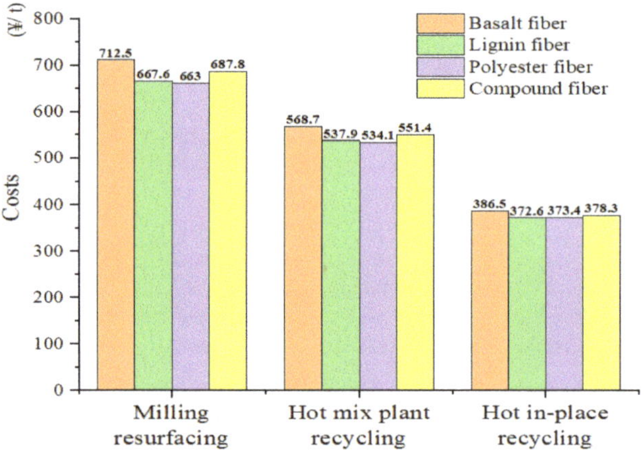

Figure 26. Maintenance cost of each maintenance process.

As can be seen from Figure 26, with the incorporation of RAP, the costs of the three maintenance processes are listed in descending order: MR > HMPR > HIPR. The cost of a new asphalt mixture in the HIPR process is only 29% of that in the MR process. The HIPR process is a little expensive in terms of equipment leasing and fuel consumption compared with the MR process. In general, the HIPR process cost only accounts for 55% of the MR process cost. The HMPR process is similar to the MR process, but due to the saving materials in the new asphalt mixture, the cost of the HMPR process is 80% of the MR process. As for fiber types, the BF is more expensive than the LF and PF when making fiber asphalt mixtures. However, the asphalt absorption for the LF is much larger than that of the BF and PF, hence the asphalt content for the LF asphalt mixture is relatively high. The cost of asphalt is greater than the cost of aggregate, making the cost of the LF asphalt mixture slightly higher than that of the PF asphalt mixture.

4.2. Comprehensive Benefit Analysis Based on the Grey Relational Method

The above study investigated the resistance to HWTT test performance and the respective economic benefit of asphalt mixture with different types of fibers incorporated in asphalt pavements under three maintenance processes. The MR process is found to be more resistant to HWTT rutting with a higher cost, and the HIPR process is found to be less resistant to moisture damage in the stripping stage but at a significant price advantage. Considering the need for test performance and economic cost in practical engineering, this study extracts the main factors affecting performance and cost through grey correlation analysis. Then, the optimal maintenance process and fiber type are selected with the highest comprehensive benefit.

The grey correlation analysis is considered as a systematic approach to analyzing finite and irregular data. It can create a grey series that gives a holistic view and comparison to determine the optimal solution. In this study, 12 kinds of asphalt mixtures with different maintenance processes and fiber types are selected as the scheme. The feature points in the HWTT curve and the cost for the corresponding maintenance process are used as the impact factors, and the optimal scheme is obtained by analyzing the influence of each factor on the comprehensive benefit. Asphalt mixtures of different types can be used to construct scheme A, where $A = \{A_1, A_1, \ldots, A_m\}$, $m = 12$. For each type of asphalt mixture, the total rut depth (TRD), N_{SN}, rut depth for high-temperature rutting effect, rut depth for stripping effect, rut depth and loading cycle at PCP, rut depth and loading cycle at SIP, and the cost of maintenance process are taken as scheme B, where $B = \{B_1, B_1, \ldots, B_n\}$, $n = 9$. Constructing matrix X from A and B, in which $X = (x_{ij})_{m \times n}$, and x_{ij} is the jth test indicator for the ith scheme. The column preference matrix X is given as,

$$X = \begin{pmatrix}
8.033 & 7238.3 & 5.001 & 2.884 & 810 & 1.95 & 14500 & 5.96 & 712.5 \\
8.284 & 6730.8 & 4.546 & 3.801 & 780 & 1.40 & 14325 & 6.11 & 667.6 \\
7.785 & 6961.8 & 4.101 & 3.899 & 790 & 1.43 & 14410 & 5.95 & 663.0 \\
6.034 & 8919.9 & 3.909 & 1.851 & 950 & 1.63 & 18120 & 5.35 & 687.8 \\
5.490 & 7446.2 & 3.623 & 2.125 & 810 & 1.16 & 14615 & 4.41 & 568.7 \\
8.104 & 7811.9 & 4.804 & 3.370 & 820 & 1.81 & 14780 & 6.41 & 537.9 \\
7.093 & 7219.1 & 3.852 & 3.424 & 800 & 1.39 & 14515 & 5.50 & 534.1 \\
5.490 & 7446.3 & 3.688 & 2.062 & 810 & 1.16 & 14610 & 4.41 & 551.4 \\
8.861 & 4827.4 & 3.084 & 6.140 & 550 & 1.00 & 13960 & 5.64 & 386.5 \\
10.92 & 4022.6 & 2.553 & 8.372 & 570 & 0.63 & 13260 & 6.17 & 372.6 \\
15.11 & 3316.7 & 3.981 & 13.41 & 505 & 0.63 & 12930 & 8.56 & 373.4 \\
16.15 & 6643.3 & 10.82 & 6.936 & 600 & 4.34 & 14530 & 13.4 & 378.3
\end{pmatrix}$$

In order to facilitate calculation and comparison, the indicators need to be dimensionless. For the larger and better indicators, the dimensionless formula is: $y_{ij} = \dfrac{max(x_j) - x_{ij}}{max(x_j) - min(x_j)}$.

For smaller and better indicators, the dimensionless formula is: $y_{ij} = \dfrac{x_{ij} - min(x_j)}{max(x_j) - min(x_j)}$, of which, $j = 1, 2, \ldots, n$. Where y_{ij} is the value of the jth index of the ith scheme, $max(x_j)$ and $min(x_j)$ represent the maximum and minimum values in indicator j, respectively. After dimensionless processing, the matrix Y can be obtained.

$$Y = \begin{pmatrix} 0.76\ 0.70\ 0.70\ 0.91\ 0.69\ 0.64\ 0.30\ 0.83\ 0.00 \\ 0.74\ 0.61\ 0.76\ 0.82\ 0.62\ 0.79\ 0.27\ 0.81\ 0.13 \\ 0.78\ 0.65\ 0.81\ 0.82\ 0.64\ 0.78\ 0.29\ 0.83\ 0.15 \\ 0.95\ 1.00\ 0.84\ 1.00\ 1.00\ 0.73\ 1.00\ 0.90\ 0.07 \\ 1.00\ 0.74\ 0.87\ 0.98\ 0.69\ 0.86\ 0.32\ 1.00\ 0.42 \\ 0.75\ 0.80\ 0.73\ 0.87\ 0.71\ 0.68\ 0.36\ 0.78\ 0.51 \\ 0.85\ 0.70\ 0.84\ 0.86\ 0.66\ 0.80\ 0.31\ 0.88\ 0.52 \\ 1.00\ 0.74\ 0.86\ 0.98\ 0.69\ 0.86\ 0.32\ 1.00\ 0.47 \\ 0.68\ 0.27\ 0.94\ 0.63\ 0.10\ 0.90\ 0.20\ 0.86\ 0.96 \\ 0.49\ 0.13\ 1.00\ 0.44\ 0.15\ 1.00\ 0.06\ 0.80\ 1.00 \\ 0.10\ 0.00\ 0.95\ 0.00\ 0.00\ 1.00\ 0.00\ 0.54\ 1.00 \\ 0.00\ 0.59\ 0.00\ 0.56\ 0.21\ 0.00\ 0.31\ 0.00\ 0.98 \end{pmatrix}$$

The H_j is defined as the entropy of the jth indicator, which can be expressed as $H_j = -\frac{\sum_{i=1}^{m} f_{ij} ln(f_{ij})}{ln\, m}$, $j = 1, 2, \ldots, n$, in which, $f_{ij} = \frac{y_{ij}}{\sum_{i=1}^{m} y_{ij}}$. In order to make sense of $ln(f_{ij})$ in the entropy formula, it is assumed that $f_{ij} = 0$, $ln(f_{ij}) = 0$. Thus, the H_j can be obtained. $H_j = [0.93\ 0.93\ 0.96\ 0.95\ 0.91\ 0.96\ 0.90\ 0.96\ 0.88]$.

The entropy weight W_j can be expressed as $W_j = \frac{1 - H_j}{\sum_{j=1}^{n}(1 - H_j)}$, and the entropy weight matrix W_j is given as,

$$W_j = diag\{0.11,\ 0.11,\ 0.06,\ 0.07,\ 0.15,\ 0.06,\ 0.17,\ 0.06,\ 0.20\}$$

Transforming matrix Y to attribute matrix R, which can be expressed as $R = Y \times W_j$.

$$R = \begin{pmatrix} 0.081\ 0.078\ 0.043\ 0.068\ 0.102\ 0.041\ 0.052\ 0.053\ 0.000 \\ 0.079\ 0.068\ 0.046\ 0.062\ 0.092\ 0.050\ 0.046\ 0.052\ 0.026 \\ 0.084\ 0.073\ 0.049\ 0.061\ 0.095\ 0.050\ 0.049\ 0.053\ 0.029 \\ 0.101\ 0.112\ 0.051\ 0.074\ 0.149\ 0.046\ 0.170\ 0.057\ 0.015 \\ 0.107\ 0.082\ 0.053\ 0.073\ 0.102\ 0.054\ 0.055\ 0.064\ 0.084 \\ 0.081\ 0.090\ 0.044\ 0.065\ 0.106\ 0.043\ 0.061\ 0.050\ 0.103 \\ 0.091\ 0.078\ 0.051\ 0.064\ 0.099\ 0.050\ 0.052\ 0.056\ 0.105 \\ 0.107\ 0.082\ 0.052\ 0.072\ 0.102\ 0.054\ 0.055\ 0.064\ 0.095 \\ 0.073\ 0.030\ 0.057\ 0.047\ 0.015\ 0.057\ 0.034\ 0.055\ 0.191 \\ 0.052\ 0.014\ 0.061\ 0.032\ 0.022\ 0.063\ 0.011\ 0.052\ 0.200 \\ 0.010\ 0.000\ 0.058\ 0.000\ 0.000\ 0.063\ 0.000\ 0.035\ 0.199 \\ 0.000\ 0.066\ 0.000\ 0.042\ 0.032\ 0.000\ 0.053\ 0.000\ 0.196 \end{pmatrix}$$

Then, the ideal point of the matrix (P) can be selected and expressed as $P = [p_1, p_2, \ldots, p_n]$. In which, $p_j = max\{r_{ij} | i = 1, 2, \ldots, m; j = 1, 2, \ldots, n\}$.

$$P = [0.107\ 0.112\ 0.061\ 0.074\ 0.149\ 0.063\ 0.170\ 0.064\ 0.200]$$

The distance to the ideal point for each scenario (L) can be calculated, where $L = [l_1, l_2, \ldots, l_m]$, of which, $l_i = \sqrt{\sum_{j=1}^{n} (r_{ij} - p_j)^2}$, $i = 1, 2, \ldots, m$.

$$L = \begin{bmatrix} 0.059\ 0.052\ 0.049\ 0.035 \\ 0.030\ 0.025\ 0.027\ 0.027 \\ 0.045\ 0.056\ 0.079\ 0.054 \end{bmatrix}$$

Therefore, the ordering of the 12 scenarios is: $L_6 < L_7 = L_8 < L_5 < L_4 < L_9 < L_3 < L_2 < L_{12} < L_{10} < L_1 < L_{11}$. According to the grey relational analysis data, it can be seen that the HMPR process has the highest comprehensive benefit among the three maintenance processes. The HMPR asphalt mixture with LF has the best performance, followed by that of the PF and BPCF. The MR fiber asphalt mixture has a similar trend. In the HIPR maintenance process, BF is the first choice. Combined with the HWTT test data, it can be seen that the HIPR maintenance process has poor stripping resistance, while the performance can be significantly improved with the addition of BF in the HIPR mixture. Therefore, the BF can be used as the preferred reinforcement and toughening material for asphalt mixtures with a high RAP dosage.

5. Conclusions

This study proposes an innovative method to analyze the performance and economic aspects of rejuvenating fiber asphalt mixtures in high-temperature moisture susceptibility. Under different maintenance processes, Hamburg wheel tracking tests are conducted to investigate the rutting and stripping effects of rejuvenated fiber asphalt mixtures. Major conclusions can be drawn as follows:

(1) The best performance in terms of resistance to hydrothermal coupling effects is achieved at the HMPR maintenance process, with the asphalt mixtures incorporating the BF and BPCF. Both of these two mixtures have a TRD of about 5.49 mm. The high-temperature rutting performance of asphalt mixtures improves with the increase in RAP content. The HIPR asphalt mixture with the LF has the best high-temperature rutting performance, with a rutting depth of 2.55 mm at a 20,000 cycle. The moisture damage resistance is negatively correlated with the high-temperature rutting performance. The asphalt mixture with the MR process is the best, and the rutting depth under the stripping effect is 1.85 mm.

(2) The HWTT curve fitting model considers the positive and negative curvature of the curve in two separate phases. The starting and ending points of the HWTT curves are connected with the intersection point to obtain two line segments. The unique solution of the PCP and SIP points is obtained by finding the maximum vertical distances between the HWTT curve and two line segments. This method can give a generalized approach to determining the value of stripping inflection points mathematically. The results show that the three feature points of the PCP, N_SN, and SIP are the earliest for the HIPR maintenance process, while these points for asphalt mixtures with the BF and BPCF under the MR maintenance process are the latest, at 950,892,018,120 loading cycles, respectively.

(3) The economic model of three maintenance processes was established. The results found that the cost of the MR maintenance process is the most expensive, while the asphalt mixtures incorporated with the BF increase the cost of maintenance to 712.5 ¥/t. The HIPR maintenance process is more expensive in terms of fuel consumption and equipment leasing, but the cost of raw materials such as aggregates and asphalt is significantly lower. The cost of the HIPR maintenance process is only 55% of the cost of the MR maintenance process.

(4) The performance and economic benefits of 12 rejuvenated asphalt mixtures were investigated by grey correlation analysis. Nine indicators including TRD, N_SN, the RD of high-temperature rutting effect, the RD of stripping effects, RD and N at PCP, RD and N at SIP, and maintenance cost are used as the impact indicators. The results indicate that the HMPR maintenance process is the most efficient within the three maintenance processes, and the HMPR maintenance process with LF is the best choice. In the HIPR maintenance process, the incorporation of BF improves the HWTT performance of the asphalt mixture.

Recycling and recovering are the main trends in the future development of the pavement maintenance industry. In this study, the HWTT curve analysis method was improved on the background of three maintenance technologies. The proposed approach can be used to identify three feature points to better control rutting resistance and moisture susceptibility. A comprehensive benefits analysis combining performance and economic benefits can be investigated. However, a full life cycle analysis is still lacking and should be focused on in future research.

Author Contributions: Conceptualization, Y.Z. and A.K.; methodology, Y.Z. and Y.W.; formal analysis, Y.Z. and Z.W. (Zhengguang Wu); validation, B.L. and C.Z.; resources, Z.W. (Zhengguang Wu), Z.W. (Zhe Wu) and A.K.; data curation, C.Z. and Y.W.; writing—original draft preparation, Y.Z.

and Y.W.; writing—review and editing, Y.Z. and B.L.; project administration, Y.Z. and B.L.; funding acquisition, Y.Z. All authors have read and agreed to the published version of the manuscript.

Funding: This research was funded by the National Natural Science Foundation of China, grant number 52108422, and the High-level Talent Introduction Project of Yangzhou University, grant number 137012062.

Institutional Review Board Statement: Not applicable.

Informed Consent Statement: Not applicable.

Data Availability Statement: Not applicable.

Conflicts of Interest: The authors declare no conflict of interest.

Abbreviations

For reading convenience, a list of abbreviations is given below.

Abbreviations	Full Form
AFM	Atomic force microscopy
BF	basalt fiber
BPCF	Basalt-polyester compound fiber
CS	Creep stage
HIPR	Hot in-place recycling
HMA	Hot mix asphalt
HMPR	Hot mix plant recycling
HRAM	Hot rejuvenated asphalt mixtures
HWTT	Hamburg wheel tracking test
LF	Lignin fiber
LVDT	Linear variable differential displacement transducer
MR	Milling resurfacing
N_{SN}	Intersection point
PCP	Post-compaction phase
PCS	Post-compaction stage
PF	Polyester fiber
RAP	Reclaimed asphalt pavement
RD	Rutting depth
RMP	Regeneration maintenance processes
RTFOT	Rolling thin film oven test
SGC	Super-pave gyratory compactor
SIP	Stripping inflection point
SS	Stripping stage
TRD	Total rut depth

References

1. Liu, S.; Shukla, A.; Nandra, T. Technological, environmental and economic aspects of Asphalt recycling for road construction. *Renew. Sustain. Energy Rev.* **2017**, *75*, 879–893. [CrossRef]
2. Zaumanis, M.; Mallick, R.B.; Frank, R. 100% recycled hot mix asphalt: A review and analysis. *Resour. Conserv. Recycl.* **2014**, *92*, 230–245. [CrossRef]
3. Yu, S.; Shen, S.; Zhou, X.; Li, X. Effect of Partial Blending on High-Content RAP Mix Design and Mixture Properties. *Transp. Res. Rec.* **2018**, *2672*, 79–87. [CrossRef]
4. Yin, F.; Chen, C.; West, R.; Martin, A.E.; Arambula-Mercado, E. Determining the Relationship Among Hamburg Wheel-Tracking Test Parameters and Correlation to Field Performance of Asphalt Pavements. *Transp. Res. Rec. J.* **2020**, *2674*, 281–291. [CrossRef]
5. Nahar, S.N.; Mohajeri, M.; Schmets, A.J.; Scarpas, A.; van de Ven, M.F.C.; Schitter, G. First Observation of Blending-Zone Morphology at Interface of Reclaimed Asphalt Binder and Virgin Bitumen. *Transp. Res. Rec.* **2013**, *2370*, 1–9. [CrossRef]
6. Zhou, Z.; Gu, X.; Dong, Q.; Ni, F.; Jiang, Y. Rutting and fatigue cracking performance of SBS-RAP blended binders with a rejuvenator. *Constr. Build. Mater.* **2019**, *203*, 294–303. [CrossRef]
7. Cheng, P.; Han, Y. Water Stability of Warm Recycled Asphalt Mixture. *J. Chongqing Jiaotong Univ.* **2015**, *34*, 60.
8. Liu, G.; Yang, T.; Li, J.; Jia, Y.; Zhao, Y.; Zhang, J. Effects of aging on rheological properties of asphalt materials and asphalt-filler interaction ability. *Constr. Build. Mater.* **2018**, *168*, 501–511. [CrossRef]

9.   Ma, T.; Huang, X.; Zhao, Y.; Zhang, Y. Evaluation of the diffusion and distribution of the rejuvenator for hot asphalt recycling. *Constr. Build. Mater.* **2015**, *98*, 530–536. [CrossRef]
10. Yue, X. Effects of Regenerant on the Properties of SBS Aged Asphalt and Thermally Regenerated Mixtures. *Road Constr.* **2016**, *41*, 7.
11. Pradhan, S.K.; Sahoo, U.C. Effectiveness of Pongamia pinnata oil as rejuvenator for higher utilization of reclaimed asphalt (RAP) material. *Innov. Infrastruct. Solut.* **2020**, *5*, 1–14. [CrossRef]
12. Li, L.; Zhang, Z.; Wang, Z.; Wu, Y.; Dong, M.; Zhang, Y. Coupled thermo-hydro-mechanical response of saturated asphalt pavement. *Constr. Build. Mater.* **2021**, *283*, 122771. [CrossRef]
13. Majidifard, H.; Jahangiri, B.; Rath, P.; Contreras, L.U.; Buttlar, W.G.; Alavi, A.H. Developing a Prediction Model for Rutting Depth of Asphalt Mixtures Using Gene Expression Programming. *Constr. Build. Mater.* **2020**, *267*, 120543. [CrossRef]
14. Zhang, Y.; Ling, M.; Kaseer, F.; Arambula, E.; Lytton, R.L.; Martin, A.E. Prediction and evaluation of rutting and moisture susceptibility in rejuvenated asphalt mixtures. *J. Clean. Prod.* **2022**, *333*, 129980. [CrossRef]
15. Chen, H.; Xu, Q.; Chen, S.; Zhang, Z. Evaluation and design of fiber-reinforced asphalt mixtures. *Mater. Des.* **2009**, *30*, 2595–2603. [CrossRef]
16. Mashaan, N.; Karim, M.; Khodary, F.; Saboo, N.; Milad, A. Bituminous Pavement Reinforcement with Fiber: A Review. *Civil Eng.* **2021**, *2*, 599–611. [CrossRef]
17. Wu, S.; Haji, A.; Adkins, I. State of art review on the incorporation of fibers in asphalt pavements. *Road Mater. Pavement Des.* **2022**. [CrossRef]
18. Shu, B.; Wu, S.; Dong, L.; Norambuena-Contreras, J.; Yang, X.; Li, C.; Liu, Q.; Wang, Q. Microfluidic synthesis of polymeric fibers containing rejuvenating agent for asphalt self-healing. *Constr. Build. Mater.* **2019**, *219*, 176–183. [CrossRef]
19. Chen, Y.; Li, Z. Study of Road Property of Basalt Fiber Asphalt Concrete: International Conference on Civil Engineering. In Proceedings of the Architecture and Sustainable Infrastructure, Zhengzhou, China, 13–15 July 2013.
20. Zhang, K.; Li, W.; Han, F. Performance deterioration mechanism and improvement techniques of asphalt mixture in salty and humid environment. *Constr. Build. Mater.* **2019**, *208*, 749–757. [CrossRef]
21. Lu, J. Research on Performance of Recycled Asphalt Mixture Based on Old Asphalt Classification and Classification Method. Master's Thesis, Yangzhou University, Yangzhou, China, 2016.
22. Yin, F.; Arambula, E.; Lytton, R.; Martin, A.E.; Cucalon, L.G. Novel Method for Moisture Susceptibility and Rutting Evaluation Using Hamburg Wheel Tracking Test. *Transp. Res. Rec.* **2014**, *2446*, 1–7. [CrossRef]
23. Nie, Y.; Kang, L.; Liu, P. Research on Economic Benefit Analysis Model of Recycling and Utilization of Hot Asphalt Pavement in Factory. *Road Constr.* **2013**, *38*, 4.

Article

Enhanced Acceptance Specification of Asphalt Binder to Drive Sustainability in the Paving Industry

Yiming Li [1,2,*] **and Simon A. M. Hesp** [2]

1   Department of Civil Engineering, Northeast Forestry University, Harbin 150040, China
2   Department of Chemistry, Queen's University, Kingston, ON K7L 3N6, Canada; simon@chem.queensu.ca
*   Correspondence: liyiming@nefu.edu.cn

Abstract: Testing small amounts of extracted and recovered asphalt binder as used in construction allows for the acceptance of materials in accordance with traffic and climate requirements. This approach facilitates the sustainable use of resources and thus prepares the paving industry for the true circular economy. Oscillatory, creep, and failure tests in a rheometer are compared for the performance grading of 32 asphalt binders extracted and recovered from real-world contract samples. Films 8 mm in diameter and 0.5 mm thick were tested from 35 to -5 °C in dynamic shear, followed by shear creep at 0 and 5 °C, and finally in tertiary tensile creep at 15 °C. The enhanced protocol uses a very small amount of material in contrast to current methods, yet it provides comparable results. Phase angle measurements appear to be optimal for performance grading, but further field study is required to determine if additional binder properties such as stiffness and/or failure strain would be required for the control of cracking.

Keywords: asphalt performance grading; thermal cracking; fatigue; phase angle; creep rate; failure strain

Citation: Li, Y.; Hesp, S.A.M. Enhanced Acceptance Specification of Asphalt Binder to Drive Sustainability in the Paving Industry. *Materials* **2021**, *14*, 6828. https://doi.org/10.3390/ma14226828

Academic Editor: Milena Pavlíková

Received: 22 October 2021
Accepted: 10 November 2021
Published: 12 November 2021

Publisher's Note: MDPI stays neutral with regard to jurisdictional claims in published maps and institutional affiliations.

1. Introduction

Optimal pavement design involves balancing material properties and structure to provide a long-life cycle with only minimal distress. It is generally accepted that rutting and moisture damage are largely controlled through the selection of appropriate aggregate types and gradation, with the addition of polymer, fiber, and/or antistrip additives when needed [1]. On the other hand, load-induced fatigue and cold temperature transverse cracking are kept in check by the selection of an appropriate pavement thickness, asphalt binder quality, and durability [1–6].

It is essential that the most accurate acceptance specification tests are conducted on carefully extracted and recovered binder, as it best reflects what is actually placed in the contract [7–10]. The presence of reclaimed asphalt pavement (RAP) in the mix, as well as associated overheating of the virgin asphalt binder during production, are factors that can have a detrimental impact on long-term performance of the pavement. Hence, these and other issues need to be accounted for in an effective quality assurance testing program.

Several Ontario municipalities have recently switched to testing of the extracted and recovered asphalt binder according to the extended bending beam rheometer (EBBR) and double-edge-notched tension (DENT) tests with promising results [7–10]. To illustrate, Figure 1a provides representative 2020 photographs for pavement on three blocks of Princess Street and King Street in downtown Kingston, Ontario, constructed before the switch. The asphalt surface was reconstructed on fresh granular base ten years ago as part of a program that replaced all downtown sewer infrastructure. The properties of the asphalt storage tank sample were used for acceptance, but—according to the current Ontario municipal asphalt specification—up to 15% RAP was allowed in the binder course and none in the surface. It is obvious that this pavement is failing through thermal cracking well before its expected design life.

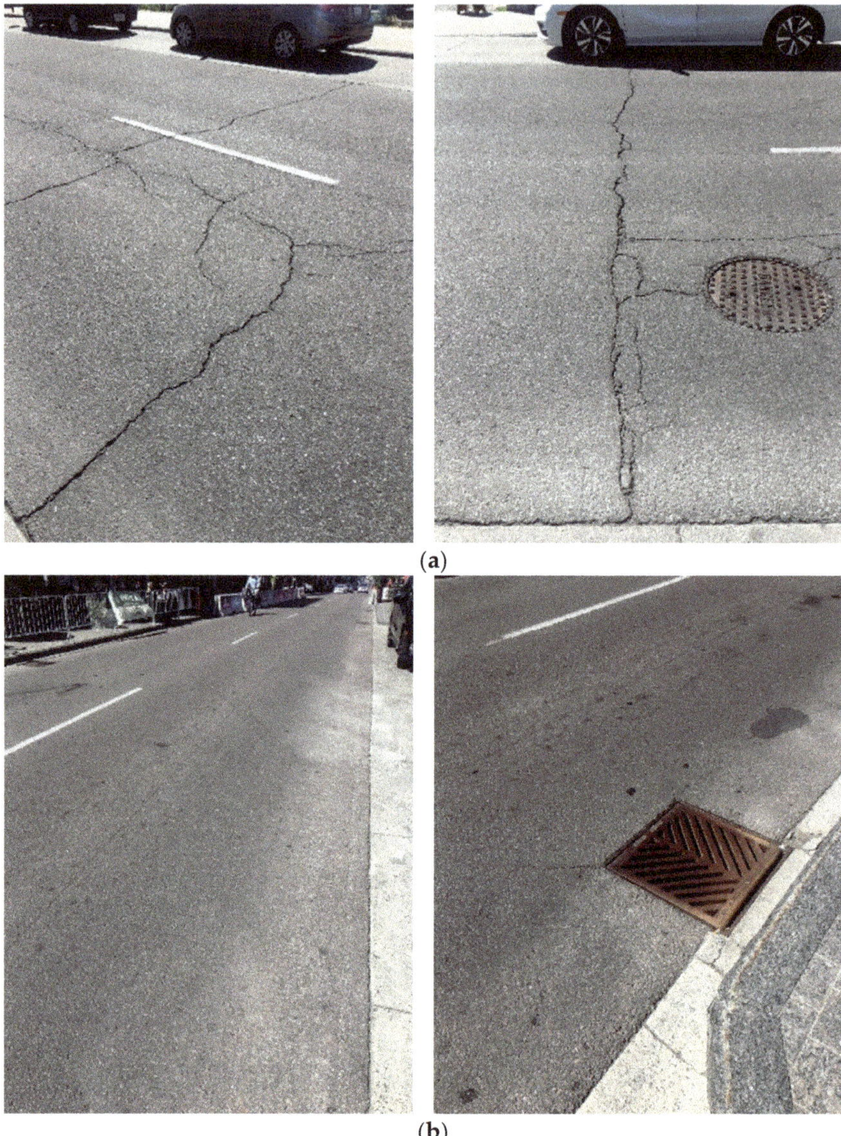

Figure 1. (**a**) Representative photographs of pavement reconstructed in 2010 on three blocks of Princess Street and King Street in downtown Kingston, Ontario, (**b**) Representative photographs of pavement reconstructed in 2011–2012 on seven adjacent blocks of Princess Street.

Figure 1b provides representative photographs for the remainder of the contract completed over the next two years on an adjacent seven blocks of Princess Street. Here, the City of Kingston had switched to acceptance of the asphalt based on properties of the extracted and recovered binder and RAP was banned from both the surface and binder courses. It is obvious that there is a stark contrast in performance after only eight to ten years of service. Hence, if the acceptance is based on extracted and recovered binder properties, then the use of RAP would be allowed as long as minimum performance is obtained in the materials as placed in the contract.

The acceptance of the asphalt for the City of Kingston has been based on extracted and recovered binder properties since 2010 and this has so far provided pavements that are meeting their design expectations [7]. While providing improved specification grading, the EBBR and DENT tests used for contract acceptance have their drawbacks. Both tests require a rather large quantity of extracted and recovered binder and take a considerable amount of time to complete. Hence, current research is focused on the development of simplified methods that use less material, take less time to complete, and with equal or better precision and accuracy [7,11–13]. Improved specification tests lower risk, which benefits both users and producers of asphalt for the betterment of the entire industry. The objective of the current research project is to develop a more practical approach for acceptance testing of extracted and recovered asphalt binder. In order to learn more about all issues involved, the research assessed a wide range of performance properties for 32 binders from commercial contracts. Future efforts will involve a detailed performance assessment of the involved pavement locations.

2. Background

It was Dow [14] of the Washington, D.C., engineering department, who in the early 1900s developed both a ductility test and an improved penetrometer for the grading of asphalt binder. With a keen eye, he had noticed that those binders that elongate when pulled by hand would perform better than those that ruptured early [15]. Both ductility and penetration tests went on to become the most widely used specification methods for straight asphalt binder and remain used with great success in many parts of the world today. In general, binder that flows well suffers little from cracking at ambient and cold temperatures. Dow [14] also commented on the fact that some binders when freshly poured perform much better compared to those that had been left to equilibrate for a time on the bench.

It was not until the publication from Hubbard and Pritchard [16] in 1916 that a quantitative assessment was made of this gradual aging phenomenon. Penetration testing showed that the consistency of binders can increase (i.e., penetration decreases) for periods of weeks and months, and that this was independent of oxidative hardening as reheating the sample could largely restore the original properties.

A series of publications by Traxler and coworkers [17–19] in the 1930s were the first to provide a comprehensive assessment of what is best described as a thermoreversible aging effect. Using tensile and shear creep experiments, major changes in rheological properties were revealed after days and weeks of isothermal storage. These authors noted that binders physically age at different rates depending on their source and production technology. They found that air oxidized binders were particularly sensitive to the effects of thermal conditioning. Filler had little effect on the degree of aging. Reheating could erase the changes. The impact of thermal equilibration was large compared to the effects of volatilization and oxidation. Finally, they described the change in consistency as a sol-to-gel transition.

Traxler and his contemporaries actively discussed the sol and gel nature of asphalt binder, as reflected by publications of Nellensteyn [20–22], Sakhanov [23], Sachanen [24], Mack [25,26], Saal [27], Pfeiffer and Van Doormaal [28], and many others. Asphalt binder is a material that is composed of a spectrum of organic molecules with molar weights that range from a few hundred to a few thousand grams per mole [29]. The individual molecules can be classified as either aliphatic (paraffins and naphthenes) or combined aliphatic and aromatic (naphthene aromatics, resins, and asphaltenes). The aliphatic fraction is defined by its molar weight and degree of branching, with those binders largely composed of linear alkanes (paraffins) providing lesser performance compared to those containing mostly branched and cyclic (napthenic) alkanes [29]. The more aromatic fractions are called asphaltenes that are typically of a higher molar weight and contain fused ring systems that could be associated with small amounts of metals such as nickel, vanadium, and iron [29].

There is a significant amount of ambiguity in the literature as the asphaltenes fraction, defined by its insolubility in n-heptane, can also be contaminated with paraffin of high enough molar weight that makes it co-precipitate [30]. The general consensus is that the asphaltenes fraction together with the paraffin slowly precipitates out into a sol-type, sol/gel-type, or gel-type structure that, depending on temperature, viscosity, and composition, takes from days to weeks or months to equilibrate [16–19,31–40].

Blokker and Van Hoorn [33] coined the term "physical hardening" and stated that it involves the rather rapid crystallization of waxes and the slower precipitation of asphaltenes. Binders with high contents of both linear paraffins (wax) and asphaltenes are most susceptible to cracking distress as, due to their gelled state at ambient and low temperatures, they are unable to relax thermal and traffic-induced stresses [38,40,41] and suffer from weak spots at the somewhat sharp interface between the crystalline and amorphous phases [42–44].

Current specifications in most of Canada and the United States are based on the work done under the U.S. Strategic Highway Research Program (SHRP) [45]. The product of SHRP was the Superpave™ specification, which grades asphalt binders at high, intermediate, and low temperatures to control rutting, fatigue, and thermal cracking distress [45].

At high temperatures, Superpave sets a lower limit on the complex modulus divided by the sine of the phase angle, $G^*/\sin\delta$, which for straight run binders is close to the complex viscosity [46]. It is generally accepted that the high temperature Superpave grade is reasonably effective at controlling rutting distress, although it should be recognized that aggregate structure and pavement thickness are two factors that can often be more important than binder properties.

At intermediate temperatures, an upper limit of 5 MPa is set on the loss modulus, $G^*\sin\delta$, for a residue aged for 20 h in a pressure aging vessel (PAV) at a temperature of 100 °C and pressure of 2.08 MPa [45]. It has been found that the intermediate temperature Superpave grade is unable to correlate well with fatigue performance largely because it confounds the beneficial effects of viscous energy dissipation and the formation of damage [47–50]. It also favors the use of binders with low phase angle and low stiffness, which are known to suffer more from oxidative age hardening, phase separation, and exudative aging [51].

At low temperatures, the Superpave specification sets limits on the creep stiffness, measured in the BBR in three-point bending at 60 s of loading, S(60 s), and creep rate (i.e., the slope of the logarithmic creep stiffness master curve) also measured after 60 s of loading, m(60). The maximum S(60 s) of 300 MPa came from a decision to increase it from 200 MPa, as otherwise, too many asphalt binders sold at the time of SHRP would not have met the specification [45]. The original 200 MPa limit was based on a single field validation study many years earlier by Readshaw [52] in British Columbia. The m-value limit of 0.300 was a late addition to the specification that was based on the average for the relaxation rate at the limiting stiffness temperature for the eight core asphalt binders used in SHRP. The m(60 s) value was intended to prevent the use of heavily air-blown binders that were known to suffer from reduced creep, reduced stress relaxation, and exudative aging.

In general, asphalt binders are highly susceptible to changes in temperature and, due to their high viscosity at and below room temperature, are often graded in a state of non-equilibrium. This problem is one that has confounded pavement design since the work of Dow [14,15], Hubbard and Pritchard [16], Traxler [17–19], and others [53–62], and was similarly of concern to SHRP researchers developing the BBR [35,36]. Hence, the SHRP program spent a considerable amount of time and resources investigating physical hardening phenomena (thermoreversible aging). An early draft of the Superpave specification contained an option to test binders after one and 24 h of conditioning at the test temperature, but for reasons that are not well documented that provision never found wide acceptance [45].

Shortly after the complete implementation of the Superpave binder specification in Ontario, the Ministry of Transportation of Ontario (MTO) initiated research projects to investigate resultant widespread premature cracking around eastern and northeastern parts of the province [3–6]. These investigations eventually resulted in an improved asphalt cement specification by incorporating both the DENT and EBBR tests [63,64].

The DENT test provides an approximate critical crack tip opening displacement (CTOD), which is a value for the strain tolerance of binders in their ductile state and is highly correlated with fatigue cracking performance [50]. The CTOD is best described as a somewhat improved measurement of ductility. It is measured under more severe constraint in deeply notched specimens and at lower temperatures compared to a regular ductility test. However, it should be noted that there is a high correlation between CTOD and conventional ductility [12,65]. The DENT test provides essential and plastic works of failure in addition to the CTOD and the relevance of those to pavement performance remains to be better understood.

The EBBR was developed to be as similar as possible to the regular BBR test. It conditions samples for 1, 24, and 72 h at T_d + 10 and T_d + 20, where T_d is the design temperature of the pavement according to climatic requirements [2–5]. The 72 h grade loss from the one-hour result at T_d + 10 (roughly equal to the AASHTO M320 grade) is calculated and serves as a measure of durability. Grade losses found for tank and recovered binders range approximately from 0 to 15 °C, depending on the quality and durability of the binder [5–7].

While both the DENT and EBBR have proven to provide much enhanced performance grading, there is a need to find simpler test methods that can be completed in less time with less material to replace these protocols for future specifications. To that end, a butt joint test (BJT) for the ductile performance grading of asphalt binders was recently proposed [13]. Introducing the BJT as an alternative method for the DENT was able to significantly reduce material requirement and testing time. Compared with the DENT test, the BJT only takes 30 min and at the same time provides precise, sensitive, and accurate ductile strain tolerances. However, the BJT test has a problem with stiff binders that can generate tensile loads that exceed the capacity of the rheometer [13]. Hence, in this research a tertiary tensile creep test is investigated, which provides a measure of strain tolerance within the load capacity of the rheometer used. In addition, dynamic and creep shear tests are conducted at ambient and just below ambient temperatures in order to provide a comparison with results obtained in the EBBR protocol at much lower temperatures.

In spite of the numerous investigations that have come to similar conclusions to those found in the seminal publications by Traxler and coworkers [17–19] on thermoreversible aging, the problems created by this gradual change in properties in specification grading remain to be effectively sorted out today. This paper focusses on the measurement of phase angle as it has been shown in previous studies to be highly correlated to EBBR limiting grades and field cracking performance [66–71]. Testing of the extracted and recovered asphalt binder allows for the proper acceptance of materials as placed in the contract, thus facilitating the proper and responsible design for a true circular economy.

3. Experimental

3.1. Materials

A total of 32 asphalt binders were used in this study from hot mix asphalt (HMA) provided by five different user agencies. Samples were shipped directly by the user agency to Queen's University by overnight courier and processed within a maximum of 1–2 months.

3.2. Methods

Binders were extracted using dichloromethylene (DCM) solvent [8,9]. All solutions were twice fed through a high-speed centrifuge (Ploog Engineering, Crown Point, IN, USA) to remove the fines. Recovery of the asphalt binder was done under a dry nitrogen gas atmosphere at moderate to high vacuum in a rotary evaporator (BUCHI Corporation, New Castle, DE, USA). Once no further DCM was visibly being distilled, the temperature of the oil bath was raised to 160 °C and the flask was subjected to vacuum below 50 mbar for an additional one hour. Asphalt binders were aged in a pressure aging vessel (PAV, Prentex, Dallas, TX, USA) according to standard procedures embodied in AASHTO R 28–09 [72] before testing at intermediate and low temperatures.

Small amounts of PAV residues were mounted in the dynamic shear rheometer (DHR-1 or DHR-2, TA Instruments, New Castle, DE, USA) at 64 °C and rapidly brought to 34 °C for testing at a film thickness of 2 mm. Samples were equilibrated for 10 min at each test temperature prior to testing at 12 °C intervals from 34 °C to −2 °C to determine the intermediate temperature complex modulus, G*, and phase angle, δ. Test frequency and strain level were kept constant at 10 rad/s and 0.1%, respectively.

The PAV residues were tested according to the DENT protocol described in AASHTO method TP 113-15 [73]. In brief, samples were poured in silicone molds with aluminum end inserts to facilitate shear transfer of the load from the test frame to the specimen. Notch depths varied to provide ligaments of 5, 10, and 15 mm in 10 mm thick specimens. The total works of failure were divided by the ligament area and plotted versus ligament length. The extrapolated intercept provided the essential work of failure, which was subsequently divided by the net section stress in the smallest ligament to determine the CTOD.

The PAV residues were tested in the EBBR protocol as described in AASTHO method TP 122-16 [74]. In brief, six samples each were conditioned at T_d +10 and T_d + 20 (where T_d is the design temperature of the pavement), for 1, 24, and 72 h. After each conditioning time, the samples were tested at T_d + 10 and T_d + 16 to determine pass and fail properties. From the individual stiffness and m-value measurements, actual grade temperatures were calculated by interpolation or extrapolation. The limiting low temperature grade (LLTG) was determined as the warmest of all limiting grade temperatures and the grade loss (GL) was determined as the difference between the 1 h limiting grade at T_d + 10 and the LLTG [74].

All PAV residues were also tested in the dynamic hybrid rheometer as 8 mm diameter by 0.5 mm thin films according to the three-in-one protocol. First, samples were mounted at 64 °C and after equilibration for 10 min at 35 °C, tested at 10 °C intervals from 35 °C to −5 °C, at frequencies of 0.1, 0.316, 1, 3.16, and 10 rad/s at a strain level of 0.1%. From these results, the temperature at which the phase angle at 10 rad/s reached 30° was calculated as a low temperature performance grade. Second, the same samples were subsequently equilibrated at 0 °C and tested for 240 s in creep shear at 1000 Pa followed by recovery for 760 s. Next, the creep test was repeated at 5 °C. From the two creep tests in shear, the temperature at which the creep rate, m, reached 0.5, was calculated according to the following equations [75]:

$$\log S'(t) = A + B[\log(t)] + C[\log(t)]^2 \tag{1}$$

$$|m| = B + 2C[\log(t)] \tag{2}$$

where $S'(t)$ is the time-dependent shear creep stiffness, m is the creep rate in shear, t is the time in seconds, A, B, and C are regression coefficients.

Finally, the same samples as measured in the first two steps of the three-in-one protocol were subsequently equilibrated at 15 °C and subjected to a tensile creep load of 8 N to failure in order to determine strain tolerance in the ductile state. Figure 2 provides a representative tertiary creep test result with an illustration of how the failure point was determined at the sudden loss of the creep load control.

The results from the three-in-one protocol to determine phase angle, creep rate and ductile failure strain were compared with findings from the standard DSR, DENT and EBBR tests. The advantage of the three-in-one protocol is obvious as it uses less than a gram of material versus approximately 150 g for the combined DENT/EBBR testing, it is automated and might therefore be more repeatable, and produces similar insights as the DENT/EBBR.

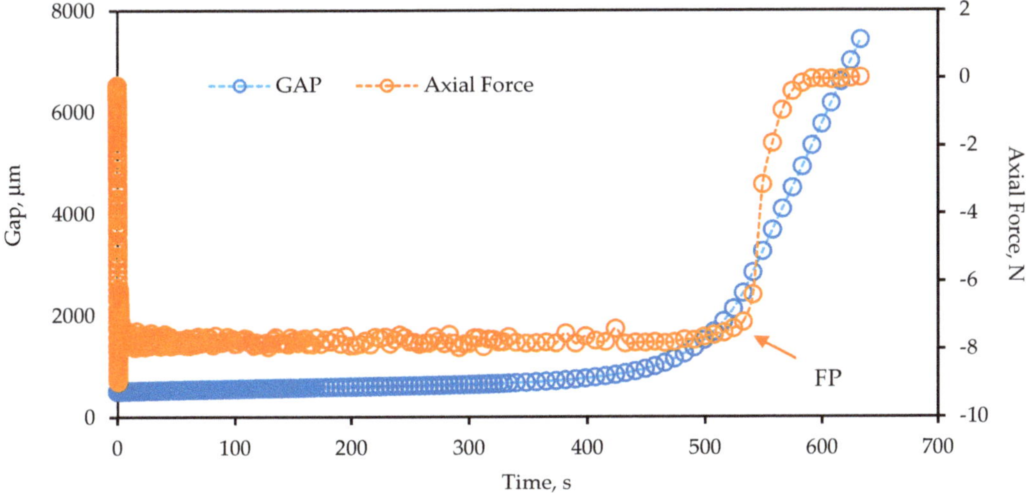

Figure 2. Tertiary creep test results on a 0.5 mm film to measure the failure point (FP).

4. Results and Discussion

4.1. Oscillatory Shear Testing

The limiting phase angle temperatures determined as part of the intermediate temperature Superpave grading and the three-in-one protocol are given in Figure 3a. It is obvious that there is a strong correlation and that the reproducibility is high (short of three outliers in squares). Tests were done by the same person several months apart. The results for the two film thicknesses are 2.63 °C apart, which is due to the thinner film being more constrained and thus less able to flow.

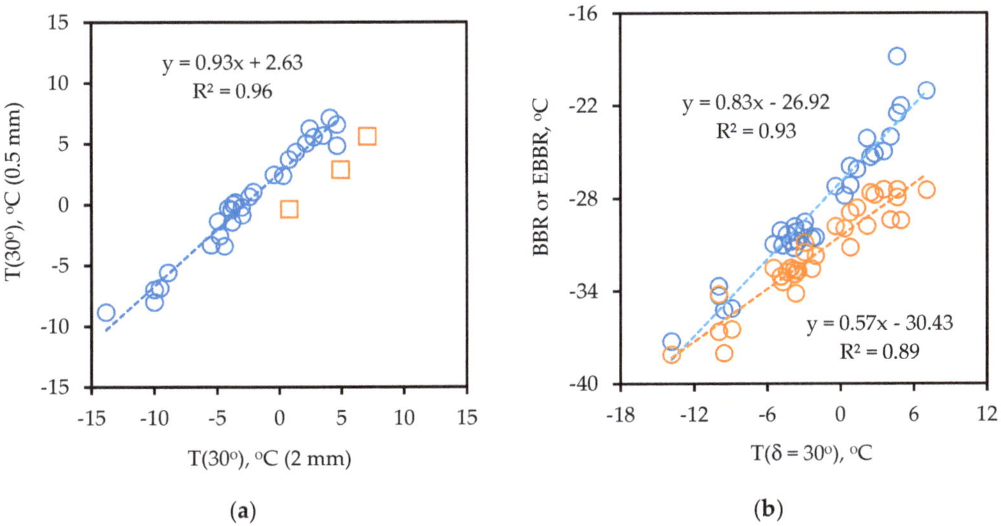

Figure 3. (**a**) Reproducibility of limiting phase angle temperature determinations using 2 and 0.5 mm thin films in the DSR; (**b**) Correlation between limiting phase angle temperature T($\delta = 30°$) (2 mm) and BBR (red symbols) and EBBR (blue symbols) limiting grades.

The correlation between the limiting phase angle temperature and the BBR and EBRR grades are in Figure 3b. These comparisons show that there are strong similarities, but that the T(δ = 30°) and EBBR have a higher sensitivity (wider range) compared to the regular BBR. The slope for the EBBR straight line fit is about 46% higher than what it is for the regular BBR (0.83 versus 0.57), which reflects the significantly improved sensitivity for the T(δ = 30°) and EBBR. The EBRR LLTG and T(δ = 30°) are most closely correlated, suggesting that cracking control may be achieved with similar efficiency when utilizing a T(δ = 30°) acceptance criterion. The difference for this limited set of data is in agreement with data from previous studies on other binders [8,11,76,77]. However, a switch from the somewhat lengthy EBBR protocol to the more practical T(δ = 30°) would mean that the information provided by the grade loss is lost. It has also been recognized that for more severely aged material, the limiting phase angle temperatures also start to suffer from the effects of thermoreversible aging [78,79].

4.2. Creep Testing

Representative creep test results for one of the binders are in Figure 4. Use of Equation (1) provides a high correlation with the raw displacement data. The correlation between T(m = 0.5) and BBR and EBBR is provided in Figure 5. As can be seen, the creep data fit Equation (1) with a high degree of accuracy. The correlation between the limiting creep rate temperature, T(m = 0.5), and the BBR and EBBR limiting temperatures is also reasonably good. As for the phase angle data in Figure 3, the range for the T(m = 0.5) at 19.6 °C is significantly wider than what it is for the BBR at 10.7 °C, somewhat wider than what it is for the EBBR at 16.5 °C, but not quite as wide as the span for the T(δ = 30°) at 20.9 °C. A wider range with equal or better precision is beneficial in a grading protocol as it allows for the better differentiation between samples.

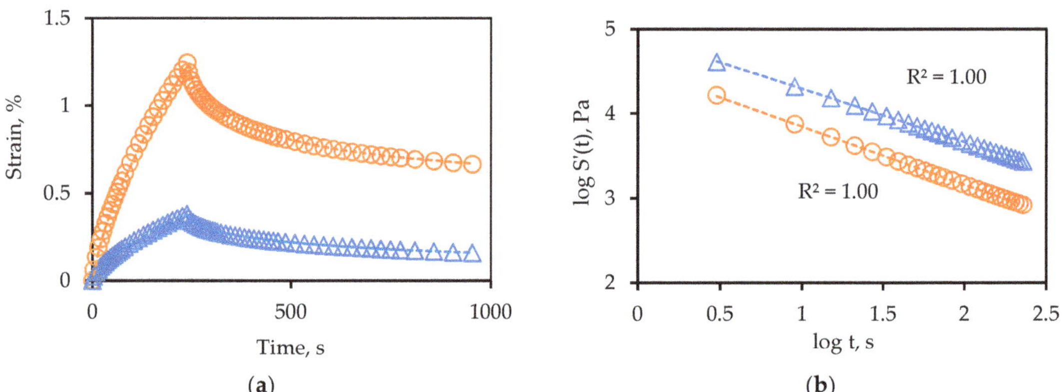

Figure 4. (**a**) Raw and (**b**) processed shear creep test results at 1000 Pa and two temperatures.

4.3. Tertiary Creep Testing

The final comparison is between the failure point in tertiary creep and the DENT CTOD as given in Figure 6a. The graph shows that there is a very high correlation and that both measurements provide nearly the same ranking. Figure 6b shows the repeatability for the tertiary creep test, which is also reasonable, although not as good as for the phase angles in Figure 3a.

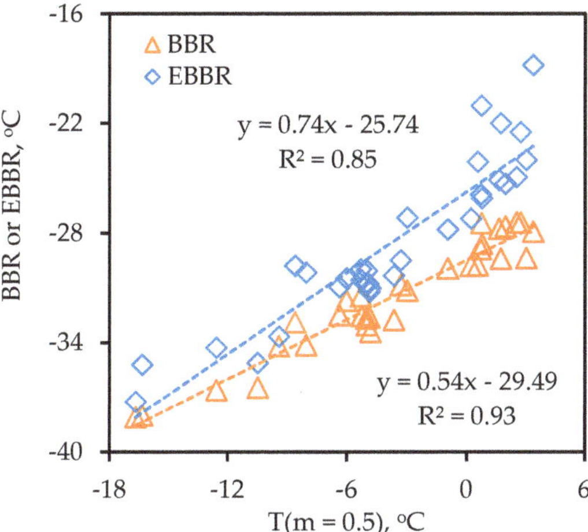

Figure 5. Correlation between T(m = 0.5) and limiting BBR and EBBR temperatures.

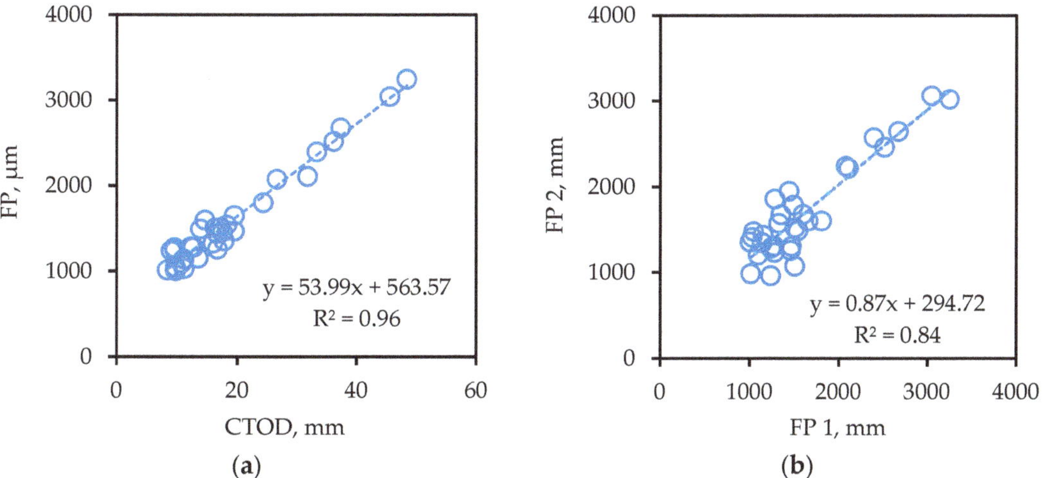

Figure 6. (**a**) Correlation between failure point (FP) and DENT CTOD; (**b**) Repeatability of the tertiary creep test (failure point FP2 is a repeat of FP1 as determined by one of the authors (Y.L.), at a time several months after the determination of FP1).

5. Summary and Conclusions

Given the results and discussion presented, the following summary and conclusions are provided:

1. Constraint increases in thinner films and the limiting phase angle temperatures increase accordingly. However, there is a strong correlation between limiting temperatures measured in films of 0.5 mm (new protocol) and 2.0 mm (AASHTO M 320 standard) thickness.

2. The limiting phase angle temperature shows a very strong correlation with the EBBR LLTG temperature ($R^2 = 0.93$), and a somewhat lesser correlation with the regular BBR temperature ($R^2 = 0.89$).

3. The ranges for limiting T($\delta = 30°$) (20.9 °C) and EBBR (16.5 °C) temperatures for this set of 32 binders were about 91 and 46% improved over the range of the regular BBR

temperature (10.7 °C). Hence, the limiting phase angle temperature is significantly more responsive to changes in binder properties than both the BBR and EBBR.

4. The phase angle reflects the binder's ability to relax thermal and traffic induced stresses and will therefore provide a good correlation with pavement cracking performance. Those binders that are of a gel type (low phase angle) are expected to perform poorly in service, while those binders that are of a sol type (high phase angle) are expected to perform well.

5. If and how a measure of binder stiffness needs to be included in the specification needs careful deliberation and further investigation through field monitoring of the investigated materials.

6. The DENT CTOD can be approximated with a high degree of accuracy by the failure point in the tertiary creep test. Whether and how this property needs to be included in future cracking specifications deserves further investigation through careful study of the long-term performance of the investigated materials.

Given the pervasiveness and seriousness of premature and excessive pavement cracking in cold climates, it is up to the user agencies to make the best use of the information provided.

Author Contributions: Conceptualization, Y.L. and S.A.M.H.; data collection and data processing, Y.L.; writing of draft manuscript, Y.L.; supervision, writing and editing of final manuscript, S.A.M.H. All authors have read and agreed to the published version of the manuscript.

Funding: The research was made possible through funding in part by Imperial Oil of Canada (URA 2018–2021) and the Ontario Ministry of Transportation (HIIFP 2018–2021). Yiming Li was generously supported by the Chinese Scholarship Council to spend one year at Queen's University in Kingston, Canada, as part of his doctoral program at Northeast Forestry University in Harbin, China.

Institutional Review Board Statement: Not applicable.

Informed Consent Statement: Not applicable.

Data Availability Statement: The raw/processed data required to reproduce these findings cannot be shared.

Acknowledgments: Yiming Li wishes to thank the China Scholarship Council (CSC) for financial support, which was provided to support his study at Queen's University. Additional appreciation goes out to the Ministry of Transportation of Ontario, Imperial Oil of Canada, and the five municipal and state user agencies for their continuing support of this research program.

Conflicts of Interest: Hesp, S.A.M. receives professional consulting fees from municipal, state and federal transportation agencies.

Disclaimer: None of the sponsors necessarily concurs with, endorses, or has adopted the findings, conclusions, or recommendations either inferred or expressly stated in the subject data developed.

References

1. The ABCs of PGAC: The Use of Performance Graded Asphalt Cements in Ontario. Available online: http://www.onasphalt.org/files/Publications/ABCs%20of%20PGAC.pdf (accessed on 20 September 2020).
2. Zhao, M.O.; Hesp, S.A.M. Performance grading of the Lamont, Alberta C-SHRP pavement trial binders. *Int. J. Pavement Eng.* **2006**, *7*, 199–211. [CrossRef]
3. Iliuta, S.; Andriescu, A.; Hesp, S.A.M.; Tam, K.K. Improved approach to low-temperature and fatigue fracture performance grading of asphalt cements. *Proc. Can. Technol. Asphalt Assoc.* **2004**, *49*, 123–158.
4. Yee, P.; Aida, B.; Hesp, S.A.M.; Marks, P.; Tam, K.K. Analysis of Premature Low-Temperature Cracking in Three Ontario, Canada, Pavements. *Transp. Res. Rec. J. Transp. Res. Board* **2006**, *1962*, 44–51. [CrossRef]
5. Hesp, S.A.M.; Soleimani, A.; Subramani, S.; Phillips, T.; Smith, D.; Marks, P.; Tam, K.K. Asphalt pavement cracking: Analysis of extraordinary life cycle variability in eastern and northeastern Ontario. *Int. J. Pavement Eng.* **2009**, *10*, 209–227. [CrossRef]
6. Hesp, S.A.M.; Shurvell, H.F. Waste engine oil residue in asphalt cement. In Proceedings of the Mairepav7: Seventh International Conference on Maintenance and Rehabilitation of Pavements and Technological Control, Auckland, New Zealand, 28–30 August 2012.
7. Ding, H.; Tetteh, N.; Hesp, S.A.M. Preliminary experience with improved asphalt cement specifications in the City of Kingston, Ontario, Canada. *Constr. Build. Mater.* **2017**, *157*, 467–475. [CrossRef]

8. Ding, H.; Gotame, Y.; Nie, Y.; Hesp, S.A.M. Acceptance testing of extracted and recovered asphalt cements for provincial and municipal paving contracts in Ontario. *Proc. Can. Technol. Asphalt Assoc.* **2018**, *63*, 43–60.
9. Akentuna, H.; Ding, A.; Khan, A.N.; Li, Y.; Hesp, S.A.M. Improved performance grading of asphalt cement and hot mix asphalt in Ontario, Canada. *Int. J. Pavement Res. Technol.* **2021**, *14*, 267–275. [CrossRef]
10. Li, Y.; Ding, H.; Nie, Y.; Hesp, S.A.M. Effective control of flexible asphalt pavement cracking through quality assurance testing of extracted and recovered binders. *Constr. Build. Mater.* **2021**, *273*, 121769. [CrossRef]
11. Angius, E.; Ding, H.; Hesp, S.A.M. Durability assessment of asphalt binder. *Constr. Build. Mater.* **2018**, *165*, 264–271. [CrossRef]
12. Campbell, S.; Ding, H.; Hesp, S.A.M. Double-edge-notched tension testing of asphalt mastics. *Constr. Build. Mater.* **2018**, *166*, 87–95. [CrossRef]
13. Diak, E.; Beyer, E.; Hesp, S.A.M. Development of a Butt Joint Test for the ductile performance grading of asphalt binders. *Constr. Build. Mater.* **2020**, *243*, 118195. [CrossRef]
14. Dow, A.W. The testing of bitumens for paving purposes. *Proc. ASTM* **1903**, *3*, 349–369.
15. Dow, A.W. Discussion on ductility test. *Proc. Assoc. Asphalt Paving Technol.* **1936**, *41*, 139–141.
16. Hubbard, P.; Pritchard, F.P. Effect of controllable variables upon the penetration test for asphalts and asphalt cements. *J. Agric. Res.* **1916**, *5*, 805–818.
17. Traxler, R.; Coombs, C. The colloidal nature of asphalt as shown by its flow properties. *J. Phys. Chem.* **1936**, *40*, 1133–1147. [CrossRef]
18. Traxler, R.N.; Schweyer, H.E. Increase in viscosity of asphalts with time. In Proceedings of the American Society for Testing Materials, Philadelphia, PA, USA, 4 March 1936; pp. 544–551.
19. Traxler, R.N.; Coombs, C.E. Development of internal structure in asphalts. In Proceedings of the American Society for Testing Materials, Philadelphia, PA, USA, 24 June–2 July 1937; pp. 549–555.
20. Nellensteyn, F.J. Bereiding en Constitutie van Asphalt. Ph.D. Thesis, Delft University Netherlands, Delft, The Netherlands, 1923.
21. Nellensteyn, F.J. The constitution of asphalt. *J. Inst. Petrol. Technol.* **1924**, *10*, 311–325.
22. Nellensteyn, F.J. Relation of the micelle to the medium in asphalt. *J. Inst. Petrol. Technol.* **1928**, *14*, 134–138.
23. Sakhanov, A.N.; Vassiliev, N.A. Solubility of solid paraffins and the solidifying temperatures of material containing them. *Neftjanoe slancevoe Chozjajstvo* **1924**, *7*, 820–837.
24. Sachanen, A. Solubility of paraffin and the solidification of paraffin containing product. *Petroleum Z.* **1925**, *21*, 735–740.
25. Mack, C. Colloid chemistry of asphalts. *J. Phys. Chem.* **1932**, *36*, 2901–2914. [CrossRef]
26. Mack, C. Physico-chemical aspects of asphalts. *Proc. Assoc. Asphalt Paving Technol.* **1933**, *5*, 40–52.
27. Saal, R.N.J.; Heukelom, W.; Blokker, P.C. Physical constants of asphaltic bitumen Part 1. *J. Inst. Petrol. Technol.* **1939**, *26*, 29–39.
28. Pfeiffer, J.P.; Van Doormaal, P.M. The rheological properties of asphaltic bitumen. *J. Inst. Pet. Technol.* **1936**, *22*, 414–440.
29. Read, J.; Whiteoak, D. *The Shell Bitumen Handbook*, 5th ed.; Thomas Telford Publishing: London, UK, 2003.
30. Redelius, P. Asphaltenes in bitumen, what they are and what they are not. Road Mater. *Pavement Design* **2009**, *10*, 25–43. [CrossRef]
31. Brown, A.B.; Sparks, J.W.; Smith, F. Steric hardening of asphalts. *Proc. Assoc. Asphalt Paving Technol.* **1957**, *26*, 486–494.
32. Brown, A.B.; Sparks, J.W. Viscoelastic properties of a penetration grade paving asphalt at winter temperatures. *Proc. Assoc. Asphalt Paving Technol.* **1958**, *27*, 35–51.
33. Blokker, P.C.; Van Hoorn, H. Durability of bitumen in theory and practice. In Proceedings of the Fifth World Petroleum Congress, New York, NY, USA, 30 May–5 June 1959.
34. Bahia, H.U.M. Low-Temperature Physical Hardening of Asphalt Cements. Ph.D. Thesis, Department of Civil Engineering, The Pennsylvania State University, State College, PA, USA, December 1991.
35. Bahia, H.U.; Anderson, D.A. Glass transition behavior and physical hardening of asphalt binders. *J. Assoc. Asphalt Paving Technol.* **1993**, *62*, 93–129.
36. Bahia, H.U.; Anderson, D.A. Isothermal low-temperature physical hardening of asphalt. *Proc. Int. Symp. Chem. Bitum.* **1991**, *10*, 114–147.
37. Kriz, P.; Stastna, J.; Zanzotto, L. Effect of thermal history on glass transition and phase compatibility of asphalts. In Proceedings of the Fourth Eurasphalt & Eurobitume Congress, Copenhagen, Denmark, 21–23 May 2008.
38. Kriz, P.; Stastna, J.; Zanzotto, L. Effects of stress on physical hardening of asphalts. In Proceedings of the Sixth International Conference Pavement Technolo, Sapporo, Japan, 1–8 July 2008.
39. Hesp, S.A.M.; Iliuta, S.; Shirokoff, J.W. Reversible aging in asphalt binders. *Energy Fuels* **2007**, *21*, 1112–1121. [CrossRef]
40. Evans, M.; Marchildon, R.; Hesp, S.A.M. Effects of physical hardening on stress relaxation in asphalt cements. *Transp. Res. Rec. J. Transp. Res. Board* **2011**, *2207*, 34–42. [CrossRef]
41. Jing, R.; Varveri, A.; Liu, X.; Scarpas, A.; Erkens, S. Rheological, fatigue and relaxation properties of aged bitumen. *Int. J. Pav. Eng.* **2019**, *21*, 1024–1033. [CrossRef]
42. Traxler, R.N.; Romberg, J.W. Asphalt, a colloidal material. *Ind. Eng. Chem.* **1952**, *44*, 155–158. [CrossRef]
43. Krenkler, K. The influence of paraffin content on the properties of bitumen. *Bitumen-Teere-Asphalte-Peche und Verwante Stoffe* **1950**, *1*, 185–191.
44. Hou, Y.; Sun, W.; Das, P.; Song, X.; Wang, L.; Ge, Z.; Huang, Y. Coupled Navier–Stokes Phase-Field Model to Evaluate the Microscopic Phase Separation in Asphalt Binder under Thermal Loading. *J. Mat. Civ. Eng.* **2016**, *28*, 04016100. [CrossRef]
45. Anderson, D.A.; Kennedy, T.W. Development of SHRP binder specification. *J. Assoc. Asphalt Paving Technol.* **1993**, *62*, 481–507.

46. Jamshidi, A.; Hamzah, M.O.; Shahadan, Z.; Yahaya, A.S. Evaluation of the rheological properties and activation energy of virgin and recovered asphalt binder blends. *J. Mater. Civ. Eng.* **2015**, *27*, 04014135. [CrossRef]
47. Andriescu, A.; Hesp, S.A.M.; Youtcheff, J.S. Essential and plastic works of ductile fracture in asphalt binders. *Transp. Res. Rec. J. Transp. Res. Board* **2004**, *1875*, 1–8. [CrossRef]
48. Andriescu, A.; Iliuta, S.; Hesp, S.A.M.; Youtcheff, J.S. Essential and plastic works of ductile fracture in asphalt binders and mixtures. *Proc. Can. Technol. Asphalt Assoc.* **2004**, *49*, 93–121.
49. Andriescu, A.; Gibson, N.; Hesp, S.A.M.; Qi, X.; Youtcheff, J.S. Validation of the essential work of fracture approach to fatigue grading of asphalt binders. *J. Assoc. Asphalt Paving Technol.* **2006**, *75E*, 1–37.
50. Gibson, N.; Qi, X.; Kutay, M.; Andriescu, A.; Stuart, K.; Youtcheff, J.; Harman, T. *Performance Testing for Superpave and Structural Validation*; Report No. FHWAHRT-11-045; FHWA, United States; Federal Highway Administration: McLean, VA, USA, 1 November 2012.
51. Kriz, J.A.P.; Noël, M.R.; Quddus, R.D.S. Rheological properties of phase-incompatible bituminous binders. In Proceedings of the Seventh Eurobitumen Eurasphalt Congress, Madrid, Spain, 15–17 June 2021.
52. Readshaw, E.E. Asphalt specifications in British Columbia for low temperature performance. In Proceedings of the Association of Asphalt Paving Technologists, Cleveland, OH, USA, 14–17 February 1972; pp. 562–581.
53. Pechenyi, B.G.; Zhelezko, E.P. Effect of thermal relaxation on asphalt properties. *Chem. Tech. Fuel. Oil.* **1973**, *9*, 767–769. [CrossRef]
54. Deme, I.J.; Young, F.D. Ste. Anne Test Road Revisited Twenty Years Later. In Proceedings of the Canadian Technical Asphalt Association, Morin Heights, QC, Canada, 30 September 1987; pp. 254–283.
55. Pechenyi, B.G.; Kuznetsov, O.I. Formation of equilibrium structures in bitumens. *Chem. Tech. Fuels. Oil.* **1990**, *26*, 372–376. [CrossRef]
56. Struik, L.C.E. *Physical Aging in Amorphous Polymers and Other Materials*; Elsevier Scientific: Amsterdam, The Netherlands, 1978.
57. Soenen, H.; De Visscher, J.; Vanelstraete, A.; Redelius, P. Influence of the thermal history on rheological properties of various bitumen. *Rheol. Acta* **2006**, *45*, 729–739. [CrossRef]
58. Togunde, O.P.; Hesp, S.A.M. Physical hardening in asphalt mixtures. *Int. J. Pavement Res. Technol.* **2012**, *5*, 46–53.
59. Soenen, H.; Ekblad, J.; Lu, X.; Redelius, P. Isothermal hardening in bitumen and in asphalt mix. In Proceedings of the Third Eurobitumen Eurasphalt Congress, Vienna, Austria, 12–14 May 2004.
60. Claudy, P.; Letoffe, J.; Rondelez, F.; Germanaud, L.; King, G.; Planche, J.-P. A new interpretation of time-dependent physical hardening in asphalt based on DSC and optical thermoanalysis. Preprint Paper. *Am. Chem. Soc., Div. Fuel Chem.* **1992**, *37*, 1408–1426.
61. Rigg, A.; Duff, A.; Nie, Y.; Somuah, M.; Tetteh, N.; Hesp, S.A.M. Non-isothermal kinetic analysis of reversible ageing in asphalt cements, Road Mater. *Pavement Des.* **2017**, *18*, 185–210. [CrossRef]
62. Ding, H.; Qiu, Y.; Rahman, A. Low-temperature reversible aging properties of selected asphalt binders based on thermal analysis. *J. Mat. Civ. Eng.* **2019**, *31*, 04018402. [CrossRef]
63. Nicol, E.; Marks, P.; Ding, H.; Hesp, S.A.M. Asphalt properties compared with performance in Ontario. *Proc. Can. Technol. Asphalt Assoc.* **2018**, *63*, 61–77.
64. Tabib, S.; Hoque, F.; Marks, P. Ontario's strategy to enhance asphalt cement quality. *Proc. Can. Technol. Asphalt Assoc.* **2015**, *60*, 325–346.
65. Khan, A.N.; Hesp, S.A.M. Comparison between thermal, rheological and failure properties for the performance grading of asphalt cements. *Constr. Build. Mater.* **2019**, *220*, 196–205. [CrossRef]
66. Goodrich, J.L. Asphalt and polymer modified asphalt properties related to the performance of asphalt concrete mixes. *Proc. Assoc. Asphalt Paving Technol.* **1991**, *60*, 117–175.
67. Button, J.W.; Hastings, C.P.; Little, D.N. *Effects of Asphalt Additives on Pavement Performance*; Report No FHWA/TX-97/187-26; Transportation Institute: College Station, TX, USA, 1996.
68. Migliori, F.; Pastor, M.; Ramond, G. Etude statistique de quelques cas de fissurations thermiques. In Proceedings of the Fifth Eurobitumen Congress, Stockholm, Sweden, 16–18 June 1993; pp. 724–728.
69. Migliori, F.; Ramond, G.; Ballie, M.; Brule, B.; Exmelin, C.; Lombardi, B.; Samanos, J.; Maia, A.F.; Such, C.; Watkins, S. Correlations between the thermal stress cracking of bituminous mixes and their binders' rheological characteristics. In Proceedings of the Eurobitume Workshop on Performance Related Properties for Bituminous Binders, Luxembourg, 3–6 May 1999; pp. 1–5.
70. Widyatmoko, I.; Heslop, M.; Elliott, R. The viscous to elastic transition temperature and the in situ performance of bituminous and asphaltic materials. *J. Inst. Asphalt Technol. Asphalt Prof.* **2005**, *14*, 3–7.
71. Soleimani, A.; Walsh, S.; Hesp, S.A.M. Asphalt cement loss tangent as surrogate performance indicator for control of thermal cracking. *Transp. Res. Rec., J. Transp. Res. Board* **2009**, *2162*, 39–46. [CrossRef]
72. AASHTO R 28. *Accelerated Aging of Asphalt Binder Using a Pressurized Aging Vessel (PAV)*; American Association of State Highway and Transportation Officials: Washington, DC, USA, 2009.
73. AASHTO TP 113–15. *Asphalt Binder Resistance to Ductile Failure Using the Double-Edge-Notched Tension (DENT)*; American Association of State Highway and Transportation Officials: Washington, DC, USA, 2015.
74. AASHTO TP 122–16. *Performance Grade of Physically Aged Asphalt Binder Using the Extended Bending Beam Rheometer (BBR) Method*; American Association of State Highway and Transportation Officials: Washington, DC, USA, 2016.
75. STP 1241. *Physical Properties of Asphalt Cement Binders*; American Society for Testing and Materials: West Conshohocken, PA, USA, 1995.
76. Khan, A.; Akentuna, M.; Pan, P.; Hesp, S.A.M. Repeatability, reproducibility, and sensitivity assessments of thermal and fatigue cracking acceptance criteria for asphalt cement. *Constr. Build. Mater.* **2020**, *243*, 117956. [CrossRef]

77. Lill, K.; Khan, A.N.; Kontson, K.; Hesp, S.A.M. Comparison of performance-based specification properties for asphalt binders sourced from around the world. *Constr. Build. Mater.* **2020**, *261*, 120552. [CrossRef]
78. Berkowitz, M.; Filipovich, M.; Baldi, A.; Hesp, S.A.M.; Aguiar-Moya, J.-P. Oxidative and thermoreversible aging effects on performance-based rheological properties of six Latin American asphalt binders. *Energy Fuels* **2019**, *33*, 2604–12013. [CrossRef]
79. Ding, H.; Zhang, H.; Liu, H.; Qiu, Y. Thermoreversible aging in model asphalt binders. *Constr. Build. Mater.* **2021**, *303*, 124355. [CrossRef]

materials

Review

Concept and Development of an Accelerated Repeated Rolling Wheel Load Simulator (ARROWS) for Fatigue Performance Characterization of Asphalt Mixture

Zeyu Zhang , Julian Kohlmeier *, Christian Schulze and Markus Oeser

Institute of Highway Engineering, RWTH Aachen University, D52074 Aachen, Germany;
zeyu.zhang@isac.rwth-aachen.de (Z.Z.); schulze@isac.rwth-aachen.de (C.S.); oeser@isac.rwth-aachen.de (M.O.)
* Correspondence: kohlmeier@isac.rwth-aachen.de

Abstract: Fatigue performance is one of the most important properties that affect the service life of asphalt mixture. Many fatigue test methods have been developed to evaluate the fatigue performance in the lab. Although these methods have contributed a lot to the fatigue performance evaluation and the development of fatigue related theory and model, their limitations should not be ignored. This paper starts by characterizing the stress state in asphalt pavement under a rolling wheel load. After that, a literature survey focusing on the experimental methods for fatigue performance evaluation is conducted. The working mechanism, applications, benefits, and limitations of each method are summarized. The literature survey results reveal that most of the lab test methods primarily focus on the fatigue performance of asphalt mixture on a material level without considering the effects of pavement structure. In addition, the stress state in the lab samples and the loading speed differ from those of asphalt mixture under rolling wheel tire load. To address these limitations, this paper proposes the concept of an innovative lab fatigue test device named Accelerated Repeated Rolling Wheel Load Simulator (ARROWS). The motivation, concept, and working mechanism of the ARROWS are introduced later in this paper. The ARROWS, which is under construction, is expected to be a feasible and effective method to simulate the repeated roll wheel load in the laboratory.

Keywords: asphalt mixture; fatigue performance; rolling load; laboratory test; device

Citation: Zhang, Z.; Kohlmeier, J.; Schulze, C.; Oeser, M. Concept and Development of an Accelerated Repeated Rolling Wheel Load Simulator (ARROWS) for Fatigue Performance Characterization of Asphalt Mixture. *Materials* **2021**, *14*, 7838. https://doi.org/10.3390/ma14247838

Academic Editor: Francesco Canestrari

Received: 29 October 2021
Accepted: 8 December 2021
Published: 17 December 2021

Publisher's Note: MDPI stays neutral with regard to jurisdictional claims in published maps and institutional affiliations.

1. Introduction

Asphalt pavement withstands millions of rolling wheel loads after opening to traffic. The rolling of wheels on the top of the pavement brings tensile, compressive, and shear stresses to the asphalt mixture. Repeating this phenomenon over the long term creates fatigue damage, degrading the integrity of asphalt mixture by creating and growing micro-cracks or cavities. As a consequence of fatigue damage accumulation, longitudinal cracks appear on pavement surface and propagate into alligator cracks, turning asphalt pavement into essentially a gravel layer as loading continues [1]. Therefore, accurately capturing and predicting the fatigue behavior of asphalt mixture is vital to prevent asphalt pavement from premature fatigue failure. For this purpose, researchers have proposed many test technologies during the past few decades, such as 4-point bending (4PB) fatigue test method [2], and indirect tensile fatigue test (ITFT) method [3]. These laboratory test methods enable the researchers to determine the influences of inherent material properties (e.g., air voids content, involved binder type, and aggregate gradation) and external service conditions (e.g., temperature, loading frequency, and load amplitude) on the fatigue response of asphalt mixtures, and to develop theories and models concerning the fatigue behavior of asphalt materials [4–6]. Although it has been extensively used, the fatigue test method is still in a dynamic development process. The overall goal of these efforts is to eliminate the differences between the lab-measured and in situ observed fatigue and cracking behavior of asphalt mixture. Specifically, researchers try to optimize the experimental setup of the fatigue tests to better represent the stress state in asphalt pavement. Bending fatigue test is designed to simulate the flexural stress at the bottom of the asphalt layer, and ITFT is expected to evaluate the fatigue response of asphalt mixture under the two-dimensional stress. The load waveform, loading frequency, and test temperature are also carefully selected to simulate the realistic service condition of asphalt pavement.

Despite the fact that much work has been done, the laboratory fatigue test technologies still have great limitations in simulating the field fatigue performance of asphalt mixture. For instance, the flexural stress at the bottom of the asphalt layer of pavement is two-dimensional. Fatigue damage occurs due to the repetition of flexural stress at both longitudinal and transverse directions caused by repeated rolling load. None of the currently available fatigue test methods can simulate this phenomenon. Hence, a paper is demanded to holistically present the development, current state of art, and limitations of the fatigue test technologies. To this purpose, this paper presents a literature survey concerning the currently available fatigue test technologies in various aspects, including sample geometry, load control mode, and load frequency. Prior to the literature survey, the stress state in asphalt mixture under the rolling tire load is discussed. Based on the literature survey results, the concept of an innovative fatigue test device named ARROWS, which is the acronym of Accelerated Repeated Rolling Wheel load Simulator, is proposed as an alternative to the current fatigue test technologies. The feasibility of the ARROWS is evaluated by comparing the stress state obtained from ARROWS with that in the multi-layered system.

2. Stress State in Asphalt Mixture under the Rolling Tire Load

The creation and accumulation of the fatigue damage in the asphalt mixture is caused by repeated rolling of tires on the surface of asphalt pavement. The good correlation between the fatigue results from the laboratory test methods and the field performance relies on how well (realistically) the repeated rolling loads have been simulated in the laboratory. Currently, it is the tensile stress at the bottom of the asphalt layer that attracts the most attention when designing asphalt pavement. Fatigue cracks are believed to be created at the bottom of the asphalt layer and propagate upwards. Given this, many studies employed cyclic one-dimensional tensile stress to evaluate the fatigue and cracking performance of the asphalt mixture. However, it is the two-dimensional tensile stress at the bottom of the asphalt layer that requires evaluation. Characterizing the tensile stress at the bottom of the asphalt layer under repeated rolling wheel loads is necessary to ensure the rationality of the fatigue test techniques.

The stress state in the asphalt mixture under repeated rolling wheel loads is analyzed in this section using the finite element (FE) method. In the FE model, a typical German pavement structure used for heavy traffic is simplified as an elastic multilayer system according to the standard RStO 12 [7]. The tire–pavement contact area and standard tire load are regarded as a circle with a diameter of 150 mm and 0.7 MPa, respectively, according to the standard RDO Asphalt 09 [8]. Table 1 details the information about pavement structure and involved materials. Figure 1 shows a schematic diagram of the pavement model. As shown, the wheel rolls towards the right side of the target point at a constant speed v. After $3/v$ seconds, the wheel rolls to directly above the target point. The stress variation caused by the rolling process is shown in Figure 2.

Table 1. Information about pavement structure and materials.

Pavement Layer	Thickness [mm]	Elastic Modulus [MPa]	Poisson's Ratio
Surface course	40	5581	0.27
Asphalt binder course	80	9686	0.27
Asphalt base course	220	6481	0.27
Frost protection layer	510	325	0.35
Subgrade	2000	50	0.40

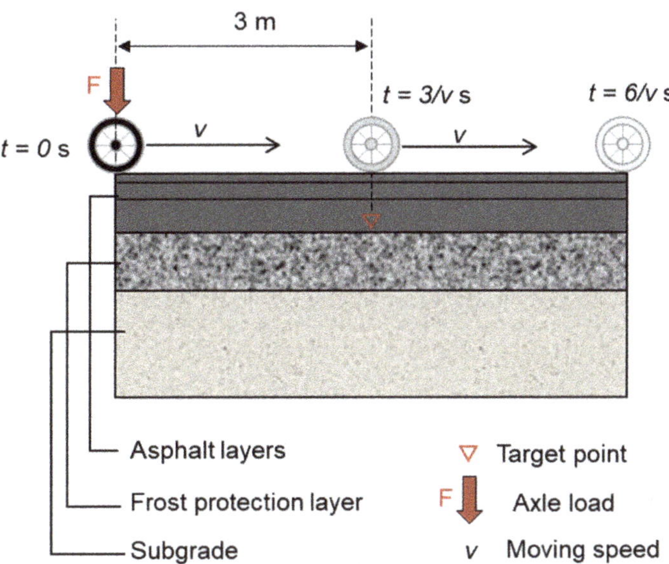

Figure 1. Schematic diagram of the pavement model.

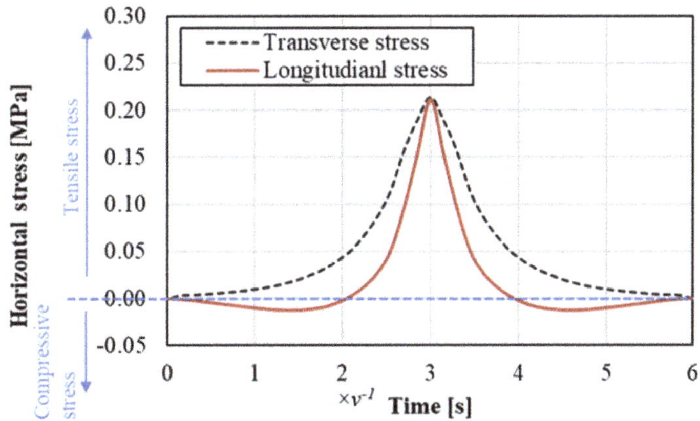

Figure 2. Stress reversal at the bottom of the asphalt layer.

Figure 2 presents the variation of horizontal stresses at the target point of asphalt mixture. As shown, both the transverse and longitudinal stresses reach the maximum when the wheel is directly above the target point. Transverse stress presents as pure tensile stress for the entire process. On the contrary, longitudinal stress experiences a compression–tension–compression process, which is called stress reversal in this paper.

An ideal fatigue test device is desired to reproduce the characteristics of tensile stresses on lab specimens. Currently, the lab test methods simplify the realistic stress waves as sinusoidal or haversine waves. By introducing a gap between two adjacent load waves, researchers can consider the self-healing performance while measuring the fatigue performance of the test specimen. However, Figure 2 indicates that the lab load on the specimen should be at least two-dimensional. Moreover, the stress at one dimension should contain a stress reversal and the stress at another dimension should be pure tensile stress.

3. Laboratory Fatigue Test Methods

The development of fatigue test devices was initially upon in the mid-1960s, where Professor Monismith et al. developed the UCB device, a 4PB fatigue test device [2]. After decades of development, there are a variety of fatigue test methods available. These methods vary from the fatigue test device and experiment setup to control methods. In general, there are two approaches to assess the fatigue properties of asphalt mixture. In the first approach, repeated loads are applied to test the specimen until the specimen fails. The responses of asphalt mixture are related to the number of loading repetitions. The second approach aims at measuring the material fundamental stress–strain relationship to formulate rigorous constitutive modes and evaluate the fatigue resistance properties of the corresponding asphalt mixture and pavement. The typical example of the second approach is the fracture mechanics-based critical strain energy release rate, Jc, derived from the notched semi-circular bending (SCB) test [9,10]. This paper primarily focuses on the test technologies employing the first approach. The involved fatigue test technologies are separated into bending fatigue test, axial fatigue test, diametral fatigue test, triaxle fatigue test, and shear fatigue test according to their loading modes.

3.1. Bending Beam Fatigue Test
3.1.1. 4PB Fatigue Test

The 4PB fatigue test is conducted on compacted beam specimens using a fixed reference point bending beam fixture. Figure 3 is the schematic diagram and the tensile stress distribution of the 4PB fatigue test. The beam specimen is fixed by two reaction clamps (clamps 1 and 4) on the outside. Repeated loads are introduced to the middle third of the specimen by two inside closed load clamps (clamps 2 and 3). The setups of the clamps allow the middle third of the prismatic specimen to work under maximum stress; the prismatic specimen is evenly distributed in the horizontal direction, and has zero moments. However, the flexural load in the specimen does not generate a homogenous stress distribution within the specimen.

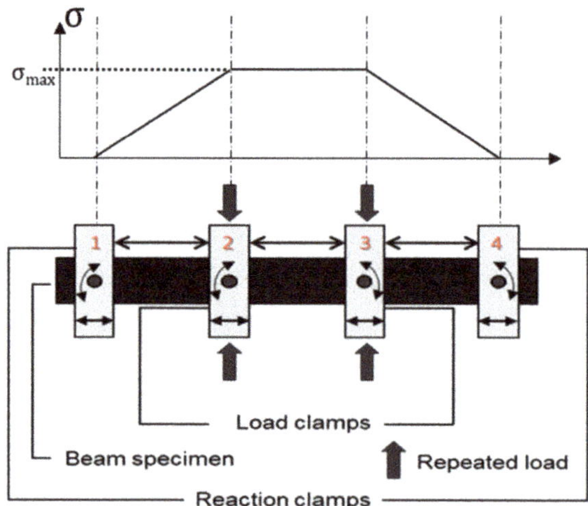

Figure 3. Scheme and tensile stress distribution of 4PB fatigue test.

In the 4PB fatigue test, the asphalt mixture specimen is subjected to repeated bending until it fails. The number of loading cycles at failure, i.e., fatigue life Nf, is recorded and plotted against the applied load value. The variation of flexural stiffness indicates the specimen's response to repeated loading. The calculation of the stiffness modulus from the deflection measurement relies on the slender beam theory, in which deflection due to shear is ignored. The Saint-Venant's principle derives the following formulas for calculating the maximum tensile stress and strain of the 4PB test where the loading span is 1/3 of the support span.

$$\sigma_t = \frac{P \times L}{2 \times b \times h^2} \tag{1}$$

$$\varepsilon_t = \frac{54 \times u \times h}{23 \times b \times h^2} \tag{2}$$

$$S = \frac{\sigma_t}{\varepsilon_t} \tag{3}$$

where

σ_t = Maximum tensile stress at the bottom of beam (Pa);
P = Peak-to-peak force applied on beam (N);
b = Average specimen breath (width) (mm);
h = Average specimen height (mm);
L = Length of beam span between outside clamps (mm);
ε_t = Maximum tensile strain at the bottom of beam (1);
u = Peak-to-peak deflection at center of beam (mm);
S = Flexural beam stiffness (Pa).

Similarly, by considering the plane hypothesis in elastic mechanics, Lv et al. proposed a methodology to calculate the compressive and tensile stress at the top and bottom of beam specimen. Using these formulas (Equations (4) and (5)), researchers are able to evaluate the compressive and tensile modulus evolution synchronously using 4PB fatigue test [11].

$$E_c = \frac{P \times L \times (\varepsilon_t + \varepsilon_c)}{2 \times b \times \varepsilon_c^2 \times h^2} \tag{4}$$

$$E_t = \frac{P \times L \times (\varepsilon_t + \varepsilon_c)}{2 \times b \times \varepsilon_t^2 \times h^2} \tag{5}$$

where

E_t = Maximum tensile modulus (Pa);
E_c = Maximum compressive modulus (Pa).

The 4PB fatigue test was first proposed by researchers from the University of California in Berkeley (UCB). Later, it was further refined during the SHRPA-003A project, completed in 1994, including the improvements to the data acquisition and control system and the use of sinusoidal loads [12]. Based on the research findings of this project, two standards, namely, ASTM D7460-08 and AASHTO T321-03, were developed. Over the years of research work, the ASTM D7460-10 and AASHTO T321-14 were released to unify the failure definition and achieve similar test results [13,14]. The main difference among these standards is the format of the repeated load waves. Because both ASTM and AASHTO methods are run in displacement-controlled mode, the format of load wave describes the waveform of the displacements imparted to the beam by the load clamps. The ASTM standard employs the cyclic haversine loads, in which the device bends the beam specimen on one side of its neutral axis [13]. The AASHTO standard recommends the use of the repeated sinusoidal loads, which means the applied displacement oscillates alternatively on both sides of the neutral axis with the same amplitude on both sides. Despite the AASHTO standard, the European standard EN 12697-24:2018 and the Australian standard AG: PT/T233-2016 suggest using cyclic sinusoidal loads as well [3,15]. Because the load waveform is not unified, cyclic loads with both sinusoidal and haversine waveform are utilized in the 4PB fatigue test. For instance, Yu et al., Dondi et al., Arsenie, et al., and Li et al. used the haversine displacement to assess the fatigue resistance of asphalt mixture [16–20]. On the other hand, Poulikakos et al., Di Benedetto et al., and Abhijith and Narayan employed sinusoidal loading in their studies [21–23]. Some concerns have been raised regarding the comparability of the 4PB fatigue test using different displacement waveforms. However, many studies reveal similar results of the 4PB fatigue test with same equivalent peak-to-peak strain in haversine and sine displacement testing modes [18,24]. A plausible explanation is due to the viscoelastic nature of asphalt mixture. Specifically, after a few loading cycles, asphalt mixture will exhibit a sinusoidal response to the applied haversine displacement, if the beam specimen is not supported by an elastic layer [18,25–27]. Given this, researchers believe it is not necessary to specifically recommend waveform of displacement in terms of the response of asphalt mixture. Nevertheless, this conclusion cannot be extrapolated to the asphalt mixtures whose nature is more elastic than viscoelastic. For instance, the strongly aged asphalt mixtures and the asphalt mixtures with harder asphalt binders. Besides, another limitation of these studies is the lack of consideration on the rest period. Mamlouk et al. induced a five-second rest period between two haversine load cycles [26]. Their findings reported the inconsistent results of 4PB fatigue test run in the haversine mode with and without rest period. Therefore, Mamlouk et al. recommended the ASTM to refine the ASTM D7460 by replacing the haversine waveform to sinusoidal waveform. It should be noted that in December 2018 the ASTM D8237-18 was published, in which the use of sinusoidal displacement waveform is recommended. Subsequently, the standard ASTM D7460-10 was withdrawn in 2019 [28].

Notably, the arguments regarding the waveform highlight the critical importance of load waveform to the fatigue response of asphalt mixture, which is also documented by the SHRP-A-003A report [29]. Besides, the rest period shows strong influences on the fatigue response of asphalt mixture by affecting the output force. Although many discussions have been conducted on the selection of load waveform, there are two consensuses: neither the sinusoidal nor the haversine waveform introduced by 4PB fatigue test realistically simulate the filed condition; and the elastic support foundation layer is important.

3.1.2. 3PB Fatigue Test

The 3PB fatigue test is also known as the center-point bending fatigue test [30]. This test is similar to the 4PB fatigue test, but the repeated loads are applied at the middle span of the specimen from a single point. Khalid believes the test duration of the 3PB fatigue test is likely to be reduced in comparison with 4PB fatigue test. This is because, for the same magnitude of applied force, the whole of the load applied by the load cell in 3PB test is transferred to the sample, whereas in the 4PB test, the applied load is split in two. However, because the tensile stress is exactly maximum under the point load, 3PB test does not allow the initiation of failure in a relatively uniform tensile stress region [29]. The scheme of the 3PB test and the distribution of the tensile stress is shown in Figure 4.

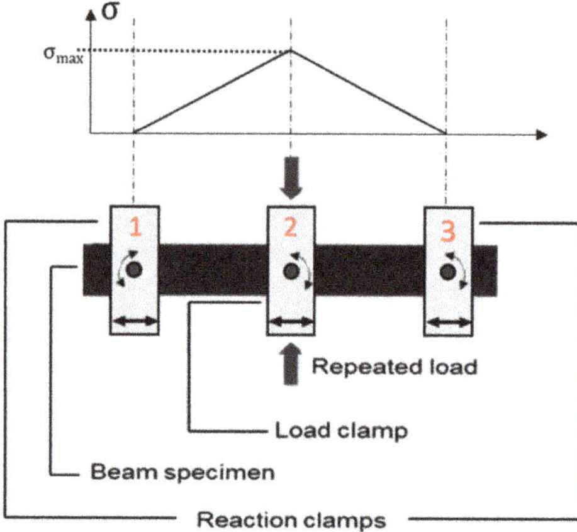

Figure 4. Schematic diagram and tensile stress distribution of 3PB fatigue test.

The maximum tensile stress of the 3PB test is mathematically expressed as below:

$$\sigma_t = \frac{3 \times P \times L}{4 \times b \times h^2}$$ (6)

3.1.3. 2PB Fatigue Test

2PB fatigue test, which was first developed in France, is widespread in European countries as a standard fatigue resistance test method [3]. This test is performed on a trapezoidal or prismatic beam set up in a test machine as a cantilever beam. The 2PB fatigue test is desired to be suitable for all types of asphalt mixture with the geometry of the sample as a function of the maximum nominal aggregate size (MNAS) of the asphalt mixture. The main result of the 2PB fatigue test is further used for pavement design. Specifically, the French pavement design principle utilizes the strain ε6, which is defined as the strain for failure at one million load repetitions using the 2PB fatigue test on the trapezoidal specimen, as the only metric of fatigue resistance.

During the 2PB fatigue test, the sample is installed vertically, its wide end is rigidly attached to a metal support, while its narrow end is subjected to repeated horizontal loads. Figure 5 shows a 2PB fatigue test device with a trapezoidal sample and the associated tensile stress distribution.

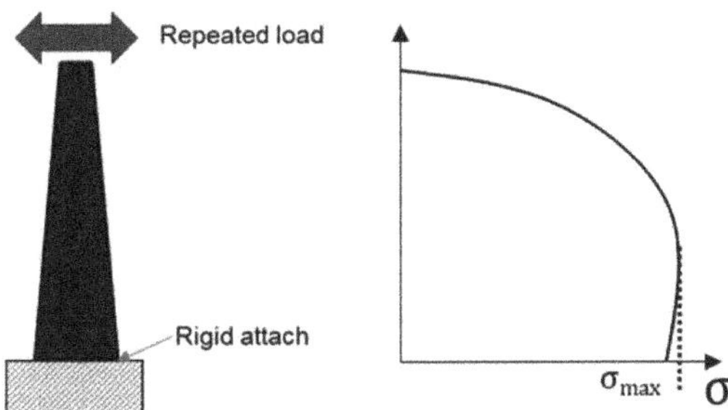

Figure 5. Schematic diagram and tensile stress distribution of 2PB fatigue test.

The applied loads are sinusoidal only but can be in either stress-controlled or strain-controlled mode [23,31]. The force–displacement used to generate the loads is recorded and used to identify fatigue failure of the sample. The specimens will fail at about mid height where the flexural stress occurs. The correlation between the flexural strain level and the displacement applied to the specimen head (narrow end of the specimen) can be described by a constant depending on the specimen geometry [3].

$$\varepsilon = K_\varepsilon \times z$$ (7)

$$K_{\varepsilon i} = \frac{(B_i - b_i)^2}{8 \times b_i \times h_i^2 \left[\frac{(b_i - B_i) \times (3B_i - b_i)}{2 \times B_i^2} + \ln \frac{B_i}{b_i} \right]}$$ (8)

where

ε = Relative strain of the specimen (-);
$K_{\varepsilon i}$ = Constant in relation to the largest strain;
z = Amplitude of the displacement at the head of the specimen;
B_i = Length of long base edge;
b_i = Length of short base edge;
h_i = Height of the specimen (mm).

The main advantages of the 2PB fatigue test are its relatively simple experimental setup compared to other fatigue tests, and the ability to include two samples simultaneously.

Several duplicate tests are required to precisely determine the Nf of asphalt mixture at a load level because the scatter of the fatigue test results is somehow inevitable. The 2PB fatigue test reduces the time needed to test the required set of samples as the 2PB fatigue test allows one to produce tests on two samples at the same time, indicating the same test condition and less scatter. As with any forms of bending fatigue tests, the weakness of the 2PB fatigue is the inhomogeneous stress distribution within the sample. In addition, Dondi et al. questioned the effects of sample gravity force on the 2PB test results. The logic behind this concern is that the direction of repeated loads employed by both 4PB and 3PB are parallel to the specimen's gravity force, while that of 2PB is perpendicular to the gravity. Consequently, they revised the classical 2PB fatigue test to a horizontal 2PB fatigue test. In the horizontal 2PB fatigue test, repeated loads are applied to the horizontally installed test specimen [24]. It should be noted that the fabrication of the high quality and precise trapezoidal sample constitutes an important challenge in this test.

Because both the 2PB and 4PB fatigue tests are the standard European test methods, their results are desired to show close agreement. However, the fatigue test results are expected to be considerably affected by test type and the mode of loading [23]. Therefore, many studies have been performed to determine the differences in the results measured by 2PB and 4PB methods. Poulikakos et al. compared the complex modulus and fatigue performance of filed aged specimens using 2PB and 4PB tests. Their study revealed that the 2PB and 4PB tests show a good linear regression for complex modulus values but dissimilar fatigue performance ranking [22]. Di Benedetto et al. reported similar results. They found the complex modulus is independent of the test method, but the fatigue performance is considerably affected by the test type [23,32]. Pronk et al. found the fatigue life obtained by 2PB test was shorter compared to that measured by 4PB test [25].

3.1.4. Loaded Wheel Fatigue Test

The loaded wheel fatigue test is derived from the wheel tracking tester, which is commonly used for evaluating the rutting resistance and moisture susceptibility of asphalt mixture. The loaded wheel fatigue test is desired to simulate the effects of a rolling wheel on the pavement by applying repeated moving wheel loads to the specimen [33,34]. To simulate the supporting effects of the base layer, in some studies, the specimen is placed on an elastic mat [35,36].

Wu et al. modified the asphalt pavement analyzer (APA) to perform an asphalt fatigue test (Figure 6) [37,38]. In their study, a beam specimen is subjected to repeated moving loads without supporting mat. A linear variable differential transformer (LVDT) was mounted on the bottom surface of the beam specimen to accurately measure the tensile strain under the moving load. To determine the tensile stress at the bottom surface of the specimen, the mechanical model of the test system was simplified as a plane-stress problem, within which, a vertical load moves on a beam.

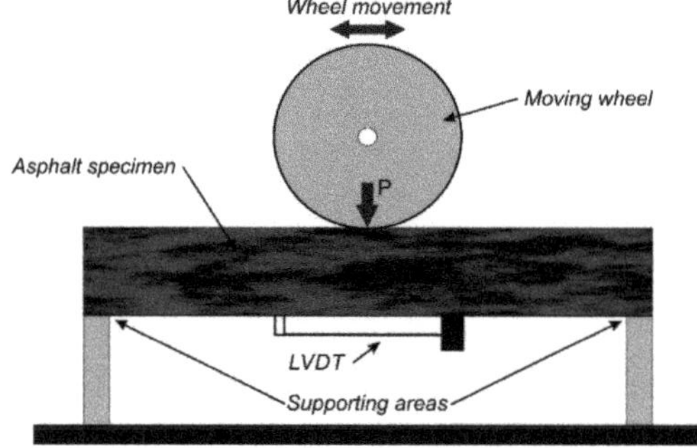

Figure 6. Scheme of the LWT fatigue test setup according to [38].

Combining this simplification with the 2D slender beam theory derives the tensile stress at the midpoint of the beam bottom:

$$\sigma_t = \sigma_{amp} \sin^2\left(\frac{2\pi}{T}t\right) = \frac{3PL}{2bh^2}\sin^2\left(\frac{2\pi}{T}t\right) \tag{9}$$

where

σ_{amp} = Amplitude of sinusoidal stress (Pa);
T = Testing cycle period (s);
t = Elapsed testing time (s);
P = Wheel load (N);
L = Length of the loading path (m).

Equations (10) and (11) yields the dynamic modulus (E^*) and phase angle (δ) of test specimen:

$$E^* = \frac{3PL}{2bh^2\varepsilon_{amp}} \tag{10}$$

$$\delta = 2\pi f \Delta t \tag{11}$$

where

ε_{amp} = Amplitude of sinusoidal strain (-);
f = Loading frequency (Hz);
Δt = Time lag between stress and strain (s).

Wu et al. evaluated the fatigue resistance of four types of asphalt mixture using modified APA test, direct tensile test, and 4PB test. The modified APA test gave the same ranking of the four asphalt mixtures in terms of fatigue resistance; the 4PB test indicated the feasibility of using loaded wheel fatigue test to differentiate between asphalt mixtures in terms of their fatigue performance [38].

However, the unsupported LWT fatigue test brings obvious vertical deformation to the test specimen, which raises concerns about the capability of the unsupported LWT fatigue test to simulate the fatigue failure process of real pavement structure [36]. Alternatively, Zhang et al. investigated the fatigue behavior of asphalt mixture using the modified Hamburg Wheel-Tracking Device (HWTD) [36]. Compared with APA modified by Wu et al., the main differences of the modified HWTD are the supporting condition and the side-confining condition. Zhang et al. utilized two 20-mm thick neoprene layers to simulate the supporting effects of the pavement base layer. The neoprene was selected as its modulus is similar to that of the pavement base course. The prismatic asphalt mixture sample was horizontally sandwiched by two wood strips, which was also used to simulate the side-confining environment. To monitor the response of the test specimen to cyclic loads, strain gauges were attached under the test specimen. Obviously, the modified HWTD system is a complex three-dimensional system, indicating the inapplicability of the slender beam system in calculating the bottom tensile stress. In their study, tensile stress is assumed as constant during the cyclic loading process. Based on this assumption, the fatigue response of asphalt mixture was analyzed using energy method.

Similarly, Nguyen and Thom characterized the fatigue performance of asphalt mixture using a Beam Wheel Tracker Fatigue Test (BWTFT) [39]. The BWTFT was developed based on the wheel tracker machine from the Nottingham Transportation Engineering Center (NTEC). In their study, two 10 mm-thick rubber mats were used to beneath the asphalt beam. Cyclic loads were applied to the beam specimen through the loaded wheel. Strain gauges were attached on the two sides of the beam, as shown in Figure 7, to monitor the response of the asphalt beam to moving wheel loads. The peak–trough strain collected using the strain gauges was plotted against the number of loading cycles as the output result of the BWTFT test.

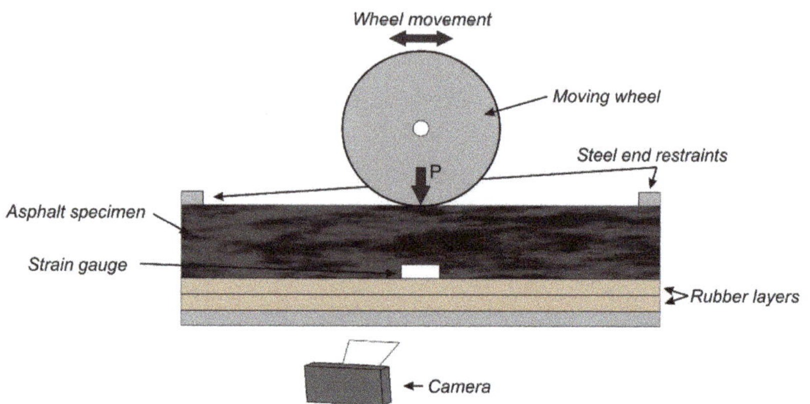

Figure 7. Scheme of experimental setup of BWTFT according to [39].

Nguyen and Thom treated the BWTFT as a beam on an elastic foundation layer. The supporting effect of the rubber mat is a function of rubber modulus and curvature, which derives the tensile stress at the midpoint of the asphalt beam:

$$\sigma_t = \frac{1.14 \times E^{0.25} \times P}{k^{0.25} \times h^{1.25}} \tag{12}$$

where

E = Modulus of asphalt (Pa);
k = Modulus of rubber (Pa);
h = Beam thickness (m).

The GoPro camera was used to capture the cracking behavior of asphalt mixture. Apart from the cracking behavior, the collected sample photos also indicated that no obvious vertical deformation occurs on the failed specimen, which is ascribed to the supporting effect of the rubber mat.

The main shortcoming of the loaded wheel fatigue test is its loading frequency. Due to the limitation of the LWT device, the loading frequency is usually low and nonadjustable. Specifically, the loading frequency used by Wu et al., Zhang et al., and Nguyen and Thom are 2 Hz, 1 Hz, and 0.1 Hz, respectively [36,38,39].

3.2. Axial Fatigue Test

The axial fatigue test refers to the test methods that apply uniaxial cyclic loads to specimen with or without stress reversal. The mechanism of the axial fatigue test brings two advantages compared with other technologies: the homogeneous distribution of applied load and ease of control and measurement. Specifically, in the axial fatigue test the load and deformation are distributed regularly over the cross-section of the specimen under loading. The axial fatigue test allows pavement engineers to directly collect the responses of test specimen in the form of force–displacement without any further process. Because of these merits, axial fatigue test plays a critical role in evaluating the fatigue performance of asphalt mixture and the development of damage theories of asphalt mixture [6,40–44].

Typical test methods of the axial fatigue test include the direct tension fatigue test (DTFT), the compression fatigue test (DCFT), and the tension compression fatigue test (TCFT). During these tests, loads are applied to the top end of the sample through the actuator. As with other test technologies, the applied cyclic loads of axial fatigue test can be controlled in force or displacement mode [45,46].

For the test methods involving uniaxial tension load, an encountered experimental problem is the end-failure of the specimen, which refers to the fatigue failure at the ends of the specimen. There are two possible reasons for the end-failure phenomenon: non-uniformity distribution of the air voids and the eccentric load [45,47]. The ends of cylinder specimen fabricated by Superpave gyratory compactor (SGC) exhibits higher air void content, increasing the likelihood of failure at the corresponding points [48–50]. Eccentric tension is believed to be another critical factor leading to the premature end-failure [51]. The maximum axial stress and corresponding strain at the sample end increase sharply with the increase in load eccentricity. The proper experimental setup is necessary for avoiding eccentric tension [52,53]. Therefore, both TCFT and DTFT require test samples to have a firm connection to the test system. Enough adhesive bond between the sample ends and the test system is necessary to avoid the adhesive failure at the sample–glue or glue–platen interface under tension loading.

CFT is relatively less used in evaluating the fatigue resistance of asphalt mixture, which may be because compression load is more related to the permanent deformation, such as rutting at high-temperatures [54–56]. However, CFT could contribute to formulating fatigue damage models for asphalt mixture. For instance, Lv et al. jointly used the CFT, DTFT, and indirect tensile fatigue test (ITFT) methods to build a normalized model of fatigue characteristics for asphalt mixture in three-dimensional stress states [57].

DTFT method employs cyclic uniaxial tension loads to determine the fatigue response of asphalt mixture, the waveform of cyclic loads can be in either sinusoidal or haversine waveform mode [58,59]. A standard test protocol (AASHTO TP107) and analytical method have been developed based on the DTFT setup [60–62]. As with other test technologies, DTFT was also able to investigate the fatigue response of asphalt mixture, considering the many variables, such as loading frequency, rest period, material properties, aging condition, and test temperature [63–66].

TCFT, as expected, contains a stress reversal for each loading cycle, because of which, TCFT can simulate the stress reversal observed in the field condition. Both strain-controlled mode and stress-controlled mode apply to the TCFT [46,67,68]. Regarding the load waveform, only the sinusoidal waveform is reported by the surveyed literature. In comparison with DTFT, TCFT offers the possibility to consider the effects of stress–strain ratio R, which is defined by Figure 8. As reported by Isailović et al., introducing a stress reversal into the loading waveform reduces the energy dissipated and partially eliminates the vertical deformation of the test sample [46].

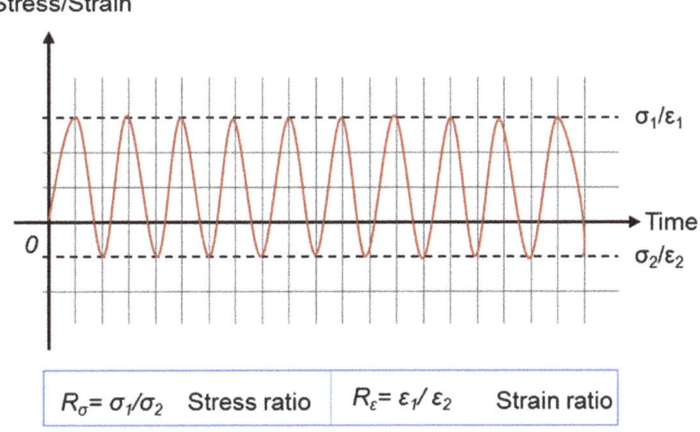

Figure 8. Definition of stress–strain ratio.

Later, Zhang and Oeser derived a damage evolution model from the continuum damage mechanics [6]. According to their model (Equation (13)), N_f is a function of material parameters α and β, and peak-to-peak stress σ:

$$N_f = \frac{\alpha}{(1+\alpha)\beta}\sigma^{-\alpha} \tag{13}$$

where $\alpha > 1$.

Replacing σ by $\sigma = \sigma(\sigma_1, R)$ yields:

$$N_f = \frac{\alpha}{(1+\alpha)\beta}\cdot\left(\frac{R_\sigma - 1}{R}\right)^\alpha \cdot \left(\frac{1}{\sigma_1}\right)^\alpha \tag{14}$$

Equation (12) indicates that N_f depends on the material parameters, stress ratio, and σ_1. For example, if the σ_1 remains the same and R_σ changes from -1 to -2, N_f will increase by $(4/3)\alpha$ times. Further studies are demanded to comprehensively investigate the effects of R on the fatigue performance of asphalt mixture.

The primary limitation of the DTFT is that the utilized pure tension loads do not necessarily represent the field conditions. In addition, DTFT under stress-controlled mode is criticized because it creates permanent vertical deformation to the specimen [46]. According to Di Benedetto, asphalt mixture exhibits two kinds of material behavior under the cyclic loading: accumulation of permanent deformation and the aggravation of fatigue damage [23,69]. These two material behaviors can be separated by the form and position of the hysteresis loop in each loading cycle, as shown in Figure 9. Rotation and expansion of the hysteresis loops with increasing number of loading cycles are primarily due to material fatigue, while the horizontal moving of the loops is ascribed to the permanent deformation [69].

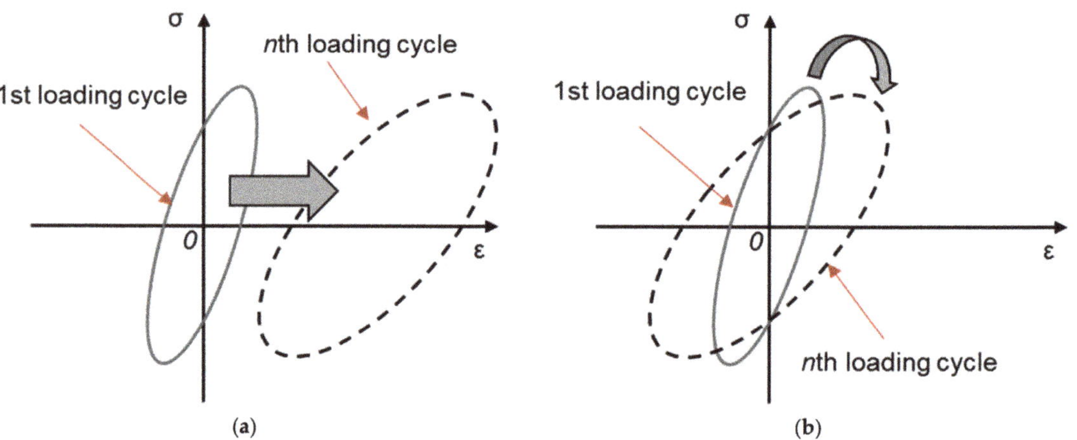

Figure 9. Changes in form and position of hysteresis loop. (a) Horizontal moving. (b) Rotation and expansion.

Isailović et al. observed the horizontal moving of the hysteresis loops of asphalt mixture under uniaxial tension stress, indicating that the failure of asphalt mixture under uniaxial tension stress is primarily associated to the accumulation of permanent deformation. Hence, they do not recommend using DTFT for fatigue analysis [46]. Similarly, in comparison with DTFT, Ashouri recommended the use of zero-mean stress-controlled TCFT instead of the zero-minimum stress-controlled DTFT to evaluate the healing behavior of asphalt mixture due to the following considerations [70]: first, TCFT can better simulate the pavement stress history than the DTFT; second, zero-mean stress-controlled TCFT makes the comparison between the fatigue test without rest periods and the healing test with reset periods easier, and third, fatigue failure is defined as the cycle when the phase angle starts to decrease. This trend is only found in the results of TCFT and not in the results of DTFT. A plausible explanation of this phenomenon is that the accumulation of viscoplastic strain prevents the material to heal.

Overall, the axial fatigue test technologies purely focus on the material property regarding fatigue performance. The loading conditions of these technologies do not necessarily represent the field conditions caused by the horizontal moving of wheel tires. Although introducing stress reversal into the loading waveform can simulate the loading pulse caused by the passing of wheel loads on pavement, TCFT does not consider the structural effect of pavement and the realistic stress–strain ratio.

3.3. Diametral Fatigue Test

Diametral fatigue tests refers to the tests that apply cyclic loads to the specimen in the diameter direction of the specimen. A typical diametral fatigue test is the indirect tension fatigue test (ITFT).

ITFT is one of the four fatigue test technologies recommended by the European standard [3]. Pavement engineers are able to design asphalt mixtures and pavements for fatigue adequacy based on the ITFT-measured fatigue response together and its correlation with the field performance.

As shown in Figure 10a, repeated vertical loads are imposed on the cylinder samples. Figure 10b illustrates that the loading configuration develops a reasonable uniform tensile stress in the specimen perpendicular to the direction of the applied load and along the vertical diametral plane. Figure 10c shows the stress distribution in the specimen on the diameter. During the ITFT, both the cyclic haversine and cyclic sinusoidal loads are employed [3,6,23,71–73]. The loading strip is employed to prevent the specimen from failing near the load line due to compression [29].

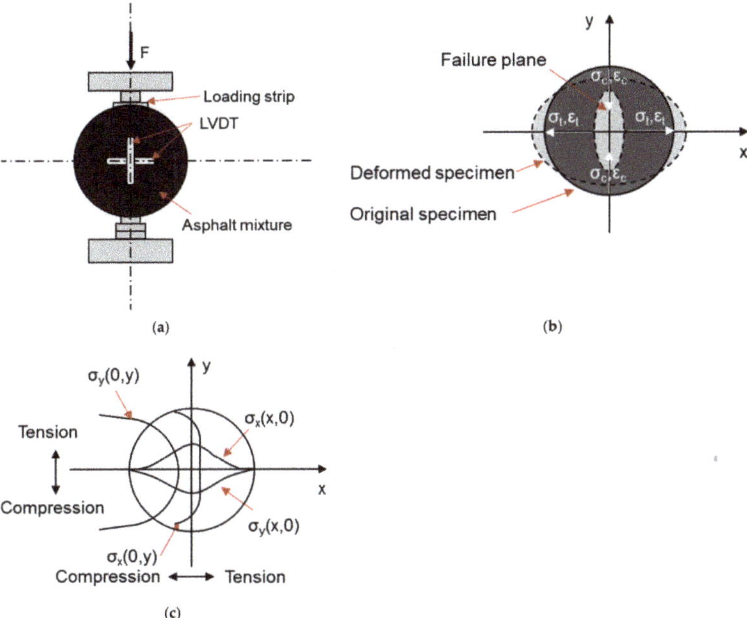

Figure 10. Schematic diagram and stress distribution of ITFT. (**a**) ITFT test setup. (**b**) Stress and strain on the diameters. (**c**) Stress distribution along horizontal and vertical diameters.

Under this configuration, the compressive and tensile stresses at the center of specimen are as follows [74]:

$$\sigma_t = \frac{2 \times P}{\pi \times a \times h} \times \sin\left(2a - \frac{a}{2R}\right) \tag{15}$$

$$\sigma_c = \frac{-6 \times P}{\pi \times a \times h} \times \sin\left(2a - \frac{a}{2R}\right) \tag{16}$$

where

σ_t = Horizontal tensile stress at the center of specimen (Pa);
σ_c = Vertical compressive stress at the center of specimen (Pa);
a = Width of loading strip (m);
h = Height of specimen (m).

Equations (15) and (16) yield that at the center of specimen, the vertical compressive stress is three times the horizontal tensile stress.

As shown in Figure 10b, the biaxial state of stress induced by ITFT is the main difference between the ITFT and the other technologies mentioned above. In the biaxial stress state, the maximum stress value is concentrated in the center of the specimen and declines outside. The biaxial stress is possibly of a type better representing the filed condition. However, because the ratio of vertical compressive stress to horizontal stress is fixed at three, the goodness of ITFT in representing the field stress state is questioned. Besides that, the current specification recommends the use of horizontal stress and strain to evaluate the fatigue performance of asphalt mixture. A plausible approach to consider the influences of the biaxial stress state is still missed. Moreover, the haversine pattern of loading raises two concerns about the ITFT technology. First, if σt is used to evaluate the fatigue response of asphalt mixture, the absence of stress reveal may compromise the correlation between experimental results and field fatigue performance. Second, this loading pattern creates a permanent deformation under the loading strips, especially at an elevated temperature [75,76]. In this case, the ITFT causes a combination of damage from fatigue and permanent deformations [31].

A number of studies have revealed that ITFT does not deliver the same results with the other test methods in terms of fatigue performance of asphalt mixture. For instance, Di Benedetto et al. reported that ITFT provides shorter fatigue lives when compared with bending tests [23]. Poulikakos and Hofko compared the fatigue lives of asphalt mixture measured by ITFT and 4PB [77]. According to them, ITFT and 4PB did not provide the same results, both in terms of value and ranking. Cheng et al. applied strain-controlled and stress-controlled cyclic haversine loads to 4PB and ITFT specimens, respectively, at different temperatures. They found that temperature shows positive influences on the Nf measured by 4PB but negative influences on the Nf measured by ITFT [78]. Yu et al. ranked the fatigue performance of asphalt binder, asphalt mortar and asphalt mixture using different test methods. They found the results of ITFT were not consistent with the results of the other methods [71].

3.4. Shear Fatigue Test

At present, several test methods are available to characterize the shear fatigue performance of asphalt mixture, and these methods can be categorized as two types: torsional and direct shear fatigue test method. Ragni et al. applied shear-torque configuration to characterize the interlayer fatigue performance of double-layered asphalt mixture specimen [79]. Ren et al. designed a uniaxial penetrating test (UPT) method to characterize the shear fatigue performance of asphalt mixture [80]. Yin et al. [81] and Rahman et al. [82] evaluated the fracture behavior of asphalt mixture under direct shear loading. The schematics of the torque- and direct-shear configurations are shown in Figure 11.

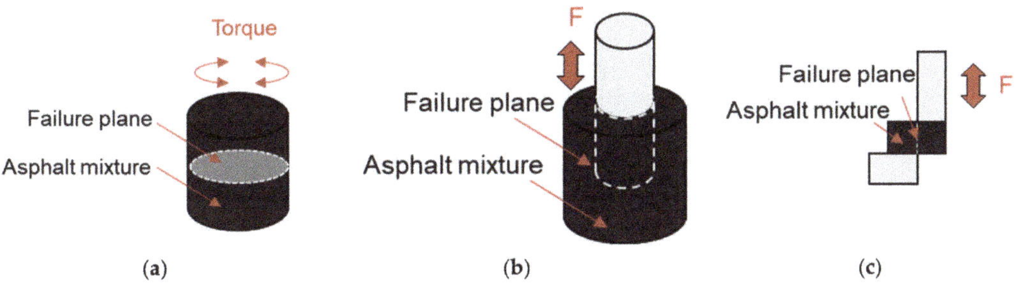

Figure 11. Schematic of shear fatigue test configurations. (**a**) Torsional shear. (**b**) Penetration shear. (**c**) Direct shear.

However, it should be noted that compared with fatigue distresses, the shear resistance of asphalt mixture is considered to be more related to the rutting of asphalt mixture [83]. This is because shear flow of asphalt mixture is ascribed as one of the major reasons leading to rutting [84].

4. Full Scale Test Method

Full-scale test method lies on the full-scale pavement structure, which is loaded by the repeated axle loads or the accelerated axle load. Because of the involved full-scale pavement structure and the realistic axle loads, the full-scale method is regarded as a reliable method to better understand and characterize the performance degradation process of asphalt pavement. The experimental data collected from these methods are widely used to verify the damage evolution models in the pavement design guide.

The typical full-scale test facilities include the AASHO road (Ottawa, Illinois, USA) [85], WesTrack (Reno, Nevada, USA) [86], NCAT test track (Opelika, Alabama, USA) [87], Mn-ROAD (Otsego, Minnesota, USA) [88], and RIOHTrack (Beijing, China) [89]. Apart from the full-scale test tracks, hydro-servo/weight loading devices are also available for evaluating the fatigue performance of pavement. In this method, repeated loads are introduced to the full-scale or model-scale pavement in a linear or circular track. The typical ring tracks are the CEDEX (Madrid, Spain) [90], CAPTIF (Christchurch, Canterbury, New Zealand) [91], LCPC (Nantes, Pays de la Loire, France) [92], ALF (Australia) [93], HVS (South Africa) [94], and MLS (United Kingdom) [95].

With the assistance of full-scale test methods, pavement researchers are able to determine the effects of traffic loads on the performance degradation of pavement structure. The advantage of these test tracks is that they can represent the real service condition of the pavement structure. However, the shortcoming is the high cost of the experiment and the large area. In addition, there is still a gap between the test loading frequency and the real traffic speed.

5. Summary and Discussion

After decades of effort, plenty of test technologies have been proposed to investigate the fatigue behavior of asphalt mixture. Although the sample geometry and load characteristics involved in these technologies vary from method to method, these technologies are the base stone of understanding and predicting the fatigue behavior of asphalt materials on different length scales. Table 2 provides an overview of the test methods and their parameters by showing their loading types, the measured properties, failure zone description, stress characteristic, and stress reversal.

Table 2. Summary of fatigue test methods.

Test	Test Specification	Loading Type	Measured Material Properties	Failure Occurs in a Uniform Stress/ Strain Zone?	Stress Characteristics	Stress Reversal?
4PB	AASHTO T321-14 ASTM D8237-18 EN 12697-24:2018	Strain-controlled sinusoidal loading wave	Stress, phase angle	Yes	1D *	Yes
3PB	-	Strain-/stress controlled haversine loading wave	Stress/strain, phase angle	No	1D	No
2PB	EN 12697-24:2018	Force-/displacement-controlled sinusoidal load	Displacement	No	1D	Yes
Axial fatigue test	AASHTO TP107	Force-/displacement-controlled	Stress/strain, phase angle	Yes	1D	Yes (TCFT)
Loaded wheel fatigue test	-	Moving wheel load	Strain at sample bottom, sample deformation	No	1D	Yes
ITFT	EN 12697-24:2018	Sinusoidal/haversine	Strain	No	2D	No
Shear fatigue test	-	Sinusoidal load	Stress/strain, phase angle	Yes	1D	Yes

- test specification is not available because the test method is not a standard fatigue test method. * D refers to dimension(s).

Despite their contributions, comparing the field service conditions with the test conditions highlights the limitations of the currently used technology: the lack of considering the influences of pavement structure and the characteristics of traffic load.

With laboratory test methods, most of them focus on the fatigue behavior of test material only without considering the influences of pavement structure. Previous studies have documented that the structural configuration can sometimes overshadow the effect of mixture property on fatigue performance, delivering different results between the laboratory fatigue performance and field fatigue performance [96,97]. During each loading cycle, the multi-layer characteristic of asphalt pavement structure resulted in a stress reversal at the longitudinal direction in the asphalt mixture, but not at the transverse direction. Stress reversal affects the fatigue performance of asphalt mixture by varying the stress ratio, however, neither the stress ratio nor the two-dimensional stress state was well simulated in the lab test methods. To omit the negative influences of the lab methods, several empirical parameters, which are determined statistically, are required to link the lab measured results to the field condition. In addition, the utilized diverse loading methods induce the non-unifying nature regarding ranking fatigue performance. When ranking the fatigue performance for a group of samples, different methods may yield different ranking results, which causes difficulty in selecting the optimal paving materials. The full-scale test methods appear to be the solution for the limitation of the lab methods. However, the high cost of the test facilities obstacles in their application.

A common shortcoming of both the lab methods and the full-scale methods is the loading speed. The importance of loading speed on the fatigue performance of asphalt materials has been reported by numerous studies. Due to their technical limitations, the maximum loading speed (approx. 20 km/h) of the hydro-servo/weight loading test devices is much slower than the real traffic speed [4,98–100], resulting in a significant difference between the test and the real service condition. The lab devices can apply load at a relatively high frequency. However, the loading frequency varies among the test standards. The ASTM D8237-18 recommends a default frequency of 10 Hz for the 4PB fatigue test [28]. The AASHTO standard requires a frequency range of 5 Hz to 25 Hz for the 4PB fatigue test [14]. The frequency for the 2PB fatigue test, however, is fixed at 25 Hz [3]. A plausible explanation is the relationship between the loading frequency and traffic speed, though this remains unclear. Table 3 lists the formulas used to converse the traffic speed v (km/h) to the lab loading frequency.

Table 3. Relationship between vehicle speed and loading time (*t*) or loading frequency (*f*).

Authors	Loading Time (t) or Loading Frequenzy (f)	Remark
NCHRP [101]	$t = L_{eff}/17.6v_s$	L_{eff} = effective length
Barksdale [102]	22.7 Hz	v = 72 km/h; depth: 30.5 cm
Mollenhauer et al. [103]	$f = 0.277\ v^{0.944}$	-

6. Concept and Development of ARROWS

6.1. Characteristics of ARROWS

Previous sections have shown the differences in the loading condition between the existing fatigue test methods and the field. These differences lead to a gap between the fatigue performance of asphalt mixture in the lab and in the field. A plausible solution for the insufficient correlation is to develop a new fatigue test device, which is able to reproduce the field loading conditions in lab. To achieve this purpose, the authors propose the concept of an innovative fatigue test device, which is called Accelerated Repeated Rolling Wheel Load Simulator (ARROWS). Because the device is under construction, this section briefly introduces the concept and mechanism of the device; practical work will be performed once the device is available.

To better reproduce the field service conditions, ARROWS is expected to have the following technical characteristics: (a) ARROWS can generate two-dimensional tensile stress at the bottom of asphalt mixture layer, (b) the loading speed should be comparable to the traffic speed, and (c) each loading pulse should consist of a loading phase and a rest phase.

The ARROWS employs a rectangle asphalt mixture slab with the dimension of 500 mm × 500 mm × 50 mm (length × width × thickness), as shown in Figure 12. A rubber mat is used to simulate the supporting effect of the base layer on the asphalt mixture, which has been documented by previous studies. The thickness of the asphalt slab and the rubber mat is adjustable to simulate different pavement structures. During the test, repeated loads are imposed on the loading paths through eleven load stamps. The contact area between the load stamps and the asphalt mixture is set as 35 mm × 35 mm and can be adjusted upon request.

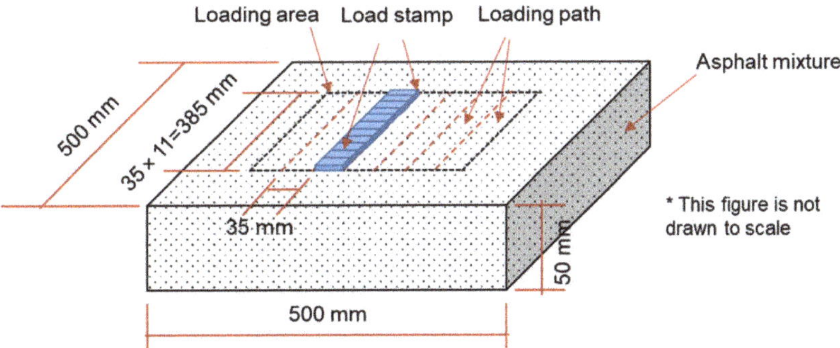

Figure 12. Geometry of asphalt mixture slab.

As mentioned in the above sections, the wheel moving speed of the APA system is much slower than the traffic speed. In order to reproduce the moving process of wheel load at traffic speed in the lab, the ARROWS involves a different loading method. Instead of applying load to the sample using a moving wheel, ARROWS applies the moving load by changing the load magnitude of stamps with time. Specifically, if the load magnitudes of all the eleven stamps at instant t can be described mathematically, for instance, the Gaussian distribution (Equation (17)).

$$F_{\mu,\sigma^2}(n) = \frac{1}{\sigma\sqrt{2\pi}}e^{-\frac{1}{2}\left(\frac{n-\mu}{\sigma}\right)^2}$$

(17)

where parameter μ is the expectation of the distribution, the parameter σ is the standard deviation, and n is the number of the load stamp.

The load profile of stamps at t(μ1), t(μ2), and t(μ3) are visualized as Figure 13. From this figure, it can be found that the load peak moves among the load stamps with time, which indicates that the repeated rolling of tire loads can be reproduced by changing the load profile with time t, and the corresponding loading speed v can be represented as:

$$v = \frac{d*(n(\mu_2) - n(\mu_1))}{t(\mu_2) - t(\mu_1)}$$

(18)

where d is the width of the load stamp.

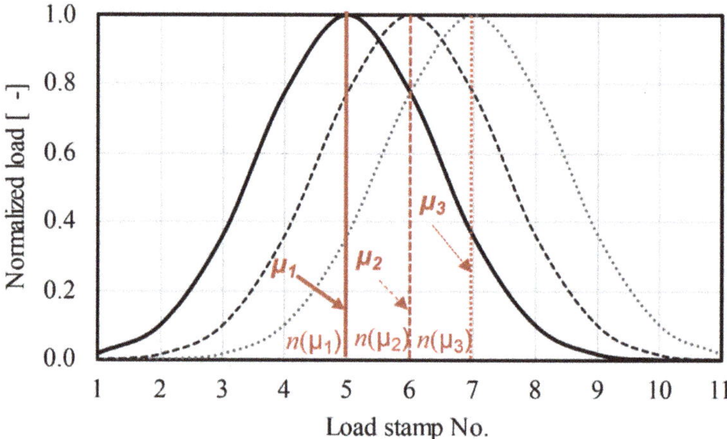

Figure 13. Load profile of stamps with different μ values.

6.2. Description of ARROWS

Stamps is used to the create load on the asphalt mixture sample. In order to reduce the necessary preload of the system, which is required to ensure contact between the load stamp contact surface and the specimen, a wheel located in the cam plate is used as a counter bearing. The setup of the counter bearing is illustrated in Figure 14b. As seen, the load stamp is connected to the cam through a spring and a piston. Vertical displacement is induced to the top of the spring, when cam contacts with the piston. The compressed spring then transfers the load to the stamp. Hence, the magnitude and waveform of the applied load are determined by the geometry of the cams.

The measurement system of the ARROWS includes the load cells, the displacement transducers, and the strain gauges. Each of the eleven load stamps is equipped with a load cell and a displacement transducer, which are operated and sampled absolutely synchronously. Figure 14b reveals the setup of the load cell and displacement transducer. The load cells and the displacement transducers are used to collect the applied vertical load and the vertical displacement of the sample, respectively. Strain gauges are attached to the selected positions on the bottom of the asphalt specimen to determine the strain evolution at the bottom of the sample.

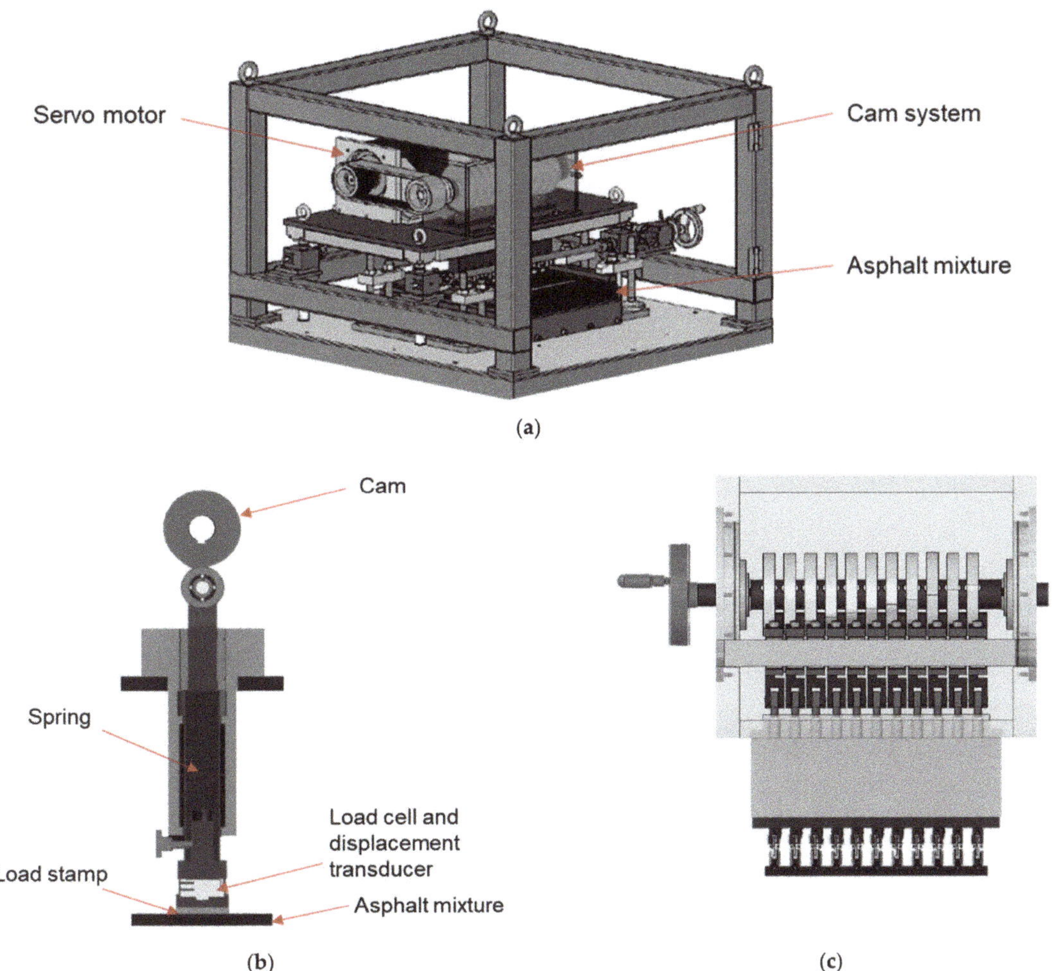

Figure 14. Details on the ARROWS. (**a**) Schematic diagram of ARROWS. (**b**) Side view of a loading unit. (**c**) Front view of the loading system.

The eleven employed cams are arranged on the shaft (Figure 14c) with an adjustable angle between them. The geometry of the cam plates and their arrangement on the shaft are strongly coupled with the loading speed, which can be controlled by the rotational speed of the shaft.

Apart from reproducing the rolling motion of a wheel, ARROWS is designed to be capable of ensuring that the introduced work remains constant, regardless of the positions of the individual cams. This requirement is met by a defined geometry of the cam discs, which is described using a Gaussian function (Equation (17)). A parameter b is introduced to describe the ratio of γ_1 to γ_2. The physical meaning of parameter b is the ratio of the duration of the loading phase to that of the rest phase of each loading pulse. Figure 15 is the schematic diagram of a cam plate.

$$r(\alpha) = E \cdot e^{f \cdot \left(\frac{\alpha}{180°} - 1 \right)^2} + \frac{d}{2} \tag{19}$$

$$f = \frac{d^2 \cdot \pi^2 \cdot b^2 \cdot 600^2}{32 \cdot \sigma^2 \cdot U^2}$$

(20)

where

α = Included angle (°);
E = Eccentricity;
d = Diameter of the cam plate;
b = Load ratio = $\gamma 1 / \gamma 2$;
σ = Standard deviation of the Gaussion distribution;
U = Circumference of the cam plate.

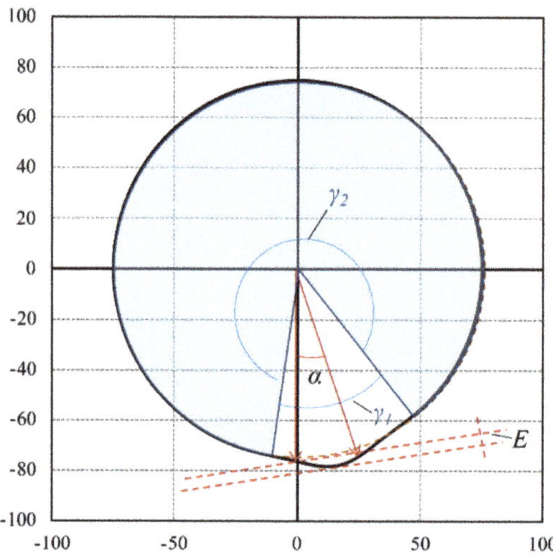

Figure 15. Schematic diagram of a cam plate.

The design of the cam geometry and arrangement allows the ARROWS to simulate the rolling motion of tire load on the pavement. Figure 15 presents the load applied to the asphalt slab at two instants, as an example. The location of all 11 cams is also shown in this figure. Figure 16a,b refer to the instant when the α of cam No. 6 rotates to 0° and 10°, respectively.

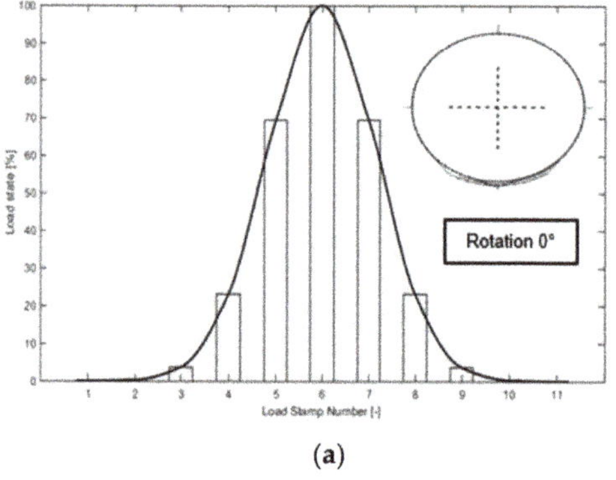

(a)

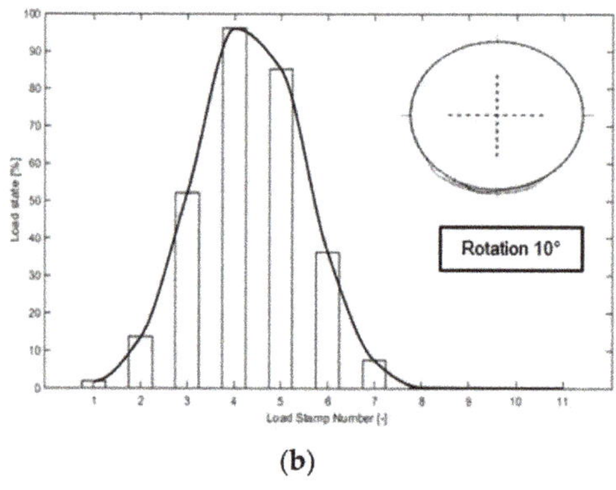

(b)

Figure 16. Load profile of load stamps at different instants. (a) $\alpha = 0°$. (b) $\alpha = 10°$.

Benefiting from the driving mode of the ARROWS, the rotation speed of the shaft can be increased to a very high level (v = 80 km/h) without compromising the output performance. This means the ARROWS allows the researchers to evaluate the fatigue behavior of asphalt mixture under a realistic loading speed. Further advantages of the ARROWS testing device are the low effort of sample preparation, the automatic test execution, as well as comparably low costs for the acquisition and operation of the test device. The preparation of the specimens for testing with the ARROWS testing device is limited to the production of the asphalt specimens of the dimensions shown in 6.1. Further processing, such as cutting, drilling, or grinding of the specimens, which is required in other test methods, is not required. The device was designed in such a way that after manual specimen insertion, a fully automatic test is carried out, so that the correct execution of the load is ensured at all times. At the latest with the multiplication of the prototype of this concept, acquisition costs are expected to be in the range of comparable mechanically-based test systems, which are significantly lower than the costs for servo-hydraulic systems, as they are used for many fatigue tests.

7. Feasibility of ARROWS

The above-calculated load profiles are inputted into a FEM model to characterize the stress states of the asphalt mixture specimen loaded by ARROWS. In the FEM model, asphalt mixture slab and the rubber mat are surrounded by a steel container. The maximum vertical stress created by the stamp was set as 0.35 MPa, the stress of other stamps is then determined accordingly. The interface between the asphalt mixture and rubber layer was considered as frictionless. The modulus of the steel is set as 20 GPa. The material properties of the asphalt mixture are the same as that of the asphalt base course listed in Table 1. Because rubber only showed a tiny strain under the loading condition, the rubber was simplified as linear viscoelastic material which had a modulus of 10 MPa and a Poisson's ratio of 0.49. Figure 17 shows the stress variation at the bottom center of the asphalt slab during a loading cycle, where a stress reversal is found in the longitudinal stress. In addition, the transverse stress is confirmed as a pure tensile stress during the loading cycle. Comparing Figure 17 with Figure 2 reveals that the ARROWS is capable of creating tensile stress, which is similar to that in the pavement structure, indicating the feasibility of using ARROWS to better simulate the realistic stress condition of asphalt mixture.

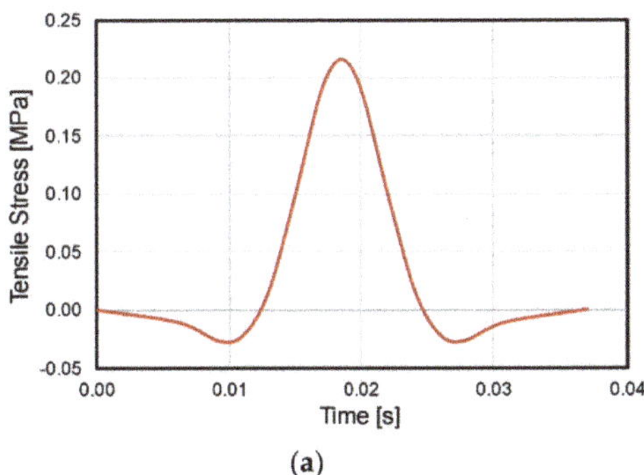

(a)

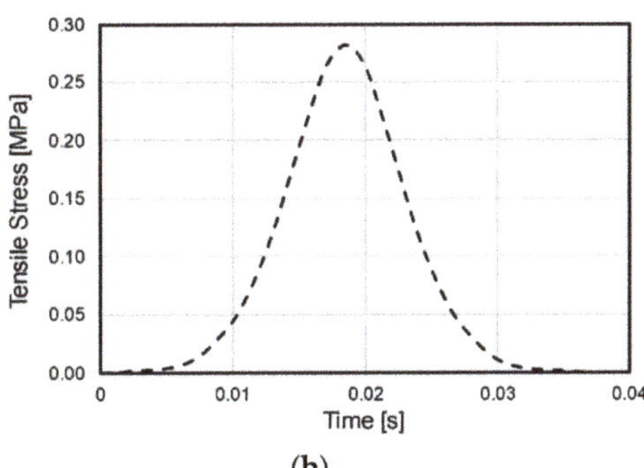

(b)

Figure 17. Horizontal stress at the bottom of asphalt slab. (**a**) Longitudinal stress. (**b**) Transverse stress.

8. Findings and Conclusions

This paper starts by characterizing the tensile stress at bottom of a typical German asphalt pavement structure. After conducting a survey of more than 100 publications, the theoretical background, working mechanisms, test standards, and applications of the currently available fatigue test methods are discussed and summarized. As a solution to these limitations, an innovative fatigue test device, named ARROWS, is proposed. The following conclusions can be drawn from this work.

(1) The stress state in the lab specimens differs from that in the pavement. The tensile stress at the bottom of the asphalt layer is two-dimensional and contains a stress reversal at longitudinal direction. However, none of the available fatigue test methods are able to reproduce this 2D tensile stress.

(2) The experimental results and the theoretical analysis reveal that the load waveform and stress ratio strongly affect the fatigue performance of asphalt mixture. This highlights the importance of reproducing field stress conditions when evaluating the fatigue performance of asphalt mixture in the lab.

(3) The ARROWS is designed to better simulate the rolling motion of the tire load in the lab with a traffic speed using a special loading system. Specifically, vertical stress is applied to the test specimen via load stamps, which is connected to a shaft through a spring and a piston.

(4) The FEM model indicates the similarity between the stress at the bottom of the asphalt slab of the ARROWS specimen and that at the asphalt layer of pavement. This indicates the feasibility and the effectiveness of using the ARROWS as an alternative to the currently available fatigue test methods.

Author Contributions: Conceptualization, J.K. and Z.Z.; methodology, J.K. and Z.Z.; software, Z.Z.; validation, J.K. and Z.Z.; writing—original draft preparation, J.K. and Z.Z.; writing—review and editing, J.K. and Z.Z.; supervision, M.O. and C.S.; project administration, C.S. and J.K.; funding acquisition, M.O. All authors have read and agreed to the published version of the manuscript.

Funding: The authors acknowledge the finical support provided by the German Research Foundation through project ARROWS (OE 514/10-1). The financial support from the China Scholarship Council (Grant No. 201908080046) is also highly appreciated by the authors.

Institutional Review Board Statement: Not applicable.

Informed Consent Statement: Not applicable.

Data Availability Statement: Not applicable.

Conflicts of Interest: The authors declare no conflict of interest.

References

1. Leng, Z.; Zhang, Z.; Zhang, Y.; Wang, Y.; Yu, H.; Ling, T. Laboratory evaluation of electromagnetic density gauges for hot-mix asphalt mixture density measurement. *Constr. Build. Mater.* **2018**, *158*, 1055–1064. [CrossRef]
2. Monismith, C.L.; Secor, K.E.; Blackmer, W.E. Asphalt mixture behaviour in repeated flexure. *J. Assoc. Asph. Paving Technol.* **1961**, *30*, 188–222.
3. DIN Deutsches Institut für Normung, e.V. *DIN EN 12697-24:2018 Asphalt—Prüfverfahren—Teil 24: Beständigkeit gegen Ermüdung;* Deutsche Fassung EN 12697-24:2018; Beuth: Berlin, Germany, 2018; p. 68. [CrossRef]
4. Kim, Y.R.; Baek, C.; Underwood, B.; Subramanian, V.; Guddati, M.N.; Lee, K. Application of viscoelastic continuum damage model based finite element analysis to predict the fatigue performance of asphalt pavements. *KSCE J. Civ. Eng.* **2008**, *12*, 109–120. [CrossRef]
5. Lv, S.; Xia, C.; Liu, C.; Zheng, J.; Zhang, F. Fatigue equation for asphalt mixture under low temperature and low loading frequency conditions. *Constr. Build. Mater.* **2019**, *211*, 1085–1093. [CrossRef]
6. Zhang, Z.; Oeser, M. Residual strength model and cumulative damage characterization of asphalt mixture subjected to repeated loading. *Int. J. Fatigue* **2020**, *135*, 105534. [CrossRef]
7. Forschungsgesellschaft für Straßen- und Verkehrswesen e. V. (FGSV). *RStO 12 Richtlinien für die Standardisierung des Oberbaus von Verkehrsflächen;* Forschungsgesellschaft für Straßen- und Verkehrswesen e. V. (FGSV): Köln, Germany, 2012.
8. Forschungsgesellschaft für Straßen- und Verkehrswesen e.V. (FGSV). *RDO Asphalt 09 Richtlinien für die Rechnerische Dimensionierung des Oberbaus von Verkehrsflächen mit Asphaltdeckschicht;* Forschungsgesellschaft für Straßen- und Verkehrswesen e. V. (FGSV): Köln, Germany, 2009.
9. Mull, M.A.; Stuart, K.; Yehia, A. Fracture resistance characterization of chemically modified crumb rubber asphalt pavement. *J. Mater. Sci.* **2002**, *37*, 557–566. [CrossRef]
10. Huang, B.; Shu, X.; Zuo, G. Using notched semi circular bending fatigue test to characterize fracture resistance of asphalt mixtures. *Eng. Fract. Mech.* **2013**, *109*, 78–88. [CrossRef]

11. Lv, S.; Liu, C.; Yao, H.; Zheng, J. Comparisons of synchronous measurement methods on various moduli of asphalt mixtures. *Constr. Build. Mater.* **2018**, *158*, 1035–1045. [CrossRef]

12. Tayebali, A.A.; Rowe, G.M.; Sousa, J.B. Fatigue response of asphalt-aggregate mixtures. *Asph. Paving Technol. Assoc. Asph. Paving Technol. Tech. Sess.* **1992**, *61*, 333–360.

13. ASTM International, D7460−10. Standard Test Method for Determining Fatigue Failure of Compacted Asphalt Concrete Subjected to Repeated Flexural Bending. *Am. Soc. Test. Mater.* **2010**, *4*, 3. [CrossRef]

14. American Association of State Highway and Transportation Officials, AASHTO T321-14. *Standard Method of Test for Determining the Fatigue Life of Compacted Asphalt Mixtures Subjected to Repeated Flexural Bending*; American Association of State Highway and Transportation Officials: Washington, DC, USA, 2014.

15. Austroads. *Austroads Test Method AGPT/T274 Characterisation of Flexural Stiffness and Fatigue Performance of Bituminous Mixes*; Austroads: Sydney, Australia, 2016.

16. Yu, H.; Zhu, Z.; Leng, Z.; Wu, C.; Zhang, Z.; Wang, D.; Oeser, M. Effect of mixing sequence on asphalt mixtures containing waste tire rubber and warm mix surfactants. *J. Clean. Prod.* **2019**, *246*, 119008. [CrossRef]

17. Yu, H.; Chen, Y.; Wu, Q.; Zhang, L.; Zhang, Z.; Zhang, J.; Miljković, M.; Oeser, M. Decision support for selecting optimal method of recycling waste tire rubber into wax-based warm mix asphalt based on fuzzy comprehensive evaluation. *J. Clean. Prod.* **2020**, *265*, 121781. [CrossRef]

18. Pronk, A.C.; Poot, M.R.; Jacobs, M.M.J.; Gelpke, R.F. Haversine fatigue testing in controlled deflection mode: Is it possible? In Proceedings of the Transportation Research Board (TRB) 89th Annual Meeting, Washington, DC, USA, 10–14 January 2010.

19. Li, N.; Molenaar, A.A.A.; van de Ven, M.F.C.; Wu, S. Characterization of fatigue performance of asphalt mixture using a new fatigue analysis approach. *Constr. Build. Mater.* **2013**, *45*, 45–52. [CrossRef]

20. Arsenie, I.M.; Chazallon, C.; Duchez, J.-L.; Hornych, P. Laboratory characterisation of the fatigue behaviour of a glass fibre grid-reinforced asphalt concrete using 4PB tests. *Road Mater. Pavement Des.* **2016**, *18*, 168–180. [CrossRef]

21. Abhijith, B.S.; Narayan, S.P.A. Evolution of the Modulus of Asphalt Concrete in Four-Point Beam Fatigue Tests. *J. Mater. Civ. Eng.* **2020**, *32*, 04020310. [CrossRef]

22. Poulikakos, L.D.; Pittet, M.; Dumont, A.-G.; Partl, M.N. Comparison of the two point bending and four point bending test methods for aged asphalt concrete field samples. *Mater. Struct.* **2014**, *48*, 2901–2913. [CrossRef]

23. Di Benedetto, H.; de La Roche, C.; Baaj, H.; Pronk, A.; Lundström, R. Fatigue of bituminous mixtures. *Mater. Struct.* **2004**, *37*, 202–216. [CrossRef]

24. Dondi, G.; Pettinari, M.; Sangiorgi, C.; Zoorob, S.E. Traditional and Dissipated Energy approaches to compare the 2PB and 4PB flexural methodologies on a Warm Mix Asphalt. *Constr. Build. Mater.* **2013**, *47*, 833–839. [CrossRef]

25. Pronk, A.C. Comparison of 2 and 4 point fatigue tests and healing in 4 point dynamic bending test based on the dissipated energy concept. In Proceedings of the Eighth International Conference on Asphalt PavementsFederal Highway Administration, Washington, DC, USA, 10–14 October 1997; Volume II, pp. 987–994.

26. Mamlouk, M.S.; Souliman, M.I.; Zeiada, W.A.; Kaloush, K.E. Refining Conditions of Fatigue Testing of Hot Mix Asphalt. *Adv. Civ. Eng. Mater.* **2012**, *1*, 20120018. [CrossRef]

27. Mateos, A.; Wu, R.; Denneman, E.; Harvey, J. Sine versus Haversine Displacement Waveform Comparison for Hot Mix Asphalt Four-Point Bending Fatigue Testing. *Transp. Res. Rec. J. Transp. Res. Board* **2018**, *2672*, 372–382. [CrossRef]

28. ASTM International, ASTM D8237-18. *Standard Test Method for Determining Fatigue Failure of Asphalt-Aggregate Mixtures with the Four-Point Beam Fatigue Device 1*; ASTM International: West Conshohocken, PA, USA, 2019.

29. Tangella, S.C.S.R.; Craus, J.; Deacon, J.A.; Monismith, C.L. *Summary Report on Fatigue Response of Asphalt Mixtures*; Institute of Transportation Studies University of California Berkeley: Berkeley, CA, USA, 1990.

30. Khalid, H.A. A comparison between bending and diametral fatigue tests for bituminous materials. *Mater. Struct.* **2000**, *33*, 457–465. [CrossRef]

31. Cocurullo, A.; Airey, G.; Collop, A.C.; Sangiorgi, C. Indirect tensile versus two-point bending fatigue testing. Proc. *Inst. Civ. Eng.-Transp.* **2008**, *161*, 207–220. [CrossRef]

32. DI Benedetto, H.; Partl, M.N.; Francken, L.; André, C.D.L.R.S. Stiffness testing for bituminous mixtures. *Mater. Struct.* **2001**, *34*, 66–70. [CrossRef]

33. Hartman, A.M.; Gilchrist, M.D.; Nolan, D. Wheeltracking Fatigue Simulation of Bituminous Mixtures. *Road Mater. Pavement Des.* **2001**, *2*, 141–160. [CrossRef]

34. Van Dijk, W. Practical fatigue characterization of bituminous mixes. *Assoc. Asph. Paving Technol.* **1975**, *44*, 38–74.

35. Liao, M.-C.; Chen, J.-S.; Tsou, K.-W. Fatigue Characteristics of Bitumen-Filler Mastics and Asphalt Mixtures. *J. Mater. Civ. Eng.* **2012**, *24*, 916–923. [CrossRef]

36. Zhang, Z.; Shen, S.; Shi, B.; Wang, H. Characterization of the fatigue behavior of asphalt mixture under full support using a Wheel-tracking Device. *Constr. Build. Mater.* **2021**, *277*, 122326. [CrossRef]

37. Wu, H.; Huang, B.; Shu, X. Characterizing viscoelastic properties of asphalt mixtures utilizing loaded wheel tester (LWT). *Road Mater. Pavement Des.* **2012**, *13*, 38–55. [CrossRef]

38. Wu, H.; Huang, B.; Shu, X. Characterizing Fatigue Behavior of Asphalt Mixtures Utilizing Loaded Wheel Tester. *J. Mater. Civ. Eng.* **2014**, *26*, 152–159. [CrossRef]

39. Nguyen, V.B.; Thom, N. Using a beam wheel tracker fatigue test to evaluate fatigue performance of asphalt mixtures. *Road Mater. Pavement Des.* **2021**, *22*, 2801–2817. [CrossRef]
40. Lv, S.; Yuan, J.; Peng, X.; Cabrera, M.B.; Liu, H.; Luo, X.; You, L. Standardization to evaluate the lasting capacity of rubberized asphalt mixtures with different testing approaches. *Constr. Build. Mater.* **2020**, *269*, 121341. [CrossRef]
41. Underwood, S.B.; Kim, R.Y. Viscoelastoplastic Continuum Damage Model for Asphalt Concrete in Tension. *J. Eng. Mech.* **2011**, *137*, 732–739. [CrossRef]
42. Luo, R.; Liu, H.; Zhang, Y. Characterization of linear viscoelastic, nonlinear viscoelastic and damage stages of asphalt mixtures. *Constr. Build. Mater.* **2016**, *125*, 72–80. [CrossRef]
43. Safaei, F.; Castorena, C.; Kim, Y.R. Linking asphalt binder fatigue to asphalt mixture fatigue performance using viscoelastic continuum damage modeling. *Mech. Time-Dependent Mater.* **2016**, *20*, 299–323. [CrossRef]
44. Luo, X.; Luo, R.; Lytton, R.L. Characterization of Fatigue Damage in Asphalt Mixtures Using Pseudostrain Energy. *J. Mater. Civ. Eng.* **2013**, *25*, 208–218. [CrossRef]
45. Lee, J.-S.; Norouzi, A.; Kim, Y.R. Determining Specimen Geometry of Cylindrical Specimens for Direct Tension Fatigue Testing of Asphalt Concrete. *J. Test. Eval.* **2016**, *45*, 20140357. [CrossRef]
46. Isailović, I.; Falchetto, A.C.; Wistuba, M.P. Energy Dissipation in Asphalt Mixtures Observed in Different Cyclic Stress-Controlled Fatigue Tests. In *8th RILEM International Symposium on Testing and Characterization of Sustainable and Innovative Bituminous Materials*; Springer: Dordrecht, The Netherlands, 2015; pp. 693–703. [CrossRef]
47. Seitllari, A.; Kutay, M. Effect of load eccentricity on uniaxial fatigue test results for asphalt concrete mixtures using FE modeling. In *Advances in Materials and Pavement Performance Prediction II*; CRC Press: Boca Raton, FL, USA, 2020; pp. 351–354. [CrossRef]
48. Xu, H.; Xing, C.; Zhang, H.; Li, H.; Tan, Y. Moisture seepage in asphalt mixture using X-ray imaging technology. *Int. J. Heat Mass Transf.* **2018**, *131*, 375–384. [CrossRef]
49. Zhang, Z.; Liu, Q.; Wu, Q.; Xu, H.; Liu, P.; Oeser, M. Damage evolution of asphalt mixture under freeze-thaw cyclic loading from a mechanical perspective. *Int. J. Fatigue* **2020**, *142*, 105923. [CrossRef]
50. Xu, H.; Guo, W.; Tan, Y. Internal structure evolution of asphalt mixtures during freeze–thaw cycles. *Mater. Des.* **2015**, *86*, 436–446. [CrossRef]
51. Zeiada, W.A.; Kaloush, K.E.; Underwood, B.S.; Mamlouk, M. Development of a Test Protocol to Measure Uniaxial Fatigue Damage and Healing. *Transp. Res. Rec. J. Transp. Res. Board.* **2016**, *2576*, 10–18. [CrossRef]
52. Zheng, M.; Li, P.; Yang, J.; Li, H.; Qiu, Y.; Zhang, Z. Fatigue character comparison between high modulus asphalt concrete and matrix asphalt concrete. *Constr. Build. Mater.* **2019**, *206*, 655–664. [CrossRef]
53. Qian, G.-P.; Liu, H.-F.; Zheng, J.-I.; Jiang, L.-J. Experiment of Tension-compression Fatigue and Damage for Asphalt Mixtures. *J. Highw. Transp. Res. Dev.* **2013**, *7*, 15–21. [CrossRef]
54. Bhairampally, R.K.; Lytton, R.L.; Little, D.N. Numerical and Graphical Method to Assess Permanent Deformation Potential for Repeated Compressive Loading of Asphalt Mixtures. *Transp. Res. Rec. J. Transp. Res. Board* **2000**, *1723*, 150–158. [CrossRef]
55. Zhang, Y.; Luo, R.; Lytton, R.L. Characterizing Permanent Deformation and Fracture of Asphalt Mixtures by Using Compressive Dynamic Modulus Tests. *J. Mater. Civ. Eng.* **2012**, *24*, 898–906. [CrossRef]
56. Zhang, Y.; Luo, R.; Lytton, R.L. Mechanistic Modeling of Fracture in Asphalt Mixtures under Compressive Loading. *J. Mater. Civ. Eng.* **2013**, *25*, 1189–1197. [CrossRef]
57. Lv, S.; Liu, C.; Chen, D.; Zheng, J.; You, Z.; You, L. Normalization of fatigue characteristics for asphalt mixtures under different stress states. *Constr. Build. Mater.* **2018**, *177*, 33–42. [CrossRef]
58. Williams, D.; Little, D.N.; Lytton, R.L.; Kim, Y.R.; Kim, Y. *Microdamage Healing in Asphalt and Asphalt Concrete, Volume II: Laboratory and Field Testing to Assess and Evaluate Microdamage and Microdamage Healing*; The National Academies of Sciences, Engineering, and Medicine: Washington, DC, USA, 2001.
59. Soltani, B.A.; Solaimanian, M.; Anderson, D.; Tate, P.S. *An Investigation of the Endurance Limit of Hot-Mix Asphalt Concrete Using a New Uniaxial Fatigue*; The National Academies of Sciences, Engineering, and Medicine: Washington, DC, USA, 2006.
60. Chehab, G.R.; O'Quinn, E.; Kim, Y.R. Specimen geometry study for direct tension test based on mechanical tests and air void variation in asphalt concrete specimens compacted by superpave gyratory compactor. *Transp. Res. Rec.* **2000**, *1723*, 125–132. [CrossRef]
61. Daniel, J.S.; Kim, Y.R.; Brown, S.; Rowe, G.; Chehab, G.; Reinke, G. Development of a simplified fatigue test and analysis procedure using a viscoelastic, continuum damage model. *Asph. Paving Technol. Assoc. Asph. Paving Technol. Tech. Sess.* **2002**, *71*, 619–650.
62. American Association of State Highway and Transportation Officials. *AASHTO TP 107 Determining the Damage Characteristic Curve of Asphalt Concrete*; American Association of State Highway and Transportation Officials: Washington, DC, USA, 2014.
63. Raithby, R.D.; Ramshaw, J.T. *Effects of Secondary Compaction on the Fatigue Performance of a Hot-Rolled Asphalt*; The National Academies of Sciences, Engineering, and Medicine: Washington, DC, USA, 1972.
64. Safaei, F.; Lee, J.-S.; Nascimento, L.A.H.D.; Hintz, C.; Kim, Y.R. Implications of warm-mix asphalt on long-term oxidative ageing and fatigue performance of asphalt binders and mixtures. *Road Mater. Pavement Des.* **2014**, *15*, 45–61. [CrossRef]
65. Lv, S.; Liu, C.; Zheng, J.; You, Z.; You, L. Viscoelastic Fatigue Damage Properties of Asphalt Mixture with Different Aging Degrees. *KSCE J. Civ. Eng.* **2018**, *22*, 2073–2081. [CrossRef]

66. Gibson, N.; Li, X. Characterizing Cracking of Asphalt Mixtures with Fiber Reinforcement. *Transp. Res. Rec. J. Transp. Res. Board* **2015**, *2507*, 57–66. [CrossRef]
67. Nguyen, Q.T.; DI Benedetto, H.; Sauzéat, C. Determination of thermal properties of asphalt mixtures as another output from cyclic tension-compression test. *Road Mater. Pavement Des.* **2012**, *13*, 85–103. [CrossRef]
68. Lundstrom, R.; Di Benedetto, H.; Isacsson, U. Influence of Asphalt Mixture Stiffness on Fatigue Failure. *J. Mater. Civ. Eng.* **2004**, *16*, 516–525. [CrossRef]
69. Di Benedetto, H. Fatigue and other Phenomena during Cyclic Loading of Bituminous Materials. In Proceedings of the Keynote Delivered at the 5th EATA Conference, European Asphalt Technology Association, Braunschweig, Germany, 3–5 June 2013.
70. Ashouri, M.; David, Y.W.; Choi, Y.-T.; Richard, Y.K. Development of healing model and simplified characterization test procedure for asphalt concrete. *Constr. Build. Mater.* **2020**, *271*, 121515. [CrossRef]
71. Yu, J.; Yu, X.; Gao, Z.; Guo, F.; Wang, D.; Yu, H. Fatigue Resistance Characterization of Warm Asphalt Rubber by Multiple Approaches. *Appl. Sci.* **2018**, *8*, 1495. [CrossRef]
72. Jiang, J.; Ni, F.; Dong, Q.; Zhao, Y.; Xu, K. Fatigue damage model of stone matrix asphalt with polymer modified binder based on tensile strain evolution and residual strength degradation using digital image correlation methods. *Meas. J. Int. Meas. Confed.* **2018**, *123*, 30–38. [CrossRef]
73. Suo, Z.; Wong, W.G. Analysis of fatigue crack growth behavior in asphalt concrete material in wearing course. *Constr. Build. Mater.* **2009**, *23*, 462–468. [CrossRef]
74. Hudson, W.R.; Kennedy, T.W. *An Indirect Tensile Test for Stabilized Materials*; Center for Highway Research, University of Texas at Austin: Austin, TX, USA, 1968.
75. Maggiore, C.; Airey, G.; Collop, A.; di Mino, G.; di Liberto, M.; Marsac, P. Fatigue resistance: Is it possible having a unique response? In Proceedings of the 3rd 4PBB Conference, 17–18 September 2012; pp. 239–249.
76. Johnson, C.M. Estimating Asphalt Binder Fatigue Resistance Using an Accelerated Test Method. Ph.D. Thesis, University of Wisconsin-Madison, Madison, WI, USA, 2010.
77. Poulikakos, L.D.; Hofko, B. A critical assessment of stiffness modulus and fatigue performance of plant produced asphalt concrete samples using various test methods. *Road Mater. Pavement Des.* **2021**, *22*, 2661–2673. [CrossRef]
78. Cheng, H.; Liu, J.; Sun, L.; Liu, L.; Zhang, Y. Fatigue behaviours of asphalt mixture at different temperatures in four-point bending and indirect tensile fatigue tests. *Constr. Build. Mater.* **2020**, *273*, 121675. [CrossRef]
79. Ragni, D.; Takarli, M.; Petit, C.; Graziani, A.; Canestrari, F. Use of acoustic techniques to analyse interlayer shear-torque fatigue test in asphalt mixtures. *Int. J. Fatigue* **2019**, *131*, 105356. [CrossRef]
80. Ren, R.; Geng, L.; An, H.; Wang, X. Experimental research on shear fatigue characteristics of asphalt mixture based on repeated uniaxial penetrating test. *Road Mater. Pavement Des.* **2014**, *16*, 459–468. [CrossRef]
81. Yin, A.; Yang, X.; Zeng, G.; Gao, H. Experimental and numerical investigation of fracture behavior of asphalt mixture under direct shear loading. *Constr. Build. Mater.* **2015**, *86*, 21–32. [CrossRef]
82. Rahman, A.; Huang, H.; Ai, C.; Ding, H.; Xin, C.; Lu, Y. Fatigue performance of interface bonding between asphalt pavement layers using four-point shear test set-up. *Int. J. Fatigue* **2018**, *121*, 181–190. [CrossRef]
83. Christensen, D.W.; Bonaquist, R. Use of strength tests for evaluating the rut resistance of asphalt concrete. *Asph. Paving Technol. Assoc. Asph. Paving Technol. Tech. Sess.* **2002**, *71*, 692–711.
84. Chen, X.; Huang, B.; Xu, Z. Uniaxial Penetration Testing for Shear Resistance of Hot-Mix Asphalt Mixtures. *Res. Rec. J. Transp. Res. Board* **2006**, *1970*, 116–125. [CrossRef]
85. Carpenter, S.H. *Load Equivalency Factors and Rutting Rates: The AASHO Road Test*; Transportation Research Board: Washington, DC, USA, 1992; pp. 31–38.
86. NCHRP Report 455. *Recommended Performance-Related Specification for Hot-Mix Asphalt Construction: Results of the Westrack Project*; Transportation Research Board: Washington DC, USA, 2002. Available online: https://trid.trb.org/view/729477 (accessed on 7 December 2021).
87. Brown, E.R.; Cooley, L.A.; Hanson, D.; Lynn, C.; Powell, B.; Prowell, B.; Watson, D. Ncat Test Track Design, Construction, and Performance. In *NCAT Report 02-12*; National Center for Asphalt Technology: Auburn, AL, USA, 2002.
88. MnROAD Minnesota's Cold Weather Pavement Testing Facility, (n.d.). Available online: https://www.dot.state.mn.us/mnroad/index.html (accessed on 7 December 2021).
89. Zhang, L.; Zhou, X.; Wang, X. Research progress of long-life asphalt pavement behavior based on the RIOHTrack full-scale accelerated loading test. *Chin. Sci. Bull.* **2020**, *65*, 3247–3258. [CrossRef]
90. Mateos, A.; Ayuso, J.P.; Jáuregui, B.C. Evaluation of Structural Response of Cracked Pavements at CEDEX Transport Research Center Test Track. *Res. Rec. J. Transp. Res. Board* **2013**, *2367*, 84–94. [CrossRef]
91. Werkmeister, S. Shakedown Analysis of Unbound Granular Materials using Accelerated Pavement Test Results from New Zealand's CAPTIF Facility. In *Pavement Mechanics and Performance*; American Society of Civil Engineers: Reston, VA, USA, 2006; pp. 220–228. [CrossRef]
92. Hornych, P.; Kerzeho, J.P.; Salasca, S. Prediction of the behaviour of a flexible pavement using finite element analysis with non-linear elastic and viscoelastic models. In Proceedings of the 9th International Conference on Asphalt Pavements, Copenhagen, Denmark, 17–22 August 2002; pp. 1–9.

93. Australian Road Research Board, Accelerated Loading Facility (ALF), (n.d.). Available online: https://www.arrb.com.au/accelerated-load-facility (accessed on 7 December 2021).

94. Viljoen, A.W.; Freeme, C.R.; Servas, V.P.; Rust, F.C. Heavy vehicle simulator aided evaluation of overlays on pavements with active cracks. In Proceedings of the 6th International Conference on the Design of Asphalt Pavements, Ann Arbor, MI, USA, 13–17 July 1987; pp. 701–709.

95. PAVETESTING, Accelerated Pavement Tester, (n.d.). Available online: https://pavetesting.com/accelerated-pavement-tester/ (accessed on 7 December 2021).

96. Sabouri, M.; Kim, Y.R. Development of a Failure Criterion for Asphalt Mixtures under Different Modes of Fatigue Loading. *Transp. Res. Rec. J. Transp. Res. Board* **2014**, *2447*, 117–125. [CrossRef]

97. Wang, Y.D.; Keshavarzi, B.; Kim, Y.R. Fatigue Performance Analysis of Pavements with RAP Using Viscoelastic Continuum Damage Theory. *KSCE J. Civ. Eng.* **2018**, *22*, 2118–2125. [CrossRef]

98. Qi, X.; Engineer, P.; Mitchell, T.; Stuart, K.; Youtcheff, J.; Petros, K.; Modeling, P.; Leader, T.; Harman, T. Strain Responses in ALF Modified-Binder Pavement Study. In Proceedings of the 2nd International Conference on Accelerated Pavement Testing, Minneapolis, MI, USA, 6–29 September 2004.

99. Liu, P.; Wang, D.; Otto, F.; Hu, J.; Oeser, M. Application of semi-analytical finite element method to evaluate asphalt pavement bearing capacity. *Int. J. Pavement Eng.* **2016**, *19*, 479–488. [CrossRef]

100. Zhang, Z.; Lu, G.; Wang, D.; Oeser, M. Performance evaluation of pervious pavement using accelerated pavement testing system. In *Airfield and Highway Pavements 2019: Design, Construction, Condition Evaluation, and Management of Pavements—Selected Paper from International Airfield Highway Pavements Conference*; American Society of Civil Engineers: Reston, VA, USA, 2019; pp. 122–130. [CrossRef]

101. National Cooperative Highway Research Program, Guide for Mechanistic–Empirical Design of New and Rehabilitated Pavement Structures. 2004. Available online: http://onlinelibrary.wiley.com/doi/10.1002/cbdv.200490137/abstract%5Cn (accessed on 7 December 2021).

102. Barksdale, R.D. Compressive stress pulse times in flexible pavements for use in dynamic testing. *Highw. Res. Rec.* **1971**, *345*, 32–44.

103. Mollenhauer, K.; Wistuba, M.; Rabe, R. Loading Frequency and Fatigue: In Situ Conditioins & Impact on Test Results. In *2nd Workshop on Four Point Bending*; Pais, J., Ed.; University of Minho: Braga, Portugal, 2009; pp. 261–276.

Article

Assessment of Aging Impact on Wax Crystallization in Selected Asphalt Binders

Wenqi Wang [1], Ali Rahman [2,3], Haibo Ding [2,3,*] and Yanjun Qiu [2,3]

1   School of Architecture and Civil Engineering, Xihua University, Chengdu 610039, China
2   School of Civil Engineering, Southwest Jiaotong University, Chengdu 610031, China
3   Highway Engineering Key Laboratory of Sichuan Province, Southwest Jiaotong University, Chengdu 610031, China
*   Correspondence: haibo.ding@swjtu.edu.cn; Tel.: +86-156-8003-1860

Abstract: For a better understanding of the changing trend in crystalline components of asphalt binders, asphalt binders originating from the SHRP Materials Reference Library with different oxidation degrees (unaged, 20 h PAV, and 60 h PAV) were prepared. The native asphalt binders and their oxidized residues were characterized by liquid-state nuclear magnetic resonance (NMR) spectroscopy and high-temperature gas chromatography (HTGC). The results showed that, compared with other carbon types, the content of internal methylene carbons of long paraffinic chains between different SHRP binders was quite different. The NMR average length of a long paraffinic internal methylene chain showed a good correlation with the wax content obtained at $-20\,^{\circ}\mathrm{C}$ using the methyl ethyl ketone (MEK) precipitation method and also the recently developed variable-temperature Fourier-transform infrared spectroscopy (VT-FTIR) method. In most cases, the average length of straight internal methylene carbons of a long paraffinic chain terminated by a methyl group increased with the oxidation of the asphalt binder. However, the difference caused by oxidation was significantly smaller than the difference caused by the source of the asphalt binder. In general, oxidation will make the n-alkanes distributed in asphalt binder fall within a narrower range. The carbon number of n-alkanes in the asphalt binder generally grew with oxidation.

Keywords: asphalt binder; aging; oxidation; wax; nuclear magnetic resonance; gas chromatography

Citation: Wang, W.; Rahman, A.; Ding, H.; Qiu, Y. Assessment of Aging Impact on Wax Crystallization in Selected Asphalt Binders. *Materials* **2022**, *15*, 8248. https://doi.org/10.3390/ma15228248

Academic Editor: Zhanping You

Received: 17 October 2022
Accepted: 18 November 2022
Published: 21 November 2022

Publisher's Note: MDPI stays neutral with regard to jurisdictional claims in published maps and institutional affiliations.

1. Introduction

The quality of asphalt binder is one of the important factors affecting the durability of asphalt pavement. In the past, researchers tried to establish a direct link between the chemical composition of asphalt and its quality; however, they failed to do so. This was mainly due to the complexity of asphalt chemical composition and a lack of in-depth understanding of the asphalt chemical composition's classification. Among the chemical components of asphalt, wax is undoubtedly an important substance affecting the low-temperature service performance of asphalt [1–4]. Consequently, wax content and its behavior in asphalt with thermal history have received considerable attention [5–7]. Wax is generally considered to be a substance that crystallizes in asphalt at 25 °C and is mainly composed of saturated long linear hydrocarbons [8]. Owing to wax self-polymorphism and the difference in the state of existence in asphalt, there is no consensus on the wax effect on asphalt performance. This is one of the reasons why countries imposed different limits on wax content in asphalt.

Separating wax is the most direct method to study wax structure in asphalt [9]. Several methods were proposed to physically separate the wax from the asphalt binder. Due to the interference of the polar components (such as asphaltenes and resins) in the asphalt, it is difficult to separate the wax directly from the asphalt binder. Therefore, distillation or solvent precipitation is usually utilized to remove asphaltenes from the asphalt binder.

The remaining components (maltenes) continue to be separated using chromatographic techniques. According to Corbett and Swarbrik's method [10], the maltenes fraction is adsorbed on a chromatographic column (alumina is used as the adsorbent phase) and sequentially desorbed with solvents of increasing polarity. Saturates, aromatics, and resins are obtained from the maltenes. Using methyl ethyl ketone-benzene as a dewaxing solvent, solid wax can be separated from saturated and aromatic fractions at low temperatures. In contrast, Rostler and White [11] separated the maltenes into nitrogen bases, first acidaffins, second acidaffins, and paraffin fractions based on their reactivity with sulfuric acid (decreasing degree of hydration). The last fraction is what is called "wax". In some studies, asphalt's neutral fraction obtained from ion exchange chromatography (IEC) is regarded as a "wax fraction" [12]. When preparative size exclusion chromatography (SEC) is utilized to separate asphalt binder, the wax component is in the SEC-II fractions (non-associating components) [13]. Waxes are mainly composed of normal alkanes and isomeric alkanes, and the former produces the most adverse effect on asphalt performance. Considering urea readily forms crystalline adducts with straight-chain hydrocarbons, Netzel et al. [14] applied this technique to determine the content of n-alkanes and regarded this as solid wax content in asphalt.

With the development of modern analytical instruments, various advanced physical and chemical characterization methods have been employed to study wax behavior in asphalt [15–18]. Differential scanning calorimetry (DSC) is a common analytical method to study the crystalline fractions in asphalt binder. The number of crystallizable fractions in asphalt binder can be quantified by the size of the endothermic peak during heating. Using a thermal analysis of asphalt components, Corbett et al. [19] concluded that the endothermic peak of asphalt was mainly caused by saturates. However, Harrison et al. [20] believed that the saturates fraction was not the main factor leading to the endothermic behavior of asphalt binder, and the average linear side chain length was the main factor that affected the size of the endothermic peak. The effects of oxidation on the thermal behavior of pure wax-doped asphalt systems were studied by Kovinich et al. [6]. They found that the size of the endothermic peaks in the heat flow curve decreased with the oxidation of the binder. Atomic force microscopy (AFM) is often applied to observe the morphology of asphalt and the "bee structures" found in asphalt are thought to be the wax. Lu et al. [21] showed that the bee structures were fewer but larger after prolonged oxidation. They attributed this phenomenon to the highly increased stiffness of aged binder and reduced compatibility between the crystalline fraction and the more polarized asphalt matrix. Recently, Ding et al. [5,22,23] developed a variable-temperature Fourier-transform infrared spectroscopy technique to study the wax in asphalt, including quantification of the wax content and the precipitation and melting process of wax in asphalt. Aging is an important factor leading to the deterioration of asphalt performance. There have been a lot of literature reports on the mechanism and chemical reaction path of asphalt aging [24–28]. However, they did not find a clear rule regarding the effects of oxidation on wax content in asphalt binder.

As can be seen from the above literature review, although the wax in asphalt has been studied to a certain degree, the effect mechanisms of oxidative aging on wax in asphalt are still unclear. In this study, various asphalt binders from different crude oil sources were oxidized to a different degree in the laboratory. High-temperature gas chromatography (HTGC) and liquid-state nuclear magnetic resonance (NMR) spectroscopy were employed to analyze the structure of asphalt binders under different aging conditions. It is expected that some new insights into oxidation's effects on the crystallizable fractions in asphalt could be provided by a series of laboratory tests.

2. Materials and Methods

2.1. Asphalt Binders

In this study, eight asphalt binders from the SHRP (Strategic Highway Research Program) materials reference library were utilized. The SHRP materials reference library was built during the SHRP and used to store the asphalt binders adopted in the implementation of the project. The physical and chemical properties of these asphalt binders have been extensively studied to reveal the mechanisms of differences in field performance. Due to the fact that the SHRP binders have different colloidal structures and crude oil sources, they are representative and generally applied as base binders for modified asphalts to validate the performance of the modifier. The short-term aging of asphalt was simulated by Rolling Thin-Film Oven (RTFO) experiments, and the long-term aging behavior of asphalt was simulated by a Pressure Aging Vessel (PAV). In addition to the traditional 20-h PAV test, 60 h oxidative aging time was also carried out to better simulate the degree of field asphalt. In a previous study [5], although the effects of oxidation on the wax contents of these SHRP binders were studied using the VT-FTIR method, no precise rule was found. This study adopts the same SHRP binders and aging modes but with different characterization methods. In this way, the obtained results could be mutually confirmed and compared. The basic properties of the selected SHRP binders can be found in the reference [23,29].

2.2. Nuclear Magnetic Resonance (NMR)

Nuclear magnetic resonance spectroscopy can provide information about the relative number of different types of hydrogen, carbon, or other atoms in a molecule, and the chemical environment in which these different types of atoms are exposed. The abscissa of the NMR spectrum is the chemical shift, and the ordinate is the intensity of the resonance absorption peak, whose peak area is proportional to the number of such nuclei in the molecule. Since the absolute value of the chemical shift of the general atomic nucleus is very small, the peak of tetramethylsilane (TMS) is usually utilized as a reference for comparison, and the relative change value is expressed as ppm (parts per million). In the case of mixtures such as asphalt, NMR provides average structure information for the entire sample. This information is useful in characterizing and differentiating the chemical properties of different asphalts as a whole. The nuclear overhauser effect (NOE) occurs due to the influence of ^{1}H adjacent to ^{13}C, which can significantly enhance the sensitivity of ^{13}C NMR. However, its influence on different types of carbon is inconsistent and could result in the spectrum's intensity not being proportional to the carbon amount. For this reason, this study adopted the measures of inverse gated decoupling and relaxation reagent to eliminate the NOE. In this way, a quantitative determination based on the integrated intensity of the spectrum could be carried out.

2.3. Gas Chromatography

Gas chromatography is used to separate the components of a sample with different partitioning coefficients between the mobile phase and stationary phase of the column [30]. In other words, the mobile phase has different adsorption values or solubility over the stationary phase. The mixture about to be separated is vaporized into a gas at the injection port, and the gas is carried by the mobile phase into the chromatographic column. Owing to the continuous flushing of the mobile phase, the components move downstream, and the components with the weakest adsorption (or dissolution) ability move downstream faster than other components. In this way, each component in the sample could flow out from respective weak to strong adsorption according to its adsorption (or dissolution) ability, so that the components can be separated in this way. The detector response versus retention time is the raw data obtained from the gas chromatography test. The time from injection to the chromatographic peak reaching the local maximum value is called the retention time of the component. Under a specific stationary phase and experimental conditions, each component has a specific retention time, which could be used as a qualitative indicator. The peak height reflects the response of the detector.

3. Results and Discussion

3.1. *^{13}C NMR Spectra of Asphalt Binders*

The ^{13}C NMR spectra of SHRP asphalt binders under different oxidation states (unaged, 20 h PAV, and 60 h PAV) are presented in Figure 1. According to previous studies, the peaks between 110–160 ppm were attributed to aromatic carbons, and the peaks between 10–40 ppm were attributed to carbons in aliphatic chains and naphthenic rings [31]. Considering the objectives of this study, only aliphatic carbons were analyzed. It can be seen from the ^{13}C NMR spectrum of unaged SHRP asphalt binders that all tested asphalts have five obvious main peaks. It was reported that the peak at 14.1 ppm was assigned to methyl carbons of aliphatic side chains [32]. The peaks at 22.7, 31.9, and 29.1 ppm correspond to carbons in methylene units successively further from the methyl group. The peak at 29.7 ppm, which is the largest peak in the aliphatic carbon region, is caused by internal methylene carbons of long paraffinic chains.

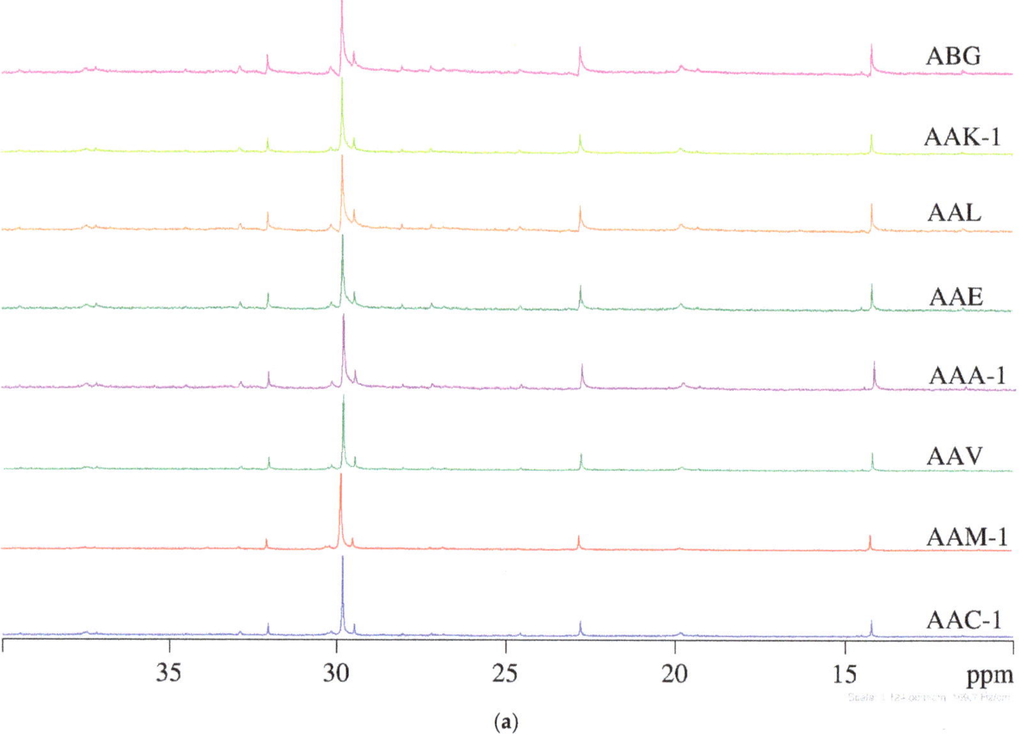

(a)

Figure 1. *Cont.*

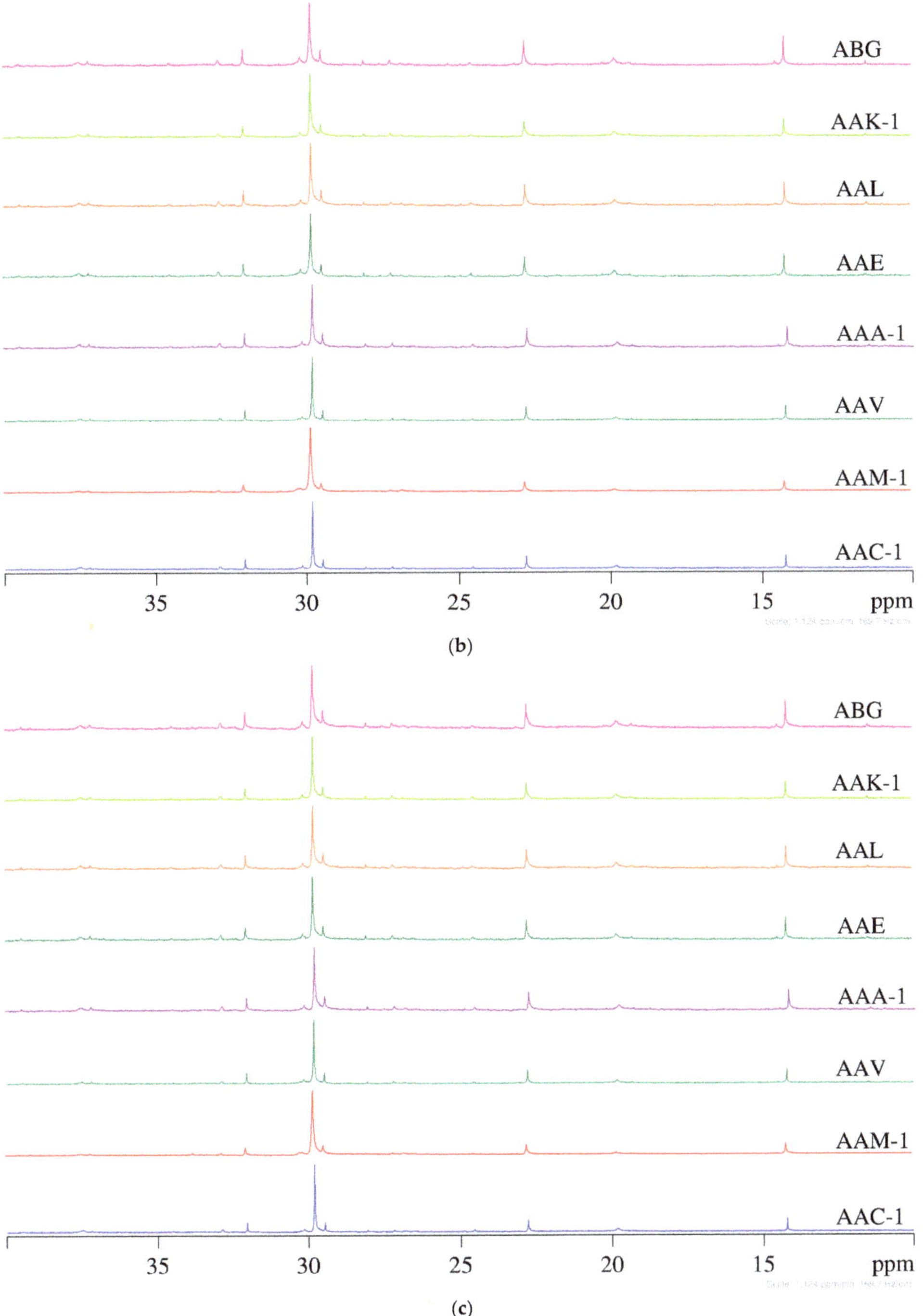

Figure 1. ^{13}C NMR spectra of SHRP asphalt binders. (a) Unaged, (b) 20 h PAV, and (c) 60 h PAV.

It is difficult to distinguish the difference between ^{13}C spectra of various SHRP asphalt binders with the naked eye. In addition, the influence of the degree of oxidative aging in SHRP binders on the ^{13}C chemical shift is also difficult to be evaluated directly. Therefore, the ratio of methylene in different positions to the 14.1 ppm methyl peak is presented in Figure 2. The ratio is an indication of the average length of a straight methylene chain terminated by a methyl group [33]. Compared with internal methylene carbons of long paraffinic chains, oxidative aging does not produce a major effect on the contents of methylene carbons adjacent to the methyl functional group in the average molecular structure of the asphalt binder. Moreover, in comparison with other carbon types, the content of internal methylene carbons of long paraffinic chains between various SHRP binders is quite different.

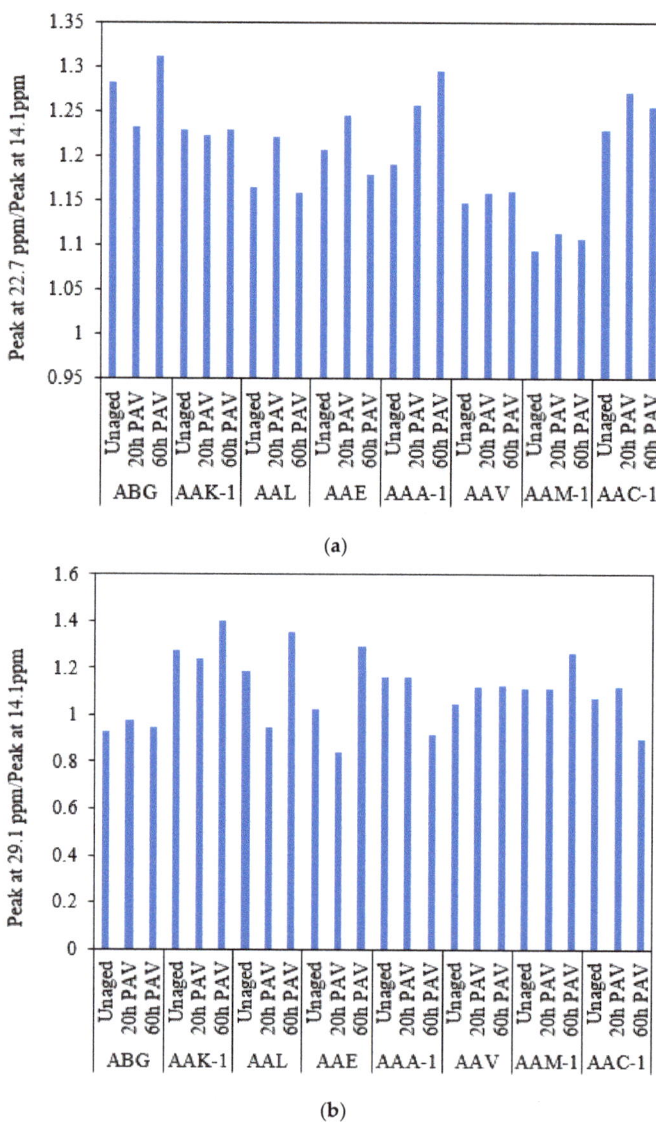

(a)

(b)

Figure 2. *Cont.*

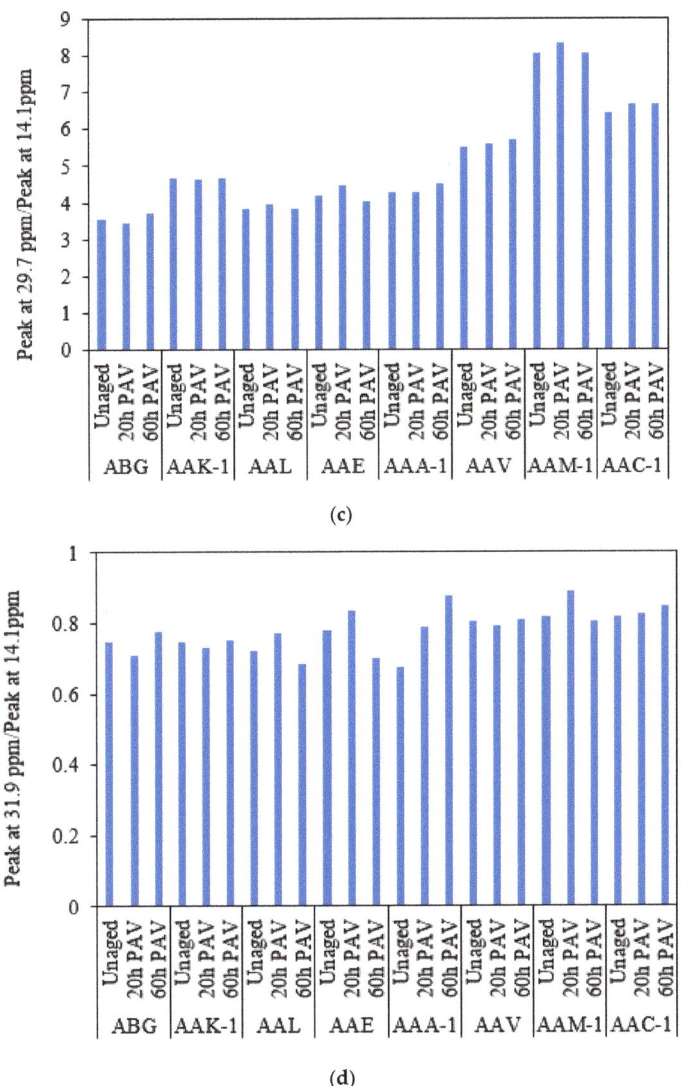

(c)

(d)

Figure 2. Percent carbons from ^{13}C NMR for SHRP asphalt binders. (**a**) Peak@22.7 ppm/Peak@14.1 ppm (**b**) Peak@29.1 ppm/Peak@14.1 ppm (**c**) Peak@29.7 ppm/Peak@14.1 ppm (**d**) Peak@31.9 ppm/Peak@14.1 ppm.

In order to determine whether the ^{13}C spectra signal can be utilized to determine the wax content in the asphalt binder, the NMR average length of a long paraffinic internal methylene chain was compared with the wax content obtained at -20 °C using the methyl ethyl ketone (MEK) precipitation method and also the recently developed variable-temperature Fourier-transform infrared spectroscopy (VT-FTIR) method, respectively. Corresponding results are illustrated in Figure 3. It can be seen that the NMR signal shows a close correlation with the other two selected methods, especially the VT-FTIR method. Compared with the commonly used chemical precipitation method to determine the wax content, the NMR method has many merits, including straightforward steps, less effect on results by the operating factors, and usage of almost no toxic chemical solvents. The most important fact is that this method can effectively distinguish asphalt binders with different wax contents from each other. However, compared with VT-FTIR, the results obtained by the NMR method are less discriminating. The difference between the highest and lowest wax content obtained using the VT-FTIR method is up to 9.07, while it is only 4.5 using the NMR method. This may be due to the fact that NMR is tested at room temperature and deuterated chloroform is used as the solvent. The presence of solvents may destroy the form of the molecular structure in the asphalt matrix. In the future, low-temperature solid-state nuclear magnetic resonance spectroscopy will be employed to further test asphalt binders and verify the accuracy of the conjecture. Similar to a previous publication by Ding et al. [5], the NMR method has not found a clear rule of the effect of oxidation on the crystallizable fractions in asphalt binder. In most cases, the average length of straight internal methylene carbons of a long paraffinic chain terminated by a methyl group increases with the oxidation of the asphalt binder. However, the difference caused by oxidation is much smaller than the source of asphalt binder.

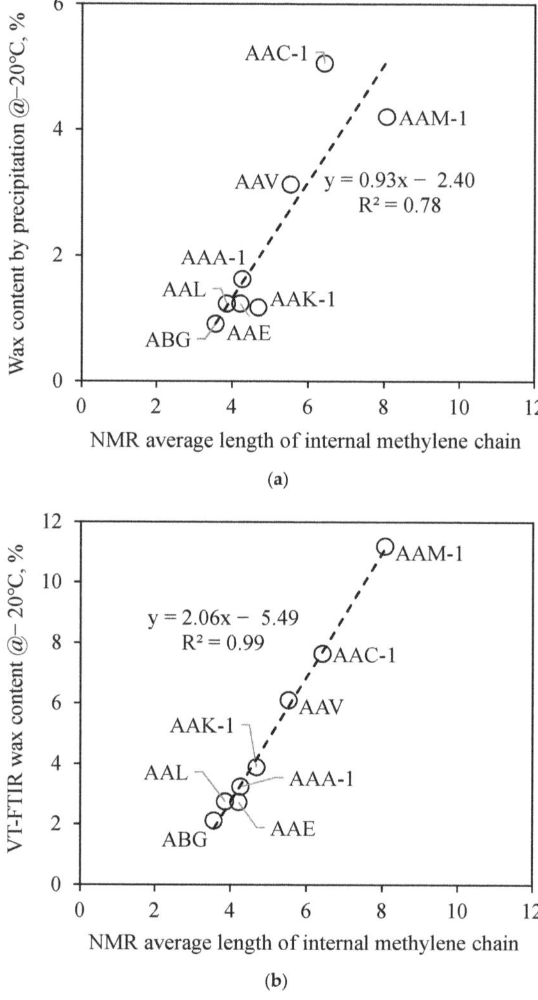

Figure 3. Comparison between the average length of straight internal methylene carbon of long paraffinic chain terminated by a methyl group and wax contents obtained using (**a**) precipitation and (**b**) VT-FTIR methods.

3.2. DEPT 135 Spectra of Asphalt Binders

The main function of using the distortionless enhancement by polarization transfer (DEPT) test is to distinguish methyl (CH_3), methylene (CH_2), and methine (CH) groups and clarify the aliphatic region of ^{13}C spectra. In general, methyl and methine carbons are displayed as upward signals in the DEPT 135 experiment, whereas the methylene carbons appear as downward signals. DEPT 135 spectra of sample AAV and AAC-1 under three oxidation states (unaged, 20 h PAV, and 60 h PAV) are shown in Figure 4. As expected, the peaks at 22.7, 31.9, 29.1, and 29.7 ppm correspond to carbons in methylene units successively further from the methyl group, and internal methylene carbons of long paraffinic chains are downward. The peaks at 14.1 ppm and 19.71 ppm correspond to methyl peaks, and methyl branches on a straight methylene backbone are upward. The aromatic/aliphatic carbons ratio in asphalt binder can also be determined using the DEPT 135 NMR experiment. Considering the main purpose of this study, the corresponding results are not shown here.

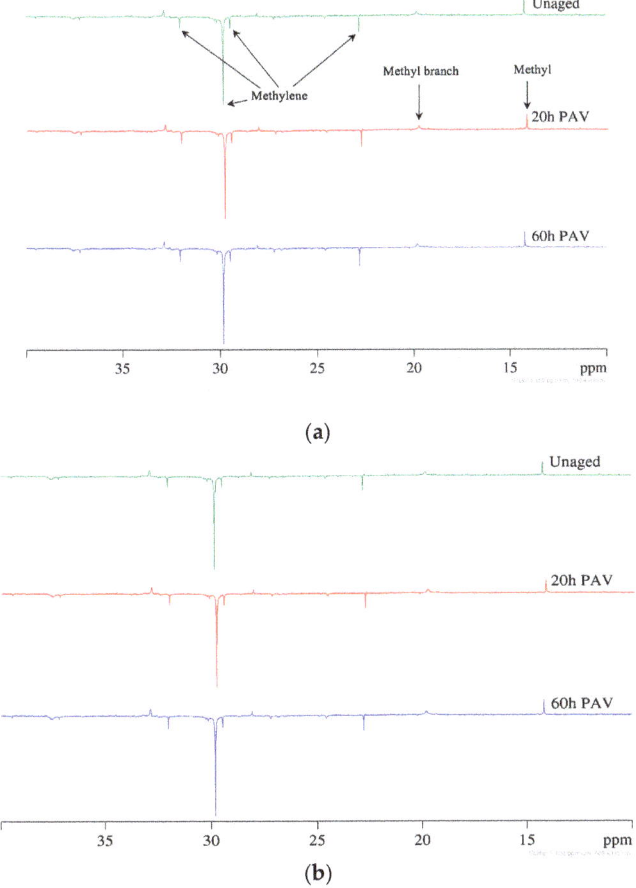

Figure 4. DEPT135 spectra of selected asphalt binders. (**a**) AAV. (**b**) AAC-1.

3.3. Effect of Oxidation on n-Alkane Distribution

Under the selected optimal chromatographic analysis conditions, the carbon disulfide solutions of asphalt with different aging states (unaged, 20 h PAV, and 60 h PAV) were analyzed and determined. The results of removing the solvent peaks are presented in Figure 5. Since the correction factors of hydrocarbon compounds on the flame ionization detector (FID) are similar, the area normalization method is directly used for quantitative analysis to obtain the content of each n-alkane. The determination results of carbon number distribution are illustrated in Figure 6. Although the effects of oxidation on the total n-alkane content in asphalt cannot be determined by gas chromatogram, some interesting observations could be made. For the unaged ABG asphalt sample, the carbon number of n-alkanes is mainly distributed in a wide range between 30 and 80. Moreover, there are very few n-alkanes with carbon numbers below 20 and above 90. With the oxidation of the binder (20 h PAV), the carbon number of n-alkanes in the asphalt is mainly concentrated in a narrow range between 35 and 50. The sample hardly contains n-alkanes with carbon numbers lower than 30 and higher than 80. By extending the binder's oxidation time (60 h PAV), the carbon number of n-alkanes in asphalt generally increased. A similar trend was found in other asphalt binders. Furthermore, the carbon number distribution of n-alkanes tends to be uniform, especially for AAM-1 and AAC-1.

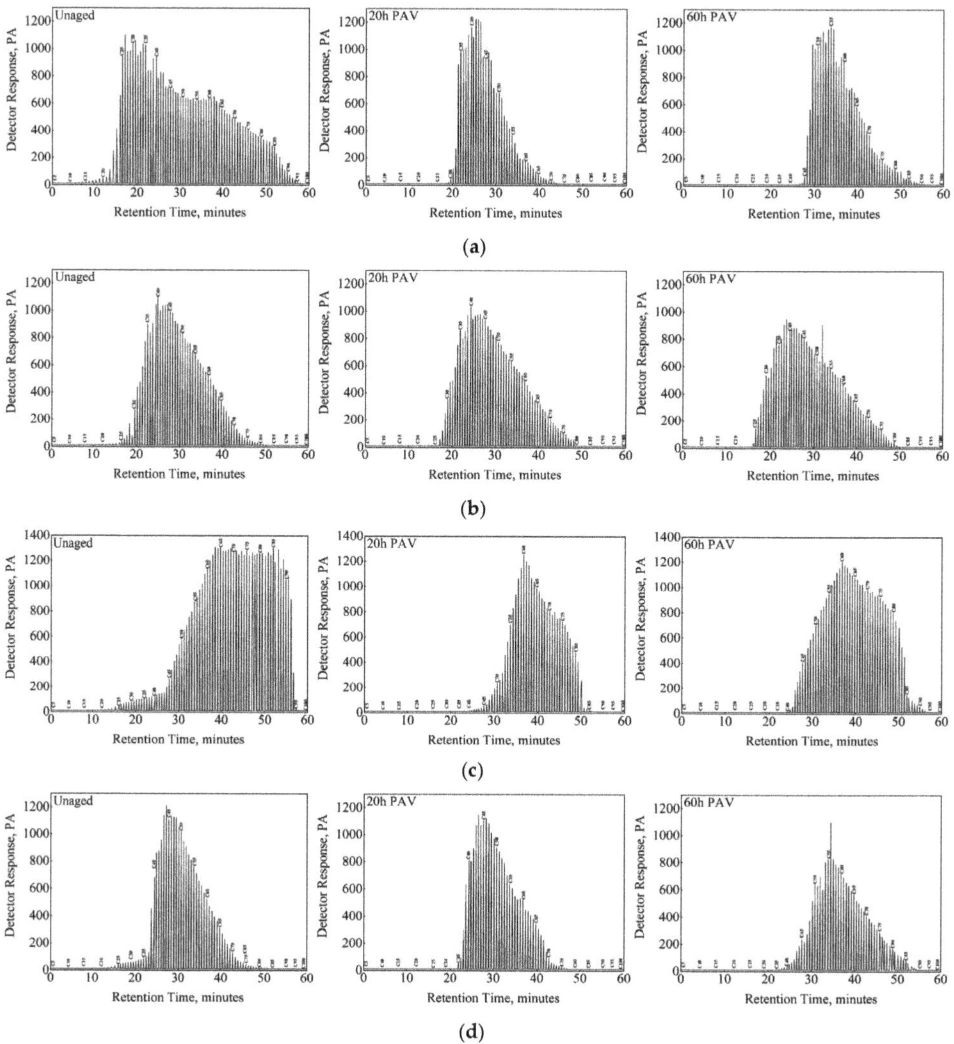

Figure 5. Gas chromatogram of asphalt binders. (**a**) ABG, (**b**) AAV, (**c**) AAM-1, and (**d**) AAC-1.

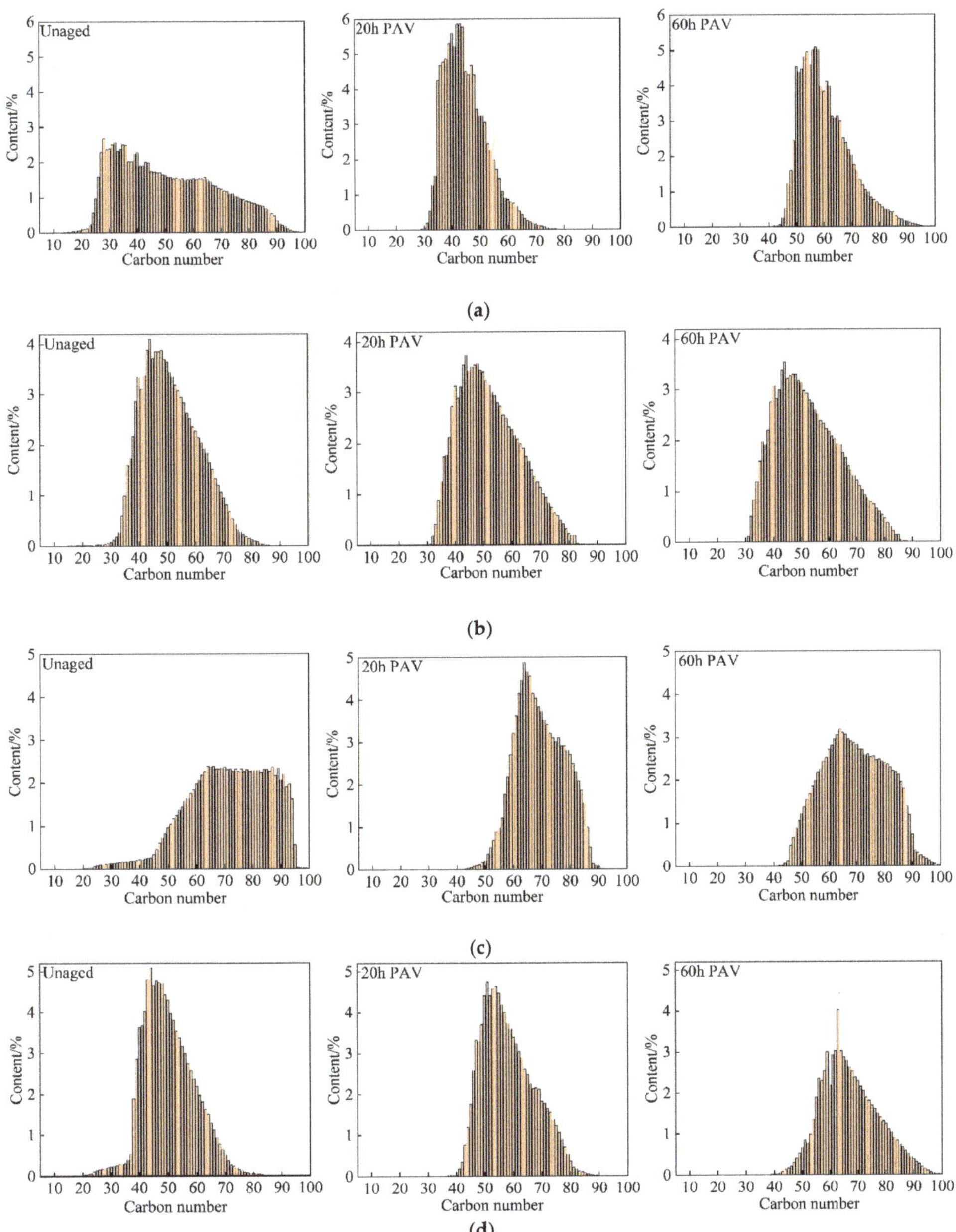

Figure 6. N-alkane distribution in asphalt binders. (**a**) ABG, (**b**) AAV, (**c**) AAM-1, and (**d**) AAC-1.

4. Summary and Conclusions

In this study, the effects of oxidation on crystallizable fractions in different asphalt binders were investigated by liquid-state nuclear magnetic resonance (NMR) spectroscopy and high-temperature gas chromatography (HTGC). The following conclusions can be drawn from a series of laboratory tests:

(1) The results of the ^{13}C NMR spectra of unaged SHRP asphalt binders revealed that all tested asphalts had five obvious main peaks. Compared with internal methylene carbons of long paraffinic chains, oxidative aging did not produce a major effect on the contents of methylene carbons adjacent to the methyl functional group in the average molecular structure of the asphalt binder.

(2) Compared with other carbon types, the content of internal methylene carbons of long paraffinic chains between different SHRP binders was quite different. The NMR average length of a long paraffinic internal methylene chain showed a close correlation with wax content obtained at $-20\ ^{\circ}$C using the methyl ethyl ketone (MEK) precipitation method and also the recently developed variable-temperature Fourier-transform infrared spectroscopy (VT-FTIR) method.

(3) The application of the NMR method could not find a clear rule in terms of the effect of oxidation on the crystallizable fractions in asphalt binder. In most cases, the average length of straight internal methylene carbons of a long paraffinic chain terminated by a methyl group increased with the oxidation of the asphalt binder. However, the difference caused by oxidation was considerably smaller than the source of asphalt binder.

(4) In general, oxidation will cause the n-alkanes distributed in asphalt binder to fall within a narrower range. The carbon number of n-alkanes in asphalt binder generally increases with oxidation.

(5) Future studies will cover the evaluation of UV effects, the short-term RTFOT test, as well as the determination of rheological properties.

Author Contributions: Conceptualization, H.D. and Y.Q.; Methodology, A.R.; Data curation, W.W. All authors have read and agreed to the published version of the manuscript.

Funding: This research was funded by 1. Talent Introduction Project of Xihua University (School Key Project) (Z202111); 2. Scientific research project of Sichuan Provincial Department of Education (16ZB0164); 3. Open Research Fund of Sichuan Provincial Key Laboratory of Road Engineering (15206569); 4. Open project of Sichuan Key Laboratory of Green Building and Energy Saving (szjj2015-074).

Institutional Review Board Statement: Not applicable.

Informed Consent Statement: Not applicable.

Data Availability Statement: The data supporting the figures in this paper, as well as other findings of this study, are available from the corresponding author upon reasonable request.

Conflicts of Interest: The authors declare no conflict of interest.

References

1. Ma, J.; Hesp, S.A.M.; Chan, S.; Li, J.Z.; Lee, S. Lessons learned from 60 years of pavement trials in continental climate regions of Canada. *Chem. Eng. J.* **2022**, *444*, 136389. [CrossRef]
2. Ding, H.; Zhang, H.; Liu, H.; Qiu, Y. Thermoreversible aging in model asphalt binders. *Constr. Build. Mater.* **2021**, *303*, 124355. [CrossRef]
3. Edwards, Y.; Isacsson, U. Wax in bitumen. *Road Mater. Pavement Des.* **2005**, *6*, 281–309. [CrossRef]
4. Ding, H.; Zhang, H.; Xie, Q.; Rahman, A.; Qiu, Y. Synthesis and characterization of nano-SiO$_2$ hybrid poly(methyl methacrylate) nanocomposites as novel wax inhibitor of asphalt binder. *Colloids Surf. A Physicochem. Eng. Asp.* **2022**, *653*, 130023. [CrossRef]
5. Ding, H.; Hesp, S. Variable-temperature Fourier-transform infrared spectroscopy study of asphalt binders from the SHRP Materials Reference Library. *Fuel* **2021**, *298*, 120819. [CrossRef]
6. Kovinich, J.; Hesp, S.; Ding, H. Modulated differential scanning calorimetry study of wax-doped asphalt binders. *Thermochim. Acta* **2021**, *699*, 178894. [CrossRef]
7. Qiu, Y.; Zhang, H.; Ding, H.; Rahman, A. Measurement and thermodynamic modeling of wax precipitation in asphalt binder. *Mater. Des.* **2022**, *221*, 110957. [CrossRef]
8. Knotnerus, J.; Krom, C.J. The constitution of wax isolated from bitumen. *Am. Chem. Soc. Div. Pet. Chem. Prepr.* **1964**, *9*, B-39.
9. Ding, H.; Zhang, H.; Zhang, H.; Liu, D.; Qiu, Y.; Hussain, A. Separation of wax fraction in asphalt binder by an improved method and determination of its molecular structure. *Fuel* **2022**, *322*, 124081. [CrossRef]
10. Thenoux, G.; Bell, C.; Wilson, J.; Eakin, D.; Schroeder, M. Experiences with the Corbett-Swarbrick procedure for separation of asphalt into four generic fractions. *Transp. Res. Rec.* **1988**, *1171*, 66–70.

11. Rostler, F.S.; White, R.M. Composition and changes in composition of highway asphalts, 85–100 penetration grade. *J. Assoc. Asph. Paving Technol.* **1962**, *31*, 35–89.

12. Redelius, P.; Lu, X.; Isacsson, U. Non-classical wax in bitumen. *Road Mater. Pavement Des.* **2002**, *3*, 7–21. [CrossRef]

13. Garrick, N.W. Use of gel-permeation chromatography in predicting properties of asphalt. *J. Mater. Civ. Eng.* **1994**, *6*, 376–389. [CrossRef]

14. Netzel, D.A.; Rovani, J.F. Direct separation and quantitative determination of (n-, iso-) alkanes in neat asphalt using urea adduction and high-temperature gas chromatography (HTGC). *Energy Fuels* **2007**, *21*, 333–338. [CrossRef]

15. Hasheminejad, N.; Pipintakos, G.; Vuye, C.; De Kerf, T.; Ghalandari, T.; Blom, J.; Van den bergh, W. Utilizing deep learning and advanced image processing techniques to investigate the microstructure of a waxy bitumen. *Constr. Build. Mater.* **2021**, *313*, 125481. [CrossRef]

16. Blom, J.; Soenen, H.; Van den Brande, N.; Van den bergh, W. New evidence on the origin of 'bee structures' on bitumen and oils, by atomic force microscopy (AFM) and confocal laser scanning microscopy (CLSM). *Fuel* **2021**, *303*, 121265. [CrossRef]

17. Ding, H.; Zhang, H.; Zhang, H.; Qiu, Y. Direct Observation of Crystalline Wax in Asphalt Binders by Variable-temperature Polarizing Microscope. *J. Mater. Civ. Eng.* **2022**, *34*, 04022244. [CrossRef]

18. Ding, H.; Zhang, H.; Zheng, X.; Zhang, C. Characterisation of crystalline wax in asphalt binder by X-ray diffraction. *Road Mater. Pavement Des.* **2022**, 1–7. [CrossRef]

19. Noel, F.; Corbett, L.W. A study of the crystalline phases in asphalts. *J. Inst. Pet.* **1970**, *56*, 261–268.

20. Harrison, I.R.; Wang, G.; Hsu, T.C. *A Differential Scanning Calorimetry Study of Asphalt Binders*; National Research Council: Washington, DC, USA, 1993.

21. Lu, X.; Sjövall, P.; Soenen, H.; Blom, J.; Makowska, M. Oxidative aging of bitumen: A structural and chemical investigation. *Road Mater. Pavement Des.* **2021**, *23*, 1091–1106. [CrossRef]

22. Ding, H.; Hesp, S. Variable-temperature Fourier-transform infrared spectroscopy study of wax precipitation and melting in Canadian and Venezuelan asphalt binders. *Constr. Build. Mater.* **2020**, *264*, 120212. [CrossRef]

23. Ding, H.; Hesp, S. Quantification of crystalline wax in asphalt binders using variable-temperature Fourier-transform infrared spectroscopy. *Fuel* **2020**, *277*, 118220. [CrossRef]

24. Tauste, R.; Moreno-Navarro, F.; Sol-Sánchez, M.; Rubio-Gámez, M.C. Understanding the bitumen ageing phenomenon: A review. *Constr. Build. Mater.* **2018**, *192*, 593–609. [CrossRef]

25. Malinowski, S.; Bandura, L.; Woszuk, A. Influence of atmospheric oxygen on the structure and electronic properties of bitumen components—A DFT study. *Fuel* **2022**, *325*, 124551. [CrossRef]

26. Hofko, B.; Maschauer, D.; Steiner, D.; Mirwald, J.; Grothe, H. Bitumen ageing—Impact of reactive oxygen species. *Case Stud. Constr. Mater.* **2020**, *13*, e00390. [CrossRef]

27. Rathore, M.; Haritonovs, V.; Zaumanis, M. Performance Evaluation of Warm Asphalt Mixtures Containing Chemical Additive and Effect of Incorporating High Reclaimed Asphalt Content. *Materials* **2021**, *14*, 3793. [CrossRef]

28. Ding, H.; Liu, H.; Qiu, Y.; Rahman, A. Effects of crystalline wax and asphaltene on thermoreversible aging of asphalt binder. *Int. J. Pavement Eng.* **2022**, *23*, 3997–4006. [CrossRef]

29. Zhang, H.; Xie, Q.; Ding, H.; Rahman, A. Spectroscopic ellipsometry studies on optical constants of crystalline wax-doped asphalt binders. *Int. J. Pavement Eng.* **2022**, 1–13. [CrossRef]

30. Gupta, A.K.; Severin, D. Characterization of petroleum waxes by high temperature gas chromatography—Correlation with physical properties. *Pet. Sci. Technol.* **1997**, *15*, 943–957. [CrossRef]

31. Siddiqui, M.N.; Ali, M.F. Investigation of chemical transformations by NMR and GPC during the laboratory aging of Arabian asphalt. *Fuel* **1999**, *78*, 1407–1416. [CrossRef]

32. Jennings, P.W.; Desando, M.A.; Raub, M.F.; Moats, R.; Mendez, T.M.; Stewart, F.F.; Hoberg, J.O.; Pribanic, J.A.S.; Smith, J.A. NMR spectroscopy in the characterization of eight selected asphalts. *Fuel Sci. Technol. Int.* **1992**, *10*, 887–907. [CrossRef]

33. Musser, B.J.; Kilpatrick, P.K. Molecular Characterization of Wax Isolated from a Variety of Crude Oils. *Energy Fuels* **1998**, *12*, 715–725. [CrossRef]

materials

Article

Preparation of Wax-Based Warm Mixture Additives from Waste Polypropylene (PP) Plastic and Their Effects on the Properties of Modified Asphalt

Gang Zhou [1,†], Chuanqiang Li [2,*,†], Haobo Wang [2], Wei Zeng [2], Tianqing Ling [1], Lin Jiang [1], Rukai Li [1], Qizheng Liu [1], Ying Cheng [1] and Dan Zhou [1]

[1] School of Civil Engineering, Chongqing Jiaotong University, Chongqing 400074, China; cjhs_2000@163.com (G.Z.); lingtq@163.com (T.L.); jianglin_cn@foxmail.com (L.J.); lirukai89@163.com (R.L.); liuqizheng_cn@163.com (Q.L.); chengying_cn@163.com (Y.C.); zhoudan_cn1@163.com (D.Z.)

[2] School of Materials Science and Engineering, Chongqing Jiaotong University, Chongqing 400074, China; wanghaobopostbox@163.com (H.W.); lzzengwei@163.com (W.Z.)

[*] Correspondence: lichuanqiang@cqjtu.edu.cn; Tel.: +86-23-6265-2389

[†] These authors contributed equally to this work.

Abstract: The production of high-performance, low-cost warm mix additives (WMa) for matrix asphalt remains a challenge. The pyrolysis method was employed to prepare wax-based WMa using waste polypropylene plastic (WPP) as the raw material in this study. Penetration, softening point, ductility, rotational viscosity, and dynamic shear rheological tests were performed to determine the physical and rheological properties of the modified asphalt. The adhesion properties were characterized using the surface free energy (SFE) method. We proved that the pyrolysis temperature and pressure play a synergistic role in the production of wax-based WMa from WPPs. The product prepared at 380 °C and 1.0 MPa (**380-1.0**) can improve the penetration of matrix asphalt by 61% and reduce the viscosity (135 °C) of matrix asphalt by 48.6%. Furthermore, the modified asphalt shows favorable elasticity, rutting resistance, and adhesion properties; thus, it serves as a promising WMa for asphalt binders.

Keywords: warm mix additive; waste polypropylene plastic; pyrolysis; PP wax; modified asphalt

Citation: Zhou, G.; Li, C.; Wang, H.; Zeng, W.; Ling, T.; Jiang, L.; Li, R.; Liu, Q.; Cheng, Y.; Zhou, D. Preparation of Wax-Based Warm Mixture Additives from Waste Polypropylene (PP) Plastic and Their Effects on the Properties of Modified Asphalt. *Materials* **2022**, *15*, 4346. https://doi.org/10.3390/ma15124346

Academic Editors: Meng Ling, Yao Zhang, Haibo Ding, Yu Chen and F. Pacheco Torgal

Received: 1 May 2022
Accepted: 2 June 2022
Published: 20 June 2022

Publisher's Note: MDPI stays neutral with regard to jurisdictional claims in published maps and institutional affiliations.

1. Introduction

In the asphalt pavement industry, the warm mix asphalt (WMA) has been introduced and used instead of the hot mix asphalt (HMA) to reduce the temperatures of mixture production and placement. This has led to reductions in fuel consumption and noxious gas emissions, which are particularly welcomed in urban environments [1–4]. A variety of asphalt binders and mixture additives are available to produce WMA mixtures among which wax-based additives are common types. For example, Sasobit can effectively reduce the viscosity of asphalt binder at high temperatures, thereby decreasing the mixing and compaction temperatures and improving the asphalt rutting resistance at usage temperatures [5–9]. Polythene wax (PEW) and polypropylene wax (PPW) are two other wax-based additives. They can also effectively reduce the construction temperature of asphalt mixtures [10–12]. However, the types of warm mixtures are still very limited, and the cost of existing warm mixes is high [13]; therefore, it is necessary to develop new low-cost warm mix additives (WMa). The production of WMa from waste materials (e.g., waste plastic) is an attractive approach.

Plastic products are essential in various fields owing to their low cost, light weight, and durability. The plastics are generally designed to be robust and durable, leading to their inherent non-degradable nature. Plastic wastes in the environment are of increasing concern because of their effects on wildlife and human. Thus, the treatment of plastic wastes has become a huge problem [14]. Most plastic wastes fall into four categories: polyester, polyolefin, polyvinyl chloride (PVC), and polystyrene (PS). Polyolefin, such as polyethylene (PE) and polypropylene (PP), with production of about 218 Mt every year, accounts for 57% of the plastic content of municipal solid waste [15]. As one of the representatives of plastics, PP plastic is extensively used due to its abundant source, low price, and chemical corrosion [16]. The widespread use of PP plastics has caused serious pollution. Therefore, there is an urgent need for the effective and environmentally friendly management of waste PP plastics [17]. Typical methods used to discard waste plastic include landfills, incineration, and mechanical recycling. However, landfill and incineration methods have the potential to pollute the surrounding soil, air and water [18]. Furthermore, the mechanical recycling method degrades the physical properties of recycled PP plastic [19]. Because of its high-hydrocarbon content, the use of waste PP plastic for the production of high-value-added products and energy recovery has attracted extensive attention among researchers, as it can be easily transformed to alkenes and alkanes via pyrolysis [17]. Because the wax can be used as WMa or can be used for improving the thermal stability of asphalt binder [20,21], the production of PP wax (PPW) should be an excellent strategy for the resource utilization of waste PP plastic. Current research has focused on the production of fuel from waste plastics via thermal pyrolysis and catalytic cracking [22–24]. Few studies have been devoted to the production of waxes from waste PP plastic, particularly those used to modify asphalt. Therefore, it is necessary to conduct a systematic study on the pyrolysis of waste PP plastics for wax production and study in depth the effect of wax on the performance of modified asphalt. In addition, few studies have been conducted on the combined effects of temperature and pressure on the pyrolysis of waste PP plastics.

In this study, a range of PPW samples was prepared by pyrolysis at different temperatures and pressures. Furthermore, the physical, viscosity-temperature, and rheological properties of the PPW-modified asphalt binders were analyzed. In addition, the surface free energy (SFE) method was used to characterize the adhesive properties of the PPW-modified binders. The results showed that the PPWs obtained at $380 \leq T \leq 400\ °C$ and $0.5 \leq p \leq 1.0$ MPa exhibited good warm mixing performances. To the best of our knowledge, this is the first study to investigate the combined effects of temperature and pressure on preparing WMa from waste PP plastic.

2. Materials and Methods

2.1. Materials

Waste PP plastic particles (WPPs) were obtained from a plastic recycling plant. Matrix asphalt (MA) with a penetration grade of 70 (penetration at 25 °C is 6.65 mm; softening point is 48.2 °C; ductility at 15 °C is greater than 100 cm; glass transition temperature (T_g) is around $-10\ °C$) was used in this work. Ethylene glycol (99+%), formamide (99+%), and deionized water were used as liquid solvents for the measurement of SFE components. All materials were of commercial origin and were used without further purification.

2.2. PPW Preparation

PPW was synthesized via pyrolysis in a pressure reactor (Figure 1). The WPP was placed in the pressure reactor, and the reactor was sealed and purged with nitrogen for 5 min to discharge the air in it. The air release valve was closed, nitrogen was injected at a specified pressure, and the air intake valve was closed. The reactor was heated to a specified temperature and then stirred at 120 revolutions per minute (rpm) for 20 min. During this period, the pressure in the reactor was maintained constant by adjusting the pressure-relief valve. Subsequently, the reactor was cooled to 120 °C, and the target PPW was poured out and collected in zip-lock bags.

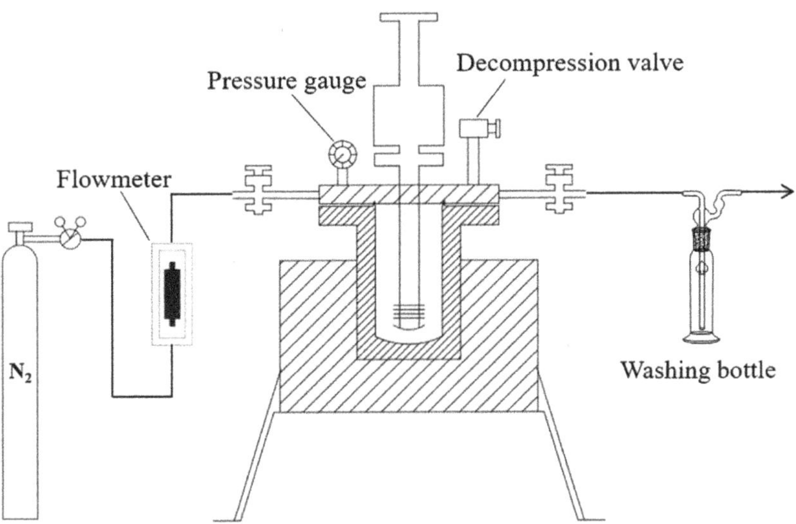

Figure 1. Schematic of pyrolysis reactor.

2.3. Preparation of PPW-Modified Asphalt (PPWA)

The PPWA was prepared by blending PPW and an asphalt binder using a high-speed shearing machine at 120 °C and 1000 rpm for 10 min. The dosage of PPW was 6% of the weight of the matrix asphalt binder. The synthetic conditions for PPW and the corresponding modified asphalt are listed in Table 1.

Table 1. Synthetic conditions of PPW and the corresponding PPWA.

PPW	Pyrolysis Conditions of PPW		PPWA
	Temperature (°C)	Pressure (MPa)	
360-0.1	360	0.1	A
360-0.5	360	0.5	B
360-1.0	360	1.0	C
380-0.1	380	0.1	D
380-0.5	380	0.5	E
380-1.0	380	1.0	F
400-0.1	400	0.1	G
400-0.5	400	0.5	H
400-1.0	400	1.0	I

2.4. Binder Tests

Penetration [25], softening point [26], and ductility tests [27] were performed to evaluate the conventional physical properties of PPWA and MA. Penetration tests were performed at three temperatures (5 °C, 10 °C, and 15 °C) to calculate the penetration index (PI) of the binder.

To identify the rotational viscosities of the binders, 10 g of MA or PPWA samples were tested using an NJD-1D Brookfield viscometer (Shanghai Changji Geological Instrument Co. Ltd., Shanghai, China) [28]. Viscosity tests were conducted from 105 to 165 °C at 15 °C intervals.

The viscoelastic properties of the binders were characterized using a dynamic shear rheometer (DSR) according to the TA DHR-3 (TA Instruments Inc., New Castle, DE, USA). The high- and intermediate-temperature performances were respectively characterized using superpave rutting and fatigue parameters [29]. Plates with specific characteristics (gap: 1 mm, diameter: 25 mm) were used for each binder to obtain the DSR value. The complex modulus (G*) and phase angle (δ) were recorded for rheological analysis.

The adhesion property of asphalt binder with aggregate was evaluated based on the SFE theory [30–32]. To calculate the surface energy, the contact angles between the asphalt and the three liquid solvents (distilled water, glycerol, and formamide) were measured using the Wilhelmy plate method [33]. First, the asphalt sample was heated to the melted state, and the cleaned glass slide was then immersed in hot asphalt for a few minutes; then, it was removed so that a layer of asphalt film with a uniform thickness was formed on the surface of the glass slide (Figure 2a). The contact angles of the three liquids with the asphalt film were measured at 20 °C using a HARKE-SPCAX3 contact-angle meter (Beijing Harke Test Instrument Factory, Beijing, China, Figure 2b). The experiment was repeated five times, and the average values were obtained.

 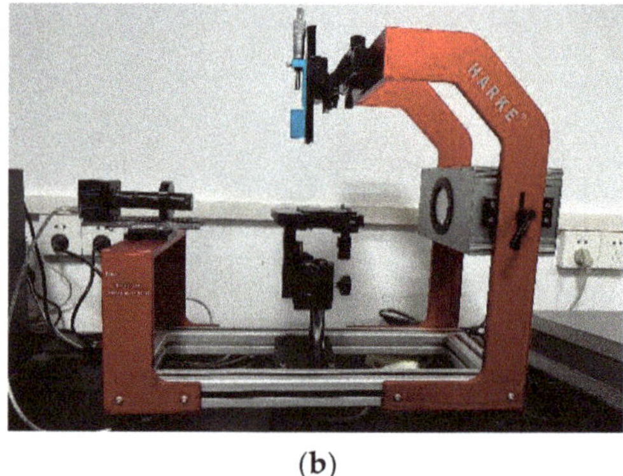

(**a**)   (**b**)

Figure 2. Asphalt film on glass slide (**a**) and contact-angle meter (**b**).

The microstructure of PPW was tested by a DM2500 optical microscope (Leica, Wetzlar, Germany). FTIR spectra of PPW and asphalt binders were tested on the Tensor II spectrometer (Bruker, Germany). DSC measurements were performed on a DSC214 System-Instrument (Netzsch Company, Selb, Germany).

3. Results and Discussion

3.1. Softening Point

The softening points of MA and PPWA are shown in Figure 3. It can be observed that the incorporation of PPW increased the softening point of MA, thus suggesting an improvement of the temperature stability of asphalt [1]. By comparing the PPWAs of **A** to **H**, it can be observed that at the same pyrolysis pressure, with an increase in pyrolysis temperature, the softening point of the obtained PPWA decreased; at the same pyrolysis temperature, the softening point of the obtained PPWA also decreased as a function of pyrolysis pressure. This indicates that an increased number of small molecules may be produced with an increase in pyrolysis temperature or pressure. However, the softening point of sample **I** is slightly higher than that of **E**, **F** and **H**. Moreover, the softening points of the four samples are not much different. This behavior suggests that increasing the pressure during pyrolysis at high temperature may have less effect on the high-temperature property of the products.

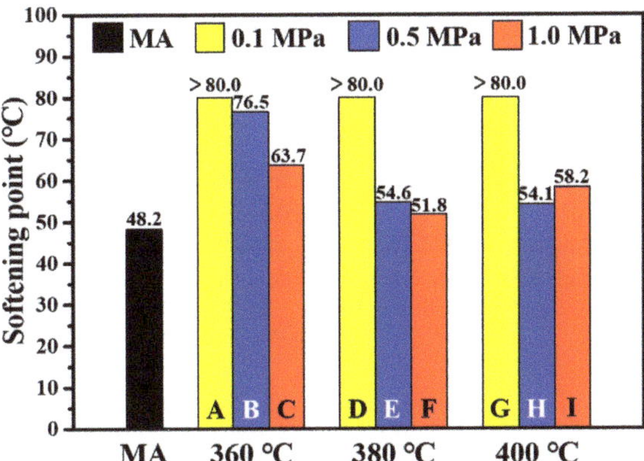

Figure 3. Softening points of matrix asphalt (MA) and PPW-modified asphalt (PPWA).

3.2. Penetration

Penetration can indicate degree of softness and the relative viscosity of asphalt binder. The higher the penetration value, the softer the asphalt binder and the lower the viscosity of asphalt [33]. Figure 4a shows the penetration values of MA and PPWA at 25 °C. It can be observed from the figure that the penetrations of **E**, **F**, **H**, and **I** are greater than that of MA, thus indicating that the corresponding PPWs can reduce the viscosity of MA. According to a previous study, the incorporation of a warm mix additive can reduce the viscosity of asphalt [34]. From this point of view, these four PPW, **380-0.5**, **380-1.0**, **400-0.5**, **400-1.0** exhibit the potential to be used as WMa. Furthermore, **380-1.0** seems to be the best one, and it can lead to 61% penetration improvement for MA. Furthermore, the penetration of **A**, **B**, **C**, **D**, **E**, and **F** increased sequentially, whereas those of **G**, **H**, and **I** first increased and then decreased. The results showed that the preparation conditions of PPW have a significant effect on the properties of PPWA. PPW produced by excessive temperature and pressure cannot reduce the viscosity of asphalt. This result is consistent with the softening point. This is an interesting phenomenon, and the reason remains to be investigated in the following chemical characterization.

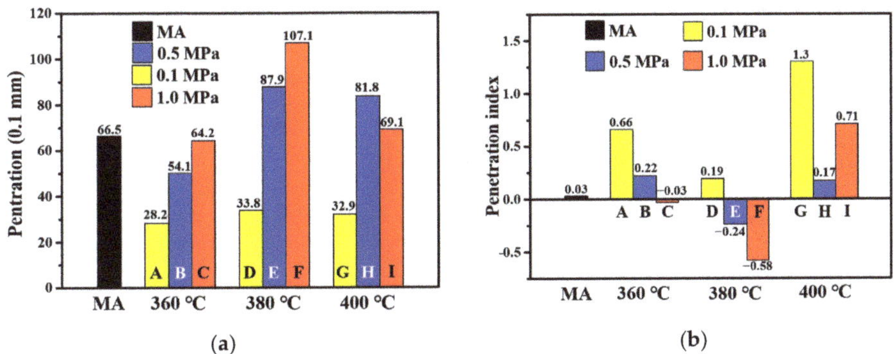

Figure 4. Penetration (**a**) and penetration index (**b**) of MA and PPWA.

The penetration index (PI) was calculated based on the penetration values of PPWA at different temperatures to evaluate the temperature sensitivity of the asphalt according to Equations (1) and (2). The results are shown in Figure 4b.

$$\lg P = K + AT \tag{1}$$

$$PI = \frac{20 - 500A}{1 + 50A} \qquad (2)$$

where T and P are the temperature and penetration, respectively, and K and A are determined from the $\lg P$–T curve.

It is known that most asphalt binders have PI values in the range of −2 to +2. Additionally, as the PI value of asphalt increases, the temperature susceptibility decreases [35,36]. In addition, the flexibility and thixotropy of the asphalt binders increased as PI increased [33]. In this study, all PPWA have PI values in the range of −2 to +2, indicating that all PPWAs are sol–gel structures, which means that adding these PPWs does not affect negatively the temperature susceptibility.

3.3. Ductility

Ductility can be used to evaluate the tensile deformation and flexibility of the asphalt binder. Figure 5 shows the ductility values of MA and PPWA at 15 °C. It can be observed that the ductility of MA decreases with the addition of PPW, thus indicating a negative effect of PPW on low-temperature flexibility and crack resistance of MA. Moreover, the ductility of PPWA increased as the corresponding pyrolysis pressure increased except for **I**. Among the nine PPWAs, **F** samples (the samples modified by PPW **380-0.1**) exhibited the best low-temperature flexibility, suggesting that the pyrolysis temperature and pressure are equally important in the production of PPW. Gao [37] proved that increasing the pressure at high pyrolysis temperature inhibited the degradation of macromolecular substances in polymers. Thereby, there might be more macromolecules in the PPW **400-1.0** than **400-0.5**, which may be the reason for the ductility of **I** sample lower than that of **H**.

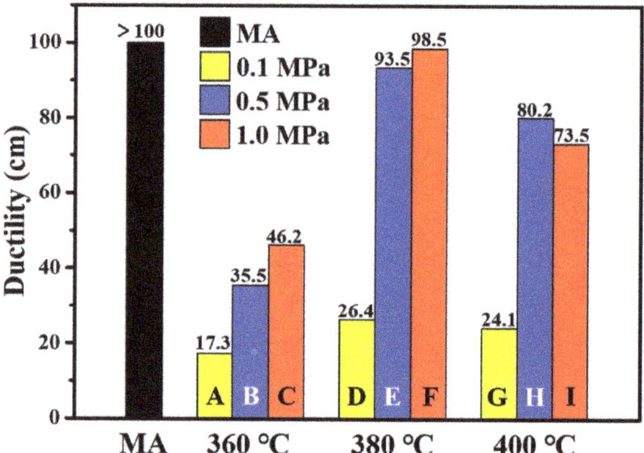

Figure 5. Ductility of MA and PPWA at 15 °C.

3.4. Rotational Viscosity

Viscosity characterizes the ability of asphalt to resist shear deformation when subjected to an external force. Figure 6 shows the viscosity–temperature curves for MA and PPWA. It can be observed that the viscosities of **E**, **F**, **H**, and **I** are lower than that of MA. The addition of **380-0.5**, **380-1.0**, **400-0.5**, and **400-1.0** to MA led to a decrease in the viscosity, which reveals that the four PPWs can decrease the mixing and compaction temperatures of the asphalt mixture; accordingly, their use may improve workability at lower temperatures [33]. The most pronounced effect was observed for **F** (the sample modified by PPW **380-1.0**), which led to a viscosity reduction of 48.6% at 135 °C compared with MA. This result is comparable to the viscosity reduction for rubberized asphalt using commercial pure PPW [12]. It is expected that the wax-based WMa may have melted at lower temperatures, which may have caused a reduction in viscosity [38].

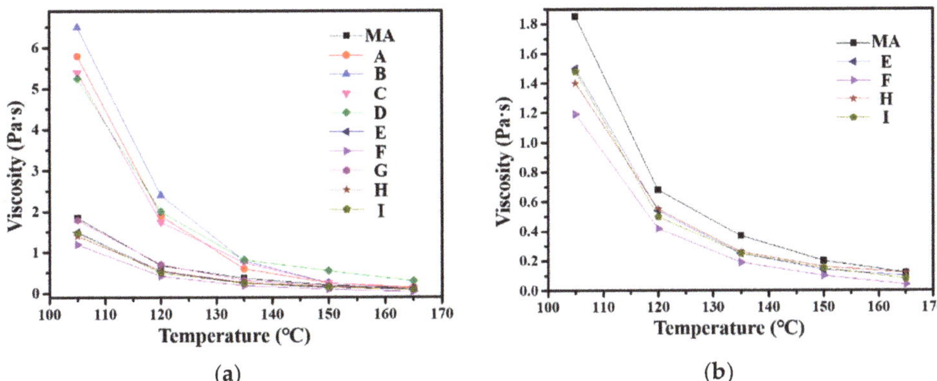

Figure 6. Viscosity–temperature curves of MA and PPWA (**a**). Magnified view of some of the curves (**b**).

3.5. Rheological Properties

The rheological properties of MA and PPWA were investigated using a DSR. The complex modulus (G*) and phase angle (δ) master curves are shown in Figure 7 over the temperature range of 52–82 °C. As shown in Figure 7a, the G* values of **A**, **B**, **C**, and **D** are higher than those of MA, which implies that the addition of PPW **360-0.1**, **360-0.5**, **360-1.0** and **380-0.1** can improve the stiffness and flow deformation resistance of the asphalt binder [39]. Figure 7b shows that the addition of PPW **380-0.5**, **380-1.0**, **400-0.1**, **400-0.5**, and **400-1.0** promoted the decrease in G* of MA. This effect is probably due to the melting of PPW, thus leading to a change in the rheological behavior of PPWA [40,41]. Among the corresponding five PPWAs, **F** exhibited the highest G* value at all test temperatures, thus suggesting its relatively high resistance to shear deformation. This might be due to the PPW **380-1.0** exhibiting better crystallinity than the other four PPWs [40].

Figure 7c shows the curves of δ for MA and PPWA. The data show that PPW **360-0.1**, **360-0.5**, **360-1.0**, and **380-0.1** can decrease considerably the δ value of MA, thus indicating an improvement in the elasticity of asphalt [42]. Interestingly, the δ values of the corresponding four PPWAs first increased and then decreased as the temperature increased. This phenomenon may be attributed to the incomplete pyrolysis of PP at low temperatures and low pressures. Additionally, the PP particles remaining in the PPW improve the high-temperature elasticity of the binder [43]. As shown in Figure 7d, the four PPWs, **380-0.5**, **400-0.1**, **400-0.5**, and **400-1.0**, decreased the values of δ at low or high temperatures. A PPW of **380-1.0** can improve slightly the elasticity of MA. Furthermore, the influence of **380-1.0** on δ values yields a linear relationship with temperature, which is consistent with Sasobit [44].

The rutting factor (RF, G*/sinδ) is related to the ability of asphalt binders to resist rutting [39]. As shown in Figure 7c, the RFs of **A**, **B**, **C**, and **D** are much larger than that of MA, thus suggesting that the addition of the corresponding four PPWs improved the rutting resistance of MA. The PPW values of **380-0.5**, **380-1.0**, **400-0.1**, **400-0.5**, **400-1.0** slightly decreased the RF of MA. However, according to the superpave specification (AASHTO: MP1), the RF value should be at least 1.0 kPa for an asphalt binder at the maximum pavement design temperature [39]. The results in Figure 7f indicate that the RF values of **E**, **F**, **G**, **H**, and **I** are all greater than 1.0 kPa at a temperature of 64 °C. The value of the **380-1.0** modified asphalt **F** is the highest among the five PPWA, indicating the best resistance against rutting ability.

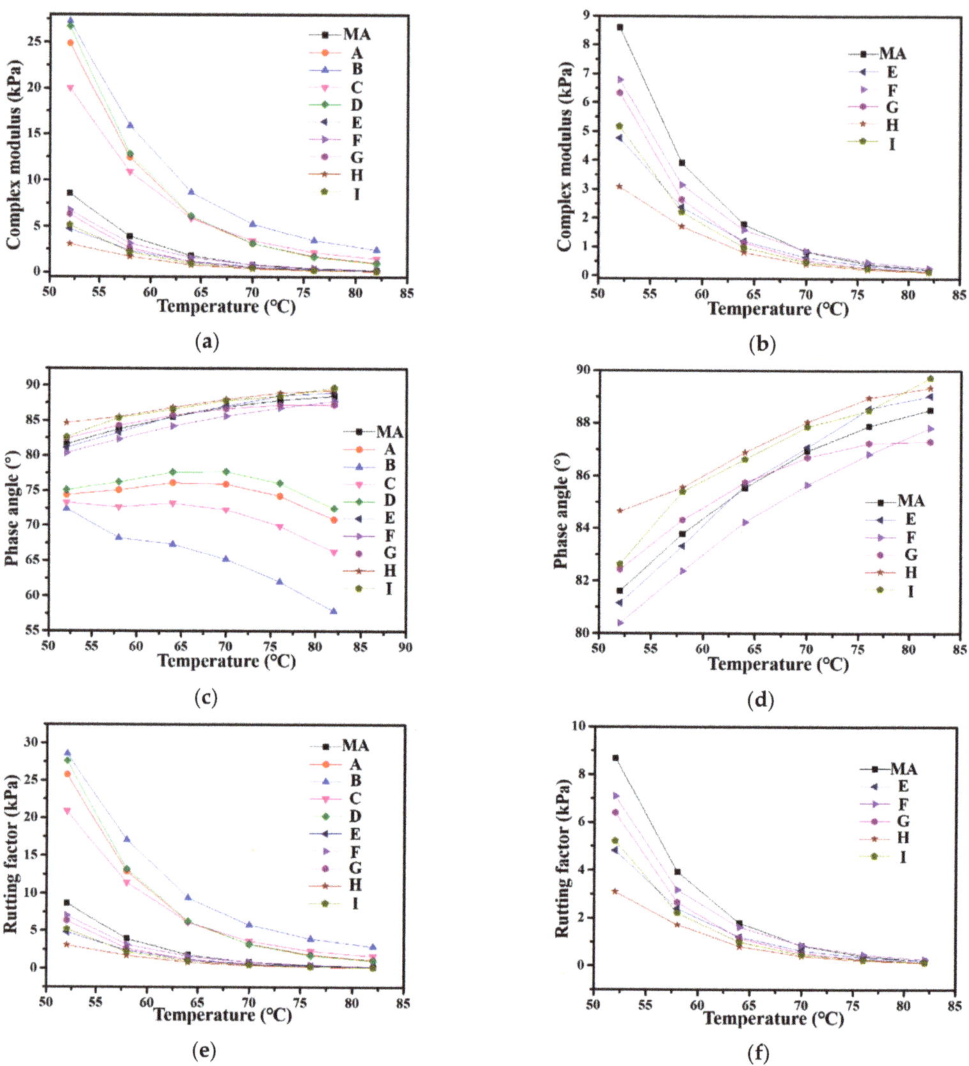

Figure 7. Rheological properties of MA and PPWA: (**a**,**b**) Complex modulus; (**c**,**d**) Phase angle; (**e**,**f**) Rutting factor.

3.6. Adhesion Properties

By comparing the nine PPWA, it can be seen that the PPW 380-1.0 has the greatest potential as a WMa. However, the use of wax-based WMa is still limited due to the concern regarding the adhesion property of modified asphalt [11,45,46]. In this study, the adhesion property of MA and **F** was investigated based on SFE theory to ensure the workability of F asphalt compared to matrix asphalt. The limestone was used as the test aggregate. Adhesion energy (ΔG_{AS}) and de-bonding energy (ΔG_{ASW}) were calculated from Equations (3) and (4), respectively.

$$\Delta G_{AS} = 2\left(\sqrt{\gamma_A^{LW}\gamma_S^{LW}} + \sqrt{\gamma_A^+\gamma_S^-} + \sqrt{\gamma_A^-\gamma_S^+} \right) \tag{3}$$

$$\Delta G_{ASW} = -2 \left[\begin{array}{l} \sqrt{\gamma_A^{LW}\gamma_W^{LW}} + \sqrt{\gamma_S^{LW}\gamma_W^{LW}} - \sqrt{\gamma_A^{LW}\gamma_S^{LW}} - \gamma_W^{LW} \\ + \sqrt{\gamma_W^+}\left(\sqrt{\gamma_A^-} + \sqrt{\gamma_S^-} - \sqrt{\gamma_W^-}\right) \\ + \sqrt{\gamma_W^-}\left(\sqrt{\gamma_A^+} + \sqrt{\gamma_S^+} - \sqrt{\gamma_W^+}\right) \\ - \sqrt{\gamma_A^+\gamma_S^-} - \sqrt{\gamma_A^-\gamma_S^+} \end{array} \right] \tag{4}$$

where $\gamma_A^{LW}/\gamma_S^{LW}/\gamma_W^{LW}$, $\gamma_A^+/\gamma_S^+/\gamma_W^+$, $\gamma_A^-/\gamma_S^-/\gamma_W^-$ are the SFE nonpolar component, polar acid component, and polar base component of asphalt/aggregate/water.

The Compatibility Ratio (CR) is used to evaluate the moisture susceptibility of asphalt. A higher value of CR implies a lower release of free energy in the presence of moisture [47]. CR was calculated as

$$CR = \frac{\Delta G_{AS}}{|\Delta G_{ASW}|} \tag{5}$$

The SFE parameters of the probe liquids and aggregate are shown in Tables 2 and 3. The SFE components of limestone have been measured by Luo et al. [31]. The contact angles measured with each probe liquids are given in Table 4. The SFE parameters of asphalt can be calculated according to the Young–Dupre equation, and the results are given in Table 5. The calculated results of ΔG_{AS}, $|\Delta G_{ASW}|$ and CR are shown in Figure 8. The results indicate that PPW **380-1.0** can increase the adhesion property and moisture susceptibility of MA. Wasiuddin et al. evaluated the adhesion property of Sasobit-modified asphalt using the SFE method [48]. The result showed that Sasobit can increase the cohesive strength slightly. From this point of this view, the PPW of **380-1.0** is similar to Sasobit.

Table 2. SFE components of the probe liquids at 20 °C.

Probe Liquids	SFE Components (mJ/m^2)				
	γ^{LW}	γ^+	γ^-	γ^{AB} [1]	γ [2]
Water	21.80	25.50	25.50	51.00	72.80
Formamide	39.00	1.92	39.60	19.00	58.00
Ethylene glycol	29.00	1.92	47.00	19.00	48.00

[1] γ^{AB} is polar acid base component; [2] γ is total SFE.

Table 3. SFE components of limestone (mJ/m^2).

Aggregate	SFE Components (mJ/m^2)				
	γ^{LW}	γ^+	γ^-	γ^{AB}	γ
Limestone	143.22	0.0023	393.68	1.89	145.11

Table 4. Contact angles (θ) and coefficient of variance (CV%) of MA and PPWA with three probe liquids.

Asphalt	Water		Formamide		Ethylene Glycol	
	θ (°)	CV (%)	θ (°)	CV (%)	θ (°)	CV (%)
MA	106.68	2.10	95.06	1.70	93.1	2.66
F	133.10	1.97	89.14	1.93	89.56	1.60

Table 5. SFE parameters of MA and F.

Asphalt	SFE Components (mJ/m^2)				
	γ^{LW}	γ^+	γ^-	γ^{AB}	γ
MA	20.08	0.42	2.71	2.15	22.23
F	33.44	0.25	6.59	2.55	35.98

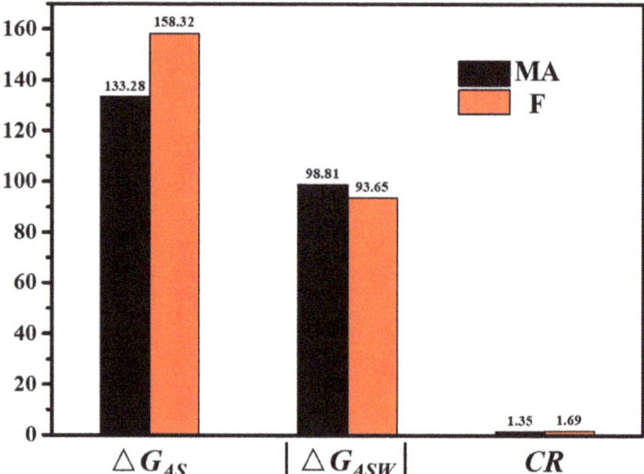

Figure 8. The ΔG_{AS}, $|\Delta G_{ASW}|$ and CR of MA and F.

3.7. Crystallinity and Composition of 380-1.0

The crystallinity of PPW 380-1.0 was measured by optical microscopy, as shown in Figure 9a. It can be seen that there are lots of crystals dispersed in PPW, which may be the reason for increasing the high-temperature performance for matrix asphalt [49]. Figure 9b shows the FTIR spectrum of 380-1.0. The absorption peaks at 2953, 2918, 1462, and 1377 cm^{-1} correspond to the -CH$_2$ asymmetric and symmetric stretching bands. The peaks at 1462 and 1377 cm^{-1} correspond to the CH$_3$ scissoring and symmetrical bending vibration. The peak at 876 cm^{-1} corresponds to the bending vibration of -CH=CH-. The FTIR results indicate that PPW 380-1.0 is mainly composed of alkanes and alkenes.

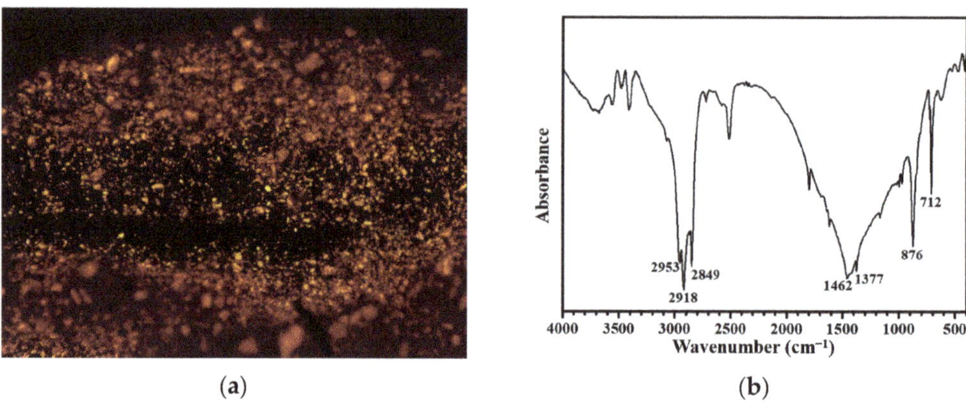

(a) (b)

Figure 9. Microphotograph (**a**) and FTIR spectrum of PPW 380-1.0 (**b**).

3.8. FTIR and DSC of PPWA F

The FTIR spectra of PPWA F and neat asphalt are shown in Figure 10a. The position of the main absorption bands of F and neat asphalt are not significantly different, which indicates that no obvious chemical reaction occurred in the modified bitumen after the addition of PPW. There are two significant exothermic peaks in the DSC curve of PPWA F. The peak at 20 °C corresponds to the phase translation of small molecules in modified asphalt. The peak at 100 °C can correspond to the melting of PPW in modified asphalt, because the melting point of matrix asphalt is generally higher than 150 °C. In addition, the T_g of PPWA F was also obtained from the DSC curve. The Tg of F is higher than that of MA, indicating that the PPW reduced the low-temperature performance of asphalt [50]. The result is consistent with the ductility result.

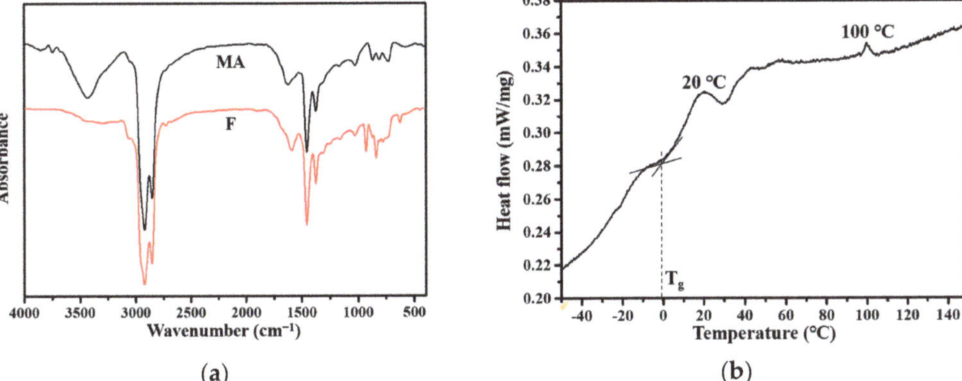

Figure 10. The FTIR spectrum (**a**) and DSC curve (**b**) of F.

4. Conclusions

In summary, nine PPW were prepared via the pyrolysis approach at different temperature and pressure in a pressure reactor. The properties of PPW modified asphalt were characterized through penetration, softening point, ductility, viscosity and rheological tests. Moreover, the adhesion property of PPW **380-1.0** modified asphalt was investigated using the SFE method based on contact angle measurements. The main conclusions are as follows:

(1) The softening point and ductility results show that the PPW prepared in this study can increase the high-temperature stability while decreasing the low-temperature flexibility of the asphalt.

(2) The penetration and rotational viscosity results indicate that PPW of **380-0.5, 380-1.0, 400-0.5,** and **400-1.0** can reduce the viscosity of asphalt and exhibit the potential to be used as WMa. The corresponding four modified PPWAs have a PI value in the range of −1 to +1, thus suggesting that the addition of the four PPWs does not affect the temperature susceptibility.

(3) The DSR results show that the addition of **360-0.1, 360-0.5, 360-1.0,** and **380-0.1** improves the flow deformation resistance and rutting resistance of the asphalt binder, whereas the other five PPWs did the opposite.

(4) The PPW **380-1.0** reduced the viscosity of asphalt by 48.6% at 135 °C, and the corresponding PPWA **F** exhibited favorable elasticity and rutting resistance characteristics. In addition, **380-0.1** improved the adhesion property and moisture susceptibility of matrix asphalt, which was measured based on the SFE theory. The PPW **380-1.0** can be considered a promising warm mix additive for asphalt binders.

(5) This study demonstrated that the temperature and pressure for the pyrolysis of PP play a synergistic role in the production of PPW-based WMa. The conditions of $380 \leq T \leq 400\ °C$ and $0.5 \leq p \leq 1.0$ MPa should be suitable.

The results obtained in this study show that the PPW prepared from waste PP plastic by the pyrolysis method can be used as WMa. Further research on the performance of PPW modified asphalt mixture is required. Furthermore, according to this work and previous research, the crystalline wax may have a negative influence on the low-temperature performance of asphalt binder [51,52]. So, the research focus is on developing the PPW WMa that can improve the low-temperature performance of asphalt, which should be carried out in the future.

Author Contributions: Conceptualization, G.Z. and C.L.; methodology, C.L.; validation, C.L., H.W. and W.Z.; formal analysis, T.L.; investigation, L.J., R.L., Q.L., Y.C. and D.Z.; data curation, W.Z.; writing—original draft preparation, G.Z.; writing—review and editing, C.L.; supervision, T.L.; funding acquisition, T.L. and C.L. All authors have read and agreed to the published version of the manuscript.

Funding: This research was funded by the Basic Science and Frontier Technology Research Program of Chongqing, grant number cstc2017jcyjAX0310 and Science and Technology Research Program of Chongqing Municipal Education Commission, grant number KJ1705124.

Institutional Review Board Statement: Not applicable.

Informed Consent Statement: Not applicable.

Data Availability Statement: Not applicable.

Conflicts of Interest: The authors declare no conflict of interest.

References

1. Rubio, M.C.; Martínez, G.; Baena, L.; Moreno, F. Warm mix asphalt: An overview. *J. Clean. Prod.* **2012**, *24*, 76–84. [CrossRef]
2. Hettiarachchi, C.; Hou, X.; Wang, J.; Xiao, F. A comprehensive review on the utilization of reclaimed asphalt material with warm mix asphalt technology. *Constr. Build. Mater.* **2019**, *227*, 117096. [CrossRef]
3. Cheraghian, G.; Falchetto, A.C.; You, Z.; Chen, S.; Kim, Y.S.; Westerhoff, J.; Moon, K.H.; Wistuba, M.P. Warm mix asphalt technology: An up to date review. *J. Clean. Prod.* **2020**, *268*, 122128. [CrossRef]
4. Sukhija, M.; Saboo, N. A comprehensive review of warm mix asphalt mixtures-laboratory to field. *Constr. Build. Mater.* **2021**, *274*, 121781. [CrossRef]
5. Liu, J.; Saboundjian, S.; Li, P.; Connor, B.; Brunette, B. Laboratory evaluation of sasobit-modified warm-mix asphalt for alaskan conditions. *J. Mater. Civ. Eng.* **2011**, *23*, 1498–1505. [CrossRef]
6. Zhao, G.; Guo, P. Workability of sasobit warm mixture asphalt. *Energy Proc.* **2012**, *16*, 1230–1236. [CrossRef]
7. Jamshidi, A.; Hamzah, M.O.; You, Z. Performance of warm mix Asphalt containing Sasobit: State-of-the-art. *Constr. Build. Mater.* **2013**, *38*, 530–553. [CrossRef]
8. Qin, Q.; Farrar, M.J.; Pauli, A.T.; Adams, J.J. Morphology, thermal analysis and rheology of Sasobit modified warm mix asphalt binders. *Fuel* **2014**, *115*, 416–425. [CrossRef]
9. Gong, J.; Han, X.; Su, W.; Xi, Z.; Cai, J.; Wang, Q.; Li, J.; Xie, H. Laboratory evaluation of warm-mix epoxy SBS modified asphalt binders containing Sasobit. *J. Build. Eng.* **2020**, *32*, 101550. [CrossRef]
10. Wen, Y.; Wang, Y.; Zhao, K.; Chong, D.; Huang, W.; Hao, G.; Mo, S. The engineering, economic, and environmental performance of terminal blend rubberized asphalt binders with wax-based warm mix additives. *J. Clean. Prod.* **2018**, *184*, 985–1001. [CrossRef]
11. Nakhaei, M.; Naderi, K.; Nasrekani, A.A.; Timm, D.H. Moisture resistance study on PE-wax and EBS-wax modified warm mix asphalt using chemical and mechanical procedures. *Constr. Build. Mater.* **2018**, *189*, 882–889. [CrossRef]
12. Ameri, M.; Afshin, A.; Shiraz, M.E.; Yazdipanah, F. Effect of wax-based warm mix additives on fatigue and rutting performance of crumb rubber modified asphalt. *Constr. Build. Mater.* **2020**, *262*, 120882. [CrossRef]
13. Vaitkus, A.; Cygas, D.; Laurinavicius, A.; Vorobjovas, V.; Perveneckas, Z. Influence of warm mix asphalt technology on asphalt physical and mechanical properties. *Constr. Build. Mater.* **2016**, *112*, 800–806. [CrossRef]
14. Burange, A.S.; Gawande, M.B.; Lam, F.L.Y.; Jayaram, R.V.; Luque, R. Heterogeneously catalyzed strategies for the deconstruction of high density polyethylene: Plastic waste valorisation to fuels. *Green Chem.* **2015**, *17*, 146–156. [CrossRef]
15. Hou, Q.; Zhen, M.; Qian, H.; Nie, Y.; Bai, X.; Xia, T.; Rehman, M.L.U.; Li, Q.; Ju, M. Upcycling and catalytic degradation of plastic wastes. *Cell Report. Phys. Sci.* **2021**, *2*, 100514. [CrossRef]
16. Aisien, E.T.; Otuya, I.C.; Aisien, F.A. Thermal and catalytic pyrolysis of waste polypropylene plastic using spent FCC catalyst. *Environ. Technol. Innov.* **2021**, *22*, 101455. [CrossRef]
17. Luo, W.; Hu, Q.; Fan, Z.; Wan, J.; He, Q.; Huang, S.; Zhou, N.; Song, M.; Zhang, J.; Zhou, Z. The effect of different particle sizes and HCl-modified kaolin on catalytic pyrolysis characteristics of reworked polypropylene plastics. *Energy* **2020**, *213*, 119080. [CrossRef]
18. Canopoli, L.; Coulon, F.; Wagland, S.T. Degradation of excavated polyethylene and polypropylene waste from landfill. *Sci. Total Environ.* **2020**, *698*, 134125. [CrossRef]
19. Hamad, K.; Kaseem, M.; Deri, F. Recycling of waste from polymer materials: An overview of the recent works. *Polym. Degrad. Stabil.* **2013**, *98*, 2801–2812. [CrossRef]
20. Ding, H.; Hesp, S.A.M. Balancing the use of wax-based warm mix additives for improved asphalt compaction with long-term pavement performance. ACS Sustain. *Chem. Eng.* **2021**, *9*, 7298–7305.
21. Ding, H.; Liu, H.; Qiu, Y.; Rahman, A. Effects of crystalline wax and asphaltene on thermoreversible aging of asphalt binder. *Int. J. Pavement. Eng.* **2021**, *2022*, 1931199. [CrossRef]

22. Xu, S.; Cao, B.; Uzoejinwa, B.B.; Odey, E.A.; Wang, S.; Shang, H.; Li, C.; Hu, Y.; Wang, Q.; Nwakaire, J.N. Synergistic effects of catalytic co-pyrolysis of macroalgae with waste plastics. *Process Saf. Environ. Protec.* **2020**, *137*, 34–48. [CrossRef]
23. Chen, W.; Jin, K.; Wang, N.L. The use of supercritical water for the liquefaction of polypropylene into oil. *ACS Sustain. Chem. Eng.* **2019**, *7*, 3749–3758. [CrossRef]
24. Tennakoon, A.; Wu, X.; Paterson, A.L.; Patnaik, S.; Pei, Y.; LaPointe, A.M.; Ammal, S.C.; Hackler, R.A.; Heyden, A.; Slowing, I.I.; et al. Catalytic upcycling of high-density polyethylene via a processive mechanism. *Nat. Catal.* **2020**, *3*, 893–901. [CrossRef]
25. *ASTM Standard D5*; Standard Test Method for Penetration of Bituminous Materials. ASTM: West Conshohocken, PA, USA, 2013.
26. *ASTM Standard D36*; Standard Test Method for Softening Point of Bitumen (Ring-and-Ball Apparatus). ASTM: West Conshohocken, PA, USA, 2006.
27. *ASTM Standard D113*; Standard Test Method for Standard Test Method for Ductility of Bituminous Materials. ASTM: West Conshohocken, PA, USA, 2007.
28. *ASTM D4402*; Standard Test Method for Viscosity Determination of Asphalt at Elevated Temperatures Using a Rotational Viscometer. ASTM: West Conshohocken, PA, USA, 2006.
29. *AASHTO Standard M320*; Standard Specification for Performance-Graded Asphalt Binder. AASHTO: Washington, DC, USA, 2010.
30. Cheng, D.; Little, D.N.; Lytton, R.L.; Holste, J.C. Surface energy measurement of asphalt and its application to predicting fatigue and healing in asphalt mixtures. *Transport. Res. Rec.* **2002**, *1810*, 44–53. [CrossRef]
31. Luo, R.; Zheng, S.; Zhang, D.; Tong, C.; Feng, G. Evaluation of adhesion property in asphalt-aggregate systems based on surface energy theory. *China J. Highw. Transp.* **2017**, *30*, 209–214.
32. Bionghi, R.; Shahraki, D.; Ameri, M.; Karimi, M.M. Correlation between bond strength and surface free energy parameters of asphalt binder-aggregate system. *Constr. Build. Mater.* **2021**, *303*, 124487. [CrossRef]
33. Chen, M.; Leng, B.; Wu, S.; Sang, Y. Physical, chemical and rheological properties of waste edible vegetable oil rejuvenated asphalt binders. *Constr. Build. Mater.* **2014**, *66*, 286–298. [CrossRef]
34. Behnood, A. A review of the warm mix asphalt (WMA) technologies: Effects on thermo-mechanical and rheological properties. *J. Clean. Prod.* **2020**, *259*, 120817. [CrossRef]
35. Al-Omari, A.A.; Khedaywi, T.S.; Khasawneh, M.A. Laboratory characterization of asphalt binders modified with waste vegetable oil using SuperPave specifications. *Int. J. Pavement Res. Technol.* **2018**, *11*, 68–76. [CrossRef]
36. Muhammad, J.; Peng, T.; Zhang, W.; Cheng, H.; Waqas, H.; Abdul, S.; Chen, K.; Zhou, Y. Moisture susceptibility and fatigue performance of asphalt binder modified by bone glue and coal fly ash. *Constr. Build. Mater.* **2021**, *308*, 125135. [CrossRef]
37. Wang, F.; Gao, N.; Magdziarz, A.; Quan, C. Co-pyrolysis of biomass and waste tires under high-pressure two-stage fixed bed reactor. *Bioresour. Technol.* **2022**, *344*, 126306. [CrossRef] [PubMed]
38. Kataware, A.V.; Singh, D. Evaluating effectiveness of WMA additives for SBS modified binder based on viscosity, Superpave PG, rutting and fatigue performance. *Constr. Build. Mater.* **2017**, *146*, 436–444. [CrossRef]
39. Ghuzlan, K.A.; Al Assi, M.O. Sasobit-modified asphalt binder rheology. *J. Mater. Civ. Eng.* **2017**, *29*, 04017142. [CrossRef]
40. Feitosa, J.P.M.; de Alencar, A.E.V.; Filho, N.W.; de Souza, J.R.R.; Branco, V.T.F.C.; Soares, J.B.; Soares, S.A.; Ricardo, N.M.P.S. Evaluation of sun-oxidized carnauba wax as warm mix asphalt additive. *Constr. Build. Mater.* **2016**, *115*, 294–298. [CrossRef]
41. Iwański, M.; Mazurek, G. Optimization of the synthetic wax content on example of bitumen 35/50. *Procedia Eng.* **2013**, *57*, 414–423. [CrossRef]
42. Fazaeli, H.; Amini, A.A.; Moghadasnejad, F.; Behbahani, H. Rheological properties of bitumen modified with a combination of FT paraffin wax (Sasobit) and other additives. *J. Civ. Eng. Manag.* **2016**, *22*, 135–145. [CrossRef]
43. Zhao, X.; Rahman, M.U.; Dissanayaka, T.; Gharagheizi, F.; Lacerda, C.; Senadheera, S.; Hedden, R.C.; Christopher, G.F. Rheological behavior of a low crystallinity polyolefin-modified asphalt binder for flexible pavements. *Case Stud. Constr. Mater.* **2021**, *15*, e00640. [CrossRef]
44. Behnood, A.; Karimi, M.M.; Cheraghian, G. Coupled effects of warm mix asphalt (WMA) additives and rheological modifiers on the properties of asphalt binders. *Clean. Eng. Technol.* **2020**, *1*, 100028. [CrossRef]
45. Lee1, H.D.; Glueckert, T.; Ahmed, T.; Kim, Y.; Baek, C.; Hwang, S. Laboratory evaluation and field implementation of polyethylene wax-based warm mix asphalt additive in USA. *Int. J. Pavement Res. Technol.* **2013**, *6*, 547–553.
46. Ahmed, T.A.; Lee, H.D.; Williams, R.C. Using a modified asphalt bond strength test to investigate the properties of asphalt binders with poly ethylene wax-based warm mix asphalt additive. *Int. J. Pavement Res. Technol.* **2018**, *11*, 28–37. [CrossRef]
47. Kakar, M.R.; Hamzah, M.O.; Akhtar, M.N.; Woodward, D. Surface free energy and moisture susceptibility evaluation of asphalt binders modified with surfactant-based chemical additive. *J. Clean. Prod.* **2016**, *112*, 2342–2353. [CrossRef]
48. Wasiuddin, N.; Saltibus, N.; Mohammad, L. Effects of a Wax-based warm mix additive on cohesive strengths of asphalt binders. In Proceedings of the Transportation and Development Institute Congress, Chicago, IL, USA, 13–16 March 2011; pp. 528–537.
49. Ding, H.; Hesp, S.A.M. Quantification of crystalline wax in asphalt binders using variabletemperature Fourier-transform infrared spectroscopy. *Fuel* **2020**, *277*, 118220. [CrossRef]
50. Ling, T.; Lu, Y.; Zhang, Z.; Li, C.; Oeser, M. Value-added application of waste rubber and waste plastic in asphalt binder as a multifunctional additive. *Materials* **2019**, *12*, 1280. [CrossRef] [PubMed]

51. Ding, H.; Zhang, H.; Zheng, X.; Zhang, C. Characterization of crystalline wax in asphalt binder by X-ray diffraction. *Road Mater. Pavement Des.* **2022**. [CrossRef]
52. Ding, H.; Zhang, H.; Zhang, H.; Liu, D.; Qiu, Y.; Hussain, A. Separation of wax fraction in asphalt binder by an improved method and determination of its molecular structure. *Fuel* **2022**, *322*, 124081. [CrossRef]

materials

MDPI

Article

Pavement Performance Investigation of Asphalt Mixtures with Plastic and Basalt Fiber Composite (PB) Modifier and Their Applications in Urban Bus Lanes Using Statics Analysis

Xueyang Jiu [1], Peng Xiao [1,2], Bo Li [1,2,*], Yu Wang [1] and Aihong Kang [1,2]

[1] College of Civil Science and Engineering, Yangzhou University, Yangzhou 225127, China; 006713@yzu.edu.cn (X.J.); xpyzu@163.com (P.X.); yuwang4012@126.com (Y.W.); kahyzu@163.com (A.K.)
[2] Research Center for Basalt Fiber Composite Construction Materials, Yangzhou 225127, China
* Correspondence: libo@yzu.edu.cn; Tel.: +86-188-5272-9286

Abstract: A new type of plastic and basalt fiber composite (PB) modifier, which is composed of waste plastic and basalt fiber using a specific process, was used for bus lanes to address severe high-temperature deformation diseases due to the heavy loads of buses. The dense gradations of asphalt mixture with a nominal maximum aggregate size of 13.2 mm (AC-13) and 19 mm (AC-20) were selected to fabricate asphalt mixtures. The impact of the modifier PB on the high-temperature rutting resistance, low-temperature crack resistance, and water damage resistance was investigated experimentally. The experimental results showed that adding the modifier PB could enhance the rutting resistance and water damage resistance of asphalt mixtures significantly while maintaining the low-temperature crack resistance. Then, PB-modified asphalt mixtures of AC-13 and AC-20 were employed into a typical pavement structure of a bus lane in Yangzhou city, China, and three types of designed pavement structures were proposed. On this basis, statics analyses of all of the designed structures were performed using the finite element method. The statics analyses revealed that, compared with the standard axle load, the actual over-loaded axle made the pavement structure of the bus lane suffer a 30% higher stress and vertical deformation, leading to accelerated rutting damage on the bus lanes. The addition of the modifier PB could make the pavement structure stronger and compensate for the negative effect caused by the heavy axle load. These findings can be used as a reference for the pavement design of urban bus lanes.

Keywords: asphalt mixture; plastic and basalt fiber composite modifier; urban bus lane; pavement performance; statics analysis

Citation: Jiu, X.; Xiao, P.; Li, B.; Wang, Y.; Kang, A. Pavement Performance Investigation of Asphalt Mixtures with Plastic and Basalt Fiber Composite (PB) Modifier and Their Applications in Urban Bus Lanes Using Statics Analysis. *Materials* **2023**, *16*, 770. https://doi.org/10.3390/ma16020770

Academic Editor: Gilda Ferrotti

Received: 6 December 2022
Revised: 6 January 2023
Accepted: 9 January 2023
Published: 12 January 2023

1. Introduction

More than 230 cities in China have created priority lanes or bus-only lanes, according to the preferential policies and measures providing priority for the development of public transport, to alleviate urban traffic congestion [1]. However, there is a lack of guidance for the design of the bus-only lanes. The current research on urban bus lanes is mainly focused on the route design, traffic flow calculation, traffic organization optimization, etc. The bus-running characteristics are seldom taken into account for the pavement structure design [2–4]. It is reported that the tire ground pressure of a bus normally exceeds 0.7 MPa, which is the standard axle load for urban pavement structure design in China according to the specification of CJJ 37-90. Subsequently, pavement distresses such as rutting, upheaval, etc., appear frequently on bus-only lanes due to the insufficient structure design [5]. Therefore, new material selection and (or) design structures of bus lanes are desired.

Waste plastics are considered as an effective modifier for asphalt since not only can they strengthen the pavement performance of asphalt mixtures [6–8], but they can also provide a solution to the environmental pollution caused by waste plastics [9,10]. Nouali et al. [11]

examined the suitability of using plastic bag waste as a bitumen modifier in order to improve the behavior of asphalt mixtures. The results show that adding plastic waste to the pure bitumen allows for reducing the void content of the mix and substantially increasing its stiffness modulus and water resistance. In Ranieri's [12] study, the rut depth values were reduced by more than 30% for waste HDPE-modified asphalt mixtures compared with conventional asphalt mixtures. Padhan's and Shahane's research [13,14] indicates that plastic gives an increase in stability, compressive strength, and split tensile strength compared to the conventional SMA/AC mixtures. It was reported that waste plastics modifiers could improve the adhesion between asphalt/aggregates and enhance the high-temperature performance of asphalt pavements [15,16]. However, adding a waste plastics modifier would also make the asphalt mixture prone to cracking, leading to a deterioration in the low-temperature performance [17,18]. Therefore, the composite modification method becomes more popular when waste plastics are used as the modifier for asphalt pavements.

Basalt fiber (BF), as a new, environmentally friendly mineral fiber [19,20], has been widely used in asphalt pavement due to its unique advantages: a wide working temperature range, excellent mechanical properties, chemical stability, anti-aging and thermal insulation properties, etc., [21]. In Celauro's and Hui's research [22,23], basalt-fiber-modified asphalt mixtures show a better high-temperature performance with reference to permanent deformation resistance when compared with the traditional mixture. Moreover, the chopped basalt fiber is combined with the asphalt well and distributed in a three-dimensional network structure in asphalt mixtures, which can reinforce the low-temperature performance of asphalt mixture significantly [24–26]. Zhu et al. [27,28] investigated the anti-fatigue property of basalt-fiber-reinforced asphalt mixture. The results present that, after adding basalt fiber, the cumulative dissipation energy is greatly improved, and then the fatigue life of the asphalt mixture is increased. The impact of the dosage and fiber length on the asphalt pavement performance was also investigated by laboratory tests, and the optimum fiber dosage and applicable fiber length for the dense graded asphalt mixture of AC-13 were recommended as 0.2~0.4% and 6 mm [29–32]. In addition, relevant specifications have also been issued to guide the application of basalt fibers in asphalt pavements [33,34]. It was pointed out that the addition of basalt fiber could enhance the high-temperature stability and low-temperature property of asphalt mixtures; in particular, the fatigue performance can be remarkably prolonged. However, the improvement effect on the high-temperature property of asphalt mixtures is inferior to that of waste plastics. If the addition of waste plastics and basalt fibers can give full play to their respective advantages in different performances of asphalt mixtures, then the waste-plastic–basalt-fiber-modified asphalt mixture can be well used to address the common pavement distresses of bus lanes.

Based on the self-developed composite modifier PB for an urban bus lane, the objective of this study was to investigate the high-temperature performance, low-temperature property, moisture stability, and dynamic modulus of PB-modified asphalt mixtures. In addition, taking the bus lanes in Yangzhou city as an example, a statics analysis of the bus lane structure was conducted using the finite element method. The findings of this study can be used as a reference for the pavement design of urban bus lanes.

2. Raw Materials and Experimental Methods

2.1. Raw Materials

2.1.1. Modifier PB

The modifier PB was made of waste plastic and basalt fiber with a weight ratio of 1:1, and manufactured using a specific process. The main source of waste plastics is waste polythene, and the components and contents are shown in Table 1. Basalt fiber is produced by Jiangsu Tianlong Basalt Fiber Co., Ltd., Yangzhou, China, and the properties are listed in Table 2. The modifier PB has colorless transparent surface and a brown metallic luster inside as flat solid particles, as shown in Figure 1c, and can be stored at ambient or low temperature. As PB contains basalt fiber component, according to the previous finding that the optimum basalt fiber length for asphalt mixtures is 6 mm [24,26], the length of the modifier PB was determined to be 6 mm in this study. Its technical indicators are shown in Table 3.

Table 1. Compositions and the contents of waste plastic.

Resin	Polythene (PE)	Ethylene Vinyl Acetate Copolymer (EVA)	Coupling Agent	Auxiliary	Melting Point/°C
10~20%	65~75%	10%	2%	3%	90~100

Table 2. The properties of basalt fiber.

Index	Fracture Strength/MPa	Elongation at Break/%	Elastic Modulus/GPa
Value	2500–5000	2.69	90–110
Requirements in T/CHTS 10016	≥2000	≥2.1	≥80

(a)   (b)   (c)

Figure 1. Illustration of modifier PB. (**a**) Waste plastic synthetic particles; (**b**) basalt fiber; (**c**) modifier PB.

Table 3. Technical indicators of the modifier PB.

Project	Appearance	Particle Size/mm	Length/mm	Density/g·cm^{-3}
PB	External colorless transparent, internal brown, flat solid particles	2.5~3.5	6	1.82~1.86

2.1.2. Aggregates and Mineral Filler

Limestone was used as the aggregates in this paper. The mineral filler selected was limestone powder, and the technical properties are shown in Table 4.

Table 4. Technical properties of the mineral filler.

Index Items		Index Results	Index Requirements
Apparent density (g/cm^3)		2.714	≥2.50
Water content (%)		0.38	≤1.0
Appearance		No clumps	No clumps
Water affinity coefficient		0.60	<1
Size range (%)	<0.6 mm	100	100
	<0.15 mm	100	90–100
	<0.075 mm	92.2	75–100

2.1.3. Asphalt

Both base asphalt and styrene butadiene styrene (SBS)-modified asphalt, produced by Tongsha asphalt factory in Jiangsu province, were adopted in this study. The technical properties of the two types of asphalts are shown in Table 5.

Table 5. Technical properties of the two types of asphalts.

Properties	Penetration (25 °C)/0.1 mm	Penetration Index PI	Softening Point /°C	Ductility (5 cm/min)/cm	Viscosity (135 °C)/Pa·s
Base asphalt	71.2	−0.8	47.1	150 (15 °C)	0.92
SBS-modified asphalt	67	0.3	78	48 (5 °C)	1.8

2.2. Gradation Design

2.2.1. Gradation Curve

Two types of dense gradations of AC-13 and AC-20 that were widely used in the bus lanes in Yangzhou city (Jiangsu Province, China) were selected in this study. The gradation curves of the two types of asphalt mixtures were illustrated according to JTG F40 [35], as shown in Figures 2 and 3.

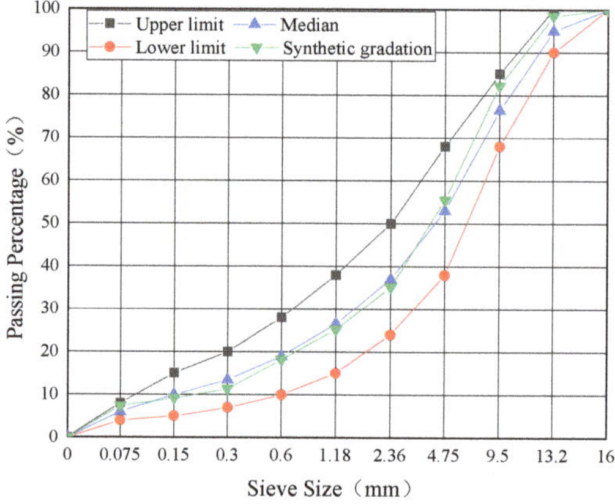

Figure 2. Gradation curve of AC-13.

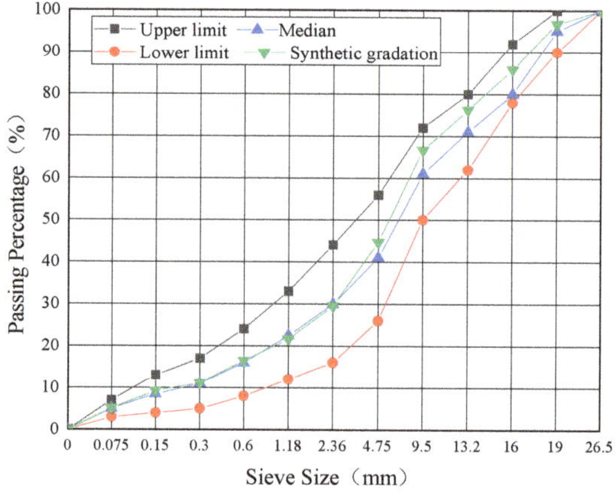

Figure 3. Gradation curve of AC-20.

2.2.2. Volumetric Properties of the Asphalt Mixtures

According to the difference in asphalt and modifier, three types of asphalt mixtures with a gradation of AC-13 were fabricated, called base asphalt + AC-13, base asphalt + AC-13 + PB, and SBS-modified asphalt + AC-13, respectively. The same three types of asphalt mixtures with a gradation of AC-20 were also fabricated. It is worth pointing out that the "dry mixing process" was adopted to fabricate the PB-modified asphalt mixtures, which means that the modifier PB was mixed with heated aggregates at a mixing temperature of 175 °C for 90 s before adding asphalt according to JTG E20-T0702 [36]. The dosage of the modifier PB in asphalt mixtures was 0.6% by the weight of aggregates. The related volumetric properties and the optimum asphalt content (OAC) of all six types of asphalt mixtures were determined by the Marshall design method in accordance with JTG F40 [35]. The results are shown in Table 6.

Table 6. Marshall test results of different asphalt mixtures.

Types of mixture	Optimum Asphalt Content (OAC)/%	Air Voids (VV)/%	Voids in Mineral Aggregate (VMA)/%	Voids Filled with Asphalt (VFA)/%	Marshall Stability/kN	Flow Value/0.1 mm
Base asphalt + AC-13	5.0	4.1	14.2	71.1	9.5	29.8
Base asphalt + AC-13 + PB	5.4	4.4	15.3	71.1	12.5	24.2
SBS-modified asphalt + AC-13	5.0	4.3	15.0	71.6	12.0	26.4
Base asphalt + AC-20	4.4	4.3	13.5	67.9	11.5	32.4
Base asphalt + AC-20 + PB	4.7	4.7	14.3	66.7	13.5	28.6
SBS-modified asphalt + AC-20	4.4	4.4	13.5	67.5	12.4	30.7

2.3. Test Methods

2.3.1. High-Temperature Stability Test

Wheel-tracking test and dynamic creep test were used to appraise the high-temperature stability of asphalt mixtures.

The wheel-tracking test was conducted in accordance with JTG E20-T0719 [36] under a test temperature of 60 °C and test tire ground pressure of 0.7 MPa. The dynamic stability (*DS*), relative deformation ratio (*RDR*), and comprehensive stability index (*CSI*) were adopted to assess the rutting resistance of asphalt mixtures, which were calculated by Equations (1)–(3), respectively. Generally speaking, higher *DS*, smaller *RDR*, and higher *CSI* values can guarantee a better rutting resistance of asphalt mixtures.

$$DS = \frac{(t_2 - t_1) \times N}{d_2 - d_1} \times C_1 \times C_2 \tag{1}$$

$$RDR = \frac{\Delta L}{D} \times 100\% \tag{2}$$

$$CSI = \frac{(t_2 - t_1) \times N}{(d_2 - d_1) \times d_1} \tag{3}$$

where: d_1 and d_2 are the deformation of the asphalt mixture at the running time t_1 (45 min), t_2 (60 min), mm; C_1 and C_2 are test coefficients, $C_1 = C_2 = 1.0$; N is the running speed of the wheel, usually 42 times/min; ΔL is the depth of rutting of the asphalt mixture under load at specific timing, mm; D is the total thickness of the specimen, mm.

In addition, a dynamic creep test was performed in accordance with the method described in NCHRP9-29. Three test temperatures of 40 °C, 50 °C, and 60 °C were selected. The test would end until the cumulative permanent strain reached 50,000 $\mu\varepsilon$ or the cumulative loading time reached 10,000 s. The test process and typical strain-loading time curve are shown in Figure 4. The flow number and creep rate were served to assess the high-temperature deformation performance of asphalt mixtures. As shown in Figure 4b, the accumulated strain can be divided into three stages: creep migration (Stage I), creep stability (Stage II), and creep failure (Stage III). Generally, the slope of the linear growth curve in Stage II is regarded as the creep rate, and the flection point between Stage II and Stage III is regarded as the flow number.

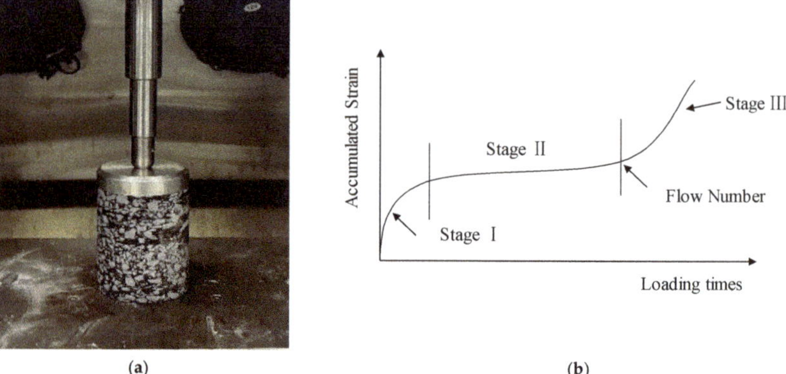

(a)   (b)

Figure 4. Dynamic creep test illustration. (**a**) Dynamic creep test process; (**b**) illustration of three stages.

2.3.2. Low-Temperature Performance Test

According to the Chinese test procedure JTG E20-T0715 [36], low-temperature bending beam test was adopted to explore the low-temperature property of asphalt mixtures. The loading speed was 50 mm/min and the test temperature was $-10\,^\circ$C. The flexural-tensile strength (R_B), failure strain (ε_B), and flexural stiffness modulus (S_B) were calculated according to Equations (4)–(6). Normally, the higher the flexural-tensile strain and lower flexural stiffness modulus are, the better the low-temperature cracking resistance of asphalt mixtures will be. The test process is illustrated in Figure 5.

$$R_B = \frac{3LP_B}{2bh^2} \tag{4}$$

$$\varepsilon_B = \frac{6hd}{L^2} \tag{5}$$

$$S_B = \frac{R_B}{\varepsilon_B} \tag{6}$$

where: b is the width of the specimen, mm; h and L are the height and span diameter of the specimen, mm; P_B is the maximum load on the specimen, N; d is the span deflection at failure point, mm.

Figure 5. Low-temperature bending beam test.

2.3.3. Water Stability Test

In accordance with the Chinese test procedure JTG E20-T0709 and T0729 [36], the water stabilities of asphalt mixtures were evaluated by immersion Marshall test and freeze–thaw splitting test. Residual stability (MS_0) and freeze–thaw splitting tensile strength ratio (TSR) were used to estimate the water damage resistance of asphalt mixtures, which are defined by Equations (7) and (8), respectively.

$$MS_0 = \frac{MS_1}{MS} \times 100 \tag{7}$$

where: MS_1 is the conditioned stability of the samples that endured hot water immersion (60 °C, 48 h), kN; MS is the unconditioned Marshall stability of the samples, kN.

$$TSR = \frac{R_{T2}}{R_{T1}} \times 100 \tag{8}$$

where: R_{T2} is the average value of the splitting strength of the conditioned samples, MPa; R_{T1} is the average value of the splitting strength of the unconditioned samples, MPa.

2.3.4. Dynamic Modulus Test

The dynamic modulus test was conducted according to the Chinese specification of JTG E20-T0738 [36]. Since dynamic modulus is temperature and frequency-related, four different test temperatures of 5 °C, 20 °C, 40 °C, and 55 °C were selected in this study. With respect to the frequency, it is reported that the frequency caused by the vehicle moving is related to its speed, pavement evenness, etc., [37]. In accordance with the data provided by the Yangzhou Passenger Transport Management Office, the average speed of the bus is approximately 30 km/h, which caused a corresponding load frequency of around 0.1 Hz [38]. Therefore, a load frequency of 0.1 Hz was adopted in this study.

For all of the above mentioned tests, the average values of three duplicate samples were used for the results analyses and discussions.

3. Results and Discussion

3.1. High-Temperature Stability

3.1.1. Wheel-Tracking Test Results

The results of the DS, RDR, and CSI of the wheel-tracking test are shown in Figure 6. It can be drawn from Figure 6 that the addition of modifier PB can improve the DS and CSI significantly, while reducing the RDR to some extent. As for the asphalt mixtures with AC-13 gradation, the DS of the PB-modified asphalt mixtures increased by 3.6 times, while the CSI increased by 8.5 times and RDR decreased by 4.7 percentage points compared with the base asphalt mixtures. Furthermore, compared with SBS-modified asphalt mixtures, the DS and CSI of the PB-modified asphalt mixtures also increased dramatically by 30.5% and 35.1%, respectively, even though the RDR remained at the same level. As for AC-20-graded asphalt mixtures, the same trends of the wheel-tracking test results could be observed when the modifier PB was used. Those findings indicate that the modifier PB possesses a superior capability to improve the high-temperature stability of asphalt mixtures, even much better than the SBS-modified asphalt. One reason could be that waste plastics can increase the viscosity of the asphalt binder, leading to an improvement in the stiffness modulus of asphalt mixtures [10,11]. The other reason could be that basalt fiber can form a three-dimensional network structure, which restricts the plastic deformation of asphalt mixtures and enhances the shear resistance of the mixtures at high temperatures [24,25].

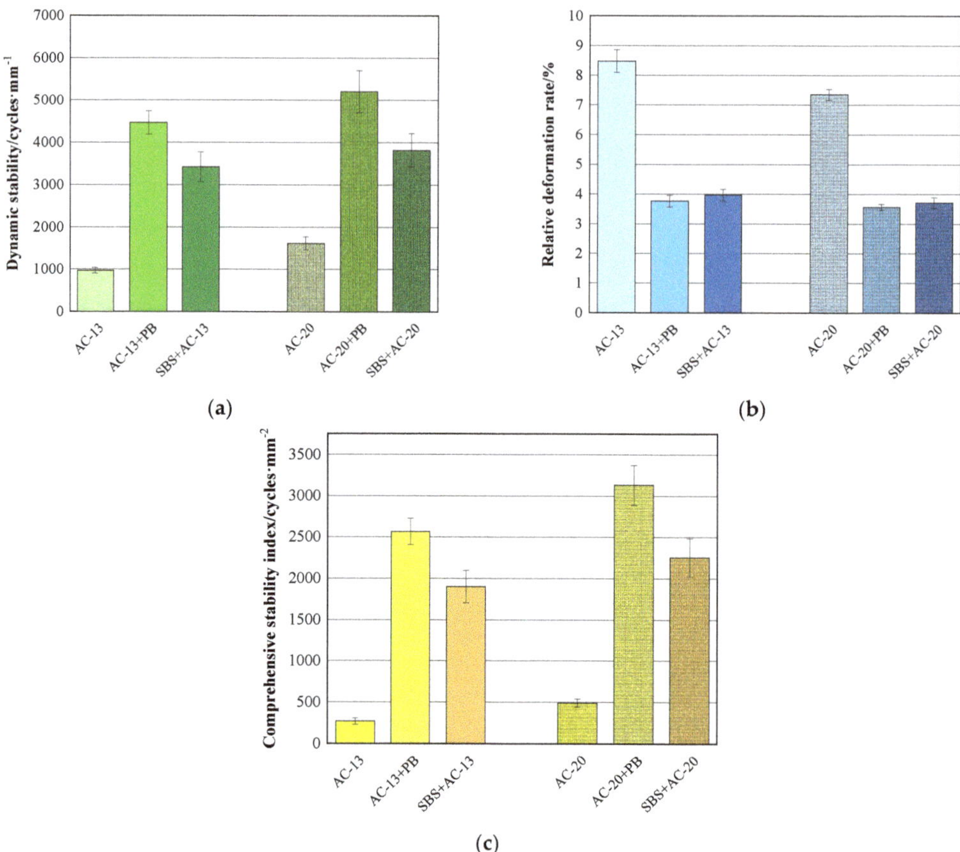

Figure 6. Results of wheel-tracking test. (**a**) Results of dynamic stability; (**b**) results of relative deformation rate; (**c**) results of comprehensive stability index.

3.1.2. Dynamic Creep Test Results

Figure 7 lists the results of the dynamic creep test. As shown in Figure 7, both the flow number and creep rate were temperature-dependent. The flow number values of all types of asphalt mixtures decreased with the increase in temperature, whereas the creep rate values presented an increasing trend. At the same temperature, for both AC-13 and AC-20 gradations, adding modifier PB could greatly enhance the flow number values of asphalt mixtures. The flow number values of different mixture gradations presented a similar trend in the following order: base asphalt mixture < SBS-modified asphalt mixture < base asphalt mixture with PB.

As for the asphalt mixtures with AC-13 gradation, the flow number values of PB-modified asphalt mixtures increased by 68.5%, 85.8%, and 89.8%, whereas the creep rate values decreased by 69.7%, 54.1%, and 47.9% at the three test temperatures of 40 °C, 50 °C, and 60 °C, respectively, when compared to those of base asphalt mixtures. Even compared with the SBS-modified asphalt mixtures, a positive improvement could also be observed in the flow number values and creep rate values of PB-modified asphalt mixtures. In addition, similar trends could be observed when modifier PB was used for the asphalt mixtures of AC-20. Compared with the base and SBS-modified asphalt mixtures, the flow number values of the PB-modified asphalt mixtures increased to as high as 92.1%, whereas the creep rate values reduced by 55.7% for the maximum.

These findings infer that adding modifier PB can significantly strengthen the permanent deformation resistance of asphalt mixtures in terms of creep deformation. The higher the temperature, the greater the enhancement achieved.

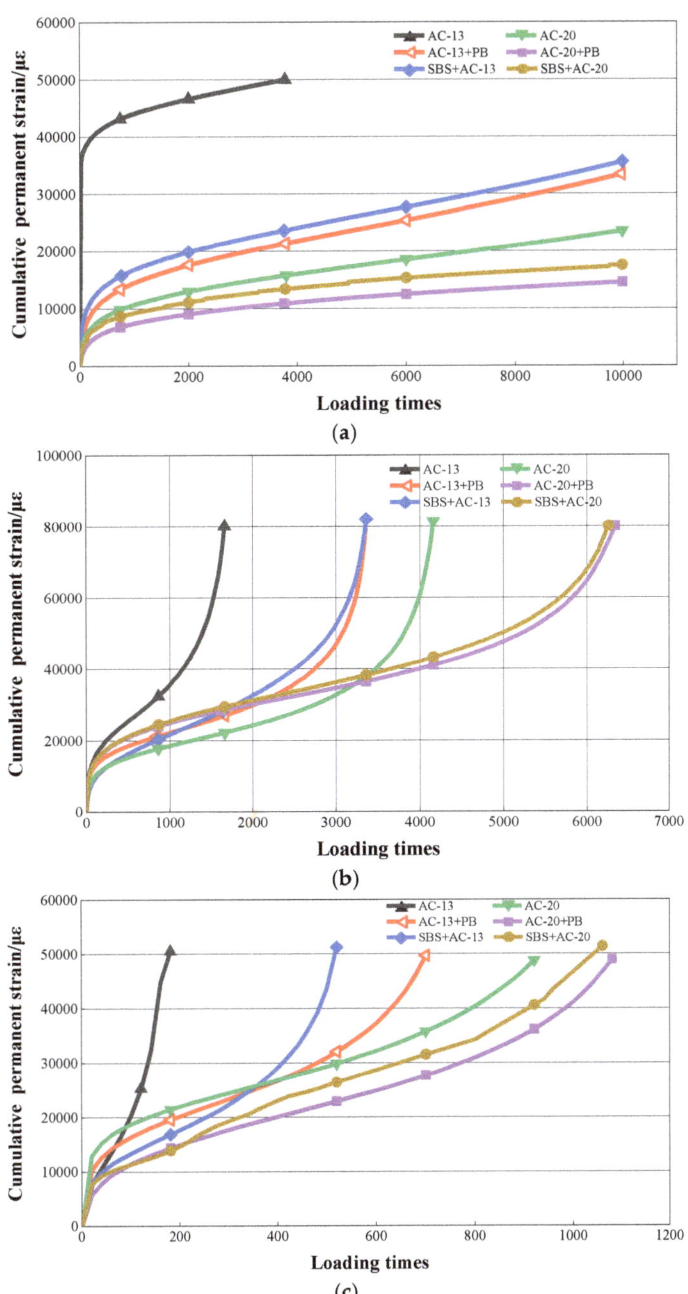

Figure 7. *Cont.*

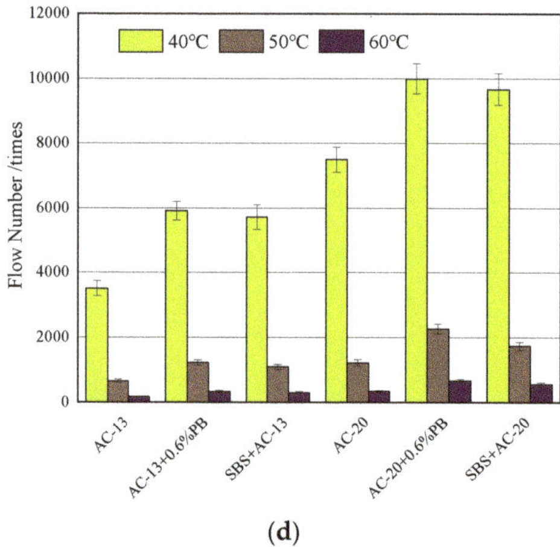

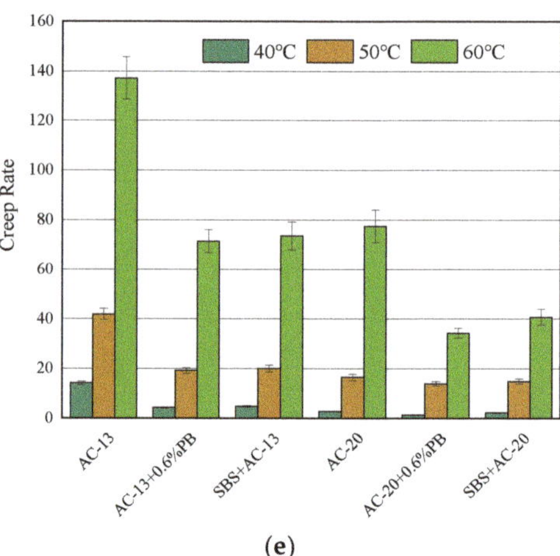

(d)

(e)

Figure 7. Results of dynamic creep test. (**a**) Cumulative permanent strain versus loading times at 40 °C; (**b**) cumulative permanent strain versus loading times at 50 °C; (**c**) cumulative permanent strain versus loading times at 60 °C; (**d**) flow number; (**e**) creep rate.

3.2. Low-Temperature Crack Resistance

The results are illustrated in Figure 8. It can be observed that, when modifier PB was used, the failure strain values of both AC-13 and AC-20 asphalt mixtures increased by 10.3% and 6.0%, respectively, compared with the corresponding base asphalt mixtures. Meanwhile, the flexural stiffness moduli presented a slightly decreasing trend. It can also be seen that the failure strain values of SBS-modified asphalt mixtures for both AC-13 and AC-20 gradations were much higher than those of PB-modified or base asphalt mixtures. These findings indicate that modifier PB does enhance the low-temperature crack resistance of the mixtures to some extent. As mentioned before, the addition of the waste plastics modifier would only make the asphalt mixtures sensitive to cracking, despite the positive impact on the high-temperature performance. By combining waste plastics with basalt fiber, the new type of modifier PB eliminates that negative effect on the low-temperature anti-cracking performance. In addition, the failure strain of the PB-modified asphalt mixture exceeded 2000 $\mu\varepsilon$, which met the requirements of JTG F40 [35].

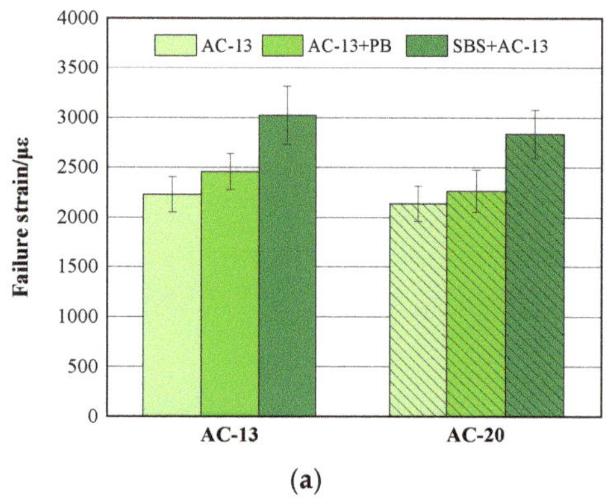

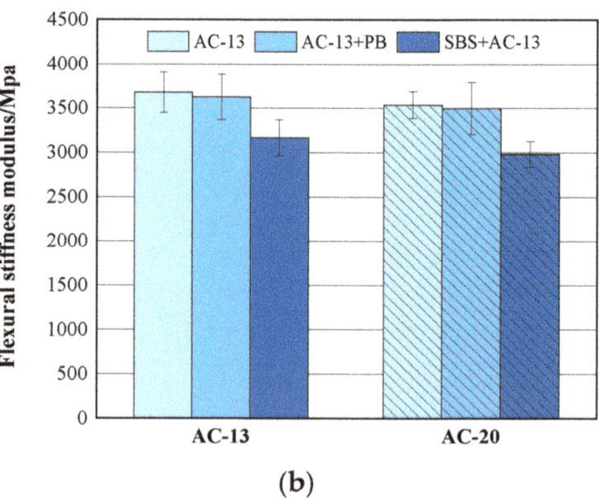

(a)

(b)

Figure 8. Low-temperature bending beam test results. (**a**) Failure strain; (**b**) flexural stiffness modulus.

3.3. Water Damage Stability

Figures 9 and 10 illustrate the results of the water stability tests. It can be seen that the MS_0 and TSR of all types of asphalt mixtures exceeded 85%, which met the requirements in the Chinese specification of JTG F40. Furthermore, adding modifier PB not only improves the MS_0 values or TSR values of mixtures but also strengthens the absolute values of the Marshall stability or splitting tensile strength.

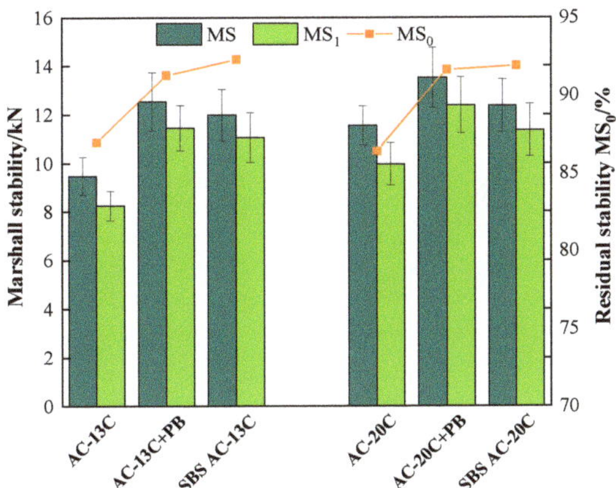

Figure 9. Results of water immersion Marshall test.

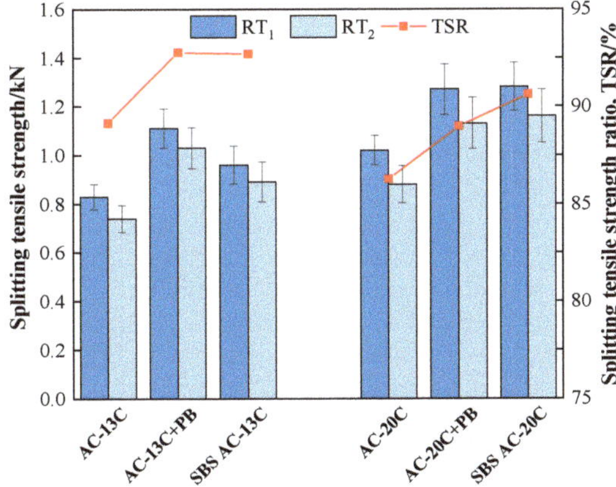

Figure 10. Results of freeze–thaw splitting test.

In terms of the mixtures of AC-13 gradation, compared with base asphalt mixtures, the MS_0 values of PB-modified mixtures increased from 87.0% to 91.3% while the TSR increased from 89.2% to 92.8%, which were comparable to those of SBS-modified mixtures. In addition, the unconditioned Marshall stability (MS) of PB-modified mixtures grew from 9.48 kN to 12.54 kN, while the unconditioned splitting tensile strength R_{T1} rose from 0.83 MPa to 1.11 MPa. In terms of the mixtures of AC-20 gradation, similar trends could be observed when modifier PB was used. These findings mean that adding modifier PB can significantly boost the water damage stability of asphalt mixtures.

3.4. Dynamic Modulus

The results of the dynamic modulus test are shown in Figure 11. It was noticeable that the dynamic moduli of the asphalt mixtures with modifier PB maintained the highest values within the test temperature range in terms of the same mixture gradation. For instance, at 55 °C, the dynamic moduli of asphalt mixtures with AC-20 gradation reached 97 MPa, 125 MPa, and 94 MPa for base asphalt mixtures, PB-modified mixtures and SBS-modified mixtures, respectively. These results indicate that the asphalt mixtures with modifier PB possess a superior deformation resistance at high temperature, which matches the results of both the wheel-tracking test and high-temperature creep test.

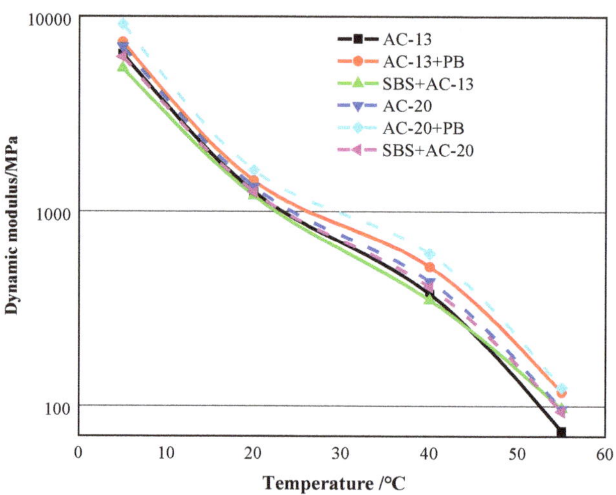

Figure 11. Results of dynamic modulus test (test frequency: 0.1 Hz).

4. Statics Analysis of the Pavement Structure of Bus Lane

4.1. Pavement Structure Design of the Bus Lane

Taking the pavement structure of the bus lane of Wenchang Road in Yangzhou City for example, the original pavement structure included five layers, namely (from top to bottom) an asphalt surface layer, asphalt binder layer, cement-stabilized gravel base layer, lime-soil base layer, and soil foundation. Asphalt mixtures with a gradation of AC-13 and AC-20 were used for the surface layer (4 cm thickness) and binder layer (8 cm thickness), respectively. PB-modified AC-13 and PB-modified AC-20 were used to replace the original materials of the asphalt surface or (and) binder layer. Therefore, three different types of pavement structures of the bus lane were proposed, as shown in Figure 12. The finite element software of ABAQUS 6.11 was employed to perform the statics analysis. The distribution of the corresponding stress and stain was analyzed for the designed structures of bus lane.

4.2. Material Parameters

The material parameters selected for the statics analysis are listed in Table 7, of which, the density and Poisson ratio were provided by Yangzhou City Municipal Administration Department. It should be noted that the dynamic modulus of the asphalt mixture at 0.1 Hz and 20 °C was used as the modulus needed in Table 7, though the static modulus of the asphalt mixture should have been used herein. This is because there is a good correlation between the dynamic modulus and static modulus of asphalt mixtures, and the values of the two indexes become comparable, especially at low-frequency conditions (such as 0.1 Hz) [37].

Asphalt surface layer(AC-13+SBS) — 4cm
Asphalt binder layer(AC-20) — 8cm
Cement stabilized gravel base layer — 40cm
Lime-soil base layer — 20cm
Soil foundation

(a)

Asphalt surface layer(AC-13+0.6%PB) — 4cm
Asphalt binder layer(AC-20) — 8cm
Cement stabilized gravel base layer — 40cm
Lime-soil base layer — 20cm
Soil foundation

(b)

Asphalt surface layer(AC-13+SBS) — 4cm
Asphalt binder layer(AC-20+0.6%PB) — 8cm
Cement stabilized gravel base layer — 40cm
Lime-soil base layer — 20cm
Soil foundation

(c)

Asphalt surface layer(AC-13+0.6%PB) — 4cm
Asphalt binder layer(AC-20+0.6%PB) — 8cm
Cement stabilized gravel base layer — 40cm
Lime-soil base layer — 20cm
Soil foundation

(d)

Figure 12. Pavement structure design of the bus lane. (**a**) Original pavement structure; (**b**) designed structure I; (**c**) designed structure II; (**d**) designed structure III.

Table 7. Material parameters for statics analysis.

Structure Sheaf	Material	Density/kg·m^{-3}	Modulus/MPa	Poisson Ratio
Surface layer	AC-13 + SBS	2360	1207	0.30
	AC-13 + 0.6% PB	2360	1438	0.30
	AC-20	2450	1340	0.30
Binder layer	AC-20 + SBS	2450	1260	0.30
	AC-20 + 0.6% PB	2450	1630	0.30
Base course	Cement-stabilized gravel	2200	1500	0.20
Sub-base	12% lime soil	2100	550	0.30
Soil base	soil	1800	45	0.40

4.3. Load Determination

The standard tire grounding pressure of 0.7 MPa is commonly used for pavement structure design in China. However, the tire ground pressure of a typical city bus normally exceeds 0.7 MPa. Therefore, the actual load was utilized for this study, along with the standard axle load. According to the data provided by Yangzhou City Bus Company, as shown in Table 8, a tire ground pressure of 0.83 MPa was considered as the actual load of a typical type of bus in Yangzhou city.

Table 8. Technical parameters of a typical type of bus in Yangzhou city.

Index	Length/mm	Width/mm	Height/mm	Curb Weight/kg	Full Quality/kg	Tire Ground Pressure/MPa
Parameter	12,000	2550	3120	11,200	17,500	0.83

Since a two-dimensional pavement model was used for the statics analysis, the applied load in the modeling process needs to be converted from the surface load (tire ground pressure) to line load. The relationship between the tire ground pressure and axle load can be expressed by Equation (9) [39,40].

$$\frac{p_i}{p} = \left(\frac{L_i}{L}\right)^{0.65} \tag{9}$$

Therefore, the axle weight of the vehicle can be calculated by Equation (10).

$$L_i = L \sqrt[13]{\left(\frac{p_i}{p}\right)^{20}} \tag{10}$$

where: p_i is the tire ground pressure, MPa; p is the standard tire ground pressure, 0.7 MPa; L_i is the axle weight of the vehicle, kN; L is the standard axle load, 100 kN.

In addition, the equivalent circle radius of the tire contacting area can be calculated by Equation (11), based on which, the line load can be calculated by Equation (12).

$$A = \frac{L_i/4}{p_i}, \; r = \sqrt{\frac{A}{\pi}} \tag{11}$$

where: A is the tire contacting area, cm^2; r is the equivalent circle radius, cm.

$$q_l = \frac{L_i/4 \times 10^3}{2r \times 10^{-2}} \tag{12}$$

where: q_l is the line load, N/m. The results are listed in Table 9.

Table 9. Results of line load conversion.

Parameter	Tire Ground Pressure/MPa	Axle Load /kN	Equivalent Circle Radius/cm	Line Load /N·m^{-1}
Standard axle load	0.7	100	10.65	117,371
Actual bus axle load	0.83	130	11.16	145,740

4.4. Model Establishing

A two-dimensional model of the cross-section of the bus lane was built for this study, with dimensions of 3.75 m (width) × 3 m (height). The CPE8R (reduced integral) unit was used. In order to accelerate the running process, the mesh was divided into a size of 0.1 m × 0.1 m for this calculation, along with meshes of 0.04 m × 0.1 m and 0.08 m × 0.1 m for the asphalt surface and binder layers, respectively. There are seven positions where the pavement is most likely to be damaged, which were marked as coordinate points in the model, namely the middle of the tire gap (Point a), the inner edge of the two tires' contacting area (Point b and b′), the middle of the two tires' contacting area (Point c and c′), and the outer edge of the two tires' contacting area (Point d and d′). The model establishment, mesh division, and loading and boundary conditions are shown in Figure 13.

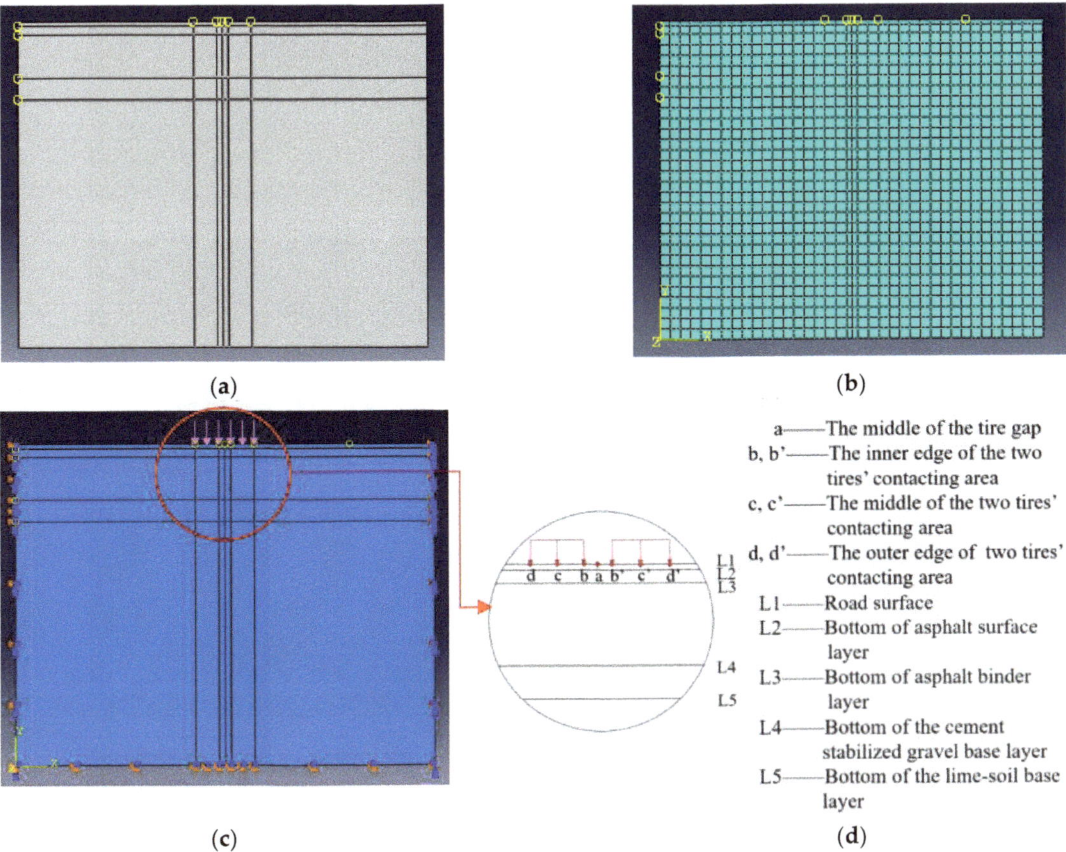

Figure 13. Model establishing process. (**a**) Modeling; (**b**) mesh division; (**c**) loading and boundary condition case; (**d**) pavement coordinate points and vertical positions.

The following assumptions were taken for this calculation: (1) the materials of each structure layer are homogeneous and uniformly continuous; (2) the material parameters keep constant with the changing of time and temperature; (3) there is no transverse displacement on the left and right sides of the model, and no transverse and vertical displacement on the bottom surface as well.

4.5. Statics Simulation Results

4.5.1. Tensile Stress Distribution

Under both standard tire ground pressure (0.7 MPa) and actual tire ground pressure (0.83 MPa), the tensile stress clouds of the original pavement structure and the three types of designed structures are illustrated in Figure 14. The maximum tensile stress of each potentially damaged point (Point a to Point d), which appears at the bottom of each structure layer, is plotted with the different vertical positions (L1 to L5) in Figure 15. Due to the symmetry of the loading and the structure, only the data of one side (Point b, c and d) were used for analysis.

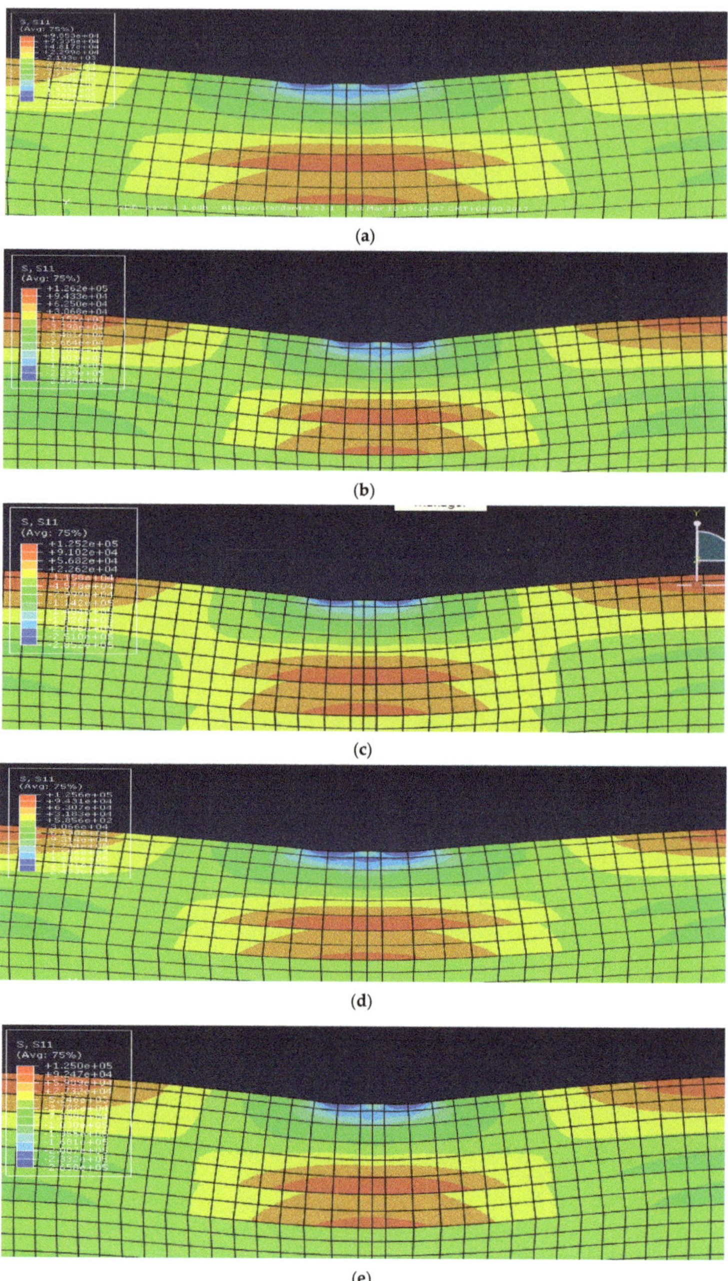

Figure 14. Tensile stress clouds of different pavement structures. (**a**) Tensile stress cloud of the original pavement structure under 0.7 MPa; (**b**) tensile stress cloud of the original pavement structure under 0.83 MPa; (**c**) tensile stress cloud of the designed structure I under 0.83 MPa; (**d**) tensile stress cloud of the designed structure II under 0.83 MPa; (**e**) tensile stress cloud of the designed structure III under 0.83 MPa.

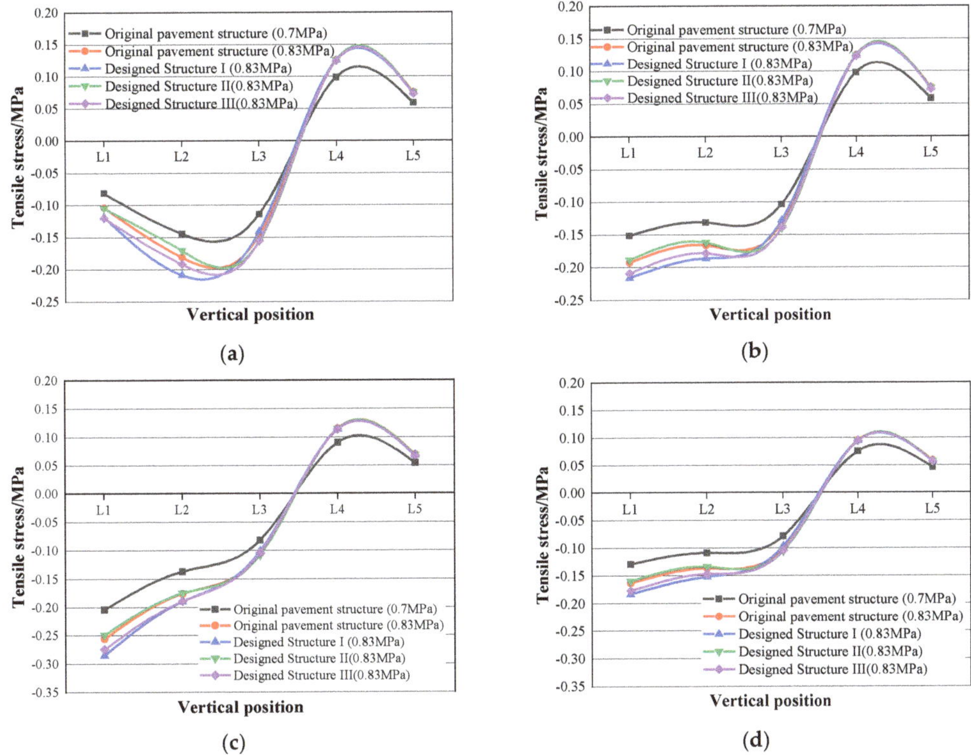

Figure 15. Maximum tensile stress of each potentially damaged point in different layers. (**a**) Maximum tensile stress at Point a; (**b**) maximum tensile stress at Point b; (**c**) maximum tensile stress at Point c; (**d**) maximum tensile stress at Point d.

As shown in Figure 15, the tensile stress values of all coordinate points were negative on L1 to L3, but turned positive on L4 and L5. These results indicate that the road surface (L1) and the bottoms of the asphalt surface layer (L2) and binder layer (L3) endure compressive stresses, whereas the bottoms of the cement-stabilized gravel base layer (L4) and lime-soil base layer (L5) suffer tensile stresses. In addition, in the middle of the tire gap (Point a), the maximum compressive stress appeared on L2, as shown in Figure 15a. The maximum compressive stress at other coordinate points occurred on L1, as shown in Figure 15b–d.

It can also be observed that the axle load causes a significant impact on the tensile stress distributions. Taking the coordinate Point a of the original structure for example, as shown in Figure 15a, when the applied axle load increased from the standard tire ground pressure of 0.7 MPa to the actual tire ground pressure of 0.83 MPa, the tensile stresses on L1 to L5 increased by 28.3%, 25.3%, 27.9%, 28.0%, and 28.5%, respectively.

Furthermore, the use of PB-modified asphalt mixtures also resulted in a fluctuation in tensile stress distributions. As shown in Figure 15, compared with the original structure, the compressive stress on L1 to L3 increased to some extent when the PB-modified surface layer or (and) binder layer were designed, whereas the tensile stresses on L4 and L5 remained as the same value. This is due to the higher moduli of the asphalt mixtures with modifier PB, leading to the PB-modified layers bearing more stress. Since asphalt mixtures possess a superior compressive strength, the increasing compressive stress does not mean an accelerated deterioration of asphalt structure layers.

4.5.2. Vertical Deformation

Under tire ground pressures of both 0.7 MPa and 0.83 MPa, the vertical deformation of all of the potentially damaged points on L1 to L5 are illustrated in Figure 16. The vertical deformation of the asphalt pavement under load could reflect the deformation resistance of a pavement.

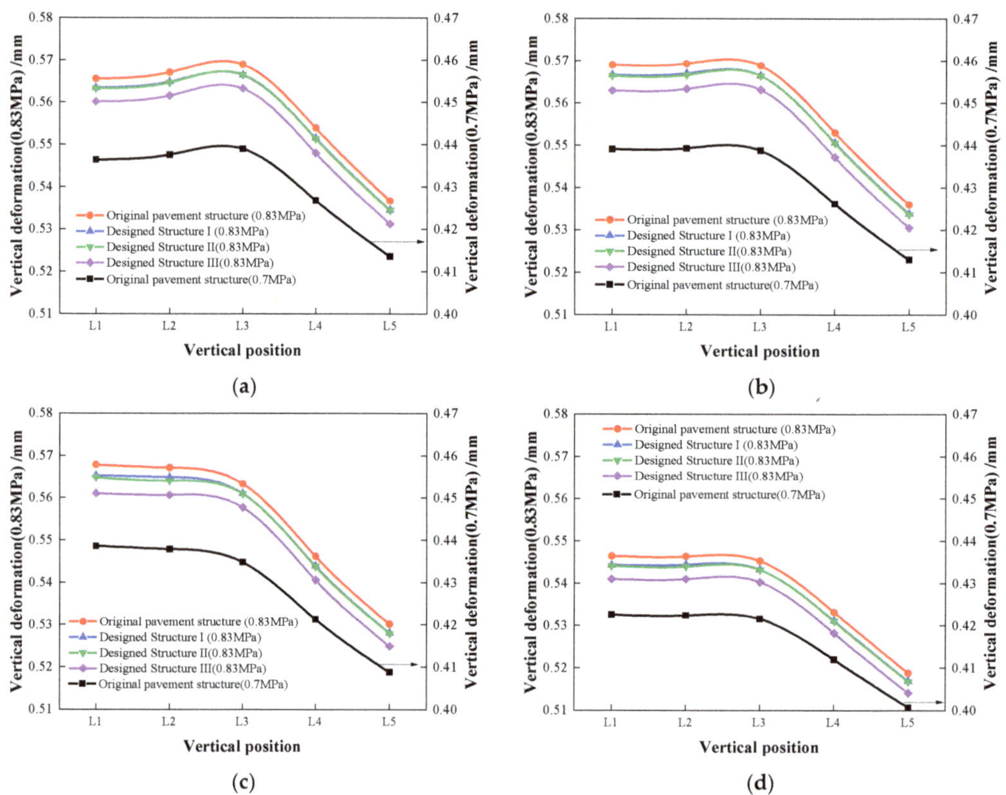

Figure 16. Vertical deformation of each potentially damaged point in different layers. (**a**) Vertical deformation at Point a; (**b**) vertical deformation at Point b; (**c**) vertical deformation at Point c; (**d**) vertical deformation at Point d.

As shown in Figure 16, the tire ground pressure impacted the vertical deformation hugely. Taking the coordinate Point a of the original structure for example, when the tire ground pressure increased from 0.7 MPa up to 0.83 MPa, the vertical deformation increased by 29.5~29.8% on L1 to L5.

In addition, the use of the PB-modified surface layer or (and) binder layer could reduce the vertical deformations effectively compared with the original structure. It was also noticeable that designed structure III presented the smallest vertical deformation, followed by designed structure II and structure I. This result indicates that structure III possesses a superior resistance to deformation.

5. Conclusions

A new type of self-developed plastic and basalt fiber composite modifier called PB modifier was used to fabricate asphalt mixtures. A series of laboratory tests, including the wheel-tracking test, dynamic creep test, low-temperature bending beam test, immersion

Marshall test, freeze–thaw splitting test, and dynamic modulus test, were adopted to explore the influence of the PB modifier on the performance of asphalt mixtures. In addition, three types of urban bus lane pavement structures were proposed by using the PB-modified asphalt mixtures for the surface layer and (or) binder layer. Statics analyses were conducted using ABAQUS 6.11 finite element software. According to the results and discussions in this study, the following conclusions can be drawn:

(1) Adding PB modifier can improve the dynamic modulus and high-temperature stability remarkably and reduce the creep rate of asphalt mixtures, presenting a superior high-temperature stability that is even better than SBS-modified asphalt mixtures.

(2) By combining with basalt fiber, the PB modifier can compensate for the adverse effect on the low-temperature crack resistance of mixtures caused by the addition of waste plastics.

(3) The PB modifier can not only improve the anti-water damage performance indexes of the residual stability and tensile strength ratio of mixtures, but can also strengthen the absolute values of the strengths, presenting a better water damage resistance.

(4) The actual axle load of a bus will cause severe tensile stress and vertical deformation compared with the standard axle load. Using PB-modified asphalt layers for bus lanes can offset the negative impact caused by a heavy axle load. Using PB-modified asphalt mixtures for both the surface layer and binder layer (designed structure III) presents the best strengthen function.

The research results of this paper provide a reference for the selection of the pavement materials and structures of the urban bus lane, which has certain theoretical significance and application value.

Author Contributions: Conceptualization, X.J. and A.K.; methodology, P.X. and A.K.; formal analysis, X.J. and Y.W.; data curation, B.L.; supervision, P.X.; visualization, Y.W.; project administration, A.K.; writing—original draft, X.J.; writing—reviewing and editing, B.L. All authors have read and agreed to the published version of the manuscript.

Funding: This research was funded by the National Natural Science Foundation of China, grant number 52178439, the Yangzhou City & University cooperation program, grant number YZ2021167, and the Yangzhou University Humanities and Social Sciences Research Fund Project, grant number xjj2020-37.

Institutional Review Board Statement: Not applicable.

Informed Consent Statement: Not applicable.

Data Availability Statement: Not applicable.

Acknowledgments: We gratefully acknowledge the financial support of the above funds and the researchers of all reports cited in our paper.

Conflicts of Interest: The authors declare no conflict of interest. The funders had no role in the design of the study; in the collection, analyses, or interpretation of data; in the writing of the manuscript; or in the decision to publish the results.

References

1. Available online: https://www.sohu.com/a/256941519_748407 (accessed on 6 December 2022).
2. Chen, Y.; Chen, G.; Wu, K. Evaluation of performance of bus lanes on urban expressway using paramics micro-simulation model. *Procedia Eng.* **2016**, *137*, 523–530. [CrossRef]
3. Ben-Dor, G.; Ben-Eliab, E.; Benensona, I. Assessing the impacts of dedicated bus lanes on urban traffic congestion and modal split with an agent-based model. *Procedia Comput. Sci.* **2018**, *130*, 824–829. [CrossRef]
4. Wei, S.; Zheng, W.; Wang, L. Understanding the configuration of bus networks in urban China from the perspective of network types and administrative division effect. *Transp. Policy* **2021**, *104*, 1–17. [CrossRef]
5. Basso, L.; Guevara, C.; Gschwender, A.; Fuster, M. Congestion pricing, transit subsidies and dedicated bus lanes: Efficient and practical solutions to congestion. *Transp. Policy* **2011**, *18*, 676–684. [CrossRef]
6. Hınıslıoğlu, S.; Ağar, E. Use of waste high density polyethylene as bitumen modifier in asphalt concrete mix. *Mater. Lett.* **2004**, *58*, 267–271. [CrossRef]

7. Heydari, S.; Hajimohammadi, A.; Javadi, N.H.S.; Khalili, N. The use of plastic waste in asphalt: A critical review on asphalt mix design and Marshall properties. *Constr. Build. Mater.* **2021**, *309*, 125185. [CrossRef]

8. Awoyera, P.; Adesina, A. Plastic wastes to construction products: Status, limitations and future perspective. *Case Stud. Constr. Mater.* **2020**, *12*, e00330. [CrossRef]

9. EI-Naga, I.A.; Ragab, M. Benefits of utilization the recycle polyethylene terephthalate waste plastic materials as a modifier to asphalt mixtures. *Constr. Build. Mater.* **2019**, *219*, 81–90. [CrossRef]

10. Yu, H.; Zhu, Z.; Zhang, Z.; Yu, J.; Oeser, M.; Wang, D. Recycling waste packaging tape into bituminous mixtures towards enhanced mechanical properties and environmental benefits. *J. Clean. Prod.* **2019**, *229*, 22–31. [CrossRef]

11. Nouali, M.; Derriche, Z.; Ghorbel, E.; Chuanqiang, L. Plastic bag waste modified bitumen a possible solution to the Algerian road pavements. *Road Mater. Pavement Des.* **2020**, *21*, 1713–1725. [CrossRef]

12. Ranieri, M.; Costa, L.; Oliveira, J.R.; Silva, H.M.; Celauro, C. Asphalt surface mixtures with improved performance using waste polymers via dry and wet processes. *J. Mater. Civ. Eng.* **2017**, *29*, 04017169. [CrossRef]

13. Padhan, R.K.; Mohanta, C.; Sreeram, A.; Gupta, A. Rheological evaluation of bitumen modified using antistripping additives synthesised from waste polyethylene terephthalate (PET). *Int. J. Pavement Eng.* **2020**, *21*, 1083–1091. [CrossRef]

14. Shahane, H.A.; Bhosale, S.S. E-Waste plastic powder modified bitumen: Rheological properties and performance study of bituminous concrete. *Road Mater. Pavement Des.* **2019**, *22*, 682–702. [CrossRef]

15. Dalhat, M.; Wahhab, H.A.-A. Performance of recycled plastic waste modified asphalt binder in Saudi Arabia. *Int. J. Pavement Eng.* **2017**, *18*, 349–357. [CrossRef]

16. Joohari, I.B.; Giustozzi, F. Chemical and high-temperature rheological properties of recycled plastics-polymer modified hybrid bitumen. *J. Clean. Prod.* **2020**, *276*, 123064. [CrossRef]

17. Abed, M.A.; Al-Tameemi, A.F.; Abed, A.H.; Wang, Y. Direct tensile test evaluation and characterization for mechanical and rheological properties of polymer modified hot mix asphalt concrete. *Polym. Compos.* **2022**, *43*, 6381–6388. [CrossRef]

18. Cheng, Y.; Han, H.; Fang, C.; Li, H.; Huang, Z.; Su, J. Preparation and properties of nano-$CaCO_3$/waste polyethylene/styrene-butadiene-styrene block polymer-modified asphalt. *Polym. Compos.* **2020**, *41*, 614–623. [CrossRef]

19. Chakartnarodom, P.; Prakaypan, W.; Ineurec, P.; Chuankrerkkul, N.; Laitila, E.A.; Kongkajun, N. Properties and performance of the basalt-fiber reinforced texture roof tiles. *Case Stud. Constr. Mater.* **2020**, *13*, e00444. [CrossRef]

20. Wang, W.; Zhu, J.; Cheng, X.; Liu, S.; Jiang, D.; Wang, W. Numerical simulation of strength of basalt fiber permeable concrete based on CT technology. *Case Stud. Constr. Mater.* **2022**, *17*, e01348. [CrossRef]

21. Xie, T.; Wang, L. Optimize the design by evaluating the performance of asphalt mastic reinforced with different basalt fiber lengths and contents. *Constr. Build. Mater.* **2023**, *363*, 129698. [CrossRef]

22. Celauro, C.; Praticò, F. Asphalt mixtures modified with basalt fibres for surface courses. *Constr. Build. Mater.* **2018**, *170*, 245–253. [CrossRef]

23. Hui, Y.; Men, G.; Xiao, P.; Tang, Q.; Han, F.; Kang, A.; Wu, Z. Recent advances in basalt fiber reinforced asphalt mixture for pavement applications. *Materials* **2022**, *15*, 6826. [CrossRef] [PubMed]

24. Wang, S.; Kang, A.; Xiao, P.; Li, B.; Fu, W. Investigating the effects of chopped basalt fiber on the performance of porous asphalt mixture. *Adv. Mater. Sci. Eng.* **2019**, *2019*, 2323761. [CrossRef]

25. Cheng, Y.; Yu, D.; Gong, Y.; Zhu, C.; Tao, J.; Wang, W. Laboratory evaluation on performance of eco-friendly basalt fiber and diatomite compound modified asphalt mixture. *Materials* **2018**, *11*, 2400. [CrossRef]

26. Lou, K.; Kang, A.; Xiao, P.; Wu, Z.; Li, B.; Wang, X. Effects of basalt fiber coated with different sizing agents on performance and microstructures of asphalt mixture. *Constr. Build. Mater.* **2021**, *266*, 121155. [CrossRef]

27. Zhu, C.; Luo, H.; Tian, W.; Teng, B.; Qian, Y.; Ai, H.; Xiao, B. Investigation on Fatigue Performance of Diatomite/Basalt Fiber Composite Modified Asphalt Mixture. *Polymers* **2022**, *14*, 414. [CrossRef]

28. Lou, K.; Xiao, P.; Kang, A.; Wu, Z.; Li, B.; Lu, P. Performance evaluation and adaptability optimization of hot mix asphalt reinforced by mixed lengths basalt fibers. *Constr. Build. Mater.* **2021**, *292*, 123373. [CrossRef]

29. Wang, W.; Cheng, Y.; Tan, G. Design optimization of SBS-modified asphalt mixture reinforced with eco-friendly basalt fiber based on response surface methodology. *Materials* **2018**, *11*, 1311. [CrossRef]

30. Li, Z.; Shen, A.; Wang, H. Effect of basalt fiber on the low-temperature performance of an asphalt mixture in a heavily frozen area. *Constr. Build. Mater.* **2020**, *253*, 119080. [CrossRef]

31. Zhang, C.; Shi, F.; Cao, P. The fracture toughness analysis on the basalt fiber reinforced asphalt concrete with prenotched three-point bending beam test. *Case Stud. Constr. Mater.* **2022**, *16*, e01079. [CrossRef]

32. Zhang, X.; Liu, J. Viscoelastic creep properties and mesostructure modeling of basalt fiber-reinforced asphalt concrete. *Constr. Build. Mater.* **2020**, *259*, 119680. [CrossRef]

33. *T/CHTS 10016-2019*; Technical Guideline for Construction of Asphalt Pavement with Basalt Fibe. China Communications Press Co., Ltd.: Beijing, China, 2019. (In Chinese)

34. *DB32/T 3710-2020*; Standard specification for construction of Asphalt Pavement with Basalt Fiber. Jiangsu Market Supervision and Administration Bureau: Beijing, China, 2020. (In Chinese)

35. *JTG F40*; Technical Specification for Construction of Highway Asphalt Pavements. Occupation Standard of the People's Republic of China: Beijing, China, 2004.

36. *JTG E20*; Standard Test Methods of Bitumen and Bituminous Mixtures for Highway Engineering. Occupation Standard of the People's Republic of China: Beijing, China, 2011.
37. Ruan, L.; Luo, R.; Hu, X.D.; Pan, P. Effect of bell-shaped loading and haversine loading on the dynamic modulus and resilient modulus of asphalt mixture. *Constr. Build. Mater.* **2018**, *161*, 124–131. [CrossRef]
38. Zhao, Y.; Pan, Y. Study on loading frequency distribution and changes with in asphalt pavements. *J. Chang. Commun. Univ.* **2007**, *23*, 7–10+17. (In Chinese)
39. Polasik, J.; Waluś, K.J.; Warguł, Ł. Experimental studies of the size contact area of a summer tire as a function of pressure and the load. *Procedia Eng.* **2017**, *177*, 347–351. [CrossRef]
40. Abinaya, L.; Thilagam, A.; Nivitha, M. Influence of tire pressure on the rutting and fatigue life of bituminous pavement. *Sadhana-Acad. Proc. Eng. Sci.* **2022**, *47*, 71. [CrossRef]

Article

Effects of Geometry and Loading Mode on the Stress State in Asphalt Mixture Cracking Tests

Yan Li [1], Weian Xuan [2], Ali Rahman [3,4,*], Haibo Ding [3,4] and Bekhzad Yusupov [3,4]

[1] Department of Civil Engineering, Ordos Institute of Technology, Ordos 017000, China; eyyly@oit.edu.cn
[2] Guangxi Key Laboratory of Road Structure and Materials, Guangxi Transportation Science and Technology Group Co., Ltd., Nanning 530029, China; 340113@my.swjtu.edu.cn
[3] School of Civil Engineering, Southwest Jiaotong University, Chengdu 610031, China; haibo.ding@swjtu.edu.cn (H.D.); bekhzad_yusupov@my.swjtu.edu.cn (B.Y.)
[4] Highway Engineering Key Laboratory of Sichuan Province, Southwest Jiaotong University, Chengdu 610031, China
* Correspondence: arahman@swjtu.edu.cn; Tel.: +86-158-8455-1418

Abstract: Currently, a variety of asphalt mixture cracking characterization tests are available as screening tools for the better selection of high-quality raw materials and also for the optimization of mixture design for different applications. However, for a same evaluation index, using different sample geometries and loading modes might lead to obtaining different values, which prevents the application of the evaluation index as a fundamental parameter in pavement design. In this paper, the effects of geometry and loading mode on the stress state in the experimental characterization of asphalt mixture cracking were discussed using numerical simulation. The results showed that applying thermally-induced load in restrained uniaxial test configuration should be considered when performing an asphalt mixture cracking test. Compared with direct tensile configuration, compressive stress clearly existed in other common test configurations, which may prevent the initiation and propagation of cracks. Moreover, it was revealed that nonuniform stress state exists in the dog-bone geometry, which makes it possible to know the failure plane in advance and place gauges at the failure plane for measuring fundamental deformation-related properties.

Keywords: geometry; mechanical load; thermal load; numerical simulation; asphalt mixture; cracking

Citation: Li, Y.; Xuan, W.; Rahman, A.; Ding, H.; Yusupov, B. Effects of Geometry and Loading Mode on the Stress State in Asphalt Mixture Cracking Tests. *Materials* **2022**, *15*, 1559. https://doi.org/10.3390/ma15041559

Academic Editor: Alessandro P. Fantilli

Received: 4 January 2022
Accepted: 16 February 2022
Published: 18 February 2022

Publisher's Note: MDPI stays neutral with regard to jurisdictional claims in published maps and institutional affiliations.

1. Introduction

Asphalt pavement has many advantages compared with concrete pavement, such as installation speed, fast usability, lower maintenance costs, and high skid resistance. In recent years, researchers have also attempted to apply asphalt concrete in trackbed due to its higher vibration attenuation capacity and waterproof function [1]. With the development of construction technology and the optimization of pavement structure, many distresses in asphalt pavement (such as segregation and rutting) can be well controlled. However, premature and excessive cracks in asphalt pavement are still inevitable, especially in cold, northern regions. Environmental temperature, mixture design, asphalt content, and properties of asphalt can all affect the cracking resistance of asphalt mixture. Among these, properties of asphalt is recognized as the most significant factor. Asphalt components have changed significantly over recent decades. On the one hand, industry has continuously increased the dosage of reclaimed asphalt pavement (RAP) for resource utilization efficiency [2]. On the other hand, a variety of waste materials and low-cost additives were used in base asphalt binder to expand its performance grade and reduce its production costs [3–8]. All of these practices raised concern about the quality of asphalt binders. Meanwhile, traditional binder characterization methods failed to rule out problematic additives or binder blends owing to the ignorance of the reversible aging phenomenon [9–11]. In order to eliminate or retard cracking, some researchers proposed using microcapsules or steel wool to enhance the cracking healing ability of asphalt pavement [12,13]. However, these technologies cannot improve chemically irreversible aging or the reversible aging resistance of asphalt binders. Thus, their long-term effectiveness is doubtful.

Irreversible aging, such as the oxidation of asphalt binders, has attracted increasing attention during recent decades, and several researchers have emphasized its factors. Reversible aging is an isothermal time-dependent hardening phenomenon which may occur at low temperature or medium–ambient temperature, and can be reversed by heating. Low-temperature reversible aging is also called physical hardening, and medium temperature reversible aging is also known as steric hardening. Free volume collapse, asphaltene aggregation, and wax crystallization were identified as the three main factors which lead to the thermal reversible process [14–17]. Researchers have also developed several empirical and fundamental theoretical models to describe and predict the reversible aging process [18–20]. The importance of reversible aging in asphalt mixture is a controversial topic. In the existing literature, few studies on the influence of reversible aging on mixture cracking test results can be found, which can be controlled by stress relaxation in a constrained state [21]. Ontario field pavement trial sections and regular contract testing results showed that low-temperature performance grade of recovered asphalt binder, after going through a physical hardening process, had a better correlation with observed field cracking [22,23]. The most pressing matter of the moment for pavement engineers is to deal with the detrimental influence of reversible aging on asphalt mixture cracking characterization. Only in this way will it be possible to establish a comprehensive theoretical model for asphalt mixture cracking in the future [24]. In addition, before performing a test, it is important to consider the configuration and procedure associated with the test in detail to ensure that the intended test conditions are realized. As the first stage of this research topic, this work was conducted to further investigate the effects of specimen geometry and loading mode on the stress state in asphalt mixture cracking tests through numerical modelling.

2. Methodology

2.1. Cracking Characterization of Asphalt Mixture

Generally, an effective approach to prevention of pavement cracking is conducting laboratory experiments to simulate the field failure mode of asphalt mixtures and rank their relative performance. After the implementation of the strategic highway research program (SHRP), a variety of test configurations, parameters, and criterions were proposed to better characterize the cracking resistance of asphalt mixture [25]. Since the main purpose of this study was to compare the effects of existing specimen geometries and loading modes on the stress state in asphalt mixture cracking test methods, the corresponding parameters and evaluation criteria are discussed in detail here. The indirect tensile (IDT) test is one of the most widely used configurations, owing to its simple geometry, low cost of test equipment, and easy implementation [26]. Recently, Zhou, Im, Sun and Scullion [27] proposed a new performance-related cracking parameter from the indirect tensile asphalt cracking test (IDEAL-CT) for asphalt mixture design and quality control (QA) purposes. Low-temperature creep compliance and strength obtained from the IDT test can also be used in the thermal cracking prediction model (TCMODEL) to perform thermal cracking analysis. In order to investigate the variation of pavement properties with depth, Velasquez, Marasteanu, Labuz and Turos [28] proposed a mixture-based bending beam rheology (BBR) test as an alternative test to IDT. For better characterizing the structural properties of asphalt mixture, four-point bending and overlay tensile tests were developed [29]. Moreover, there were some other unnotched testing configurations reported in the literature, such as restrained cooling, dog-bone shape direct tensile, and hollow cylinder tensile tests [30]. The features of unnotched testing configurations are summarized in Table 1. Most of the previous mentioned unnotched tests are based on the linear viscoelastic analysis of creep and strength data. These approaches do not take into account the evolution of cracks with time and are not sensitive to polymer modification type or level [27–30]; thus, testing on a prenotched sample was proposed. Seven common and representative asphalt mixture fracture characterization methods on notched samples include the single-edge notched beam [31], semi-circular bending [32], the semi-circular tensile [33], the disk-shaped compact tensile [34], the indirect ring tensile [35], the double-edge notched tensile [36], and the restrained notched ring tests [37]. The advantages and disadvantages of these fracture mechanical-based test methods on notched samples are listed in Table 2.

Table 1. Specimen geometry and features of common unnotched asphalt mixture fracture test configurations (tests on unnotched samples).

Specimen Geometries	Features
Indirect tensile [26,27]	• Only applicable for mechanical loading. • Extremely easy to perform. • The internal stress distribution is complex. • Not "real" tension values obtained.
Mixture BBR [28]	• Only applicable for mechanical loading. • Small thickness of beam allows analysis of the effect of aging at very small pavement depths. • Binder properties can be back-calculated using appropriate composite material model. • The volume of material tested may not be representative.
Four-point bending [29]	• Only applicable for cyclic mechanical loading. • Cyclic load needs longer time. • Results depend on a combination of structural and materials properties.
Overlay tensile [29]	• Only applicable for mechanical loading. • A structural test to simulate reflective cracking. • No fundamental material property is related
Restrained cooling [30]	• Applicable for mechanical and thermal loading. • Possibility of load eccentricity due to end fixtures. • Misalignment may cause bending. • The position of specimen failure is unknown and in case of failure at the caps the stress distribution is unknown.

Table 1. *Cont.*

Specimen Geometries	Features
Dog-bone shape direct tensile [30] 	<ul><li>Applicable for mechanical and thermal loading.</li><li>Subjected to a known and simple state of stress.</li><li>Misalignment may cause bending.</li><li>Sample failures at a known position.</li></ul>
Hollow cylinder tensile [30] 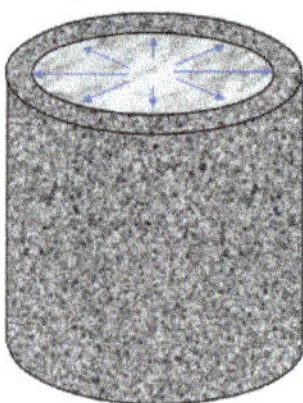	<ul><li>Only applicable for mechanical loading.</li><li>Complex special devices are needed.</li><li>Can obtain fundamental material properties.</li><li>No stress concentration at the point of load application.</li></ul>

Table 2. Specimen geometries and features of common notched asphalt mixture fracture test configurations (test on notched sample).

Specimen Geometry	Features
Single-edge notched beam [31] 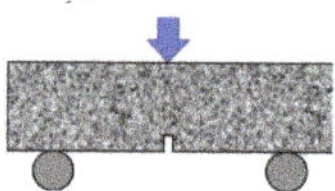	<ul><li>Constrained for crack propagation.</li><li>Higher fracture surface area.</li><li>Only applicable for mechanical loading.</li></ul>
Semi-circular bending [32] 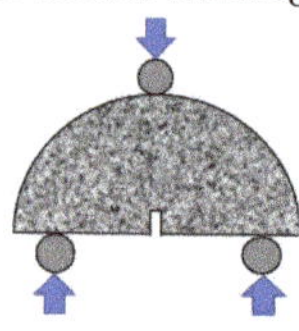	<ul><li>Complicated stress distribution.</li><li>Constrained for crack propagation.</li><li>Smaller fracture surface area.</li><li>Only applicable for mechanical loading.</li></ul>
Semi-circular tensile [33] 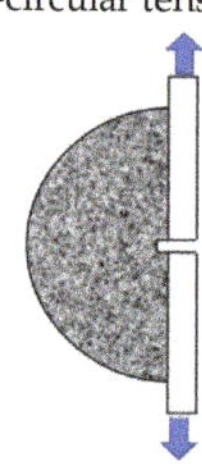	<ul><li>Applicable for mechanical and thermal loading.</li><li>Generates tensile stresses around the cracking area.</li><li>Cracking can easily propagate.</li><li>Similar to semi-circular bending (SCB) test.</li></ul>

Table 2. *Cont.*

Specimen Geometry	Features
Disk-shaped compact tensile [34] 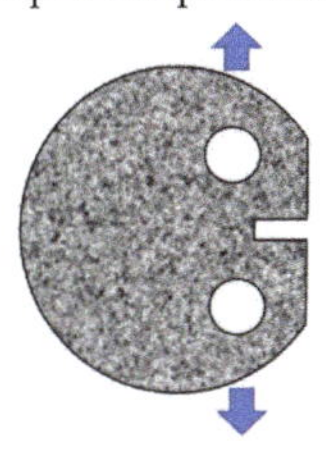	<ul><li>Complex stress distribution.</li><li>Failure around the loading holes.</li><li>Larger fracture surface area.</li><li>Applicable for mechanical and thermal loading.</li></ul>
Indirect ring tensile [35] 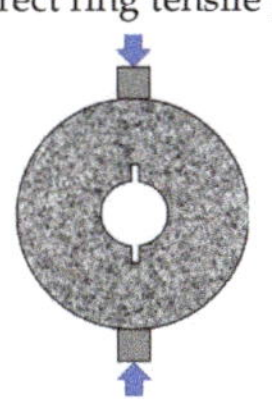	<ul><li>Higher fracture surface area.</li><li>Only applicable for mechanical loading.</li><li>Similar test frame as indirect tensile test (IDT).</li><li>Pure tensile state along the crack propagation line.</li></ul>
Double-edge notched tensile [36] 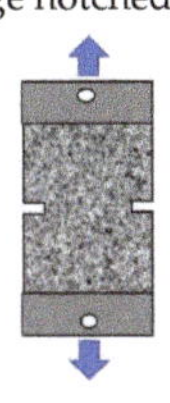	<ul><li>Applicable for mechanical and thermal loading.</li><li>Pure tension state along the crack propagation line.</li><li>Results can be analyzed using essential work of fracture.</li></ul>
Restrained notched ring [37] 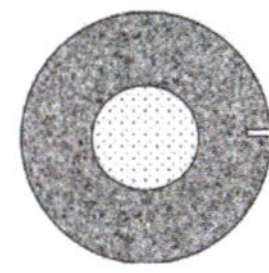	<ul><li>Only applicable for thermal loading.</li><li>Sample compaction needs special equipment.</li><li>Directly measure the cracking resistance under field-like conditions.</li></ul>

2.2. Importance of Thermally-Induced Load on Cracking Behavior of Asphalt Mixture

The main thermal cracking mechanism of asphalt pavement can be described as when the asphalt surface layer is subjected to continuously cooling events, and thermal contraction stress generates within asphalt mixture due to being restrained by the base layer and road shoulder. If the thermal shrinkage stress is higher than the tensile strength of material itself, pavement fracture would be inevitable. Based on this principle, the thermal stress restrained specimen test (TSRST) was developed by Jung and Vinson [25] during SHRP program. However, this test method was not widely used in regular quality assurance (QA) and quality control (QC) procedure owing to the relative complexity and longer time required to apply the thermal cooling in a laboratory. Consequently, many researchers and road agencies apply monotonic mechanical loading in place of thermal loading. Undoubtedly, it is a fast, simple, and convenient approach to rank the relative cracking performance of asphalt mixtures under a set of standard conditions. Furthermore, it can also simulate mechanical loading associated with cracking phenomena, such as surface-initiated longitudinal wheel path cracks which are usually caused by the higher tire pressures of truck wheels [38]. However, these laboratory tests cannot reflect some important factors that could lead to the observed cracking in the field, such as cooling rate [39], glass transition temperature (Tg) [40], thermal contraction coefficient [41], and reversible aging [42]. In addition, compared with mechanical load-induced local cracking, thermally induced shrinkage stress may result in more serious transverse cracking. Field monitoring and comparison of thermal- and mechanical-load-induced strains in asphalt pavement showed that although the frequency of thermal-induced strain was lower, resulting damage caused by thermal-induced strains could be more than that of mechanical load-induced strains, due to its higher amplitude [43]. From the perspective of prevention, thermally induced loading is hardly affected by human factors compared with mechanical-induced loading, which can be easily controlled by limiting the wheel load of trucks. Therefore, the asphalt thermal cracking analyzer (ATCA) test [44], uniaxial thermal stress and strain test (UTSST) [45], and asphalt concrete cracking device (ACCD) test [37] were developed for the optimized selection of materials with improved resistance to thermal shrinkage cracking. Since cracking propagation is an important property for understanding the cracking mechanism of pavements, Mandal and Bahia [46] also proposed to test notched samples in a restrained configuration.

3. FE Model Development

3.1. Geometry and Structure of Model

In order to evaluate the influence of sample geometry, two kinds of load, mechanical and thermally induced loads, were applied on the finite element model (FEM) of samples with different geometries using finite element simulation software ABAQUS® ver.6.12 [47]. The boundary conditions played a major role in predicting the response of the model. Consequently, fixed boundary conditions at top and bottom surfaces, and fixed boundary conditions only at the bottom surface were applied for thermal load samples and mechanical load samples, respectively. Due to the importance of the meshing step of the model in obtaining the most accurate results, many trials were undertaken during simulation to determine the best mesh size. The six-node wedge element (C3D6) was utilized to mesh the model, as shown in Figure 1. Through setting a correct mesh size and step analysis, a simulation with higher accuracy and smaller computational time can be obtained.

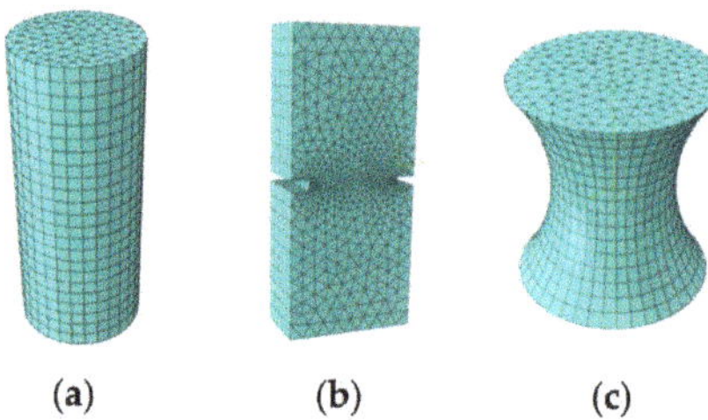

Figure 1. Three-dimensional FE model of selected sample geometries after meshing. (**a**) Cylindrical geometry; (**b**) double-notched geometry; and (**c**) dog-bone geometry.

3.2. Material Characterization

In this study, asphalt mixture type was a dense-graded asphalt concrete (AC-20) with nominal maximum aggregate size (NMAS) of 20 mm which was characterized as a linear viscoelastic material. Hooke's law was utilized to describe the behavior of linear elastic material. The elastic properties of the asphalt material are given in Table 3.

Table 3. Elastic material properties [48,49].

Material	Moduli (MPa)	Poisson's Ratio	Density (kg/m^3)
AC-20	14,500	0.35	2400

The Burgers model is usually employed by researchers to describe the viscoelastic properties of asphalt mixture. However, the Burgers model can only reflect the deformation characteristics of asphalt mixture in the short term. Thus, the generalized Maxwell model was developed by assembling several Maxwell models in parallel to better describe the relaxation performance of asphalt mixture in the long term [50]. The relaxation model adopted in ABAQUS® finite element software is the Prony model, which has the same mathematical expression as the generalized Maxwell model. A 9-element Prony's series was used to define the relaxation modulus at the reference temperature of 20 °C. The Prony's series model parameters are shown in Table 4. The Williams–Landel–Ferry (WLF) equation parameters were obtained from conducting dynamic modulus tests on the same asphalt materials at the same reference temperature and are presented in Table 5 [49].

Table 4. Prony series model parameters (E_0 = 14,500 MPa).

N (Number of Maxwell Units)	G (Ratio of Normal Modulus)	K (Ratio of Tangent Modulus)	τ (Relaxation Time)
1	0.1679	0	0.00001
2	0.1793	0	0.0001
3	0.2555	0	0.001
4	0.2194	0	0.01
5	0.1049	0	0.1
6	0.0368	0	1
7	0.0129	0	10
8	0.0048	0	100
9	0.0022	0	1000

Table 5. WLF equation constants of asphalt materials [49].

Mixture Type	C_1	C_2
AC-20	33.5	284.9

In order to consider the temperature field transmission process from outside to the central section of the model, the coefficients of thermal contraction at different temperatures are provided in Table 6. The values of the thermal properties of asphalt mixture were acquired from previous works in the literature [51]. The thermal conductivity and the specific heat of the asphalt mixture used in this study were 0.74 W/(m·°C) and 880 J/(kg·°C), respectively.

Table 6. Thermal contraction and expansion coefficients.

N	Contraction and Expansion Coefficients	Temperature/°C
1	1.00×10^{-5}	40
2	1.20×10^{-5}	30
3	1.50×10^{-5}	25
4	1.80×10^{-5}	20
5	2.10×10^{-5}	15
6	2.40×10^{-5}	10
7	2.60×10^{-5}	0
8	2.10×10^{-5}	−10
9	1.60×10^{-5}	−20
10	1.30×10^{-5}	−26

4. Results and Discussion

4.1. Effects of Sample Geometry on Stress State in Asphalt Mixture

Indirect tensile test configuration is widely utilized to evaluate the cracking resistance of asphalt mixture because it is simple to carry out without requiring cumbersome cutting or gluing procedures. Using theoretical mechanics, it is straightforward to obtain the stress distribution in testing specimens under mechanical loading. Figure 2 shows the stress distribution along the vertical and horizontal directions of the asphalt mixture sample. It can be seen that the internal stress distribution is complex, and the values obtained from the test are not "real" tension values. If the horizontal and vertical strains at the center of the specimen are measured, a reliable Poisson's ratio can be determined. The stiffness values are always rough approximations, because there exists no direct relationship between the applied force and the stress at the center as a result of the complicated internal stress distribution [52]. Therefore, the observed response is related to a particular state of stress and cannot be generalized in other situations.

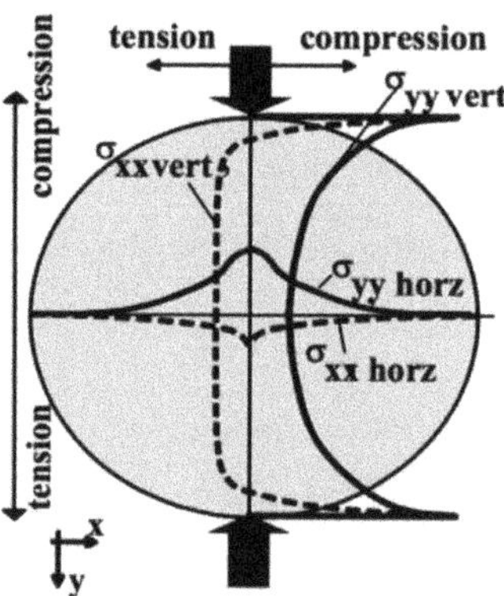

Figure 2. Stress distribution along the vertical and horizontal directions using indirect tensile test.

Compared with direct tensile configuration, compressive stress clearly exists in other common test configurations such as three-point bending, single-edge notched beam, semi-circular bending/tensile, and disk-shaped compact tensile tests. It is well recognized that asphalt pavement cracking performance in the field mainly depends on the tensile properties of asphalt binders. The ideal stress state in a tested sample is a simple tensile. Except for the binder, there are also many factors which would affect the compression behavior of the asphalt mixture, such as aggregate gradation, aggregate strength, and interlock effects. Therefore, the final tensile failure strain would be affected by the abovementioned factors, which are not consistent with the real field conditions. Based on this analysis, only three direct tensile test configurations were considered for numerical simulation analysis, as shown in Figure 3. It is evident that an almost uniform stress state existed for the cylindrical geometry. Moreover, there was an obvious stress concentration in both the notched sample and the dog-bone sample. The advantages of this stress concentration are that, firstly, the failure plane is known in advance, and secondly, it is possible to install gauges at the failure plane for measuring deformation-related properties.

The middle cross section of the stress states for the three configurations are shown in Figure 4. It is evident that stress on the edge of the notched sample was significantly higher than the stress in the middle part, thereby suggesting that the crack propagation in notched geometry is straightforward. It is a controversial topic as to whether a notched asphalt mixture sample should be utilized for testing or not. Some researchers hold the opinion that tests performed on a notched sample can better differentiate materials with different compositions. In addition, it is believed that it would be beneficial to use a notched specimen when measuring the temperature at which the mixture becomes vulnerable to cracking [53]. However, it should be noted that the notches have an extremely high possibility of passing through the aggregates, owing to the high proportion of aggregates in the mixture, and that is not consistent with what occurs in the field. Furthermore, it is difficult to control the crack initiation, which could lead to uncertainty in the test.

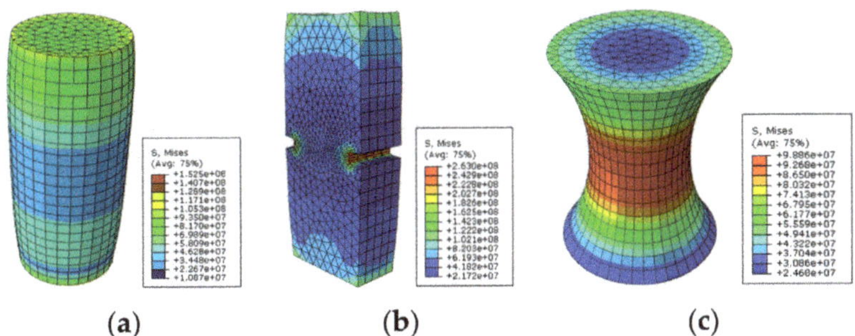

Figure 3. Stress state in direct tensile test configurations. (**a**) Cylindrical geometry; (**b**) double-notched geometry; and (**c**) dog-bone geometry.

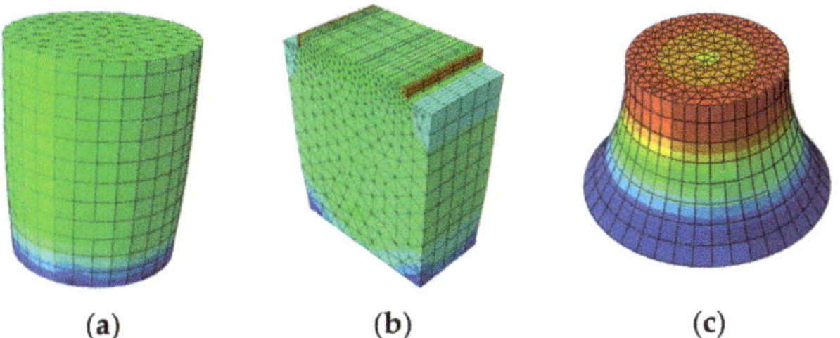

Figure 4. The stress states of middle cross sections: (**a**) cylindrical geometry; (**b**) double-notched geometry; and (**c**) dog-bone geometry.

4.2. Comparison between Thermal and Mechanical Load Modes

In order to realize the difference between the influence of the thermal load and the mechanical load on mixture response, thermal load was also applied to asphalt mixtures with three different direct tensile configurations. The numerical results are shown in Figure 5. It can obviously be seen that thermal load had a symmetrical effect on the mechanical response. For the notched sample, it is certain that the failure plane appeared in the middle cross section due to a higher degree of stress concentration. However, local stress concentration appeared at both ends for the other two geometries, resulting in localized damage and fracture near the loading platens. This phenomenon can be attributed to the fully constrained boundary conditions. One special treatment could be the application of the glue at both ends of the sample before performing the test. Similar to mechanical loading, the stress around the specimen was higher than that at the core of the sample. Some researchers applied a compressive mechanical loading with the rate of 1.27 mm/min in indirect tensile test (IDT) to represent thermally induced tensile loading rate [54]. However, if it is assumed that the temperature decreases at a rate of 10 °C per hour, as usually applied in TSRST, and a typical contraction coefficient for asphalt mixture is approximately taken as 2×10^{-5} per degree, it will result in a loading rate of 0.001 mm/min. That is three orders of magnitude less than the mechanical load, with a loading rate of 1.27 mm/min. Different loading rates may change the failure mode in the asphalt mixture, e.g., a higher loading rate may readily cause the fracture of coarse aggregate [36].

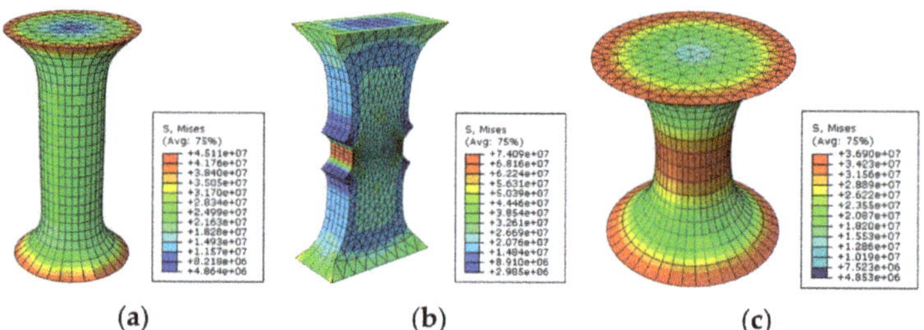

Figure 5. Effects of thermal load on asphalt mixture response direct tensile test configurations. (**a**) Cylindrical geometry; (**b**) double-notched geometry; and (**c**) dog-bone geometry.

5. Conclusions

This numerical study aimed to investigate the effects of geometry and loading mode on the stress state of asphalt mixture cracking tests. Considering the results presented in this paper, the following conclusions can be obtained:

(1) The application of thermally induced load in a restrained uniaxial test configuration should be considered when asphalt mixture cracking characterization is carried out. Consequently, important factors, such as cooling rate, glass transition temperature, thermal contraction coefficient, and reversible aging, which influence the observed cracking in the field, are of great importance during testing;

(2) Compared with direct tensile configuration, compressive stress clearly exists in other common test configurations, such as three-point bending, single-edge notched beam, semi-circular bending/tensile, and disk-shaped compact tensile tests, which may prevent the initiation and propagation of cracks;

(3) Although the test on a notched sample can better characterize the cracking propagation process, it should be noted that notches have an extremely high possibility of passing through the aggregates due to their high proportion in the mixture, and that is not consistent with what happens in the field. In addition, it is difficult to control the crack initiation which can lead to uncertainty in the test;

(4) A uniform stress state almost exists in cylindrical geometry. Moreover, there is an obvious stress concentration for notched and dog-bone samples. The advantages of this stress concentration are that, firstly, failure plane is known in advance, and secondly, it is possible to install gauges at the failure plane for measuring deformation-related properties.

6. Future Work

The future studies on this research topic could be continued in the following two directions:

(1) In order to simplify the analysis of numerical simulation, asphalt mixture, in this study, was considered as a single-phase material with a single constitutive relationship. In fact, at least two phase materials (binder and coarse aggregate) exist in the mixture. For thinner asphalt mixture samples or a notched sample, size effects should not be ignored. Furthermore, aggregate and binder or mastic have totally different thermal contraction coefficients and moduli. Thus, micromechanical analysis of thermal stresses and strains with multi-phase models from the image scanning of a sample cross section is imperative;

(2) Laboratory restrained notched cooling tests could be performed on asphalt mixtures containing asphalt binders with totally different reversible aging trends to observe if reversible aging in asphalt binder will transfer into the properties of asphalt mixture. Since laboratory asphalt mixture aging procedures may not fully consider the field condition [55,56], more harsh asphalt mixture aging protocol should be considered before running cracking tests.

Author Contributions: Conceptualization: A.R. and H.D.; methodology: Y.L.; software: Y.L. and B.Y.; formal analysis: Y.L. and W.X.; investigation: Y.L. and B.Y.; writing—original draft preparation: Y.L.; writing—review and editing: A.R. All authors have read and agreed to the published version of the manuscript.

Funding: The research was funded by Fundamental Research Funds for the Central Universities, Southwest Jiaotong University, grant number XJ2021004701.

Institutional Review Board Statement: Not applicable.

Informed Consent Statement: Not applicable.

Data Availability Statement: Not applicable.

Acknowledgments: The authors appreciate the Analytical and Testing Center of Southwest Jiaotong University for the testing and data analysis.

Conflicts of Interest: The authors declare no conflict of interest.

References

1. Luo, Q.; Fu, H.; Liu, K. Monitoring of train-induced responses at asphalt support layer of a high-speed ballasted track. *Constr. Build. Mater.* **2021**, *298*, 123909. [CrossRef]
2. Tran, N.; West, R.; Taylor, A.; Willis, R. Evaluation of moderate and high RAP mixtures at laboratory and pavement scales. *Int. J. Pavement Eng.* **2017**, *18*, 851–858. [CrossRef]
3. Fini, E.H.; Kalberer, E.W.; Shahbazi, A.; Basti, M.; You, Z.; Ozer, H.; Aurangzeb, Q. Chemical characterization of biobinder from swine manure: Sustainable modifier for asphalt binder. *J. Mater. Civ. Eng.* **2011**, *23*, 1506–1513. [CrossRef]
4. Colbert, B.W.; You, Z. Properties of modified asphalt Binders blended with electronic waste powders. *J. Mater. Civ. Eng.* **2012**, *24*, 1261–1267. [CrossRef]
5. Kaskow, J.; Poppelen, S.V.; Hesp, S. Methods for the quantification of recycled engine oil bottoms in performance-graded asphalt cement. *J. Mater. Civ. Eng.* **2018**, *30*, 04017269. [CrossRef]
6. Wu, S.; Feng, Y.; Wong, A. Selected rheological properties of tall oil pitch binder for asphaltic road pavement construction. *Int. J. Pavement Eng.* **2004**, *5*, 175–182. [CrossRef]
7. Dalhat, M.A.; Al-Abdul Wahhab, H.I. Performance of recycled plastic waste modified asphalt binder in Saudi Arabia. *Int. J. Pavement Eng.* **2017**, *18*, 349–357. [CrossRef]
8. Yu, X.; Dong, F.; Liang, X.; Ji, Z. Rheological properties and microstructure of printed circuit boards modified asphalt. *China Pet. Process. Petrochem. Technol.* **2017**, *19*, 72–80.
9. Ding, H.; Qiu, Y.; Rahman, A.; Wang, W. Low-temperature reversible aging properties of coal liquefaction residue modified asphalt. *Mater. Struct.* **2018**, *51*, 51–63. [CrossRef]
10. Johnson, K.; Hesp, S. Effect of waste engine oil residue on quality and durability of SHRP materials reference library binders. *J. Transp. Res. Board* **2014**, *2444*, 102–109. [CrossRef]
11. Kodrat, I.; Sohn, D.; Hesp, S. Comparison of polyphosphoric acid–modified asphalt binders with straight and polymer-modified materials. *J. Transp. Res. Board* **2007**, *1998*, 45–55. [CrossRef]
12. Sun, D.; Li, B.; Ye, F.; Zhu, X.; Lu, T.; Tian, Y. Fatigue behavior of microcapsule-induced self-healing asphalt concrete. *J. Clean. Prod.* **2018**, *188*, 466–476. [CrossRef]
13. Liu, Q.; Li, B.; Schlangen, E.; Sun, Y.; Wu, S. Research on the Mechanical, Thermal, Induction Heating and Healing Properties of Steel Slag/Steel Fibers Composite Asphalt Mixture. *Appl. Sci.* **2017**, *7*, 1088. [CrossRef]
14. Hesp, S.; Iliuta, S.; Shirokoff, J.W. Reversible aging in asphalt binders. *Energy Fuels* **2007**, *21*, 1112–1121. [CrossRef]
15. Ding, H.; Zhang, H.; Zheng, X.; Zhang, C. Characterisation of crystalline wax in asphalt binder by X-ray diffraction. *Road Mater. Pavement Des.* **2022**, 1–17. [CrossRef]
16. Ding, H.; Liu, H.; Qiu, Y.; Rahman, A. Effects of crystalline wax and asphaltene on thermoreversible aging of asphalt binder. *Int. J. Pavement Eng.* **2021**, *51*, 1–10. [CrossRef]
17. Ding, H.; Zhang, H.; Liu, H.; Qiu, Y. Thermoreversible aging in model asphalt binders. *Constr. Build. Mater.* **2021**, *303*, 124355. [CrossRef]
18. Freeston, J.-L.; Gillespie, G.; Hesp, S.; Paliukaite, M.; Taylor, R. Physical hardening in asphalt. In Proceedings of the 60th Annual Conference of the Canadian Technical Asphalt Association (CTAA), CTAA, Ottawa, ON, Canada, 15–18 November 2015.
19. Rigg, A.; Duff, A.; Nie, Y.; Somuah, M.; Tetteh, N.; Hesp, S. Non-isothermal kinetic analysis of reversible ageing in asphalt cements. *Road Mater. Pavement Des.* **2017**, *8*, 1–26. [CrossRef]
20. Tabatabaee, H.; Velasquez, R.; Bahia, H. Predicting low temperature physical hardening in asphalt binders. *Constr. Build. Mater.* **2012**, *34*, 162–169. [CrossRef]
21. Shenoy, A. Stress relaxation can perturb and prevent physical hardening in a constrained binder at low temperatures. *Road Mater. Pavement Des.* **2002**, *3*, 87–94. [CrossRef]

22. Ding, H.; Tetteh, N.; Hesp, S. Preliminary experience with improved asphalt cement specifications in the City of Kingston, Ontario, Canada. *Constr. Build. Mater.* **2017**, *157*, 467–475. [CrossRef]
23. Tabib, S.; Khuskivadze, O.; Marks, P.; Nicol, E.; Ding, H.; Hesp, S. Pavement performance compared with asphalt properties for five contracts in Ontario. *Constr. Build. Mater.* **2018**, *171*, 719–725. [CrossRef]
24. Hesp, S.; Soleimani, A.; Subramani, S.; Phillips, T.; Smith, D.; Marks, P.; Tam, K. Asphalt pavement cracking: Analysis of extraordinary life cycle variability in eastern and northeastern Ontario. *Int. J. Pavement Eng.* **2009**, *10*, 209–227. [CrossRef]
25. Jung, D.; Vinson, T.S. Low temperature cracking resistance of asphalt concrete mixtures. *J. Assoc. Asph. Paving Technol.* **1993**, *62*, 54–92.
26. Roque, R.; Buttlar, W.G. The development of a measurement and analysis system to accurately determine asphalt concrete properties using the indirect tensile mode. *J. Assoc. Asph. Paving Technol.* **1992**, *61*, 304–332.
27. Zhou, F.; Im, S.; Sun, L.; Scullion, T. Development of an IDEAL cracking test for asphalt mix design and QC/QA. *Road Mater. Pavement Des.* **2017**, *18*, 405–427. [CrossRef]
28. Velasquez, R.; Marasteanu, M.; Labuz, J.; Turos, M. Evaluation of bending beam rheometer for characterization of asphalt mixtures. *J. Assoc. Asph. Paving Technol.* **2010**, *29*, 295–324.
29. Zhou, F.; Scullion, T. Overlay Tester: A simple performance test for thermal reflective cracking. *J. Assoc. Asph. Paving Technol.* **2005**, *74*, 443–484.
30. Buttlar, W.G.; Wagoner, M.P.; You, Z.; Brovold, S.T. Simplifying the hollow cylinder tensile test procedure through volume-based strain. *J. Assoc. Asph. Paving Technol.* **2004**, *73*, 367–399.
31. Kim, H.; Wagoner, M.P.; Buttlar, W.G. Micromechanical fracture modeling of asphalt concrete using a single-edge notched beam test. *Mater. Struct.* **2008**, *42*, 677. [CrossRef]
32. Elseifi, M.A.; Mohammad, L.N.; Ying, H.; Cooper, S. Modeling and evaluation of the cracking resistance of asphalt mixtures using the semi-circular bending test at intermediate temperatures. *Road Mater. Pavement Des.* **2012**, *13*, 124–139. [CrossRef]
33. Pérez-Jiménez, F.; Valdés, G.; Miró, R.; Martínez, A.; Botella, R. Fénix Test: Development of a new test procedure for evaluating cracking resistance in bituminous mixtures. *Transp. Res. Rec. J. Transp. Res. Board* **2010**, *2181*, 36–43. [CrossRef]
34. Chiangmai, C.N.; Buttlar, W.G. Cyclic loading behavior of asphalt concrete mixture using disk-shaped compact tension(DCT) test and released energy approach. *J. Assoc. Asph. Paving Technol.* **2015**, *84*, 593–614.
35. Zeinali, A.; Mahboub, K.C.; Blankenship, P.B. Development of the indirect ring tension fracture test for hot-mix asphalt. *Road Mater. Pavement Des.* **2014**, *15*, 146–171. [CrossRef]
36. Andriescu, A.; Iliuta, S.; Hesp, S.; Youtcheff, J.S. Essential and plastic works of ductile fracture in asphalt binders and mixtures. In Proceedings of the 49th Annual Conference of Canadian Technical Asphalt Association(CTAA), Montreal, QC, Canada, 21–24 November 2004; pp. 93–121.
37. Akentuna, M.; Kim, S.S.; Nazzal, M.; Abbas, A.R.; Arefin, M.S. Study of the thermal stress development of asphalt mixtures using the Asphalt Concrete Cracking Device (ACCD). *Constr. Build. Mater.* **2016**, *114*, 416–422. [CrossRef]
38. Myers, L.A.; Roque, R.; Ruth, B.E. Mechanisms of surface-initiated longitudinal wheel path cracks in high-type bituminous pavements. *J. Assoc. Asph. Paving Technol.* **1998**, *67*, 401–432.
39. Apeagyei, A.K.; Dave, E.V.; Buttlar, W.G. Effect of cooling rate on thermal cracking of asphalt concrete pavements. *J. Assoc. Asph. Paving Technol.* **2008**, *77*, 709–738.
40. Tabatabaee, H.; Velasquez, R.; Bahia, H. Modeling thermal stress in asphalt mixtures undergoing glass transition and physical hardening. *J. Transp. Res. Board* **2012**, *2296*, 106–114. [CrossRef]
41. Das, P.K.; Jelagin, D.; Birgisson, B. Importance of thermal contraction coefficient in low temperature cracking of asphalt concrete. In Proceedings of the 59th Annual Conference of the Canadian Technical Asphalt Association (CTAA), Winnipeg, MB, Canada, 11–14 November 2014.
42. Yee, P.; Aida, B.; Hesp, S.; Marks, P.; Tam, K. Analysis of premature low-temperature cracking in three Ontario, Canada, pavements. *J. Transp. Res. Board* **2006**, *1962*, 44–51. [CrossRef]
43. Bayat, A.; Knight, M.A.; Soleymani, H.R. Field monitoring and comparison of thermal- and load-induced strains in asphalt pavement. *Int. J. Pavement Eng.* **2012**, *13*, 508–514. [CrossRef]
44. Bahia, H.; Tabatabaee, H.; Velasquez, R. Asphalt thermal cracking analyser (ATCA). In Proceedings of the 7th RILEM International Conference on Cracking in Pavements, Delft, The Netherlands, 20–22 June 2012; pp. 147–156. [CrossRef]
45. Alavi, M.; Hajj, E.Y.; Sebaaly, P.E. A comprehensive model for predicting thermal cracking events in asphalt pavements. *Int. J. Pavement Eng.* **2017**, *18*, 871–885. [CrossRef]
46. Mandal, T.; Bahia, H. Measuring crack propagation resistance of asphalt mixtures using notched samples in the TSRST. In Proceedings of the 62th Annual Conference of the Canadian Technical Asphalt Association (CTAA), Halifax, NS, Canada, 12–15 November 2017.
47. Dassault Systémes Simulia Corporation. *Abaqus 6.12 Analysis User's Manual*; Dassault Systémes Simulia Corporation: Johnston, RI, USA, 2012.
48. Qadir, A.; Guler, M. Finite element modelling of thermal stress restrained specimen test. In Proceedings of the 5th Eurasphalt & Eurobitumen Congress, Eurasphalt & Eurobitumen Congress, Istanbul, Turkey, 13–15 June 2012.
49. Li, C.; Niu, K. Simulation of asphalt concrete cracking using Cohesive Zone Model. *Constr. Build. Mater.* **2013**, *38*, 1097–1106. [CrossRef]

50. Zheng, J.; Qian, G.; Ying, R. Testing thermalviscoelastic constitutive relation of asphalt mixtures and its mechanical applications. *Eng. Mech.* **2008**, *25*, 34–41.
51. Gajewski, M.; Langlois, P.-A. Prediction of asphalt concrete low-temperature cracking resistance on the basis of different constitutive models. *Procedia Eng.* **2014**, *91*, 81–86. [CrossRef]
52. Mirza, M.W.; Graul, R.A.; Groeger, J.L.; Lopez, A. Theoretical Evaluation of Poisson's Ratio and Elastic Modulus Using Indirect Tensile Test with Emphasis on Bituminous Mixtures. *Transp. Res. Rec.* **1997**, *1590*, 34–44. [CrossRef]
53. Hesp, S.; Terlouw, T.; Vonk, W. Low temperature performance of SBS-modified asphalt mixes. *J. Transp. Res. Rec.* **2000**, *69*, 540–574.
54. Qian, G.P.; Guo, Z.Y.; Zheng, J.L.; Zhou, Z.G. Calculation for thermal stresses of asphalt pavement under environmental conditions based on thermalviscoelasticity theory. *J. Tongji Univ.* **2003**, *18*, 12–19.
55. Elwardany, M.D.; Rad, F.Y.; Castorena, C.; Kim, Y.R. Evaluation of asphalt mixture laboratory long-term ageing methods for performance testing and prediction. *Road Mater. Pavement Des.* **2016**, *18*, 28–61. [CrossRef]
56. Mensching, D.J.; Andriescu, A.; DeCarlo, C.; Li, X.; Youtcheff, J.S. Effect of extended aging on asphalt materials containing re-refined engine oil bottoms. *Transp. Res. Rec. J. Transp. Res. Board* **2017**, *2632*, 60–69. [CrossRef]

materials

MDPI

Article

Long-Term Performance Evolution of RIOHTrack Pavement Surface Layer Based on DMA Method

Zhimin Ma [1,2], Xudong Wang [1,2,*], Yanzhu Wang [3], Xingye Zhou [1,2] and Yang Wu [1,2]

[1] Research Institute of Highway Ministry of Transport, Beijing 100088, China
[2] National Observation and Research Station of Corrosion of Road Materials and Engineering Safety in Dadushe Beijing, Research Institute of Highway Ministry of Transport, Beijing 100088, China
[3] School of Transportation Science and Engineering, Harbin Institute of Technology, Harbin 150090, China
* Correspondence: xd.wang@rioh.cn

Abstract: Asphalt mixture is a typical viscoelastic material, and its road performance will change with the action of environment and load during actual service. This study conducted experimental research on the surface course asphalt mixture of three categories and six typical structures of RIOHTrack based on the Dynamic Mechanical Analysis method. Moreover, this study explored the performance evolution law of asphalt mixture under the coupling action of load and environment in the process of loading from 0 million to 54 million standard axle times. Results demonstrated that the phase transition characteristic temperature of the surface course materials of the three types of typical structures showed a trend of first increasing and then decreasing with the accumulation of load and environmental effects, indicating the presence of two stages of the dual coupling effect of environmental aging and load rolling on the asphalt mixture during service. In addition, the results suggested that the phase transition characteristic temperature, modulus, and phase angle of the surface layer materials have obvious material differences and structure dependencies.

Keywords: asphalt mixture; phase transformation; characteristic temperature; dynamic mechanical analysis; load; environment

Citation: Ma, Z.; Wang, X.; Wang, Y.; Zhou, X.; Wu, Y. Long-Term Performance Evolution of RIOHTrack Pavement Surface Layer Based on DMA Method. *Materials* **2022**, *15*, 6461. https://doi.org/10.3390/ma15186461

Academic Editors: Meng Ling, Yao Zhang, Haibo Ding and Yu Chen

Received: 12 August 2022
Accepted: 13 September 2022
Published: 17 September 2022

Publisher's Note: MDPI stays neutral with regard to jurisdictional claims in published maps and institutional affiliations.

1. Introduction

Asphalt pavement in the whole life cycle of service process degrades by the environment and vehicle load coupling. The service performance of each structural layer shows a gradual decay until the state of destruction. The surface condition, skid resistance, and structural strength of the pavement are constantly attenuated with the continuous action of the load and the environment, resulting in a gradual decline in the bearing capacity of the pavement and eventually failure and damage. Some asphalt pavements have exceeded their design life without structural damage in the actual service process. Meanwhile, other pavements have been damaged before reaching the design lifetime [1].

The construction of more durable long-life asphalt pavements has been an important development direction in the field of road engineering. The construction of long-life asphalt pavement construction and maintenance technology system includes several aspects of research: (1) research on the damage mechanism of pavement materials and balanced design methods; (2) research on the structural design methods and indicators of long-life asphalt pavement; (3) research on the integration of asphalt pavement design and construction technology. The evaluation of the service performance of in-service pavements and the study of the mechanical state damage evolution law have been the main bottleneck that restricts the construction of long-life asphalt pavement design methods. The performance evolution law of asphalt pavement in service is very complex, subject to both environmental and loading effects, and at the same time closely related to the physical and mechanical properties of each structural layer material. The same material under different temperatures and loads exhibits very different mechanical response states, which in turn makes the overall service state of asphalt pavements vary significantly. Therefore, understanding how to evaluate the bearing capacity and service status of the active pavement and how to reveal the damage evolution law of pavement materials and structures is an important prerequisite and theoretical foundation for asphalt pavement to achieve long-life goals.

In order to achieve the evaluation of in-service pavements in service condition, researchers have conducted a large number of tests on asphalt pavements for a long time. These tests are roughly divided into two categories. The first category is outdoor tests represented by Benkelman beam deflectometers, drop-weight deflectometers, and load-bearing plate tests, which are often used to evaluate the bearing capacity of pavement structures on site. Refs. [2–5] used a portable falling weight deflectometer (PFWD) to study the relationship between dynamic elastic modulus and compactness and water content. They found that the correlation coefficient between dynamic elastic modulus and static elastic modulus, compactness, and water content was greater than 0.85, and the maximum value of dynamic elastic modulus was close to the optimal water content, which is consistent with the actual situation. The results showed that PFWD can be used for the rapid detection and evaluation of subgrade resilience. Zha Xudong [6,7] determined the data analysis method of PFWD rapid detection of subgrade modulus by analyzing the dynamic test principle of portable drop weight deflector and established the correlation between PFWD modulus and bearing plate resilience modulus, Beckmann beam deflection, and FWD modulus through a large number of comparative tests of subgrade on site. The results showed that PFWD had good correlation with other detection methods. However, this type of test method represents the bearing capacity of the overall structure from the subgrade, the base layer to the surface layer, and cannot reflect its real bearing capacity for a specific layer [8]. The second category is to conduct laboratory tests on asphalt mixtures obtained by drilling cores in the field and characterize the service performance of specific layers through the test results of the materials. The most common mechanism is to use UTM and MTS to apply sinusoidal load to the material and measure its dynamic modulus, loss modulus, and phase angle to evaluate the mechanical properties of the material. However, the limitation of this method is that the required size of the test piece is particularly large. When evaluating the service state of the active pavement, the damage to the pavement is large, which is not conducive to long-term testing. Neither of these two types of methods can achieve a long-term study of the service condition of in-service pavements due to the limitations mentioned above.

Dynamic mechanical analysis as an important means of testing the properties of viscoelastic polymer materials, not only the required size of the specimen is small, but also can simulate a variety of loading and temperature environments, so the method is more widely used in the field of polymer materials, especially composite materials. Asphalt mixture is a typical viscoelastic material that exhibits distinct mechanical properties under various external conditions and viscoelastic properties under different temperatures and loading modes. Asphalt mixtures can be considered to be close to elastic under high-speed vehicle loads even in summer, while slow temperature shrinkage exhibits viscous flow properties in winter [9]. The phase state of the asphalt mixture changed from a glassy state to a highly elastic state with the gradual increase in temperature, eventually reaching a viscous flow state. Relevant studies have shown that the cracking, rutting, fatigue, and other damage problems of asphalt pavement are closely related to the phase transition and viscoelasticity of asphalt mixtures [10]. Therefore, the study of the phase state of the asphalt mixture is important for characterizing the service state of the pavement. Therefore, in this study, the test method of dynamic mechanical analysis was chosen to investigate the service condition of in-service pavement materials in relation to the service condition [11].

For the reasons stated above, this study relies on the Research Institute of Highway Ministry of transport, RIOHTrack, to carry out pavement coring for the surface layer of the three types of track structures from 0 loading, with 6 and 8 million times of standard axle load as the cycles. A continuous temperature sweep test based on a fixed frequency and a heating rate was carried out using the DMA slice test method, and the change law of the characteristic temperature of the phase transition of the asphalt mixture with the number of loads and the material and structural differences of the characteristic temperature were analyzed.

2. Materials and Tests

2.1. Materials

2.1.1. Road Coring

This research is based on the first full-scale track in the field of road engineering in China, the Research Institute of Highway Ministry of transport (RIOHTrack), which is located in the road traffic test site of the Ministry of Transport, Majuqiao, Tongzhou District, Beijing. The RIOHTrack test road is an elliptical closed curve composed of a straight line and a circular curve. This road runs north–south and is symmetrically arranged, with a total length of 2039 m. According to the linear characteristics of the test road and the purpose of the test, RIOHTrack has paved a total of 38 pavement structures, of which 19 types of asphalt pavement structures are found in seven categories. In this study, six typical structures of three categories were selected for research.

RIOHTrack has been loaded since November 2016. According to the load level, the corresponding action times converted to standard axle load (0.7 MPa) have been loaded for more than 60 million times, which is equivalent to the load level of China's heavy haul traffic for 30 years. This study took 6 and 8 million times of standard axle loads as the cycles to carry out pavement coring to study the service performance evolution of asphalt mixture in the whole process under the coupling effect of environmental load. The coring information is illustrated in Table 1, and the obtained core samples (part) are shown in Figure 1.

Table 1. RIOHTrack coring information.

Frequency	Coring Time	Loading Axle Time	Core State
1	31 December 2018	8 million times	
2	23 July 2019	16 million times	
3	14 January 2020	24 million times	
4	20 June 2020	30 million times	Carriageway (intact)
5	3 November 2020	36 million times	
6	24 February 2021	42 million times	
7	23 June 2021	48 million times	
8	19 October 2021	54 million times	

Figure 1. Pavement core sample.

2.1.2. Material Parameters

This study focuses on RIOHTrack's three major categories of six types of pavement structure surface layers, namely, semi-rigid base asphalt surface layer (STR1, 3), flip-chip structure (STR10, 12), and thick asphalt surface layer structure (STR18, 19). The surface layer material of STR1/3 and 10/12 is AC-13, and that of STR18/19 is SMA-13. The asphalt used in both materials is SBS modified asphalt. The schematic of each structure and the basic technical parameters of the two materials are shown in Figure 2 and Tables 2 and 3.

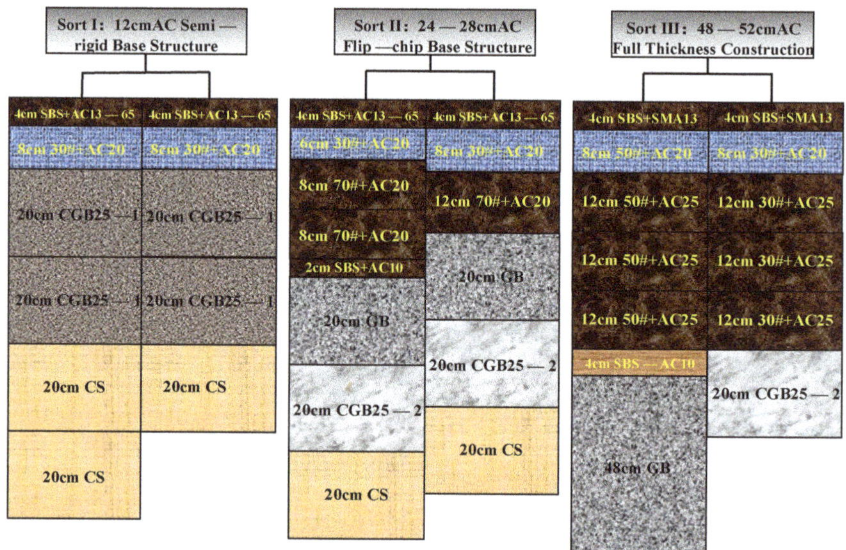

Figure 2. Schematic of each structure.

Table 2. Basic performance data of asphalt.

Asphalt	Ductility/cm (10 °C)	Ductility/cm (5 °C)	Penetration/0.1 mm (25 °C)	Softening Point/°C
SBS modified asphalt	48.7	28.5	63.4	72.7

Table 3. Material gradation of the surface layer of each structure.

Structure	Material	Pass Rate of Each Grade of Sieve Hole (%)									
		16	13.2	9.5	4.75	2.36	1.18	0.6	0.3	0.15	0.075
STR1/3 STR10/12	AC13-65	100	98.0	60.6	34.8	25.4	17.7	13.4	9.9	8.8	7.1
STR18/19	SMA13-75	100	97.7	54.7	24.9	16.7	13.9	12.5	11.4	11.0	9.7

2.2. Test

2.2.1. Specimen Preparation

The core samples were drilled on site as cylinders with a diameter of 15 cm. In this study, the dynamic mechanical test method was used, and the test piece was a sheet-like body with a 60 mm length, 15 mm width, and 3.5 mm thickness, as shown in Figure 3. The cutting method of the test piece is as follows: first, the core sample is cut according to the structure of each layer of the asphalt pavement to obtain the cylindrical core sample of the upper layer, the middle surface layer, and the lower layer structure; a precision cutting machine is then used to cut it into the test piece required for the DMA test. The process is shown in Figure 4 [12–14].

Figure 3. DMA specimen.

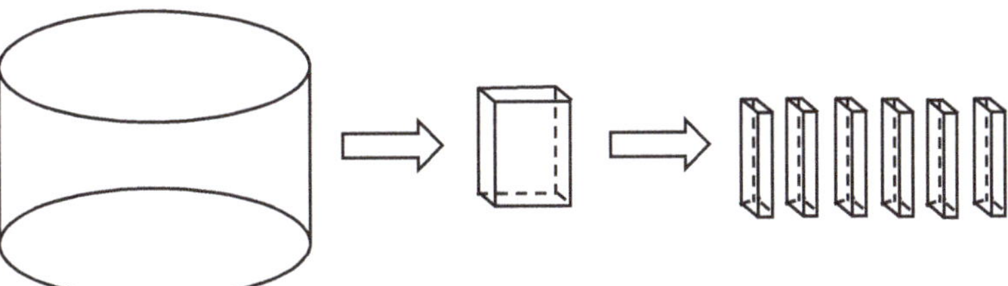

Figure 4. Diagram of specimen cutting.

2.2.2. Test Principle

The dynamic mechanical behavior of a material refers to the strain (or stress) response of the material under the action of alternating stress (or strain). The most commonly used alternating stress in the dynamic mechanical test method is the sinusoidal stress, which can be expressed as follows:

$$\tau(t) = \tau_0 sin\omega t \tag{1}$$

where τ_0 is the stress amplitude, ω is the angular frequency (unit: radians), and the strain response of the specimen under sinusoidal alternating stress varies with the material properties.

Asphalt materials are viscoelastic materials, and the strain will lag behind the stress by a phase angle δ ($0° < \delta < 90°$).

$$\gamma(t) = \gamma_0 \sin(\omega t - \delta) \tag{2}$$

Expanding Formula (2) yields

$$\gamma t = \gamma_0 (cos\delta sin\omega t - sin\delta cos\omega t) \tag{3}$$

The strain response includes two items: the first term is in phase with the stress, reflecting the elasticity of the material; the second term is 90° behind the stress, reflecting the viscosity of the material.

If a sinusoidal strain is applied to the viscoelastic specimen:

$$\gamma(t) = \gamma_0 \sin(\omega t) \tag{4}$$

then the stress response of the specimen will lead the strain by a phase angle δ:

$$\tau(t) = \tau_0 sin\omega(\omega t + \delta) \tag{5}$$

The modulus of a material is the ratio of stress to strain, and the resulting modulus should be a complex number due to the phase difference between stress and strain in viscoelastic materials. The stress–strain function is written in complex form for convenience of calculation:

$$\gamma(t) = \gamma_0 \exp(i\omega t) \tag{6}$$

and

$$\tau t = \tau_0 \exp[i(\omega t + \delta)] \tag{7}$$

Thus, complex modulus E^* is:

$$E^* = \frac{\tau(t)}{\gamma(t)} = \frac{\tau_0}{\gamma_0} e^{i\delta} = \frac{\tau_0}{\gamma_0}(cos\delta + isin\delta) \tag{8}$$

that is,

$$E^* = |E^*|(cos\delta + isin\delta) = E' + iE'' \tag{9}$$

where

$$E\prime = |E^*| \cos \delta = \frac{\tau_0}{\gamma_0} cos\delta \tag{10}$$

$$E'' = |E^*| \sin \delta = \frac{\tau_0}{\gamma_0} sin\delta \tag{11}$$

$$E^* = \sqrt{E'^2 + E''^2} \tag{12}$$

In the formula, the real number of complex modulus E' represents the energy stored by the material due to elastic deformation during the deformation, which is called the storage modulus. Imaginary number E'' characterizes the energy lost in the form of heat due to viscous deformation during the deformation of the material, which is called dissipation energy [15–18].

2.2.3. Experimental Equipment and Methods

In this study, the double cantilever fixture of the dynamic mechanical analyzer (DMAQ800) produced by TA Company (Boston, MA, USA) was used to conduct the test, as shown in Figure 5. Asphalt mixture is a temperature (frequency) sensitive material. Accordingly, the test conditions have a significant effect on the results. After conducting several tests, the research group determined the standard test conditions for the phase temperature of asphalt mixture based on DMA: the strain level was 50 $\mu\varepsilon$, the loading frequency was 1 Hz, the temperature sweep range was $-30\,°C$–$70\,°C$, and the heating rate was $2\,°C/min$.

To ensure the reliability of the test, a force measuring wrench was used to apply the same tightening force, and the test piece cut was fixed to the fixture in accordance with Section 2.2.1. The furnace was closed, the strain and frequency were set, the nitrogen temperature was controlled to $-30\,°C$ and kept at a constant temperature for 20 min, and a dynamic sine wave flexural tensile load was applied to the middle of the sliced specimen through the loading axis to carry out the corresponding tests. Five parallel specimens were set for each group of tests.

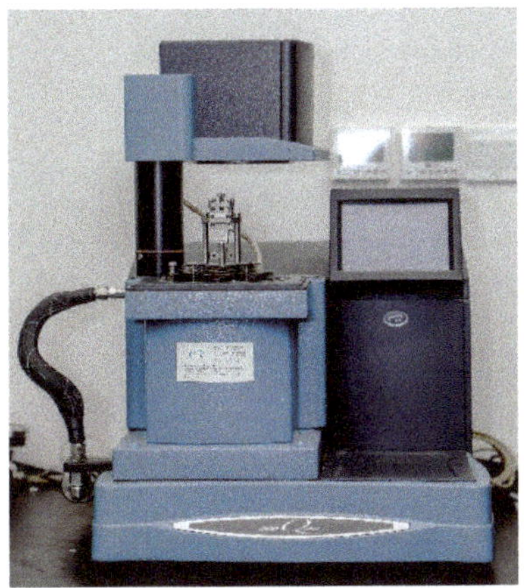

Figure 5. Test apparatus: DMAQ800.

2.2.4. Analysis Method

The complex modulus E^* of the asphalt mixture specimen, the loss modulus E'', and the tangent value of the phase angle $\tan \varphi$ can be obtained according to the abovementioned test steps. After the outliers were removed, the data of the parallel specimens were averaged, and the obtained complex modulus curve was inverse S shape. According to the characteristics of the curve, the Boltzmann function (Formula (13)) was used for fitting, the temperature corresponding to the inflection point was taken as the characteristic temperature of phase transition T_1, and the tangent line was taken through this point. Moreover, the temperatures corresponding to the intersection of the two progressive lines were characteristic temperatures T_2 and T_3, as shown in Figure 6.

$$y = \frac{A_1 - A_2}{1 + e^{(x-x_0)/dx}} + A_2 \tag{13}$$

where A_1, A_2, x_0, and dx are parameters.

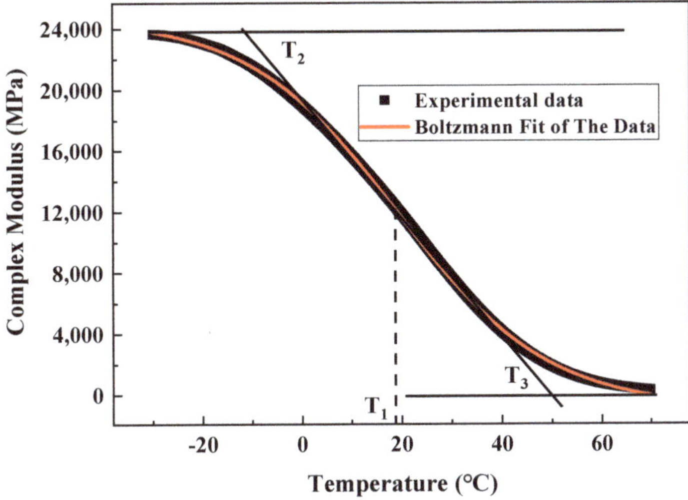

Figure 6. Schematic of complex modulus feature points.

The loss modulus curve obtained from the test had a peak point, and the curves on both sides of the peak point were asymmetrical. Accordingly, the Bigaussian function (Formula (14)) was used for fitting, and the temperature value corresponding to the peak point was taken as T_4, as shown in Figure 7. The positive curve of the phase angle had a peak point, and the curves on both sides of the peak point were symmetrical. The Gauss function (Formula (15)) was used for fitting, and the temperature value corresponding to the peak point was taken as T_5, as shown in Figure 8.

$$y = y_0 + He^{-0.5\left(\frac{x-x_c}{\omega_1}\right)^2}(x < x_c); \; y = y_0 + He^{-0.5\left(\frac{x-x_c}{\omega_2}\right)^2}(x > x_c) \tag{14}$$

$$y = y_0 + \frac{A}{\omega\sqrt{\pi/2}}e^{-2\frac{(x-x_c)^2}{\omega^2}} \tag{15}$$

where y_0, H, x_c, ω_1, and ω_2 are parameters.

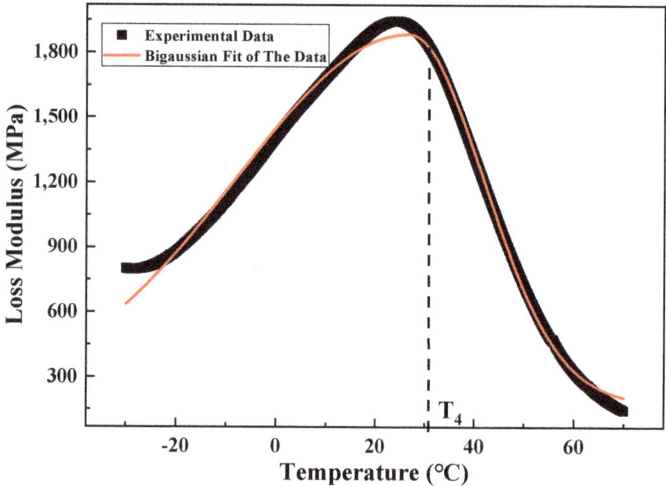

Figure 7. Schematic of the characteristic points of loss modulus.

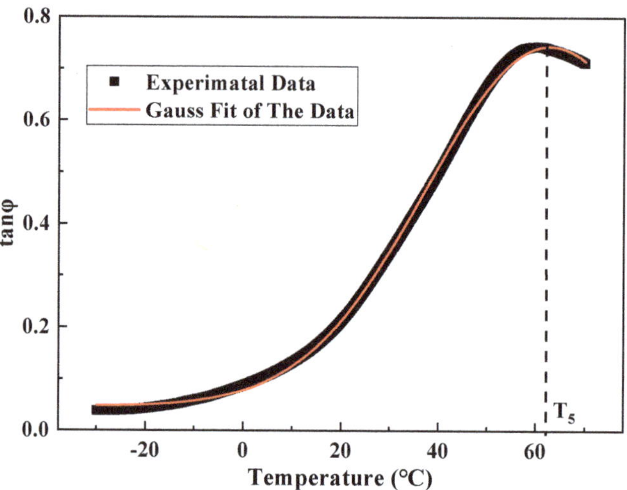

Figure 8. Schematic of the phase angle feature points.

Asphalt mixture is a typical viscoelastic material whose pavement properties change with load and temperature. In this study, the core samples of RIOHTrack with different loading cycles were drilled, and the surface layers of three types and six structures were tested by using the test method described in Section 2.2.3. Five characteristic temperatures T_1–T_5 were obtained by using the above-mentioned data processing and analysis methods. T_2, T_4, and T_5 were used to characterize the low, medium, and high temperature properties of the asphalt mixture, respectively. The long-term performance evolution law of the asphalt pavement surface layer under the influence of environment and load was characterized by analyzing the change trend of the characteristic temperature with the increase in load times [19].

3. Results and Discussion

3.1. Effect of Load-Environment Coupling on the Phase Characteristic Temperature of Asphalt Mixtures

Under the long-term action of environment and load, the service performance of asphalt mixture will change accordingly. The T_2, T_4 and T_5, results of the three major types of structures are calculated and plotted as shown in Figures 9–11 to analyze the long-term performance evolution law of asphalt mixture.

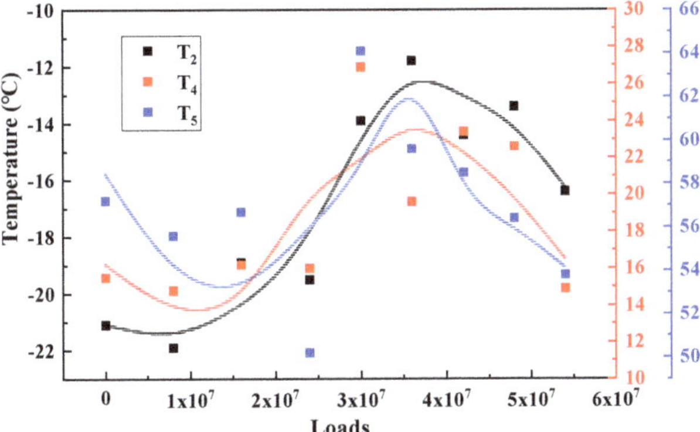

Figure 9. Characteristic temperature variation of STR1/3.

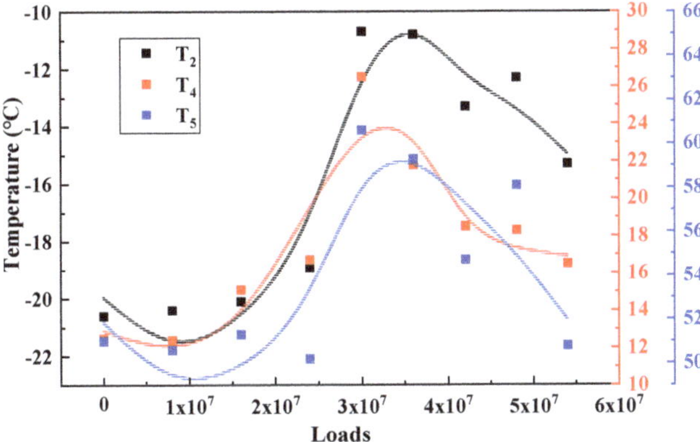

Figure 10. Characteristic temperature variation of STR10/12.

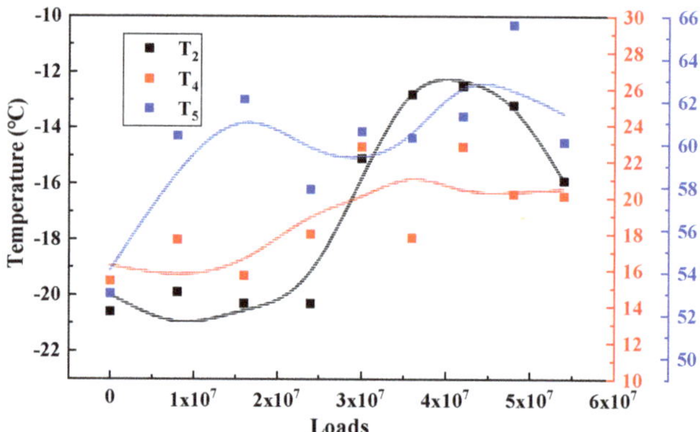

Figure 11. Characteristic temperature variation of STR18/19.

T_2 is the characteristic temperature to characterize the low temperature performance of asphalt mixture, as shown in Figures 9–11. With the accumulation of load action, T_2 of all three types of structures showed a trend of increasing and then decreasing and reached the peak in the range of 30–40 million load actions. This indicates that, in the surface layer of asphalt mixture in the coupling of load and environment, the material from the glassy state into the high elasticity of the degree of difficulty first increased and then decreased, that is, the low temperature performance of the asphalt mixture first deteriorated and then recovered. Analysis of the possible causes make the author believe that: At the beginning of the phase, the effect of the environment is weaker than the effect of the load action, and the pavement material gradually develops fatigue damage, making the viscous specific gravity of the asphalt mixture increase, which is manifested by an increase in the characteristic temperature of the phase transformation. As the load and the environment continue to act, the influence of the environment on the material gradually increases, and the material generates a large number of unsaturated bonds due to aging, which further leads to the occurrence of polymerization reactions, resulting in an increase in the molecular weight of the material, slowing down the movement of the molecular chain, which is manifested as a decrease in the characteristic temperature of the phase transition, and the performance of the pavement in terms of low temperature performance [20].

T_4 is the characteristic temperature to characterize the medium temperature crack resistance of asphalt mixture. The higher the temperature, the harder the material is, and the more likely it is to produce fatigue cracks. As shown in Figures 9–11, T_4 also shows a trend of increasing and then decreasing under load. This indicates that, during long-term service, there is also a dynamic change process of the medium-temperature crack resistance of the surface layer that is first better and then worse.

T_5 is the peak temperature of the phase angle tangent. This temperature is the maximum viscous ratio of the material, used to characterize the high temperature performance of the asphalt mixture. The higher the T_5 temperature, the higher the temperature required to achieve the maximum viscous and elastic ratio of the asphalt mixture at high temperature, that is, the better the high temperature performance of the material. From Figures 9–11 can be seen, T_5 with the cumulative effect of the load, showing a trend of increasing and then decreasing, and T_2, T_4 the same evolutionary law, reflecting a good correlation. This indicates that the temperature required to reach the maximum viscous ratio of asphalt mixture first increases and then decreases, in terms of road performance in high temperature service to produce permanent deformation of the degree of difficulty first increases and then decreases, that is, its high temperature performance first becomes better and then worse.

3.2. Long-Term Evolution of Modulus and Phase Angle of Asphalt Mixtures

From the analysis in Section 3.1, it can be seen that, in the asphalt mixture in the long-term service process, subject to the cumulative effect of the environment and load, its low-temperature crack resistance first becomes worse and then recovers, while the high-temperature rutting resistance first becomes better and then gradually deteriorates, showing a contradictory relationship, which is consistent with our knowledge. On the other hand, this also reflects that the performance of asphalt pavement during service does not show a constant decay until destruction, but a dynamic change due to the different main influencing factors in the early and late stages, with an inflection point of performance evolution in the middle of service. In order to describe this phenomenon more intuitively, it is necessary to analyze the dynamic modulus and phase angle changes of the asphalt mixture during the whole process of service. Due to the differences in characteristic temperatures, it is not reasonable to select the modulus values corresponding to the characteristic temperatures for comparison, so the complex modulus values corresponding to $-20\ ^\circ\mathrm{C}$, $20\ ^\circ\mathrm{C}$, $60\ ^\circ\mathrm{C}$ and the phase angle tangent values are selected for analysis in this study, and the results are shown in Figures 12–14.

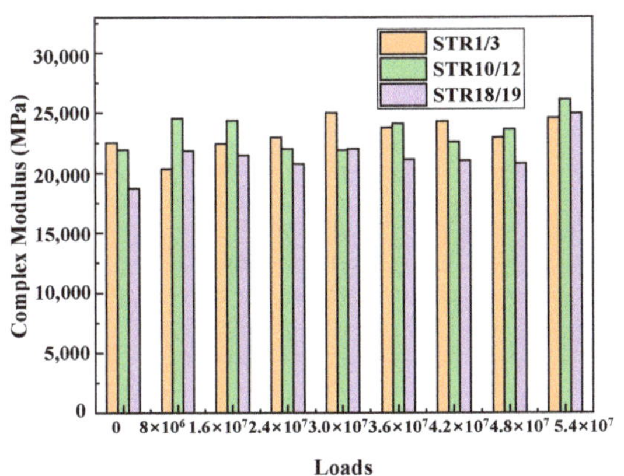

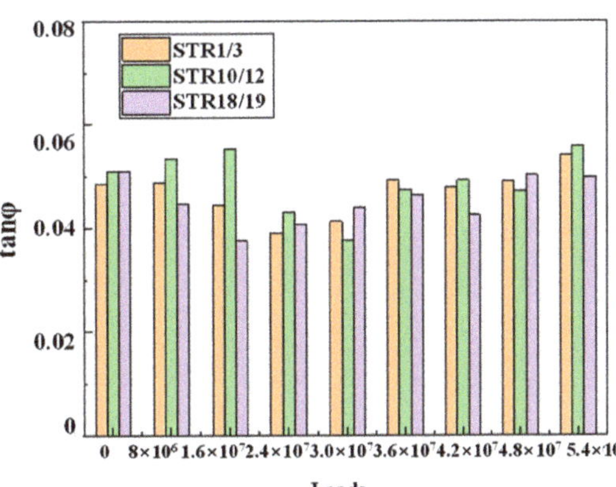

Figure 12. Complex modulus and phase angle at $-20\ ^\circ\mathrm{C}$ for different loading cycles.

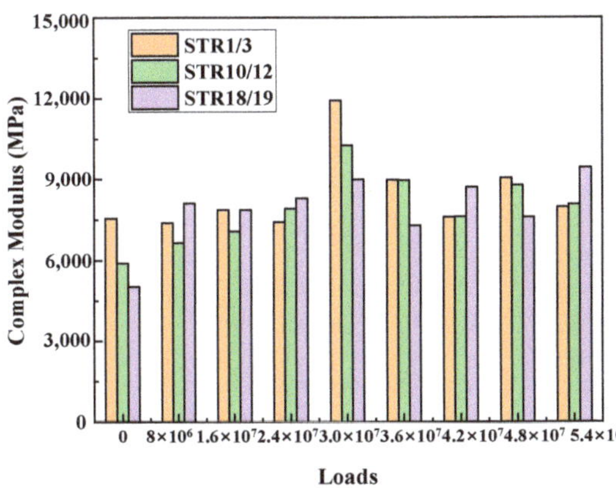

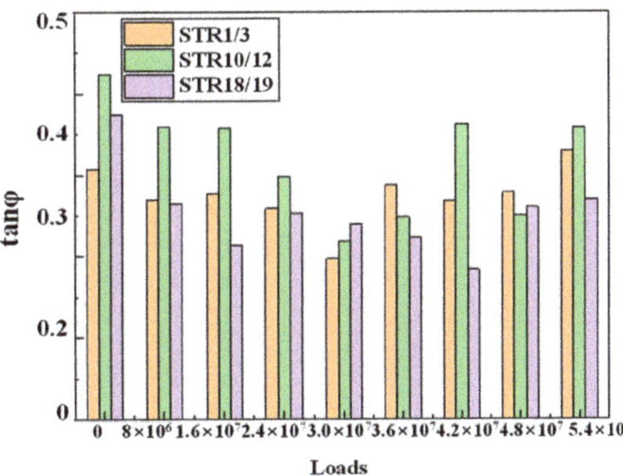

Figure 13. Complex modulus and phase angle at $20\ ^\circ\mathrm{C}$ for different loading cycles.

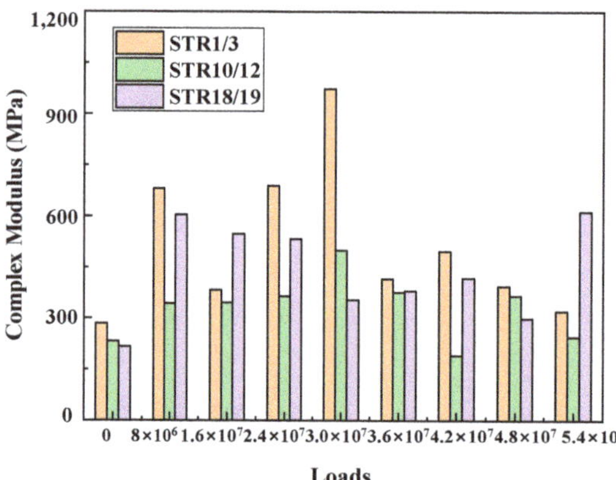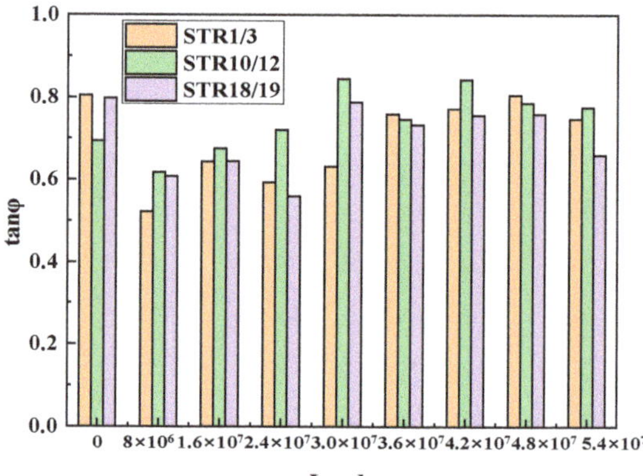

Figure 14. Complex modulus and phase angle at 60 °C for different loading cycles.

As shown in Figures 12–14, under the continuous action of load and environment, both in the low and medium temperature zone and high temperature zone, the complex modulus of asphalt mixture shows a trend of increasing and then fluctuating decrease, indicating that the surface layer of asphalt mixture in service is subjected to environmental aging and the double coupling effect of load crushing. Aging hardens the material. In load rolling there are two stages: when the rutting is small, in the state of compression density is dominant, the material hardens; when the rutting is larger, the mixture produces "flow" deformation. The material becomes soft, which is also a fatigue damage performance. In the long-term service process, the phase angle tangent value of the surface layer of asphalt mixture shows a trend of first decreasing and then increasing, indicating that the asphalt mixture in the early service of the elastic proportion is increasing and viscous proportion is decreasing. In about 30 million loadings the elastic proportion reaches the last stage, after which the elastic part of the proportion is decreasing and viscous proportion is increasing. These phenomena are consistent with the conclusions obtained in Section 3.1. Therefore, it can be considered that, for the mixture in the service process, its performance evolution exhibits two stages: one is the hardening stage, the second is the fluctuation of the soft stage, which for the study of the surface layer of asphalt mixture in service during the mechanical decay law has a certain guiding significance.

3.3. Material Variability of Characteristic Temperatures

From the analysis of Section 3.2, it can be seen that the surface layers of the three types of structures show similar evolutionary laws under the action of environmental loading, but there are some differences among them because the surface layer materials of the three types of structures are different. The phase transition characteristic temperatures T_2, T_4, and T_5 of the three types of structures are obtained by fitting and calculating according to the aforementioned experimental analysis method, respectively, and they are plotted as scatter plots and the corresponding trend lines are added, and the results are shown in Figures 15–17.

As shown in Figure 15, T_2 is the characteristic temperature to characterize the low temperature performance of asphalt mixture. The results show that under low temperature conditions, whether it is used for STR1/3 and STR10/12 surface layer of AC-13 or for STR18/19 SMA-13, the trend of the characteristic temperature T_2 is very close. The author believes that this is because in the low temperature zone, the two asphalt mixtures are in the glassy state, and the material has a high modulus. In the mechanical properties of the performance of the elasticity of the main, i.e., the long-term role of the load, the two performance evolution laws do not show significant differences.

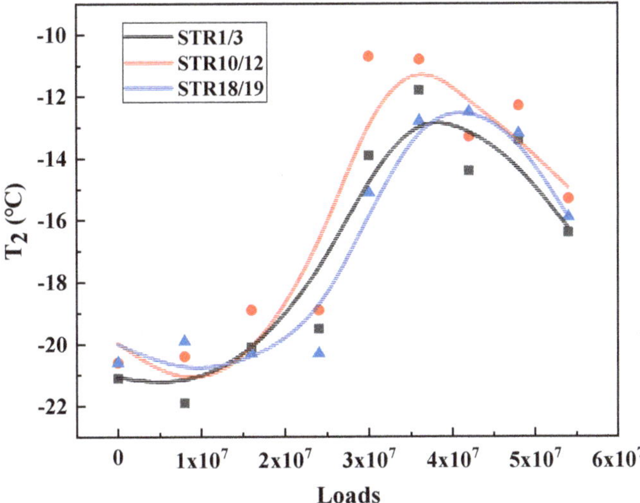

Figure 15. Characteristic temperature T_2 for the three types of structures.

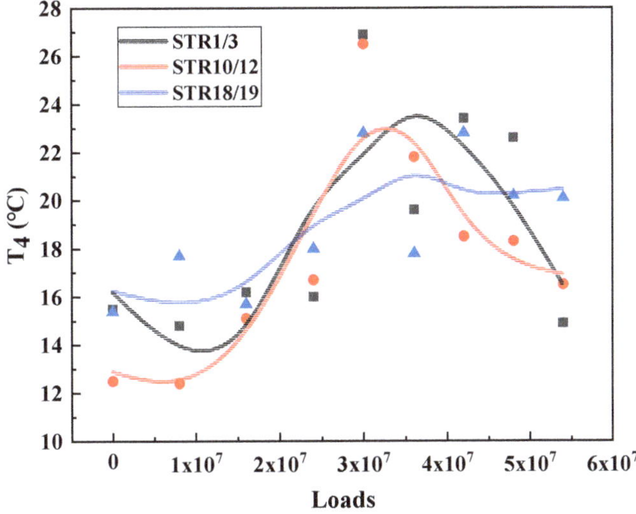

Figure 16. Characteristic temperature T_2 for the three types of structures.

As shown in Figure 16, T_4 is the characteristic temperature to characterize the medium temperature performance of the asphalt mixture, and the results show that AC-13 and SMA-13 exhibit very different long-term performance evolution patterns in the medium temperature region. The characteristic temperature of AC-13 is lower than that of SMA-13 at the beginning of the loading period, while the characteristic temperature of AC-13 tends to increase with the increase of loading times, while SMA-13 also increases, but its growth rate is much less than that of AC-13. After reaching the peak, both of them show a decreasing trend, and AC-13 also shows a faster decline. This indicates that SMA-13 maintains a more stable medium-temperature performance during the long-term service cycle, while AC-13 is more influenced by the environmental loading. Analyzing the reason is that SMA-13 coarse aggregate content is high, in the mix, the particle surface and surface direct contact, mutual embeddings constitute the skeleton, and improve the temperature sensitivity of the asphalt mixture. At the same time, the higher content of mineral powder and asphalt formation of the material has a high cohesion, to ensure the stability of its overall mechanical properties, so compared with AC-13, SMA-13 by environmental loading. Therefore, SMA-13 performs more consistently than AC-13 over the long-term service cycle under the coupling effect of the loading environment.

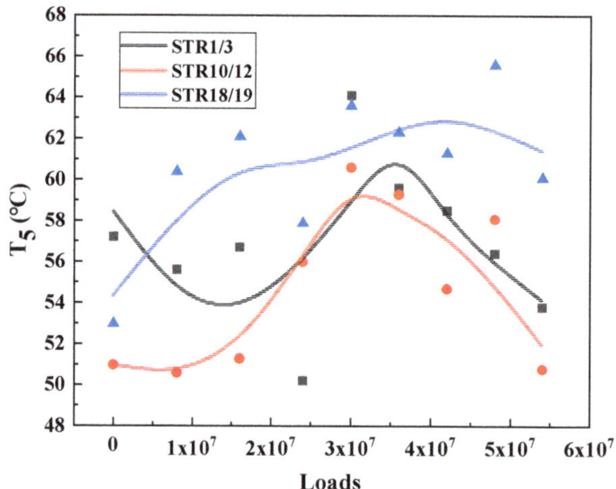

Figure 17. Characteristic temperature T_2 for the three types of structures.

As shown in Figure 17, in the high temperature region, SMA-13 has a higher characteristic temperature T_5 than AC-13 throughout the service cycle and maintains a relatively stable performance. This indicates that the high temperature performance of SMA-13 is always better than that of AC-13 during the full service cycle of asphalt pavement and is less affected by the cumulative effect of environmental load coupling. In addition to the above-mentioned coarse aggregate skeleton embedding, mineral powder and asphalt, the increase of asphalt dosage, and the addition of fibers also played a role in improving the high temperature stability of SMA-13.

3.4. Structural Dependence of the Characteristic Temperature

As described in Section 3.2, it can be seen that the trend of the phase transition characteristic temperature is basically the same for STR1/3 and STR10/12 throughout the service cycle because they use the same surface layer material, but there are some differences in the values due to the differences in the pavement structure. In order to analyze the differences, the differences between the three characteristic temperatures of STR1/3 and STR10/12 were calculated as shown in Tables 4–6.

Table 4. Comparison of STR1/3 and STR10/12 characteristic temperature T_2.

Loading Axle Time (Million)	STR1/3	STR10/12	Difference
8	−21.1	−20.6	−0.5
16	−21.9	−20.4	−1.5
24	−20.1	−18.9	−1.2
30	−19.5	−18.9	−0.6
36	−13.9	−10.7	−3.2
42	−11.8	−10.8	−1
48	−14.4	−13.3	−1.1
54	−13.4	−12.3	−1.1
	−16.4	−15.3	−1.1

Table 5. Comparison of STR1/3 and STR10/12 characteristic temperature T_4.

Loading Axle Time (Million)	STR1/3	STR10/12	Difference
0	15.5	12.5	3
800	14.8	12.4	2.4
1600	16.2	15.1	1.1
2400	16.0	16.7	−0.7
3000	26.9	26.5	0.4
3600	19.6	21.8	−2.2
4200	23.4	18.5	4.9
4800	22.6	18.3	4.3
5400	14.9	16.5	−1.6

Table 6. Comparison of STR1/3 and STR10/12 characteristic temperature T_5.

Loading Axle Time (Million)	STR1/3	STR10/12	Difference
0	57.2	51.0	6.2
800	55.6	50.6	5
1600	56.7	51.3	5.4
2400	56	50.2	5.8
3000	64.1	60.6	3.5
3600	59.6	59.3	0.3
4200	58.5	54.7	3.8
4800	56.4	58.1	−1.7
5400	53.8	50.8	3

As seen in Table 4, the characteristic temperature T_2 of STR1/3 is about 1–2 °C lower than that of STR10/12 in the pre-service period. While in the middle of service, i.e., when the characteristic temperature reaches its peak, the difference increases with a difference of about 3 °C, and in the late service period, the difference resumes with about 1 °C. This indicates that the structural form of STR1/3 has some advantages in reducing the low temperature cracking performance of the asphalt pavement surface layer.

As shown in Tables 5 and 6, the characteristic temperature of the thin asphalt surface structure (STR1/3) is higher than that of the thick asphalt surface structure (STR10/12) during the whole service life in the medium and high temperature zone, excluding individual loading stages, which indicates that the thin asphalt surface structure (STR1/3) performs better in terms of high temperature rutting resistance. Combined with the above analysis, it can be judged that STR1/3 has better performance than STR10/12 in both low temperature cracking and high temperature rutting resistance, and the cost of STR1/3 is much lower than STR10/12.

4. Conclusions

In this study, the DMA test method was used to test the surface layer asphalt mixes of six asphalt pavement structures in three major categories of RIOHTrack, and the complex modulus, loss modulus, and phase angle of the materials at different loading cycles were obtained and the five-phase transformation characteristic temperatures were obtained by fitting the corresponding functions. The performance evolution law of the in-service pavement surface layer asphalt mixture for the whole process of 0–54 million standard axial loadings was obtained by analyzing the test results, which are summarized as follows.

(1) Under the coupling effect of loading environment, the five-phase transformation characteristic temperatures of the surface layer asphalt mixture all show a trend of first increasing and then decreasing, indicating that with the increase of service time of the pavement, the low temperature crack resistance of the surface layer asphalt

mixture first becomes better and then worse, while the high temperature performance first becomes worse and then recovers.

(2) With the coupling of load environment, the modulus of asphalt pavement surface layer asphalt mixture first increases and then gradually decreases, while the phase angle first decreases and then gradually increases, indicating that the asphalt mixture in the actual service process, there are two stages of compression density hardening and then fluctuating softness.

(3) During the long-term service, the phase transition characteristics of the surface layer asphalt mixture temperature shows obvious material variability and structural dependence, SMA-13 shows more stability than AC-13, and is less affected by the environment and loading, while in terms of structure, the semi-rigid base layer thin asphalt surface structure represented by STR1/3 shows better performance than the inverted structure represented by STR10/12. better road performance.

Author Contributions: Conceptualization, Z.M. and X.W.; methodology, Z.M. and Y.W. (Yanzhu Wang); formal analysis, Z.M. and X.Z.; investigation, Z.M. and Y.W. (Yanzhu Wang); resources, Y.W. (Yanzhu Wang); data curation, X.Z.; writing—original draft preparation, Z.M. and Y.W. (Yang Wu); writing—review and editing, Z.M. and Y.W. (Yang Wu). All authors have read and agreed to the published version of the manuscript.

Funding: This work was funded by the National Key of Research and Development Plan under Grant number 2020YFA0714300.

Institutional Review Board Statement: Not applicable.

Informed Consent Statement: Not applicable.

Data Availability Statement: Not applicable.

Conflicts of Interest: The authors declare no conflict of interest.

References

1. Gao, Y.; Geng, D.; Huang, X.; Li, G. Degradation evaluation index of asphalt pavement based on mechanical performance of asphalt mixture. *Constr. Build. Mater.* **2017**, *140*, 75–81. [CrossRef]
2. Zhou, F.; Scullion, T. Guidelines for Evaluation of Existing Pavements for HMA Overlay. *Concr. Pavements* 2007.
3. Seo, J.-W.; Kim, S.-I.; Choi, J.-S.; Park, D.-W. Evaluation of layer properties of flexible pave-ment using a pseudo-static analysis procedure of Falling Weight Deflectometer. *Constr. Build. Mater.* **2009**, *23*, 3206–3213. [CrossRef]
4. Nega, A.; Nikraz, H.; Al-Qadi, I.L. Dynamic analysis of falling weight deflectometer. *J. Traffic Transp. Eng.* **2016**, *3*, 427–437. [CrossRef]
5. Sebaaly, B.; Mamlouk, M.; Davies, T. Dynamic Analysis of Falling Weight Deflectometer Data. *Transp. Res. Rec.* **1986**, *1070*, 63–68.
6. Duan, D.; Zha, X.; Zhang, Q. Subgrade resilience modulus measuring by portable falling weight deflectometer(PFWD). *J. Traffic Transp. Eng.* **2004**, *4*, 10–12. [CrossRef]
7. Zha, X. Study of Rapid Test of Subgrade Modulus with PFWD. *J. Highw. Transp. Res. Dev.* **2009**, *26*, 26–30. [CrossRef]
8. Lee, K.; Pape, S.; Castorena, C.; Kim, Y.R. Evaluation of Small Specimen Geometries for Asphalt Mixture Performance Testing and Pavement Performance Prediction. *Transp. Res. Rec. J. Transp. Res. Board* **2017**, *2631*, 74–82. [CrossRef]
9. Yuan, Q.; Liu, W.; Pan, Y.; Deng, D.; Liu, Z. Characterization of Cement Asphalt Mortar for Slab Track by Dynamic Mechanical Thermoanalysis. *J. Mater. Civ. Eng.* **2016**, *28*, 04015154.1–04015154.8. [CrossRef]
10. Kim, H.; Wagoner, M.; Buttlar, W. Micromechanical fracture modeling of asphalt concrete using a single-edge notched beam test. *Mater. Struct.* **2008**, *42*, 677–689. [CrossRef]
11. Wang, Y.; Wang, X.; Ma, Z.; Shan, L.; Zhang, C. Evaluation of the High- and Low-Temperature Performance of Asphalt Mortar Based on the DMA Method. *Materials* **2022**, *15*, 3341. [CrossRef] [PubMed]
12. Chen, D.-H.; Bilyeu, J.; Scullion, T.; Nazarian, S.; Chiu, C.-T. Failure Investigation of a Foamed-Asphalt Highway Project. *J. Infrastruct. Syst.* **2006**, *12*, 33–40. [CrossRef]
13. Tarefder, R.; Ahmad, M. Evaluating the Relationship between Permeability and Moisture Damage of Asphalt Concrete Pavements. *J. Mater. Civ. Eng.* **2015**, *27*, 04014172. [CrossRef]
14. Gu, F.; Luo, X.; Zhang, Y.; Lytton, R.L. Using overlay test to evaluate fracture properties of field-aged asphalt concrete. *Constr. Build. Mater.* **2015**, *101*, 1059–1068. [CrossRef]
15. Soliman, H.; Shalaby, A. Glass Transition Temperature and Low-Temperature Stiffness of Hot-Pour Bituminous Sealants. In Proceedings of the Transportation Research Board Meeting, Washington, DC, USA, 13–17 January 2008.

16. Lei, Z.; Tan, Y.; Bahia, H. Relationship between glass transition temperature and low temperature properties of oil modified binders. *Constr. Build. Mater.* **2016**, *104*, 92–98. [CrossRef]
17. Jin, L.; Qian, Z.; Zheng, Y. High temperature performance and evaluation index of gussasphalt mortar based on DMA method. *J. Southeast Univ. Nat. Sci. Ed.* **2014**, *44*, 1062–1067. [CrossRef]
18. Yang, G.; Wang, X.; Zhou, X.; Wang, Y. Experimental Study on the Phase Transition Characteristics of Asphalt Mixture for Stress Absorbing Membrane Interlayer. *Materials* **2020**, *13*, 474. [CrossRef] [PubMed]
19. Zhang, C.; Ren, Q.; Qian, Z.; Wang, X. Evaluating the Effects of High RAP Content and Rejuvenating Agents on Fatigue Performance of Fine Aggregate Matrix through DMA Flexural Bending Test. *Materials* **2019**, *12*, 1508. [CrossRef] [PubMed]
20. Gedik, A. A Literature Review of Bitumen Hardening from Past to Present: Aging Mechanism, Indicators, Test Methods, and Retarders. In Proceedings of the International Airfield and Highway Pavements Conference 2021, Austin, TX, USA, 8–10 June 2021. [CrossRef]

MDPI

Article

Pyrolysis Combustion Characteristics of Epoxy Asphalt Based on TG-MS and Cone Calorimeter Test

Xiaolong Li [1,2], Junan Shen [3,*], Tianqing Ling [1] and Qingbin Mei [2]

1   School of Civil Engineering, Chongqing Jiaotong University, Chongqing 400074, China; xiaolongli1989@foxmail.com (X.L.); lingtq@163.com (T.L.)
2   Yunnan Research Institute of Highway Science and Technology, Kunming 650051, China; meiqb@126.com
3   Civil Engineering and Construction Management, Georgia Southern University, Statesboro, GA 30458, USA
*   Correspondence: jshen@georgiasouthern.edu; Tel.: +86-13669711440

Abstract: To examine the pyrolysis and combustion characteristics of epoxy asphalt, the heat and smoke release characteristics were analyzed via TG-MS and cone calorimeter tests, and the surface morphology of residual carbon after pyrolysis and combustion was observed via scanning electron microscopy. The results showed that the smoke produce rate of epoxy asphalt was high in the early stage, and then sharply decreased. Moreover, the total smoke produced was close to that of base asphalt, and the surface of residual carbon presented an irregular network structure, which was rough and loose, and had few holes, however most of them existed in the form of embedded nonpenetration. The heat and smoke release characteristics of epoxy asphalt showed that it is not a simple fusion of base asphalt and epoxy resin. Instead, they promote, interact with, and affect each other, and the influence of epoxy resin was greater than that of base asphalt.

Keywords: epoxy asphalt; TG-MS; cone calorimeter test; pyrolysis combustion characteristics

Citation: Li, X.; Shen, J.; Ling, T.; Mei, Q. Pyrolysis Combustion Characteristics of Epoxy Asphalt Based on TG-MS and Cone Calorimeter Test. *Materials* **2022**, *15*, 4973. https://doi.org/10.3390/ma15144973

Academic Editors: Francesco Canestrari and Simon Hesp

Received: 20 June 2022
Accepted: 11 July 2022
Published: 17 July 2022

Publisher's Note: MDPI stays neutral with regard to jurisdictional claims in published maps and institutional affiliations.

1. Introduction

At present, asphalt pavement is widely used in highway construction in China, and various asphalt modification technologies have been developed. Among them, epoxy asphalt is used for constructing steel bridge surfaces and tunnel pavement because of its high strength, high temperature resistance, and fatigue resistance [1–3]. Epoxy asphalt is synthesized from asphalt and epoxy resin. Asphalt is a complex mixture of hydrocarbon and nonhydrocarbon derivatives, and is mainly composed of carbon, hydrogen, and other elements. Epoxy resin (EP) is a polymer material with two or more epoxy groups on the molecular chain and is a widely used thermosetting plastic [4]. Both asphalt and EP are combustible. In case of a fire caused due to fuel leakage, a large amount of heat and toxic smokes are released [5], which severely affect the structural safety of bridges and tunnels [6–8], endanger personal safety, and pollute the environment [9–11]. Therefore, examining the pyrolysis and combustion characteristics of epoxy asphalt is important for road fire safety and methods of flame retardant and smoke suppression.

Wu et al. [12] used the Rosemount NGA2000 analyzer to analyze the combustion characteristics of asphalt at a high-temperature rise rate and examined the release law of gaseous products of asphalt and its slurry combustion. Zhu et al. [13] studied the combustion mechanism and gaseous products of asphalt binder at five different oxygen concentrations (21%, 18%, 15%, 12%, and 10%) by thermogravimetry-Fourier transform infrared spectroscopy (TG-FTIR), and results show that the influence on the combustion of heavy components is more significant. Shi et al. [14] analyzed the combustion characteristics of asphalt via a thermogravimetry–infrared combined experiment and discussed the dynamic evolution law of asphalt along with a four-component combustion process. Xia et al. [15] studied the composition of gaseous products of four-component combustion of asphalt via thermogravimetry and mass spectrometry (TG-MS). Xia et al. [16] analyzed the thermogravimetric curves of asphalt pyrolysis under different heating rates and found that the heating rate can affect the pyrolysis mass loss, morphology, and total number of products. Yang et al. [17] evaluated the composition and distribution of pyrolysis products of four components in the pyrolysis process by pyrolysis gas chromatograph coupled with mass spectrometry (PY-GC–MS). Huang et al. [18] reported that the mixing speed, heating temperature, and heating duration of asphalt are the main factors that change the asphalt smoke emission. Wang et al. [19] summarized the test methods of fire effluents produced by a bitumen and asphalt mixture after combustion and determined the influencing factors of fume concentration and composition. Mouritz [20] summarized the key problems of thermosetting matrix composites' fire behavior of epoxy and phenolic resins, including combustion mechanism, flame retardant, and fire reaction characteristics. Wen et al. [21] studied the pyrolysis characteristics of epoxy resin in SF_6/N_2 environment via molecular dynamics. Li et al. [22] studied the heat release rate (HRR), total heat release (THR), smoke production rate (SPR), and total smoke release (TSR) of epoxy resin based on the cone calorimeter method. Xu et al. [23,24] studied the combustion pyrolysis characteristics of carbon fiber/epoxy composites via thermogravimetric analysis and the cone calorimeter method. Zhong et al. [25] used Amsterdam density functional (ADF) software to simulate the correlation between the formation of oxygen-containing small molecules and the number of hydroxyl radicals at different temperatures during resin pyrolysis. It can be seen that researchers all around the world have explored the pyrolysis and combustion properties of asphalt and epoxy resin; however, these were all independent analyses of materials, and there was little discussion on the synthesized epoxy asphalt. Therefore, further research is needed.

According to the composition characteristics of epoxy asphalt, the heat and smoke release characteristics of epoxy asphalt are analyzed via TG-MS and cone calorimeter tests, and the surface morphology of residual carbon after pyrolysis and combustion can be observed via scanning electron microscopy (SEM), which provides a theoretical basis for further analysis of the pyrolysis and combustion characteristics of epoxy asphalt.

2. Materials and Methods

2.1. Raw Materials

The base asphalt adopts Shell Pen 70, and the main performance indexes are listed in Table 1. The test results meet the specification requirements [26]. Epoxy asphalt was prepared by premixing component A (E-51 bisphenol A type) and component B (modified aromatic amine curing agent) in a ratio of 60:40 (mass ratio) at 60 °C to first form the epoxy resin and was then poured into equal mass base asphalt at 150 °C for 4min of shear mixing. The main performance indexes of component A and epoxy asphalt are listed in Tables 2 and 3, respectively. The test methods, conditions, and technical requirements for performance indexes of epoxy resin and epoxy asphalt shall be implemented in accordance with the specification [27]. The samples of epoxy resin and epoxy asphalt were first cured in an oven at 150 °C for 3 h, followed by curing in an oven at 60 °C for 4 days. It was finally placed at 25 °C for 1 day before testing.

Table 1. Performance indexes of base asphalt.

Test Items	Pen. /0.1 mm	R&B/°C	Ductility (10 °C) /cm	RTFOT		
				Quality Change/%	Residual Penetration Ratio/%	Residual Ductility (10 °C)/cm
Test value	72	47.5	33	0.48	68.5	9
Technical requirement	60–80	≥45	≥15	≤±0.8	≥61	≥6

Table 2. Performance indexes of epoxy resin component A.

Test Items	Viscosity (23 °C) /(Pa·s)	Epoxy Equivalent /(g·mol^{-1})	Water Content /%	Flash Point/°C	Density /(g·cm^{-3})	Appearance
Test value	13.278	189	0.02	281	1.164	Transparent
Technical requirement	11–15	185–192	≤0.05	≥200	1.16–1.17	Transparent

Table 3. Performance indexes of epoxy asphalt.

Test Items	Tensile Strength (23 °C)/MPa	Fracture Elongation (23 °C)/%	Thermoset (300 °C)	Residence Time /min
Test value	4.3	221	Un-melted	98
Technical requirement	≥1.5	≥200	300°C Un-melted	≥40

2.2. Test Methods

(1) TG-MS

In this study, the pyrolysis characteristics of base asphalt, epoxy resin, and epoxy asphalt were examined using the Japanese physio-thermo plus EV2/thermo mass photo TG-MS technology under a 70 eV electron ionization (EI) source. The heating range of TG ranged from 25 to 800 °C and the heating rate was 10 °C/min. The thermogravimetric analyzer and mass spectrometer were used simultaneously, and the scanning mode was ion scanning, simulating air atmosphere.

(2) Cone Calorimeter Test

The cone calorimeter test is one of the most commonly used test instruments to study the combustion performance of materials. It is capable of examining time to ignition (TTI), HRR, THR, SPR, TSR, and gas parameters under different oxygen mole fractions, and is widely used in the evaluation of the combustion performance of materials. In this test, the FTT0402 cone calorimeter manufactured by the Fire Testing Technology Company in the UK was used for the mesoscale combustion test. The tests conducted were in accordance with ISO 5660-1 and ASTM E1354 standards. Each sample was wrapped with aluminum foil paper and was tested under 50 kW/m^2 of thermal radiation. The sample specification was $100 \times 100 \times 3$ mm^3.

3. Results and Analysis

3.1. TG-MS Analysis

3.1.1. TG Analysis

TG and differential thermal gravity (DTG) of epoxy asphalt are shown in Figure 1.

It can be seen from the figure that the weight loss of epoxy asphalt mainly occurs at 232.1–579.9 °C. Moreover, the weight loss rate is 96.13% and the maximum weight loss rate is −6.093 %/min. According to the weight loss rate curve, the pyrolysis weight loss process of epoxy asphalt can be divided into three consecutive stages. The temperature in the first stage ranges from 232.1 to 387.9 °C and the corresponding T_{max1} is observed at 342.8 °C, and the weight loss is approximately 38.1%. In this stage, the weight loss mainly occurs due to the fracture and degradation of the network structure of epoxy resin, the large amount of pyrolysis weight loss of saturated components, and participation of some small molecules of volatile aromatic components during pyrolysis. The saturated fraction and aromatic fraction are decomposed into alkanes and cycloalkanes under high-temperature conditions. With the further increase in temperature, the alkyl side chain breaks to generate some short-chain alkanes and gaseous products [28]. The temperature in the second stage ranges from 387.9 to 467.3 °C and the corresponding T_{max2} is observed at 440.8 °C, and the weight loss is approximately 23.8%. This stage mainly involves the decomposition of compounds and colloids formed in the previous stage. At higher temperatures, some relatively stable chemical structures begin to undergo cracking reaction and generate a large number of free radicals and a large number of gaseous products. Moreover, dehydrogenation polymerization of residual substances generates more stable macromolecular substances. The temperature in the third stage ranges from 467.3 to 579.9 °C and the corresponding T_{max3} is observed at 510.9 °C, and the weight loss is approximately 33.7%. This stage is mainly caused by the further degradation of semicoke polymer formed in the first stage of epoxy resin and the pyrolysis of heavy components in asphalt. In this stage, the carbonized layer generated by alkyl side-chain dehydrogenation polymerization in the first and second stages begins to crack and burn, and the heavy substance, asphaltene, in asphalt begins to burn [29].

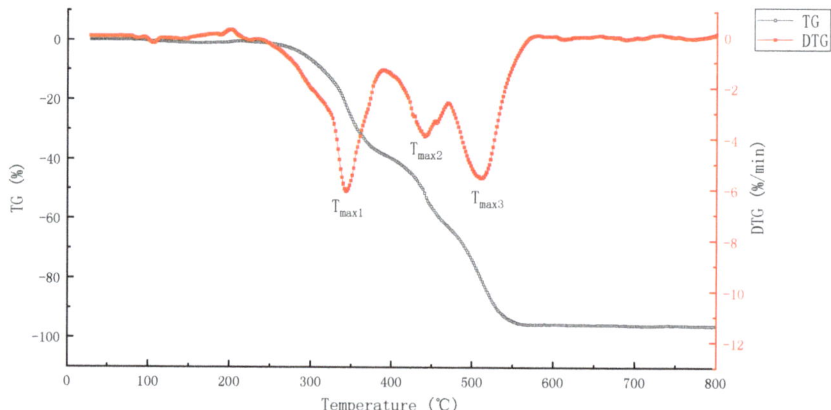

Figure 1. TG and DTG rate curve of epoxy asphalt.

The pyrolysis process of epoxy asphalt shows obvious multi-stage characteristics, indicating that the pyrolysis reaction process of epoxy asphalt is very complex. This is mainly because the base asphalt and epoxy resin are complex mixtures, and each component has different thermal stability. During the heating process, they gradually start pyrolysis, combustion, absorb, or release different amounts of heat, resulting in the multi-stage characteristics of the DTG curve.

3.1.2. MS Analysis

Figure 2 shows the MS-DTG curve of pyrolysis volatiles of epoxy asphalt, in which the DTG curve provides a reference for analyzing the precipitation process of volatiles. Table 4 lists the distribution of pyrolysis volatiles of epoxy asphalt at each stage. Figures 3–5 show the mass spectrum of volatile matter from pyrolysis combustion of epoxy asphalt at characteristic temperatures.

According to the test results of the MS analysis, in the first stage of pyrolysis, the percentage of weight loss was the largest; however, the amount of volatile released was insufficient and mainly included carbon monoxide, water, carbon dioxide, hydrogen peroxide, acetaldehyde, and propane. This is because in the first stage, the fracture and degradation of the epoxy resin network structure occurred with large weight loss; however, the composition of epoxy resin was relatively uncomplicated and less volatile. The third stage comprised the most volatile components, because with the gradual increase in temperature, the analysis of the light component in the saturated component showed that the alkyl side chain broke, there were more high molecular weight components, and the generated coke began to burn and release more volatiles.

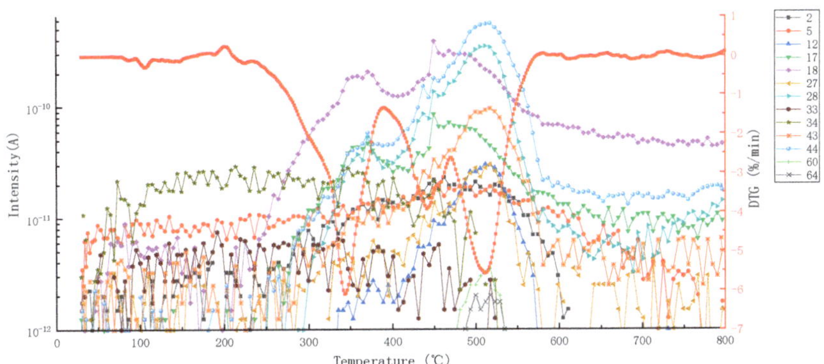

Figure 2. MS-DTG curve of main volatiles from pyrolysis combustion of epoxy asphalt.

Table 4. Distribution of volatile matter from pyrolysis combustion of epoxy asphalt.

Pyrolysis Combustion Stage	Temperature Range/°C	Thermal Weightlessness/%	Main Volatiles (Characteristic Products)
Phase I	232.1–387.9	38.1	H_2(2), C ion fragment(12), NH_3(17), H_2O(18), CNH(27), N_2(28), CO(28), C_2H_4(28), HS-ion fragment(33), H_2S(34), H_2O_2(34), N_3H(43), CH_2O_2(44), C_3H_8(44), CO_2(44), N_2O(44)
Phase II	387.9–467.3	23.8	H_2(2), C ion fragment(12), NH_3(17), H_2O(18), CNH(27), N_2(28), CO(28), C_2H_4(28), HS-ion fragment(33), H_2S(34), H_2O_2(34), N_3H(43), CO_2(44), C_2H_4O(44), C_3H_8(44), N_2O(44)
Phase III	467.3–579.9	33.7	H_2(2), C ion fragment(12), NH_3(17), H_2O(18), CNH(27), N_2(28), CO(28), C_2H_4(28), N_3H(43), CO_2(44), C_2H_4O(44), N_2O(44), C_3H_8(44), C_3H_8O(60), $C_2H_4O_2$(60), $C_2H_8N_2$(60), C_2H_4S(60), SO_2(64)

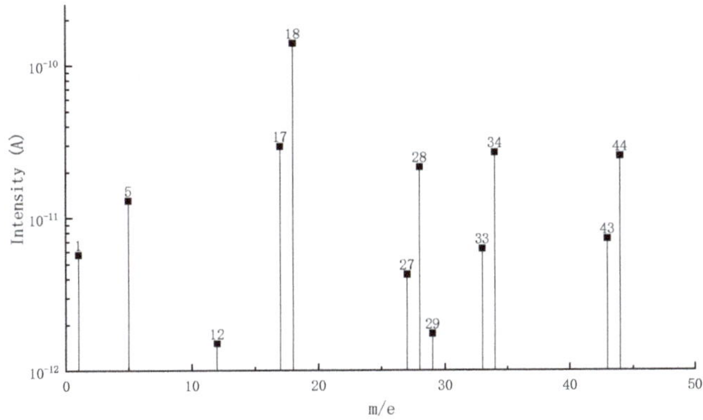

Figure 3. Mass spectrum of volatile matter from pyrolysis combustion of epoxy asphalt at characteristic temperature (342.8 °C).

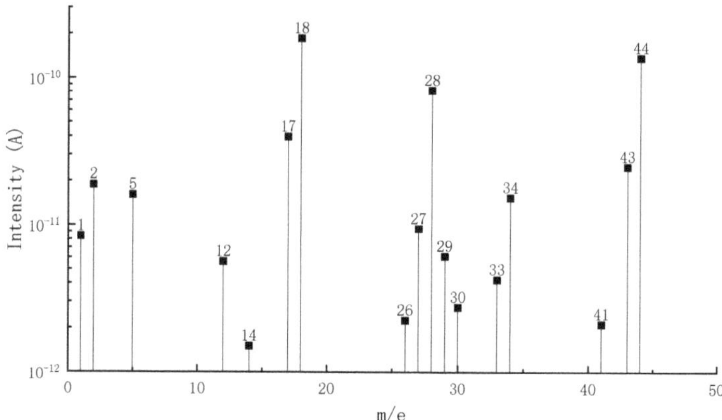

Figure 4. Mass spectrum of volatile matter from pyrolysis combustion of epoxy asphalt at characteristic temperature (440.8 °C).

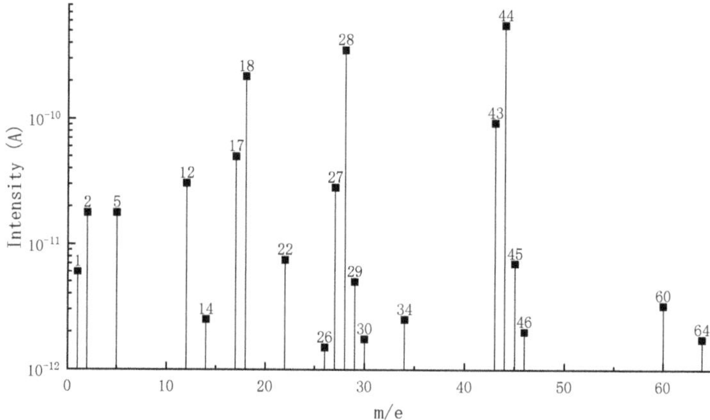

Figure 5. Mass spectrum of volatile matter from pyrolysis combustion of epoxy asphalt at characteristic temperature (510.9 °C).

It can be seen from Figure 2 that the peak value of the ion flow curve appears when it is close to the peak temperature of each pyrolysis stage, indicating that the volatile release is the largest at this time. Figures 3–5 show that the high-content ions in the gas volatiles at the peak temperature in the three stages of pyrolysis are the same, that is, H_2O with a mass charge ratio of 18, N_2, CO, and C_2H_4 with a mass charge ratio of 28, and CO_2, C_2H_4O, N_2O, and C_3H_8 with a mass charge ratio of 44. Compared with the second stage, the overall ion current intensity in the third stage is larger, and C_3H_8O, $C_2H_4O_2$, $C_2H_8N_2$, and C_2H_4S with a mass charge ratio of 60 and SO_2 with a mass charge ratio of 64 are also produced.

3.2. Cone Calorimeter Test Analysis

The FTT0402 cone calorimeter was used for the mesoscale combustion test. The test results of different materials are listed in Table 5, and the residue after the test is shown in Figure 6.

Table 5. Test results of the cone calorimeter test.

Sample	TTI (s)	HRR (kW/m^2)	THR (MJ/m^2)	SPR (m^2/s)	TSR (m^2/m^2)	COP (g/s)	CO$_2$P (g/s)
Base asphalt	36	328.8	50.5	0.145	2413.9	0.00635	0.1639
Epoxy resin	30	1015.7	104.5	0.203	2995.2	0.01728	0.5845
Epoxy asphalt	31	910.5	85.2	0.220	2492.8	0.01776	0.4991

(**a**)   (**b**)   (**c**)

Figure 6. Images of residue after the cone calorimeter test using different materials: (**a**) Base asphalt, (**b**) epoxy resin, and(**c**) epoxy asphalt.

Under strong thermal radiation, all types of materials quickly burned. The HRR reached the peak rapidly after ignition, and then abruptly dropped. The insufficient combustion of the polymer generated a large amount of black smoke and had a pungent smell. Within 500 s, all materials burned. Among them, the residual carbon remaining after pyrolysis and combustion of epoxy resin burned the least, and the residual carbon rate was less than 1%, which is consistent with the test results of thermogravimetric analysis. It was speculated that most substances in the materials were decomposed into gas components. Although base asphalt and epoxy asphalt had some residual carbon after combustion, the residual carbon was relatively broken and did not form a complete carbon layer.

3.2.1. Combustion Heat Release Performance of Different Materials

The HRR and THR of different materials are shown in Figures 7 and 8, respectively.

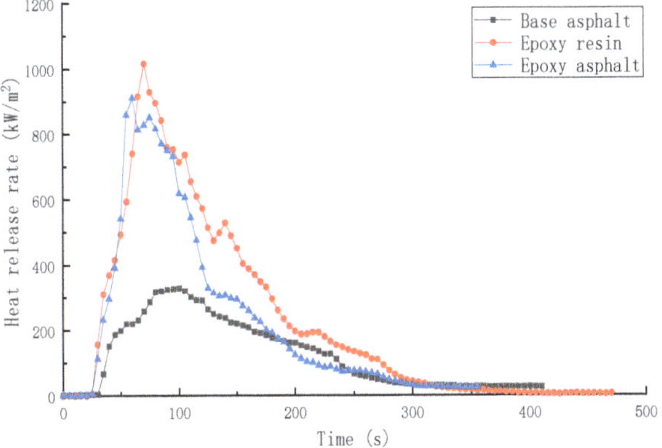

Figure 7. HRR of different materials.

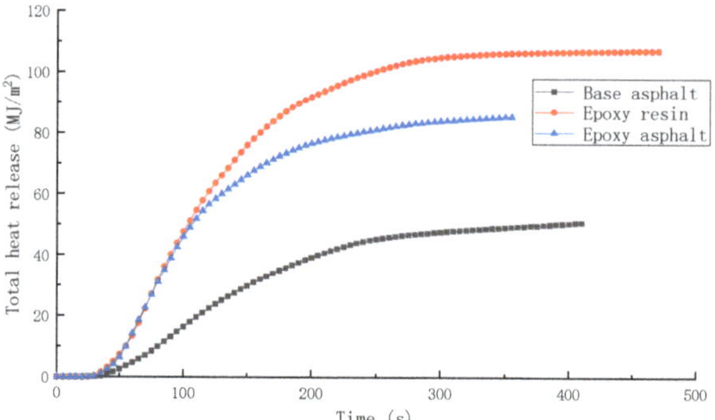

Figure 8. THR of different materials.

TTI determined the flammability of polymer under actual fire conditions. The ignition time of each material ranged between 31 and 36 s. Moreover, resin materials were easier to ignite, and it took 6–8 min for each material to burn completely until it self-extinguished.

HRR and THR are important for estimating the combustion safety of polymers. Both these parameters represent the thermal feedback generated in the material combustion process. The larger the HRR and THR, the more intense the pyrolysis process; however, it also results in worsening the safety. According to the test results, the time for base asphalt, epoxy resin, and epoxy asphalt to reach the peak of HRR was 100, 70, and 60 s, respectively. It was easier for the resin materials to reach the combustion conditions, and the peak heat release rate (pkHRR) and THR of epoxy resin were the highest, reaching 1015.7 kW/m^2 and 104.5 MJ/m^2, respectively. The HRR curve presented sharp peaks. Although the pkHRR and THR of base asphalt were the lowest, that is, 328.8 kW/m^2 and 50.5 MJ/m^2, respectively, the combustion curve was relatively smooth, indicating that the epoxy resin combustion was more intense, and the base asphalt combustion was less intense.

3.2.2. Combustion Smoke Release Performance of Different Materials

SPR, TSR, CO production rate (COP), and CO_2 production rate (CO_2P) of different materials are shown in Figures 9–12.

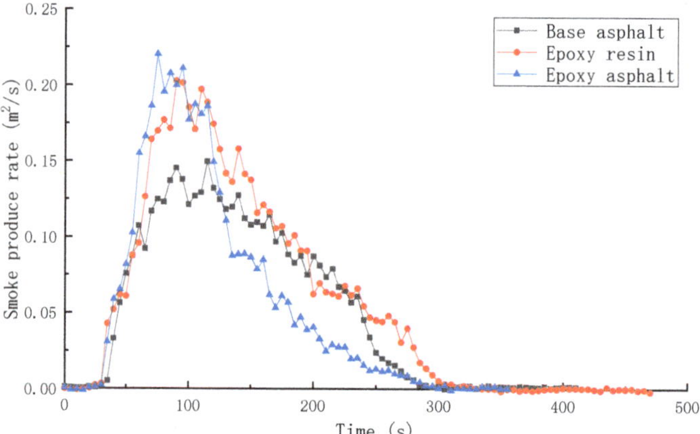

Figure 9. SPR of different materials.

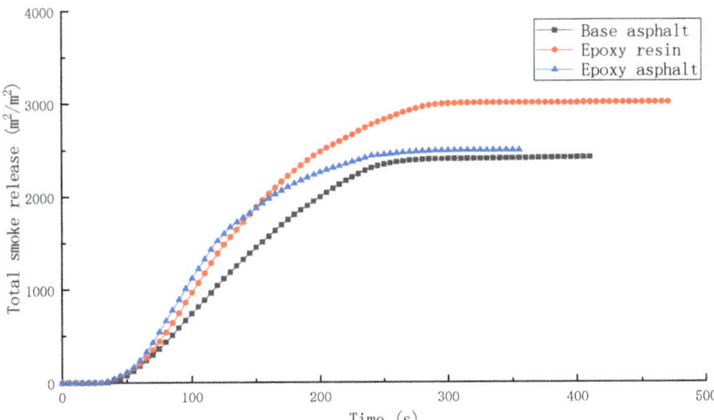

Figure 10. TSR of different materials.

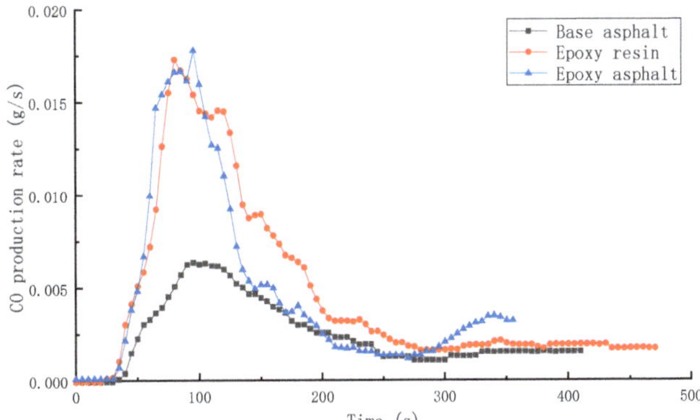

Figure 11. CO production rate (COP) of different materials.

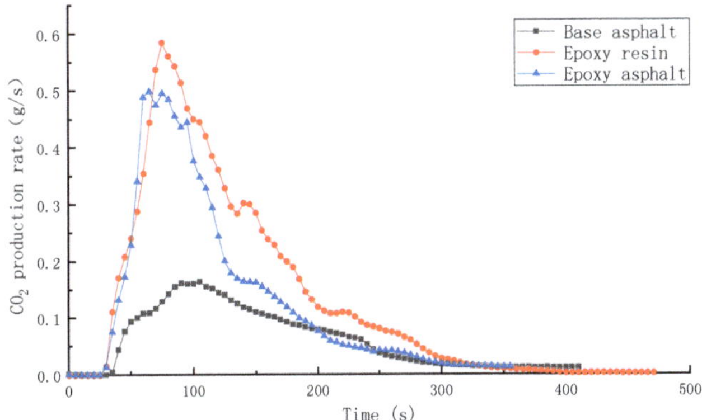

Figure 12. CO_2 production rate (CO_2P) of different materials.

Overall, the trend of SPR was consistent with that of HRR. When the HRR reached the peak, the SPR was the highest. The SPR and TSR of epoxy resin were higher during the test, while the SPR of epoxy asphalt was higher in the early stage, and then sharply decreased. The TSR was close to that of base asphalt, which was approximately 2400 m^2/m^2, indicating that the introduction of base asphalt into epoxy resin in the form of an interpenetrating network reduced the smoke generation of resin to a certain extent.

The values of COP and CO_2P of base asphalt were the lowest, that is, 0.00635 and 0.1639 g/s, respectively. In contrast, the values of COP and CO_2P of resin materials were higher, which was mainly caused by high-temperature fracture and combustion of the polymer chain.

According to the results of heat release and smoke release tests, the combustion characteristics of epoxy asphalt were not between those of base asphalt and epoxy resin; instead, they promoted, interacted with, and influenced each other. In addition, although epoxy asphalt was the equal mass synthesis of base asphalt and epoxy resin, the pyrolysis and combustion characteristics of epoxy asphalt were closer to that of epoxy resin, indicating that the influence of epoxy resin was greater than that of base asphalt.

3.3. SEM Analysis

To further understand the combustion mechanism of different materials, the residual carbon was examined using the cone calorimeter method and was sprayed with gold, and SEM (manufactured by ZEISS Sigma 300, Jena, Germany) was used. The accelerating voltage was 20 kV, amplified 50, 200, and 2000 times to observe the surface morphology. The morphology of residual carbon is shown in Figures 13–15.

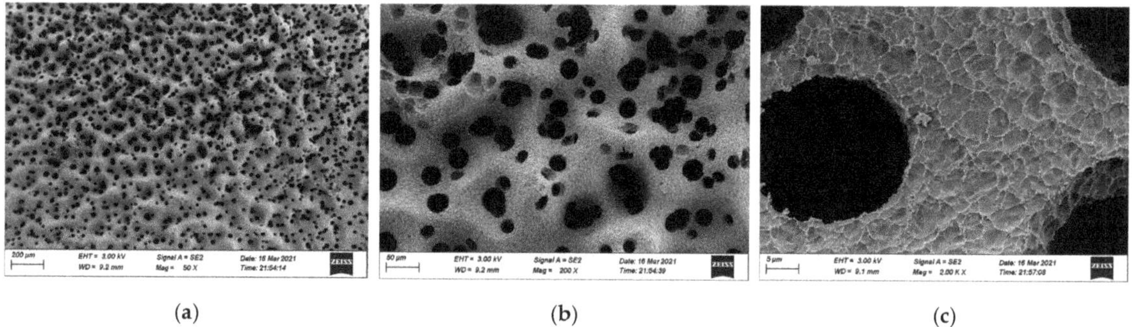

(a)   (b)   (c)

Figure 13. Morphology of residual carbon after the cone calorimeter test of base asphalt: (**a**) 50×, (**b**) 200×, and (**c**) 2000× magnification.

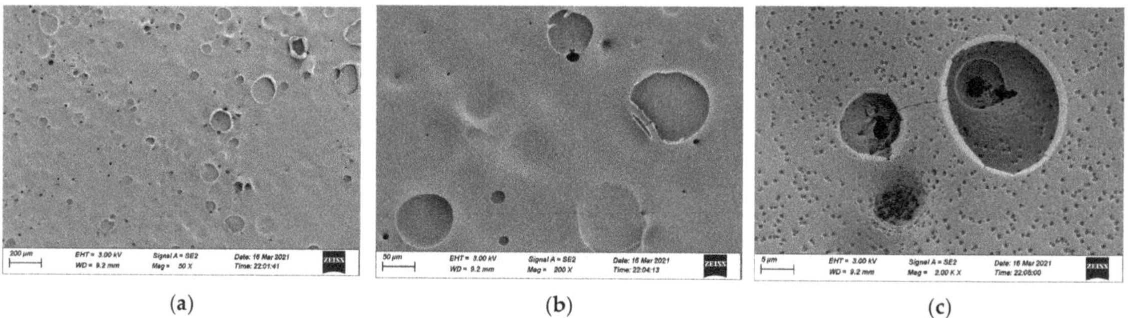

(a)   (b)   (c)

Figure 14. Morphology of residual carbon after the cone calorimeter test of epoxy resin: (**a**) 50×, (**b**) 200×, and (**c**) 2000× magnification.

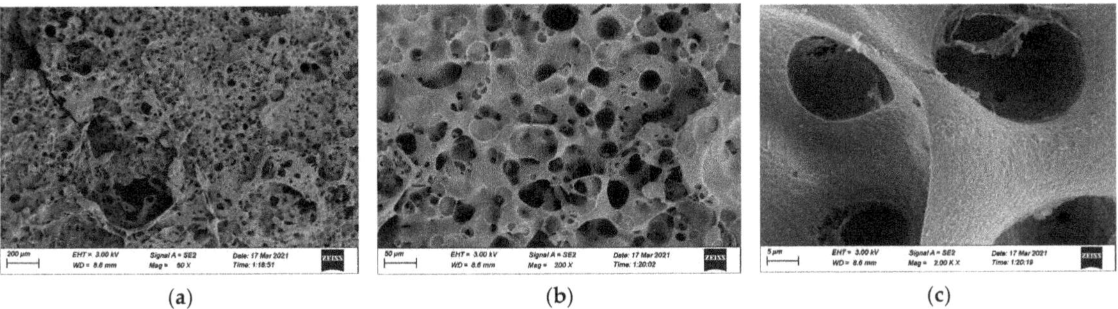

Figure 15. Morphology of carbon residue after the cone calorimeter test of epoxy asphalt: (**a**) 50×, (**b**) 200×, and (**c**) 2000× magnification.

At high temperature, the surface of base asphalt will flow and agglomerate, resulting in molten droplets, which accelerates the endothermic decomposition of asphalt. After combustion, the surface of residual carbon presented a uniform high-low fluctuation shape, which is a skeleton pore structure and has many through holes with a large pore diameter and loose structure. This is mainly caused by the precipitation of internal decomposition gas breaking through the barrier layer during asphalt combustion. Notably, more through holes indicate more volatile matter in gas phase. Under a high-power microscope (2000× magnification), it can be observed that the residue had a fine-scale structure, which may be formed by stacking collapsed structures after decomposition.

The carbon residue rate of epoxy resin after combustion was low, but the carbon residue surface was dense and flat with few through holes. The main reason for this is that the volatile content was relatively small, and the Figure 14c 2000× magnification figure shows that the combustion form is deepening combustion layer-by-layer at the hole-forming place.

It can be seen from Figure 15 that after the combustion of epoxy asphalt, the surface of residual carbon presented uneven high and low fluctuation, and the whole sample was an irregular network structure. The structure was rough and loose and had some holes with a large pore diameter; however, most of them existed in embedded nonpenetrating form, which is mainly caused by noncombustible gases such as N_2 and CO_2 released by polymers after combustion on the carbon layer.

4. Conclusions

In this study, the pyrolysis and combustion characteristics of epoxy asphalt were examined using TG-MS, cone calorimeter, and SEM. The main conclusions are as follows:

(1) Epoxy asphalt exhibited different characteristics in pyrolysis and combustion stages. When it was close to the peak temperature of each stage, a peak value was observed in the ion flow curve, and the release of volatiles was the largest.

(2) The peak HRR, THR, SPR, and TSR of epoxy resin were high, reaching 1015.7 kW/m^2, 104.5 MJ/m^2, 0.203m^2/s, and 2995.2 m^2/m^2, respectively, and the combustion was intense. However, the smoke produced from epoxy resin reduced to a certain extent after the base asphalt was added in the epoxy resin in the form of an interpenetrating network. The SPR of epoxy asphalt was high in the early stage, up to 0.220 m^2/s, then sharply decreased. Moreover, the TSR was close to that of base asphalt.

(3) The residual carbon rate was low, and the volatile content was relatively small after the combustion of epoxy resin. Its combustion form is to deepen the combustion layer-by-layer at the pore-forming place. Therefore, the residual carbon surface was dense and flat with few through holes. However, the carbon residue surface of epoxy asphalt after combustion exhibited an irregular network structure, which was rough and loose, with some holes; nevertheless, most of them existed in the form of embedded

nonpenetration, which is mainly caused by the impact of noncombustible gases such as N_2 and CO_2 released by polymers after combustion on the carbon layer.

(4)   The results of heat release and smoke release tests showed that epoxy asphalt is not a simple fusion of base asphalt and epoxy resin. Instead, they promote, interact with, and affect each other, and the influence of epoxy resin on the pyrolysis and combustion characteristics of epoxy asphalt was greater than that of base asphalt.

Author Contributions: J.S. conceived and designed the experiments; X.L. and Q.M. conducted the experiments; X.L., T.L. and J.S. analyzed the data; X.L. wrote the paper. All authors have read and agreed to the published version of the manuscript.

Funding: This research received no external funding.

Institutional Review Board Statement: Not applicable.

Informed Consent Statement: Not applicable.

Data Availability Statement: Not applicable.

Acknowledgments: This research was supported by the Yunnan Research Institute of Highway Science and Technology.

Conflicts of Interest: The authors declare no conflict of interest regarding the publication of this paper.

References

1. Huang, H.M.; Zeng, G.D.; Xu, W.; Li, S.J.; Zhou, Z.G. Study of curing reaction mechanism and construction control performance of epoxy asphalt. *J. Build. Mater.* **2020**, *23*, 941–947. [CrossRef]
2. Zhou, X.-Y.; Bo, Y.-Z.; Duan, S.-Y. Study on Low Temperature and High Temperature Performance of Epoxy Resin Particles Used for Asphalt Pavement. *DEStech Trans. Mater. Sci. Eng.* **2017**, *448*, 10861. [CrossRef]
3. Guo, P.C.; Yao, B.; Li, M.Z. Fracture characteristics and numerical simulation of epoxy resin mixture. *J. Build. Mater.* **2020**, *23*, 1160–1166,1176. [CrossRef]
4. Xu, M.-J.; Xu, G.-R.; Leng, Y.; Li, B. Synthesis of a novel flame retardant based on cyclotriphosphazene and DOPO groups and its application in epoxy resins. *Polym. Degrad. Stab.* **2016**, *123*, 105–114. [CrossRef]
5. Androjić, I.; Dimter, S. Laboratory evaluation of the physical properties of hot mix asphalt exposed to combustion. *Constr. Build. Mater.* **2022**, *323*, 126569. [CrossRef]
6. Puente, E.; Lázaro, D.; Alvear, D. Study of tunnel pavements behaviour in fire by using coupled cone calorimeter—FTIR analysis. *Fire Saf. J.* **2016**, *81*, 1–7. [CrossRef]
7. Zhu, K.; Huang, Z.Y.; Wu, K.; Wu, B.; Zhang, X.; Zhang, C. Hydrated lime modification of asphalt mixtures with improved fire performance. *J. Zhejiang Univ. Sci.* **2015**, *49*, 963–968. [CrossRef]
8. Wang, X.; Hu, Y.; Song, L.; Xing, W.; Lu, H.; Lv, P.; Jie, G. Flame retardancy and thermal degradation mechanism of epoxy resin composites based on a DOPO substituted organophosphorus oligomer. *Polymer.* **2010**, *51*, 2435–2445. [CrossRef]
9. Wang, J.; Qian, L.; Huang, Z.; Fang, Y.; Qiu, Y. Synergistic flame-retardant behavior and mechanisms of aluminum poly-hexamethylenephosphinate and phosphaphenanthrene in epoxy resin. *Polym. Degrad. Stab.* **2016**, *130*, 173–181. [CrossRef]
10. Yang, S.; Wang, J.; Huo, S.; Wang, M.; Wang, J.; Zhang, B. Synergistic flame-retardant effect of expandable graphite and phosphorus-containing compounds for epoxy resin: Strong bonding of different carbon residues. *Polym. Degrad. Stab.* **2016**, *128*, 89–98. [CrossRef]
11. Qiu, J.; Yang, T.; Wang, X.; Wang, L.; Zhang, G. Review of the flame retardancy on highway tunnel asphalt pavement. *Constr. Build. Mater.* **2019**, *195*, 468–482. [CrossRef]
12. Wu, K.; Zhu, K.; Huang, Z.Y.; Wang, J.C.; Yang, Q.M.; Liang, P. Research on the gas based on infrared spectral analysis. *Spectrosc. Spectr. Anal.* **2012**, *32*, 2089–2094. [CrossRef]
13. Zhu, K.; Qin, X.; Wang, Y.; Lin, C.; Wang, Q.; Wu, K. Effect of the oxygen concentration on the combustion of asphalt binder. *J. Anal. Appl. Pyrolysis.* **2021**, *160*, 105370. [CrossRef]
14. Shi, H.; Xu, T.; Jiang, R. Combustion mechanism of four components separated from asphalt binder. *Fuel* **2017**, *192*, 18–26. [CrossRef]
15. Xia, W.; Xu, T.; Wang, H. Thermal behaviors and harmful volatile constituents released from asphalt components at high temperature. *J. Hazard. Mater.* **2019**, *373*, 741–752. [CrossRef]
16. Xia, W.; Xu, T.; Wang, S.; Wang, H. Mass loss evolution of bituminous fractions at different heating rates and constituent conformation of emitted volatiles. *Energy Sci. Eng.* **2019**, *7*, 2782–2796. [CrossRef]
17. Yang, D.; Yang, C.; Xie, J.; Wu, S.; Ye, Q.; Hu, R. Combusting behavior and pyrolysis products evaluation of SARA components separated from bitumen. *Constr. Build. Mater.* **2020**, *244*, 118401. [CrossRef]

18. Huang, G.; He, Z.Y.; Huang, T. Laboratory measurement and analysis of effect factor on asphalt fume under the elevated temperature. *J. Build. Mater.* **2015**, *18*, 322–327. [CrossRef]
19. Wang, W.Z.; Shen, A.Q.; Wang, L.S.; Liu, H.C. Measurements, emission characteristics and control methods of fire effluents from tunnel asphalt pavement during fire: A review. *Res. Square.* **2022**, 480339. [CrossRef]
20. Mouritz, A.P.; Gibson, A.G. *Fire Properties of Polymer Composite Materials*; Springer: Dordrecht, The Netherlands, 2006.
21. Wen, H.; Zhang, X.; Xia, R.; Yang, Z.; Wu, Y. Thermal Decomposition Properties of Epoxy Resin in SF6/N2 Mixture. *Materials* **2018**, *12*, 75. [CrossRef] [PubMed]
22. Li, S.; Zhang, G.X.; Wang, W.B.; Wang, J.; Yang, Y.J.; Fan, H.J.; Tan, L. Research on the combustion performance of epoxy resin based on the cone calorimeter. *Chem. Adhes.* **2019**, *41*, 457–459,483. [CrossRef]
23. Xu, Y.; Yang, Y.; Shen, R.; Parker, T.; Zhang, Y.; Wang, Z.; Wang, Q. Thermal behavior and kinetics study of carbon/epoxy resin composites. *Polym. Compos.* **2019**, *40*, 4530–4546. [CrossRef]
24. Xu, Y.; Lv, C.; Shen, R.; Wang, Z.; Wang, Q. Experimental investigation of thermal properties and fire behavior of carbon/epoxy laminate and its foam core sandwich composite. *J. Therm. Anal. Calorim.* **2019**, *136*, 1237–1247. [CrossRef]
25. Zhong, Y.; Jing, X.; Wang, S.; Jia, Q.-X. Behavior investigation of phenolic hydroxyl groups during the pyrolysis of cured phenolic resin via molecular dynamics simulation. *Polym. Degrad. Stab.* **2016**, *125*, 97–104. [CrossRef]
26. Ministry of Transport of the People's Republic of China. *Technical Specifications for Construction of Highway Asphalt Pavements*; JTG F40–2014; Ministry of Transport of the People's Republic of China: Beijing, China, 2014.
27. General Administration of Quality Supervision, Inspection and Quarantine of the People's Republic of China. *General Specifications of Epoxy Asphalt Materials for Paving Roads and Bridges*; GB/T 30598–2014; General Administration of Quality Supervision, Inspection and Quarantine of the People's Republic of China: Beijing, China, 2014.
28. Shi, H.; Xu, T.; Zhou, P.; Jiang, R. Combustion properties of saturates, aromatics, resins, and asphaltenes in asphalt binder. *Constr. Build. Mater.* **2017**, *136*, 515–523. [CrossRef]
29. Xu, T.; Huang, X. Study on combustion mechanism of asphalt binder by using TG–FTIR technique. *Fuel* **2010**, *89*, 2185–2190. [CrossRef]

Article

Evaluation of the High- and Low-Temperature Performance of Asphalt Mortar Based on the DMA Method

Yanzhu Wang [1], **Xudong Wang** [1,2,*], **Zhimin Ma** [2], **Lingyan Shan** [2] and **Chao Zhang** [1]

[1] School of Transportation Science and Engineering, Harbin Institute of Technology, Harbin 150090, China; wangyzh0102@163.com (Y.W.); 18b332002@stu.hit.edu.cn (C.Z.)

[2] Research Institute of Highway, Ministry of Transport, Beijing 100088, China; mzm1755709496@163.com (Z.M.); ly.shan@rioh.cn (L.S.)

* Correspondence: xd.wang@rioh.cn

Abstract: Asphalt mortar is a typical temperature-sensitive material that plays a crucial role in the performance of asphalt mixture. This study evaluates the high- and low-temperature performance of asphalt mortar based on the dynamic mechanical analysis (DMA) method. Temperature-sweep tests of asphalt mortars were conducted using the DMA method under fixed strain level, frequency, and heating rate conditions. The dynamic mechanical response curves, characteristic temperature, and other indices were obtained and used to investigate the high- and low-temperature performance of asphalt mortar. The results showed that the phase transition temperatures T_1, T_0, and T_g can be used to evaluate the low-temperature performance of asphalt mortar. Additionally, they had a good linear relationship, and the evaluation results were consistent. Meanwhile, T_2, E_{60}, and $\tan(\delta)_{max}$ indicators can effectively evaluate the high-temperature performance of asphalt mortar. Asphalt plays a key role in the performance of asphalt mortar. Mortars with neat asphalt A70 and modified asphalt AR had the worst and best high- and low-temperature performances, respectively. Furthermore, the finer gradation improved the low-temperature performance of asphalt mortar, while the coarser gradation improved the high-temperature properties of modified asphalt mortars but had the opposite effect on neat asphalt A70.

Keywords: asphalt mortar; dynamic mechanical analysis; phase transition; high- and low-temperature performance

Citation: Wang, Y.; Wang, X.; Ma, Z.; Shan, L.; Zhang, C. Evaluation of the High- and Low-Temperature Performance of Asphalt Mortar Based on the DMA Method. *Materials* **2022**, *15*, 3341. https://doi.org/10.3390/ma15093341

Academic Editor: Gabriele Milani

Received: 11 April 2022
Accepted: 2 May 2022
Published: 6 May 2022

Publisher's Note: MDPI stays neutral with regard to jurisdictional claims in published maps and institutional affiliations.

1. Introduction

Asphalt mixtures are widely used in highway construction, and their performance can be studied from various scales including asphalt binder, mastic, mortar (or fine aggregate matrix, FAM), and mixtures [1]. Asphalt mortar is a soft phase filled in the coarse aggregate particles of the asphalt mixture. It is composed of asphalt binder, voids, filler particles, and fine aggregate particles with specific gradation [2,3]. Asphalt mortar significantly impacts the road performance and service life of asphalt mixtures [4–7]. Therefore, research on the performance of asphalt mortar is useful to further study the material characteristics of asphalt mixtures.

Asphalt mortar is a homogeneous material because its composition is relatively simple. However, some studies have shown that additives, such as composite additives, synthetic fibers, and mineral fillers [8–10], can improve the performance of asphalt mortar, and the asphalt, gradation, and asphalt–aggregate ratio are the main factors determining the viscoelastic properties of mortar [11,12]. Researchers hope to gain a better understanding of the road performance of asphalt mixtures by evaluating asphalt mortar, because performance tests on asphalt mortar are more efficient and repeatable [3,13].

In previous research, various test procedures were adopted by different researchers to evaluate the properties of asphalt mortar. Yang et al. [7], Im et al. [14], and Aragão et al. [15]

evaluated the low-temperature performance and fracture properties of asphalt mortar usings semicircular bending tests. Meanwhile, Gong et al. [16] used a bending beam rheometer to effectively investigate the low-temperature property of asphalt mortar. Zhang et al. [17] investigated the flexural–tensile rheological behavior of asphalt mortar by the beam bending creep test. The stress relaxation behavior of cement and asphalt mortar under different initial strain levels was investigated by Fu [18]. Zhu et al. [19] evaluated the fatigue performance of fine aggregate matrix containing recycled asphalt shingles using the conventional time sweep test and a proposed strain sweep test. Furthermore, the dynamic mechanical analysis method (DMA) has widely been used recently to analyze the viscoelastic properties of asphalt mortar. Fu et al. [20] evaluated the viscoelastic behavior of cement and asphalt mortar using the DMA method at a temperature range of -40 to $60\ ^\circ$C. Yu et al. [21] evaluated the effect of ultraviolet aging on the dynamic mechanical properties of SBS-modified asphalt mortar using the bending and dynamic moduli. They found that the bending modulus of SBS-modified asphalt mortar increased with the increase in UV intensity and aging time. Additionally, they found that the storage and complex moduli gradually increased with the intensity of aging, but the aging impact of the loss modulus and factor were not obvious. Fang et al. [22] studied the effects of mix parameters on the dynamic mechanical properties of the asphalt mortar using the DMA method. They quantitatively established the relationships between dynamic mechanical properties and the volume fraction of phases in asphalt mortar. In sum, testing and evaluating viscoelastic properties of asphalt mortar using the DMA method is popular because of its simplicity, reproducibility, and flexibility.

The viscoelasticity of asphalt mortar exhibits different phase states over a wide temperature range, including glass brittle solid, elastic rubber, or a viscous fluid, which is directly related to the road performance of asphalt mixture [23]. The DMA method can be used to study the phase transition of viscoelastic materials [24]. The characteristic temperature corresponding to the transition of a material from one phase state to another is called the phase transition temperature, which can be obtained from DMA testing by measuring the complex modulus of the viscoelastic material at different temperatures. When a certain viscoelastic material is subjected to sinusoidal stress, it exhibits delayed strain. The complex modulus, storage modulus, loss modulus, and phase angle tangent value curve can be obtained using the dynamic mechanical test method. The relationship between them is as follows:

$$E' = E\cos(\delta) \tag{1}$$

$$E'' = E\sin(\delta) \tag{2}$$

$$\tan(\delta) = \frac{E''}{E'} \tag{3}$$

where E is the complex modulus; E' is the storage modulus; E'' is the loss modulus; δ is the phase angle.

The temperature corresponding to the inflection point of the complex modulus curve or the temperature corresponding to the peak point of the loss modulus curve and tangent of the phase angle curve is called the phase transition temperature or glass transition temperature [23–26]. The phase transition temperature of a viscoelastic material is an important criterion in predicting its field performance. However, although many studies on the viscoelastic properties of asphalt mortar used the DMA method, few evaluations on the phase transition characteristics and phase transition temperature of asphalt mortar exist.

In this study, six asphalt mortars with three asphalt and two gradations were prepared and tested using the DMA method. The phase transition temperature and other characteristic indices of asphalt mortar were obtained through the mechanical response curves, and the high- and low-temperature performance of asphalt mortars were evaluated.

2. Materials and Methods

2.1. Materials

In this study, neat asphalt A70, SBS-modified asphalt, and rubber asphalt AR with 22 wt% crumb-rubber-powder content were used. Table 1 shows the basic technical indices of asphalt. Limestone aggregate and mineral powder were used to make asphalt mortar, and its technical indices met the technical specifications for asphalt pavement construction. Asphalt mortar was formed using two grading methods: G1 and G2. The maximum particle size of the aggregate used was less than or equal to 2.36 mm, and Table 2 shows the specific gradation. The specimens tested in this work were produced by the Superpave gyratory compactor, and the air void level was measured as shown in Table 3. As shown in Figure 1, the cylindrical specimen was cut with a length, width, and height of 60, 13, and 3 mm, respectively.

Table 1. Basic technical indices of asphalt.

Asphalt	Penetration (25 °C)/0.1 mm	Softening Point/°C	Ductility (5 cm/min, 10 °C)	PG Grade
A70	67	49.0	55	PG64-22
SBS	63	72.7	49	PG82-22
AR	39	72.6	-	PG88-28

Table 2. Gradation of aggregates used in the study.

Percent Passing /%wt	Sieve Size/mm						
	4.75	2.36	1.18	0.6	0.3	0.15	0.075
G1	100	67.8	46.2	31.7	21.6	14.7	10
G2	100	84.8	69.8	55.1	40.1	25	10

Table 3. Scheme of asphalt mortar molding.

Asphalt Mortar	Asphalt	Gradation	Asphalt–Aggregate Ratio/%	Porosity/%
G1-A70	A70	G1	6.5	3.6
G1-SBS	SBS	G1	6.5	3.7
G1-AR	AR	G1	6.5	3.4
G2-A70	A70	G2	6.5	2.8
G2-SBS	SBS	G2	6.5	2.8
G2-AR	AR	G2	6.5	2.7

Figure 1. Asphalt mortar samples.

2.2. Methods

The DMA test was conducted using a TA instrument DMA Q-800 apparatus (New Castle, DE, USA) and dual cantilever loading clamp as shown in Figure 2. The temperature-sweep mode was used with a fixed frequency of 1 Hz and strain of 50 $\mu\varepsilon$. The criteria for using a strain of 50 $\mu\varepsilon$ were based on ensuring that the materials remained in the linear viscoelastic zone and avoiding the influence of noise caused by too small strain on the test accuracy [25]. The test was conducted at a temperature ranging from −35 to 75 °C and a rate of heating of 2 °C/min. Liquid nitrogen was used for temperature control during the test. Each specimen was fixed on the double cantilever clamp with the force of fixed torque. The furnace was turned off, and the temperature was cooled down to −35 °C and held constant for 10 min. Then, the temperature was raised at a heating rate of 2 °C/min. At the same time, the dynamic load was applied to the specimen, and the complex modulus, storage modulus, loss modulus, phase angle, and other information were collected. Two specimens were tested from each asphalt mortar, and the results were averaged.

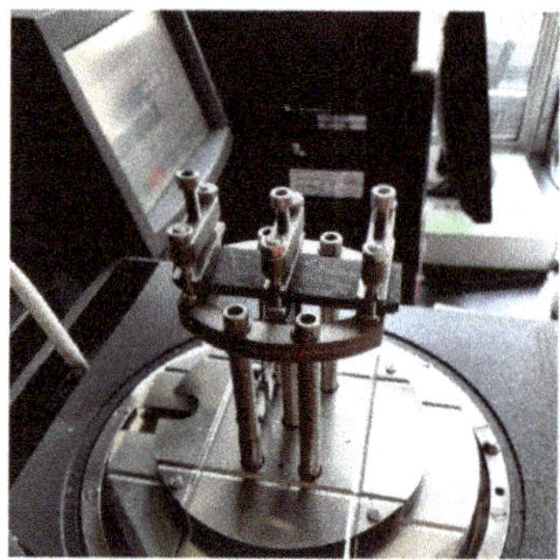

Figure 2. DMA instrument and dual cantilever clamp.

Figure 3 shows the complex modulus, loss modulus, and tan(δ) curves obtained from the temperature-sweep test. The curve of the complex modulus, varying with the temperature, shows an inverse S-shape. In this study, the modulus curve was fitted using the Boltzmann function [24] in the temperature range of −30 to 70 °C as shown in Equation (4).

$$y = \frac{A_1 - A_2}{1 + e^{(T-x_0)/dx}} + A_2 \tag{4}$$

where A_1 and A_2 are the maximum and minimum modulus, respectively; x_0 and d_x are the shape parameters of the modulus curve.

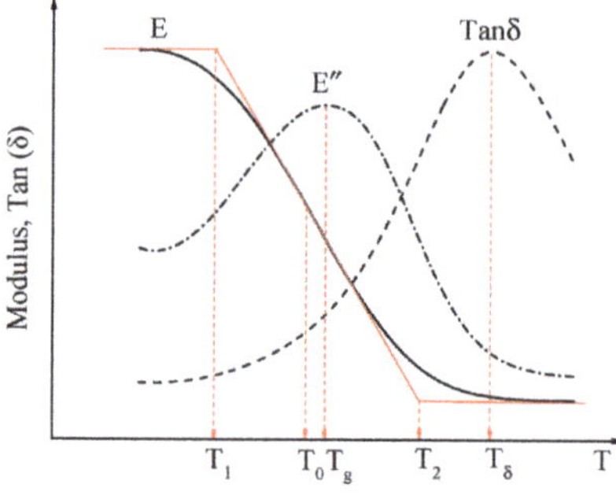

Figure 3. Determination of the glass transition temperature from dynamic testing.

According to the characteristics of the complex modulus curve and the fitting parameters of the Boltzmann function, three-phase transition temperature points can be obtained from the complex modulus curve to describe the phase transition characteristics of asphalt mortar. The three-phase transition temperatures are as follows:

- The temperature corresponding to the midpoint (x_0, y_0) of the complex modulus curve is defined as T_0, $T_0 = x_0$. This is also the temperature point where the complex modulus decreases the fastest with increasing temperature;
- The temperature corresponding to the intersection of the midpoint tangent of the complex modulus curve and the asymptote in the low-temperature zone is defined as T_1, $T_1 = x_0 - 2d_x$. T_1 is the temperature corresponding to the change in asphalt mortar from a glassy to a rubbery state, which can be considered the glass transition temperature of the asphalt mortar;
- The temperature corresponding to the intersection of the midpoint tangent of the complex modulus curve and the asymptote in the high-temperature zone is defined as T_2, $T_2 = x_0 + 2d_x$. This temperature reflects the phase behavior characteristics of asphalt mortar in the high-temperature zone. The loss modulus and $\tan(\delta)$ curves have distinct peak points, and the temperatures corresponding to the peak points are denoted as T_g and T_δ, respectively, which can be referred to as the glass transition temperature [23,25].

This study obtained five-phase transition temperatures through the DMA test as shown in Figure 3. The phase transition temperatures were used to evaluate the high- and low-temperature performance of asphalt mortar.

3. Results and Discussion

3.1. Complex Modulus

Figure 4 shows the complex modulus for the tested asphalt mortar drawn on normal and logarithmic scales. In the range of temperature less than T_1, the rate of the modulus increasing with decreasing temperature decreased until the modulus remained constant. At this temperature, the asphalt mortar was close to the elastomer. The complex modulus decreased rapidly with increasing temperature in the temperature range T_1–T_2. Although the value of complex modulus was small in the high-temperature region where the temperature was greater than T_2, it can be seen in the logarithmic scale (Figure 4b) that the change rate of the complex modulus of different types of asphalt mortar showed obvious variability with the increase in temperature. This difference was mainly due to the different types of asphalt and the minor impact of mineral aggregate gradation. The complex modulus of the A70 asphalt mortar was the smallest, and the rate of modulus decrease with increasing temperature was the largest, whereas that of the AR asphalt mortar was the opposite.

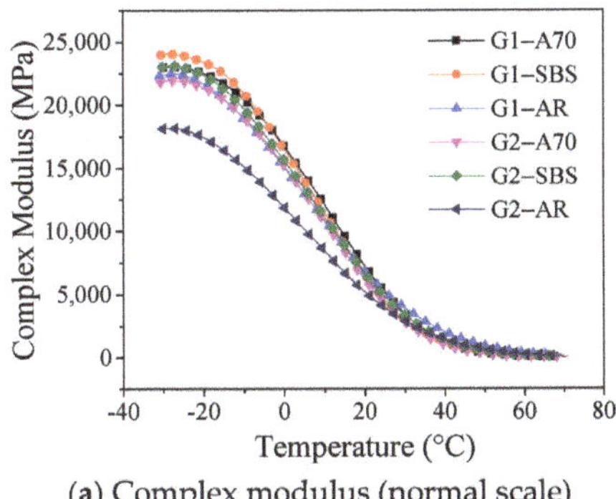

(**a**) Complex modulus (normal scale)

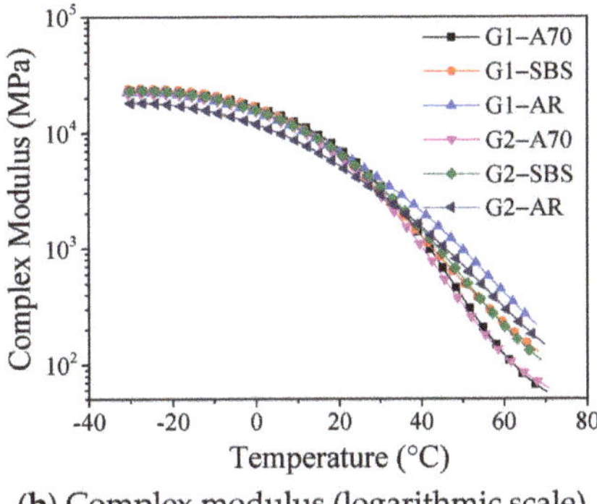

(**b**) Complex modulus (logarithmic scale)

Figure 4. Complex modulus curve (normal and logarithmic scales).

The Boltzmann function model was used to fit the complex modulus curve, and the fitting coefficients R^2 were greater than 0.999. The phase transition temperatures T_0, T_1, and T_2 of the asphalt mortar were calculated according to the fitting parameters. Figure 5a shows the phase transition temperature T_1 of six kinds of asphalt mortar. The highest temperature T_1 of G1-A70 was -12.9 °C, and the lowest temperature T_1 of G2-AR was -22.2 °C. When the aggregate gradation was the same, the phase transition temperature T_1 of the three samples were A70, SBS, and AR from high to low. Due to the different types of asphalt, the maximum and minimum phase transition temperatures T_1 of the mortar were nearly 10 °C. The characteristic temperature T_1 of the asphalt mortar with G1 gradation was higher than G2 when the asphalt was the same. Aggregate gradation G1 was coarser than that of G2, resulting in a difference in asphalt film thickness and porosity, which affects the transition temperature of the mortar. In this study, the transition temperature T_1 of asphalt mortar with coarse gradation, thick asphalt film, and large porosity was relatively higher. However, the maximum difference in T_1 caused by different aggregate gradations was less than 3 °C, indicating that asphalt was still the main factor influencing the transition temperature T_1 of asphalt mortar.

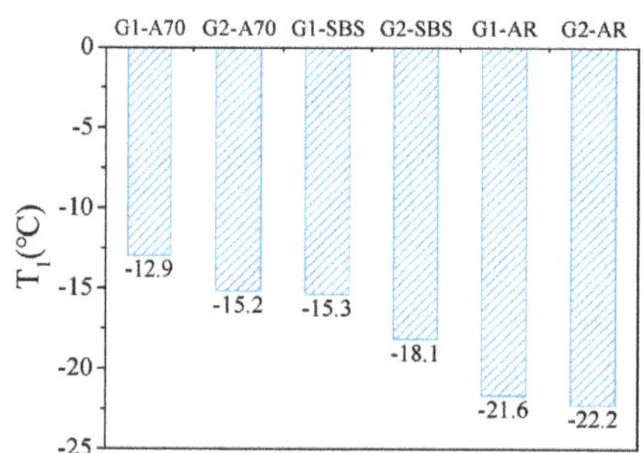

(**a**) Phase transition temperature T_1

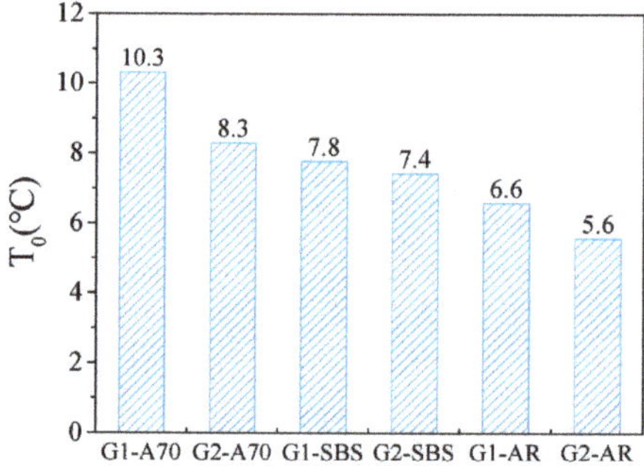

(**b**) Phase transition temperature T_0

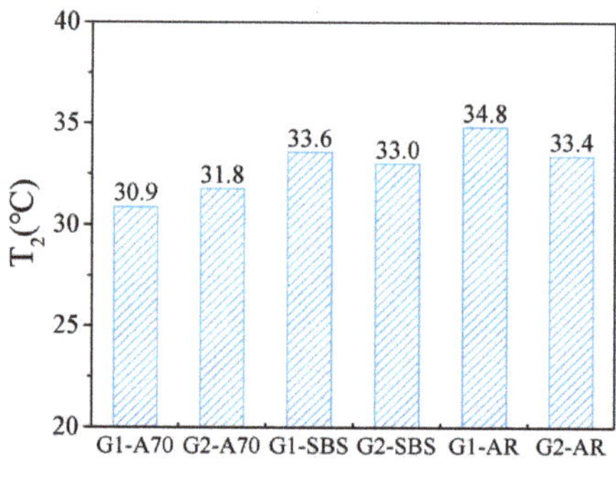

(**c**) Phase transition temperature T_2

Figure 5. Phase transition temperature obtained from complex modulus (i.e., T_1, T_0, and T_2).

Figure 5b shows the phase transition temperature (T_0) of asphalt mortar where T_0 and T_1 had the same change trend. When combined with Figure 6, it is clear that T_0 and T_1 had a good linear relationship with a fitting coefficient R^2 greater than 0.86. Due to the fact of this good relationship, it can be deduced that the phase transition temperatures T_1 and T_0 can be used to evaluate the low-temperature performance of asphalt mortar. Furthermore, the lower the phase transition temperature, the better the low-temperature crack resistance of asphalt mortar. In this study, the order of the low-temperature performance of the mortar prepared usings three kinds of asphalt was AR > SBS > A70.

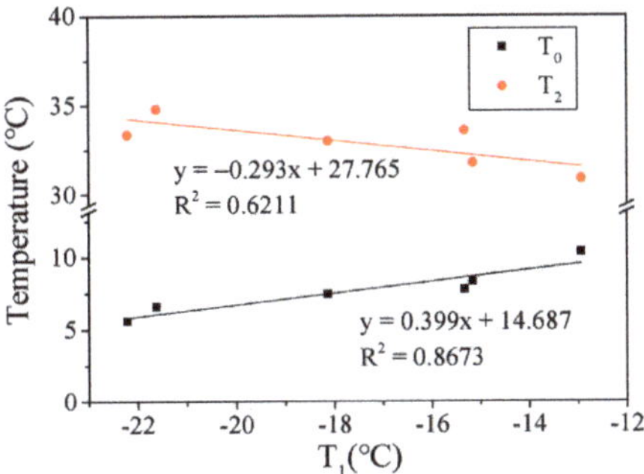

Figure 6. Relationship between T_1 and T_0 and T_2.

Figure 5c shows the phase transition temperature T_2 of asphalt mortar. Phase transition temperature T_2 can be considered the characteristic temperature at which asphalt mortar changes from a rubbery to a viscous state. The higher the temperature, the better the high-temperature performance of the mortar. Figure 6 shows that temperature T_2 had a poor linear correlation with T_1, with the fitting coefficient R^2 = 0.62. The maximum phase transition temperature T_2 of G1-AR asphalt mortar was 34.8 °C, and the minimum T_2 of G1-A70 asphalt mortar was 30.9 °C. The T_2 of modified asphalt mortar with coarse grading (G1) was relatively high. Meanwhile, the three kinds of asphalt had high-temperature performance in the order AR > SBS > A70. In this study, the phase transition temperature T_2 of six kinds of asphalt mortars had small variability, indicating that the asphalt type and aggregate gradation have little influence on the characteristic temperature T_2 of the mortar.

Figure 4b shows that the modulus of asphalt mortar varied significantly in the high-temperature range where the temperature was greater than 40 °C. The index usually evaluates the high-temperature performance of pavement materials under 60 °C. The complex modulus of asphalt mortar at 60 °C was calculated in this study using the parameters of the Boltzmann fitting equation as shown in Table 4. The E_{60} value of mortar with the same asphalt was close, but there was a significant difference when the asphalt was different. The asphalt mortar G1-A70 had the worst performance at a high temperature with the smallest E_{60} value of 33 MPa. The asphalt mortar G1-AR had the best high-temperature performance with the largest E_{60} value of 450 MPa. T_2 and E_{60} were used as indices to evaluate the high-temperature performance of six mortars, and the results were consistent.

Table 4. The complex modulus at 60 °C.

Asphalt Mortar	E_{60}/MPa
G1-A70	33
G2-A70	41
G1-SBS	157
G2-SBS	150
G1-AR	450
G2-AR	330

3.2. Loss Modulus

The loss modulus is the amount of energy dissipated due to the irreversible viscous deformation when a material deforms under stress, and it reflects the material's viscosity. The temperature corresponding to the peak point of the loss modulus curve can be used as the glass transition temperature to evaluate the low-temperature properties of materials.

Figure 7 shows the loss modulus for the tested asphalt mortar drawn on normal and logarithmic scales. It can be observed from this figure that asphalt type and aggregate gradation cause significant differences in loss modulus across a wide temperature range. Figure 8 shows the phase transition temperature T_g of asphalt mortar obtained from the loss modulus temperature curve. The highest phase transition temperature T_g of asphalt mortar G1-A70 was 17.2 °C, and the lowest phase transition temperature of G2-AR was 6.9 °C. With the same asphalt, the phase transition temperature T_g of mortar with an aggregate gradation of G1 was higher than that of G2.

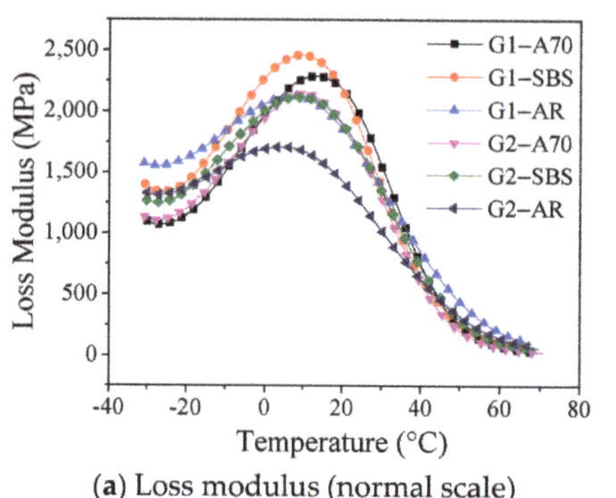

(**a**) Loss modulus (normal scale)

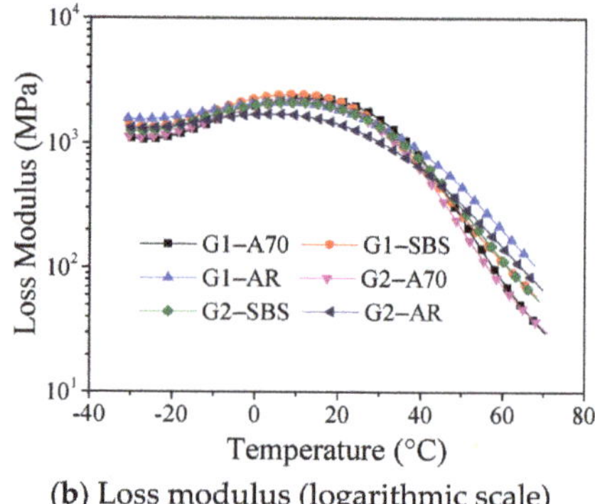

(**b**) Loss modulus (logarithmic scale)

Figure 7. Loss modulus curve (normal and logarithmic scales).

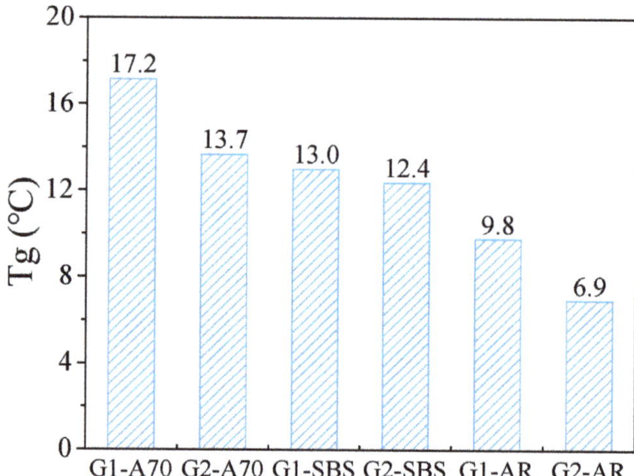

Figure 8. Phase transition temperature T_g.

The glass transition temperature is an important parameter to evaluate the low-temperature performance of materials. T_1, T_0, and T_g obtained using the DMA method can be used as glass transition temperatures to evaluate the low-temperature performance of asphalt mortar. Figures 6 and 9 show a good linear relationship between T_1, T_0, and T_g, and the evaluation results of the low-temperature performance of six kinds of asphalt mortar were consistent.

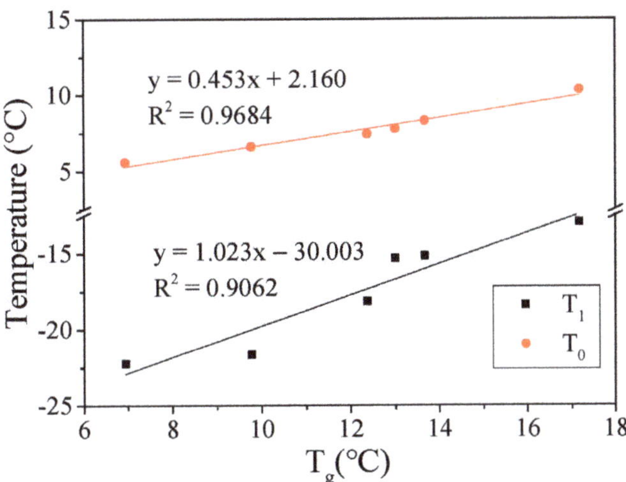

Figure 9. Relationship between T_g and T_1 and T_0.

3.3. Tan(δ) (E″/E′)

Figure 10 shows the tan(δ) curve of asphalt mortar. In the test temperature range, the tan(δ) value of the asphalt mortar first increased and then decreased. The tan(δ) value gradually increased as the temperature rose, and the rate of increase was slow in the low-temperature range (less than 0 °C). When the temperature exceeds 20 °C, the tan(δ) value rapidly increased and decreased after reaching its maximum. In the low-temperature region, the tan(δ) values of the six kinds of asphalt mortars had little difference. Still, in the temperature range of 20–70 °C, the tan(δ) temperature curves of the different kinds of asphalt mortar varied significantly. When the temperature exceeded 20 °C, the tan(δ) value and its increasing rate of A70 asphalt mortar were the largest, while the AR asphalt mortar was the opposite. The tan(δ) curves of the two aggregate gradation mortars were similar when the asphalt was the same, which indicates that the asphalt type was the main factor affecting the tan(δ) curve of the asphalt the mortar.

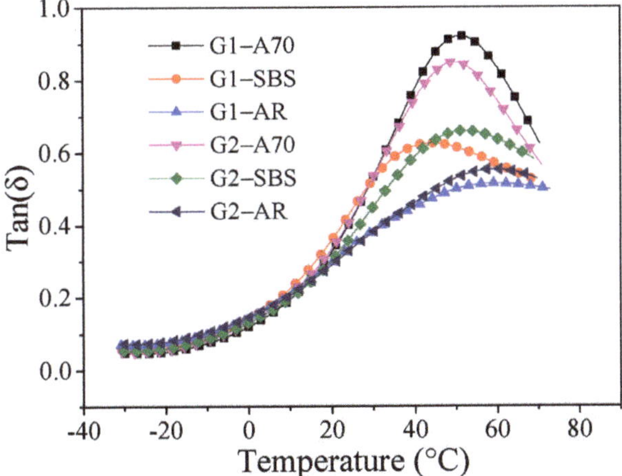

Figure 10. The curve of tan(δ).

Tan(δ) is the ratio of loss modulus to storage modulus. The larger the tan(δ) value, the larger the proportion of loss modulus, the greater the irreversible deformation capacity of the material under the action of external force, and the closer the material is to the viscous phase. In the high-temperature region, the higher the tan(δ) value of asphalt mortar, the more likely the irreversible viscous deformation of asphalt mortar will occur when the asphalt mortar is subjected to external force, indicating that the high-temperature deformation resistance of asphalt mortar is worse. Meanwhile, the lower the peak temperature of the tan(δ) value curve, the worse the high-temperature performance of mortar.

Figure 11 shows the maximum tan(δ) value and the corresponding temperature T_δ. Asphalt mortar G1-SBS had the lowest phase transition temperature T_δ of 43.8 °C, whereas asphalt mortar G2-AR had the highest at 59.3 °C. This result was inconsistent with the evaluation results of indicators T_2 and E_{60}. This is because asphalt mortar is not a uniformly simple material composition, and its phase transformation characteristics under a high-temperature range are more complex. However, asphalt mortar G1-A70 had the largest Tan(δ)$_{max}$ value of 0.92, and asphalt mortar G1-AR had the smallest Tan(δ)$_{max}$ value of 0.51. The Tan(δ)$_{max}$ values of two mortars with the same asphalt were similar, and the order of the high-temperature performance of the asphalt mortar prepared with the same aggregate gradation of three kinds of asphalt was AR > SBS > A70. This was different from the evaluation result of phase transition temperature T_δ but consistent with the evaluation results of indicators T_2 and E_{60}, and it was also consistent with the high-temperature grade (PG grade) of asphalt as shown in Table 1. Therefore, Tan(δ)$_{max}$ can more effectively evaluate the high-temperature performance of asphalt mortar.

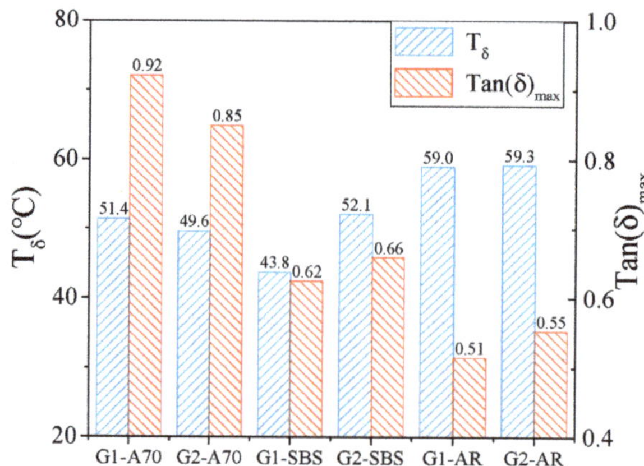

Figure 11. Maximum tan(δ) and corresponding temperature T_δ.

Asphalt plays a key role in the high- and low-temperature performance of asphalt mortar. The mortars formed by asphalt AR and SBS had the better high- and low-temperature performance. This was mainly because polymer SBS and crumb rubber powder absorbed the light component of neat asphalt, resulting in swelling and changing the phase characteristics of asphalt. The interactions between modifier and asphalt were strengthened, leading to the improved high-temperature performance and toughness. The swelled SBS and crumb rubber particles made the asphalt more elastic and flexible; therefore, the elasticity and low-temperature performance of the asphalt were also improved.

4. Conclusions

In this study, three kinds of asphalt and two kinds of aggregate gradation were used to form six kinds of asphalt mortar samples. The viscoelastic properties of six kinds of asphalt mortars were tested using the DMA method. The temperature-sweep of asphalt mortar was conducted with a dual cantilever clamp under the conditions of a fixed strain level, frequency, and heating rate. The dynamic response curves of asphalt mortar under flexural and tensile modes were obtained including complex modulus, storage modulus, loss modulus, and tan(δ).

The Boltzmann function model was used to fit the complex modulus curve, and the phase transition temperatures T_1, T_0, and T_2 of asphalt mortar were obtained according to the fitting parameters. The loss modulus and tan(δ) curves had clear peak points, and the phase transition temperatures T_g and T_δ corresponding to peak points were obtained. Meanwhile, the complex modulus at 60 °C (E_{60}) and the maximum tan(δ) ($Tan(\delta)_{max}$) were used as indicators to evaluate the high-temperature performance of asphalt mortar. The following conclusions were drawn:

a) The phase transition temperatures T_1, T_0, and T_g can be used as the glass transition temperature of asphalt mortar, and they had a good linear relationship. T_1, T_0, and T_g effectively evaluated the low-temperature performance of asphalt mortar, and the evaluation results of the three indices were consistent. Mortars with neat asphalt A70 had a significantly higher T_1, T_0, and T_g and poor low-temperature performance, while mortars with modified asphalt AR had the best low-temperature performance. The finer gradation had a positive effect on the low-temperature performance of asphalt mortar.

b) Phase transition temperature T_2, complex modulus at 60 °C (E_{60}), and the maximum tan(δ) ($Tan(\delta)_{max}$) can be used to evaluate the high-temperature performance of asphalt mortar. The mortar formed by asphalt AR had the best high-temperature performance, followed by asphalt SBS and A70. Coarser gradation had a good effect on the high-temperature performance of the modified asphalt, but it had the opposite effect on the neat asphalt A70.

c) Asphalt plays a key role in the high- and low-temperature performance of asphalt mortar. In this study, the order of high- and low-temperature performance was AR > SBS > A70.

The temperature-sweep test based on the DMA method can effectively evaluate the phase transition characteristics of asphalt mortar in a wide temperature range, and the significance of the evaluation indicators are clear. The reliability, effectiveness, and practicability of the method for evaluating the high- and low-temperature performance of asphalt mortar using phase transition temperature will be further validated in future research by combining with the road performance research of asphalt mixture.

Author Contributions: Conceptualization, Y.W. and X.W.; methodology, Y.W. and Z.M.; formal analysis, Y.W. and L.S.; investigation, Y.W. and Z.M.; resources, L.S.; data curation, C.Z.; writing—original draft preparation, Y.W. and Z.M.; writing—review and editing, Y.W. and C.Z. All authors have read and agreed to the published version of the manuscript.

Funding: This work was funded by the National Key of Research and Development Plan (grant number: 2020YFA0714300).

Institutional Review Board Statement: Not applicable.

Informed Consent Statement: Not applicable.

Data Availability Statement: Not applicable.

Conflicts of Interest: The authors declare no conflict of interest.

References

1. Jiang, J.; Ni, F.; Gu, X.; Yao, L.; Dong, Q. Evaluation of aggregate packing based on thickness distribution of asphalt binder, mastic and mortar within asphalt mixtures using multiscale methods. *Constr. Build. Mater.* **2019**, *222*, 717–730. [CrossRef]
2. Arshadi, A.; Bahia, H. Development of an image-based multi-scale finite-element approach to predict mechanical response of asphalt mixtures. *Road Mater. Pavement Des.* **2015**, *16*, 214–229. [CrossRef]
3. Suresha, S.; Ningappa, A. Recent trends and laboratory performance studies on FAM mixtures: A state-of-the-art review. *Constr. Build. Mater.* **2018**, *174*, 496–506. [CrossRef]

4. Li, Q.; Chen, X.; Li, G.; Zhang, S. Fatigue resistance investigation of warm-mix recycled asphalt binder, mastic, and fine aggregate matrix. *Fatigue Fract. Eng. Mater. Struct.* **2017**, *41*, 400–411. [CrossRef]

5. Nejad, F.M.; Habibi, M.; Hosseini, P.; Jahanbakhsh, H. Investigating the mechanical and fatigue properties of sustainable cement emulsified asphalt mortar. *J. Clean. Prod.* **2017**, *156*, 717–728. [CrossRef]

6. Chen, M.; Javilla, B.; Hong, W.; Pan, C.; Riara, M.; Mo, L.; Guo, M. Rheological and Interaction Analysis of Asphalt Binder, Mastic and Mortar. *Materials* **2019**, *12*, 128. [CrossRef]

7. Yang, S.; Jiang, J.; Leng, Z.; Ni, F. Feasibility and performance of the Semi-circular Bending test in evaluating the low-temperature performance of asphalt mortar. *Constr. Build. Mater.* **2020**, *269*, 121305. [CrossRef]

8. Niu, D.; Chen, H.; Kim, Y.R.; Sheng, Y.; Geng, J.; Guan, B.; Xiong, R.; Yang, Z. Damage assessment of asphalt concrete with composite additives at the FAM–coarse aggregate interfacial zone. *Constr. Build. Mater.* **2019**, *198*, 587–596. [CrossRef]

9. Apostolidis, P.; Liu, X.; Daniel, G.C.; Erkens, S.; Scarpas, T. Effect of synthetic fibres on fracture performance of asphalt mortar. *Road Mater. Pavement Des.* **2019**, *21*, 1918–1931. [CrossRef]

10. Chuanfeng, Z.; Yupeng, F.; Zhuang, M.; Xue, Y. Influence of mineral filler on the low-temperature cohesive strength of asphalt mortar. *Cold Reg. Sci. Technol.* **2017**, *133*, 1–6. [CrossRef]

11. Underwood, B.S.; Kim, Y.R. Effect of volumetric factors on the mechanical behavior of asphalt fine aggregate matrix and the relationship to asphalt mixture properties. *Constr. Build. Mater.* **2013**, *49*, 672–681. [CrossRef]

12. Osmari, P.H.; de Souza, R.C.; Nascimento, L.A.H.D.; Aragão, F.T.S. Evaluation of the relationship between the fatigue performance of FAM and AC mixtures based on volumetric characteristics and on the S-VECD theory. *Constr. Build. Mater.* **2020**, *265*, 120294. [CrossRef]

13. Espinosa, L.; Caro, S.; Wills, J. Study of the influence of the loading rate on the fracture behaviour of asphalt mixtures and asphalt mortars. *Constr. Build. Mater.* **2020**, *262*, 120037. [CrossRef]

14. Im, S.; Ban, H.; Kim, Y.-R. Characterization of mode-I and mode-II fracture properties of fine aggregate matrix using a semicircular specimen geometry. *Constr. Build. Mater.* **2014**, *52*, 413–421. [CrossRef]

15. Aragão, F.T.S.; Badilla-Vargas, G.A.; Hartmann, D.A.; de Oliveira, A.D.; Kim, Y.-R. Characterization of temperature- and rate-dependent fracture properties of fine aggregate bituminous mixtures using an integrated numerical-experimental approach. *Eng. Fract. Mech.* **2017**, *180*, 195–212. [CrossRef]

16. Gong, X.; Romero, P.; Dong, Z.; Li, Y. Investigation on the low temperature property of asphalt fine aggregate matrix and asphalt mixture including the environmental factors. *Constr. Build. Mater.* **2017**, *156*, 56–62. [CrossRef]

17. Zhang, X.; Xu, L.; Lv, J. Investigation on the Flexural–Tensile Rheological Behavior and Its Influence Factors of Fiber-reinforced Asphalt Mortar. *Polymers* **2020**, *12*, 1970. [CrossRef]

18. Fu, Q.; Xie, Y.-J.; Niu, D.-T.; Long, G.; Luo, D.-M.; Yuan, Q.; Song, H. Integrated experimental measurement and computational analysis of relaxation behavior of cement and asphalt mortar. *Constr. Build. Mater.* **2016**, *120*, 137–146. [CrossRef]

19. Zhu, J.; Alavi, M.Z.; Harvey, J.; Sun, L.; He, Y. Evaluating fatigue performance of fine aggregate matrix of asphalt mix containing recycled asphalt shingles. *Constr. Build. Mater.* **2017**, *139*, 203–211. [CrossRef]

20. Fu, Q.; Xie, Y.; Long, G.; Niu, D.; Song, H. Dynamic mechanical thermo-analysis of cement and asphalt mortar. *Powder Technol.* **2017**, *313*, 36–43. [CrossRef]

21. Yu, H.; Yao, D.; Qian, G.; Cai, J.; Gong, X.; Cheng, L. Effect of ultraviolet aging on dynamic mechanical properties of SBS modified asphalt mortar. *Constr. Build. Mater.* **2021**, *281*, 122328. [CrossRef]

22. Fang, L.; Yuan, Q.; Deng, D.; Pan, Y.; Wang, Y. Effect of Mix Parameters on the Dynamic Mechanical Properties of Cement Asphalt Mortar. *J. Mater. Civ. Eng.* **2017**, *29*, 04017080. [CrossRef]

23. Tan, Y.; Guo, M. Study on the phase behavior of asphalt mastic. *Constr. Build. Mater.* **2013**, *47*, 311–317. [CrossRef]

24. Yang, G.; Wang, X.; Zhou, X.; Wang, Y. Experimental Study on the Phase Transition Characteristics of Asphalt Mixture for Stress Absorbing Membrane Interlayer. *Materials* **2020**, *13*, 474. [CrossRef]

25. Soliman, H.; Shalaby, A. Characterizing the Low-Temperature Performance of Hot-Pour Bituminous Sealants Using Glass Transition Temperature and Dynamic Stiffness Modulus. *J. Mater. Civ. Eng.* **2009**, *21*, 688–693. [CrossRef]

26. Liu, H.; Luo, R.; Lv, H. Establishing continuous relaxation spectrum based on complex modulus tests to construct relaxation modulus master curves in compliance with linear viscoelastic theory. *Constr. Build. Mater.* **2018**, *165*, 372–384. [CrossRef]

 *materials*

Article

The Relationship between Poisson's Ratio Index and Deformation Behavior of Asphalt Mixtures Tested through an Optical Fiber Bragg Grating Strain Sensor

Xu Liu [1,2], Mo Zhang [3,*] and Wanqiu Liu [4]

[1] Fundamental Research Innovation Center, Research Institute of Highway, Ministry of Transport, 8 Xitucheng Road, Beijing 100088, China; sikui2003@outlook.com
[2] Department of Civil Engineering, Tsinghua University, 30 Shuangqing Road, Beijing 100084, China
[3] School of Civil and Transportation Engineering, Hebei University of Technology, 5340 Xiping Road, Tianjin 300401, China
[4] College of Civil Engineering and Architecture, Hainan University, 58 Renmin Avenue, Haikou 570228, China; liuwanqiu@dlut.edu.cn
[*] Correspondence: mozhang@hebut.edu.cn

Citation: Liu, X.; Zhang, M.; Liu, W. The Relationship between Poisson's Ratio Index and Deformation Behavior of Asphalt Mixtures Tested through an Optical Fiber Bragg Grating Strain Sensor. *Materials* **2022**, *15*, 1882. https://doi.org/10.3390/ma15051882

Academic Editor: Meng Ling

Received: 11 January 2022
Accepted: 28 February 2022
Published: 3 March 2022

Publisher's Note: MDPI stays neutral with regard to jurisdictional claims in published maps and institutional affiliations.

Abstract: Flow-rutting is the main distress leading asphalt pavement to undergo premature maintenance, and is produced by the rapid accumulation of shear deformation in asphalt layers under high temperature and heavy loads. The excessive permanent deformation of the asphalt mixture at high temperature is related to the decrease of the material's stability during the temperature increase and an unfavorable stress state, e.g., low confining pressure and high shear stress, which eventually leads to significant nonlinear viscoplastic behavior. In this research, dynamic modulus tests and repeated loading tests were carried out at 35 °C and 50 °C to analyze the deformation response of materials under a strain amplitude of <200 με and 400~500 μεs, respectively. Based on the in-lab repeated loading tests, the total deformation of the asphalt mixture in each loading and rest cycle was divided into three parts, being elastic, viscoelastic, and viscoplastic strain, and the measurement of the axial and lateral strain of cylindrical samples was realized with the aid of optical fiber Bragg grating strain sensors. It was found that the experimental index of the ratio between lateral strain and longitudinal strain (RLSLS), derived, but distinguished, from Poisson's ratio defined limited in elastic strain, can characterize the deformation in viscoelastic and viscoplastic behaviors of the mixes. Furthermore, the indices of dynamic modulus, phase angle, complex Poisson's ratio, stiffness, and creep rate of four types of mixes containing different volcanic ash fillers and asphalt binders at 35 °C and 50 °C were systematically analyzed by the jointed experiments of modified dynamic modulus tests and repeated loading tests, and their consistent trending to the RLSLS index was obtained.

Keywords: asphalt mixture; flow-rutting; axial and lateral deformation; optical fiber Bragg grating

1. Introduction

Flow-rutting (FR) is one of the most harmful forms of distress on high-grade asphalt pavement. It is a type of unrecoverable shear deformation resulting from the creeping flow of the asphalt binder and the rearrangement of aggregate particles in the mesoscale of the mix [1,2], which accumulates rapidly in the asphalt layer under repeated vehicle loads. FR is highly dependent on the properties of the material and greatly impacted by environmental temperatures and traffic loading. In particular, the FR disease becomes more significant under conditions of high temperature and heavy wheel loads, which can result in destructive impacts on the service lives of pavements. Due to serious rutting induced by overload states, many asphalt pavements in China have to be greatly repaired or rebuilt within two to three years, far from the designed service life of approximately 15 to 20 years [3]. From the perspective of pavement design, this phenomenon is due mainly to the mismatch between the evaluation method of material performance and the formation mechanism of FR.

Based on the relationship between the material and structure of asphalt pavement, the permanent shear deformation in asphalt layers is usually introduced by the coupling effects of the following factors: (1) high-temperature conditions that induce more significant viscoplastic behavior of the asphalt mixture and results in higher permanent deformation; (2) heavy wheel loads that can lead to larger shear stress that then intensify the viscoplastic flow; and (3) the stiffness constitution in the structural layers of pavement can result in an adverse combination of principal stresses (e.g., low average stress or large deviator stress) and decrease the plastic yielding threshold of hot mixed asphalt (HMA).

In respect to material performance improvement, many scholars have tried to introduce new materials into the mixture design to comprehensively improve the high temperature performance of asphalt mixtures [4–11]. However, precise evaluation methods and indicative indices are still needed to reflect the development of permanent deformation within asphalt pavement under high temperature and complex loading conditions to further improve and predict the performance and stability of an asphalt mixture. Empirical and theoretical assessments are the two major classes of evaluation methods of rutting. The Marshall test and wheel tracking test are the most widely applied empirical laboratory testing methods to reflect the high-temperature failure of HMA at a specific temperature and loading mode [12–15]. However, they cannot completely simulate the working states of real pavement, such as continuously varied loads and temperatures. Therefore, field accelerated loading tests, such as accelerated loading facility and accelerated pavement testing, are used to investigate permanent deformation and establish a life-cycle prediction model of rutting [16–18]. However, theses field accelerated loading tests would be hugely expensive to perform for large-scale road tests and the testing can only be carried out for certain structural compositions of pavement.

Furthermore, there is always a theoretical gap between empirical evaluation indices and HMA performances, leading to difficulties in performing an accurate analysis. Mechanical models and testing indices are important for evaluating deformation properties of asphalt binders and mixtures. To better understand the states of permanent deformation, Lagos-Varas et al. developed a viscoelastic model using derivatives of fractional order, which can describe the creep, recovery, and relaxation behaviors in an asphalt mixture [19,20]. With the increasing application of viscoelastic and viscoplastic theories in the mechanical analysis of asphalt mixtures, temperature-controlled dynamic triaxial tests are becoming important evaluation methods [21], and the related repeated load permanent deformation (RLPD) test is the most recommended evaluation method [22,23]. The periodical loads that consist of a half-sine loading period and a rest period in each cycle can well represent the complex loading condition, and the calculated flow number (FN) and the deductive FN index can effectively evaluate the permanent deformation of the HMA [22–25]. The mixture samples have to be continuously loaded until the accelerated shear failure stage to obtain the FN value, which needs a long loading period that is sometimes not achievable. Rather than a slow and constant development, the corresponding FR permanent deformation is a fast increase of shear deflection, which is generated by nonlinear mechanical behaviors, mainly viscoplastic and plastic deformation, which are hard to quantitatively evaluate by the FN or FN index.

In this study, the ratio of lateral strain to longitudinal strain (RLSLS) tested in the RLPD test was adopted to elucidate the significance of viscoplastic and plastic behavior of HMA in the FR. RLSLS was proposed based on the modification of Poisson's Ratio (PR), which is a traditional concept in elastic theory. As a more generic index, RLSLS can better reflect the complex elastic-visco-plastic behavior of HMA, while the accurate measurement of lateral and longitudinal strain is necessary. However, the common deformation sensor, linear variable differential transformer (LVDT) that adhered to HMA samples, is prone to soften under high temperatures, which affects the testing accuracy. To overcome this problem, an optical fiber Bragg grating (OFBG) strain sensor was adopted to measure the lateral and longitudinal strain of cylindrical HMA specimens in this study. The standard dynamic modulus (DM) tests, modified DM tests and RLPD tests were applied at 35°C and 50°C, determined by consideration of test procedures and sensor range, in order to analyze the change characteristics of the mechanical parameters of the asphalt mixture when the ambient temperature was transformed from a medium temperature to a higher temperature. First, four types of HMA with two types of VA fillers and two types of asphalt binder, i.e., base asphalt and SBS-modified asphalt, were investigated in this study. The HMA samples were tested by applying a haversine load in an uniaxial direction at the loading frequencies of 1, 0.5, and 0.1 Hz to (i) explore the feasibility of the OFBG strain sensor in the measurement of complex Poisson's ratio (CPR) in modified DM tests; and (ii) analyze the trend of CPR to evaluate the high-temperature stability of HMA. Second, based on RLPD testing results, different types of HMA were compared with respect to the stiffness and rate of permanent strain (RPS). In addition, the total strain of HMA in a single loading-unloading-rest period was divided into three parts: instantaneous recoverable strain, delayed recoverable strain, and residual strain. By calculating the RLSLS corresponding to these strains, the effects of different types of VA and asphalt binders on deformation behavior were analyzed and the correlation between RLSLS and the shear permanent deformation of the HMA was discussed.

2. Methodology

2.1. Materials

The research group of this study used porous volcanic ash (VA) to replace traditional mineral powder as a fine filler in the asphalt mixture. It was found that the mastics and HMA with the combination of VA-and SBS-modified asphalt had better performance on both high- and low-temperature properties [10,11]. The representative types of HMA and their compositions are detailed as follows:

2.1.1. Volcanic Ash Fine Filler and Asphalt Binder

The two types of VA were collected from the Lazihe and Bahaozha areas in the Jilin province of China, and are denoted by LA and BA, respectively (as shown in Figure 1). They were ground to particle sizes smaller than 0.075 mm and used as fine fillers in the HMA, of which the physical properties are listed in Table 1. LA and BA have similar apparent densities, while LA had a larger average particle size and higher specific surface area than BA. According to a previous study [11], the LA particles have a richer porous structure than BA. In addition, BA has a smoother particle surface than LA, based on scanning electron microscopy (SEM) characterization.

On the other hand, base asphalt and 5 wt% of SBS-modified asphalt with a penetration level of 90 (0.1 mm) were adopted as two types of asphalt binders, which were denoted by P and S, respectively. The properties of the base asphalt were measured following the ASTM Standards, as shown in Table 2.

Figure 1. The volcanic ash fine filler of Bahaozha (BA) and Lazihe (LA).

Table 1. Physical properties of BA and LA volcanic ashes.

Filler	BA	LA
Average particle diameter (nm)	453	1901
Diameter (nm)/Proportion of the dominant particles	294/98.2%	1963/98.8%
Bulk specific surface area (m^2/cm^3)	4.94	114.88
Apparent density (g/cm^3)	2.40	2.55

Table 2. The physical and mechanical properties of base asphalt.

Index	Results (units)	Specification
Penetration (25 °C) [1]	99.0 (0.1 mm)	ASTM D5
Softening point	45.0 (°C)	ASTM D36
Ductility (10 °C, 5 cm/min) [2]	78.0 (cm)	ASTM D113

Note: [1] The same batch of base asphalt was used for the SBS-modified asphalt binder. Penetration tests were performed at 25 °C. [2] The ductility of asphalt was taken at 10 °C instead of 25 °C as stipulated in ASTM D113 due to the low annual average temperature in Jilin area (China), between 40° and 44° north latitude.

2.1.2. Asphalt Mixture

In this study, four types of HMA with different combinations of the two types of asphalt binder (base/SBS) and two types of volcanic ash fine fillers (LA/BA) were investigated, as listed in Table 3. The aggregate gradation in Table 4 was used for all of the HMA samples to eliminate the impacts of aggregates. An asphalt/aggregate mass ratio of 4.6%, determined by the Marshall method [26], and a fine filler to asphalt binder ratio of 1:1 were used for all the sample sets. The samples were designated by "type of VA-type asphalt binder", which were LA-S, LA-P, BA-S, and BA-P, respectively. Cylindrical HMA samples, with a diameter of 100 mm and height of 150 mm, were prepared with a gyratory compactor for the standard DM tests, modified DM tests, and the RLPD tests in this study. All the specimens were cured in an environmental chamber at a specific temperature for 4 h before testing.

Table 3. The composition of mixtures with different fine fillers and asphalt binders.

Asphalt Mixture	Volcanic Ash Fine Filler	Asphalt Binder
LA-S		5 wt% SBS modified asphalt
LA-P	LA	Base asphalt
BA-S		5 wt% SBS modified asphalt
BA-P	BA	Base asphalt

Table 4. The grading of aggregates.

Sieve Size (mm)	Percent Passing by Weight
19	100
16	92.6
13.2	82.7
9.5	68.2
4.75	49.1
2.36	32.4
1.18	23.6
0.6	17
0.3	12.7
0.15	9.0
0.075	6.6

2.2. Optical Fiber Bragg Grating Strain Sensor

The optical fiber Bragg grating (OFBG) sensing technique has been used in experimental research on pavement material and structure, which can bear a high compaction force and high temperatures during the testing process [27,28]. In the modified DM tests and RLPD tests, an OFBG strain sensor was adopted to measure the longitudinal and lateral strain of HMA cylindrical samples, with the highest sampling frequency of 300 Hz and largest strain capacity of 10,000 $\mu\varepsilon$. Due to its high ductility and highest accuracy of 1 $\mu\varepsilon$, OFBG was allowed to attach to the surrounding cylinders with the closest contact to measure the lateral strain of the HMA under dynamic loading.

With particularly fabricated fastening equipment, three sets of OFBG sensors were aligned surrounding the cylindrical samples, of which two were aligned in the vertical direction and one was aligned in the horizontal direction, as schematically shown in Figure 2. At the two sides of the grating section in each OFBG sensor, two epoxy resin blocks were fixed and connected to the adjusting screws, with which the OFBG was stretched during testing to ensure consistent deformation between the sensor and samples. The OFBG was stretched to a strain value slightly higher than 50 $\mu\varepsilon$ in the horizontal direction and approximately 5000 $\mu\varepsilon$ in the vertical direction. The average reading of the two vertical sensors was used to calculate the longitudinal strain, while that of the horizontal sensor was the circumferential strain used to obtain the lateral strain.

The linear measurement of the strain by the OFBG was realized based on the reflecting wavelength change in the grating section of optical fiber [27], of which the strain can be calculated by Equation (1):

$$\Delta\varepsilon = \frac{1}{1 - P_e}\left(\frac{\Delta\lambda_1}{\lambda_1} - \frac{\Delta\lambda_2}{\lambda_2}\right) \tag{1}$$

where $\Delta\varepsilon$ is the strain of the OFBG, P_e is the stress optical coefficient, λ_1 is the Bragg wavelength of grating areas when stress and temperature both changed, and λ_2 is the Bragg wavelength of grating areas when only temperature is changed.

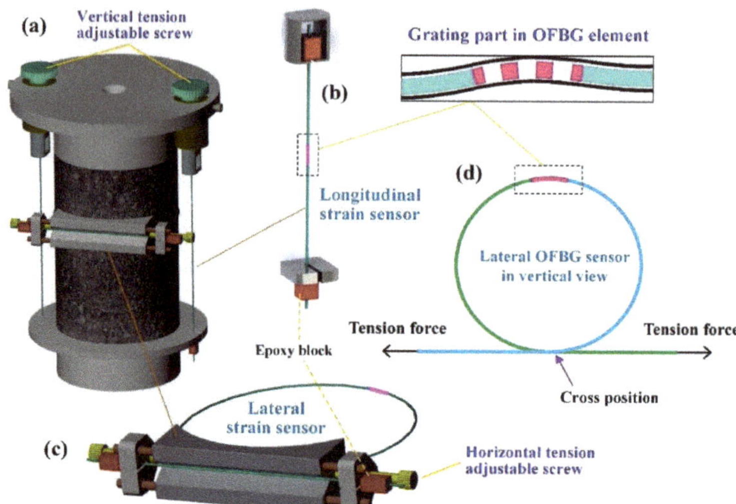

Figure 2. Schematic setup of (**a**) the cylindrical HMA sample in DM tests with two (**b**) vertical OFBG sensors and one (**c**) horizontal OFBG sensor, in which the (**d**) OFBG element was fixed with two epoxy resin blocks and stretched with two screwing nuts.

The longitudinal and lateral strain of the cylindrical sample were converted from the sensor reading according to the geometric correlation, as shown in Equations (2) and (3):

$$\varepsilon_{long} = \frac{\varepsilon_{olong} \cdot l_o}{h} \tag{2}$$

$$\varepsilon_{lat} = \frac{\varepsilon_{olat} \cdot l_o}{\pi d} \tag{3}$$

where ε_{long} and ε_{lat} are the longitudinal and lateral strain of the sample, respectively, ε_{olong} and ε_{olat} are the longitudinal and lateral strain of the OFBG, respectively, in mm/mm, l_o is the length of the OFBG between the two epoxy resin blocks, in mm, and h and d are the height and diameter of the sample, respectively, in mm.

2.3. Dynamic Modulus Tests

2.3.1. Standard Dynamic Modulus Test

The dynamic modulus (DM) and phase angle (PA) of the HMA at temperatures of 35 °C and 50 °C, and frequencies of 25 Hz, 20 Hz, 10 Hz, 5 Hz, 1 Hz, 0.5 Hz, and 0.1 Hz were tested with a standard dynamic modulus test to evaluate the medium- to high-temperature performance of the materials. The 35 °C and 50 °C conditions were two representative temperatures selected from the middle-to high-temperature section in the temperature range of −10 °C~54 °C, specified in AASHTO TP62-2007 [29], used to analyze the change characteristics of material parameters in the process of temperature increase. For each type of asphalt mixture, four groups of parallel specimens were prepared for testing, and the mean value of the results was taken as the representative value of the material. The loading was conducted using a universal testing machine (UTM) with a maximum load capacity of 100 kN. During testing, the contact pressure of 5 N was retained between the UTM loading cell and the top surface of the cylindrical sample. The testing temperature was pre-adjusted to, and retained at, 35 °C and 50 °C with the embedded environmental chamber.

2.3.2. Modified Dynamic Modulus Test

The modified DM test is a test method for the exploration of applying the OFBG as a deformation sensor based on the standard DM test. The traditional LVDT strain measuring apparatus in the standard DM test was replaced by an OFBG strain sensor in the modified DM test to simultaneously measure the axial and lateral strain of the cylindrical samples, while the testing temperature, materials, and loading equipment remained the same. The material properties at higher temperatures can be characterized at low testing frequencies according to the time-temperature superposition principle (TTSP) [30]. Therefore, the relatively low testing frequencies of 1 Hz, 0.5 Hz, and 0.1 Hz were adopted for the performance analysis of the mixtures, at which permanent deformation was more prone to develop. In this part, two sets of parallel specimens were tested for each type of HMA.

The amplitude of the strain wave of the HMA specimen was limited in the linear viscoelastic range (<200 µε) during testing. The longitudinal strain was negative since the specimen was compressed, while the lateral strain was positive. As shown in Figure 3, the strain–time curves in the last 5 loading cycles were obtained and fit using Equations (4) and (5). The curve fitting process and parameter calculation were accomplished with MATLAB code.

$$\varepsilon_{long}(t) = (-1) \times \left[\varepsilon_{0,long} \sin\left(2\pi ft + \varphi_{long}\right) + a_{c,long}(t) + Y_{long} \right] \tag{4}$$

$$\varepsilon_{lat}(t) = \varepsilon_{0,lat} \sin(2\pi ft + \varphi_{lat}) + a_{c,lat}(t) + Y_{lat} \tag{5}$$

where f is the loading frequency, t is the testing time, ε_0 and φ are the amplitude and phase angle of the strain curve, respectively, while ε_0 is positive, Y is the vertical compensation during curve fitting, $a_c(t)$ is the creeping rate, and the subscript *long* and *lat* represent the longitudinal and lateral directions, respectively.

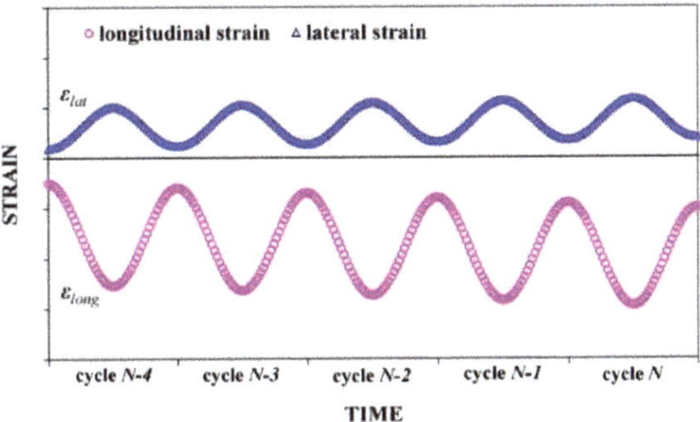

Figure 3. Longitudinal and lateral strain in the last five cycles obtained by OFBG sensors.

In addition, the complex Poisson's ratio (CPR) and related mechanical parameters were calculated by using Equations (6) to (8):

$$v^* = \frac{\varepsilon_{lat}(t)}{\varepsilon_{long}(t)} = |v^*| \cdot e^{-i\delta_v} \tag{6}$$

$$|v^*| = \frac{\varepsilon_{0,lat}}{\varepsilon_{0,long}} \tag{7}$$

$$\delta_v = \varphi_{long} - \varphi_{lat} \tag{8}$$

where v^* is the CPR, $|v^*|$ is the normal of v^*, which was named the dynamic Poisson's ratio (DPR), δ_v is the phase lag of lateral strain compared to longitudinal strain, and $0 \leq \delta_v \leq \frac{\pi}{2}$.

2.4. Repeated Load Permanent Deformation (RLPD) Tests

2.4.1. Testing Procedure

The RLPD tests were conducted to investigate the development of axial and lateral deformations of the HMA specimens under complex loading conditions by using different loading/rest time combinations at the two temperatures of 35 °C and 50 °C in order to compare with the results of the DM and modified DM tests. Limited by the strain range of the OFBG sensor, the 60 °C condition was not adopted as the experimental temperature, which was usually used in the traditional high-temperature performance evaluation test of asphalt mixtures. At 60 °C, the permanent deformation of HMA develops more rapidly than at lower temperatures, and the test data of enough load cycles cannot be obtained. Therefore, 50 °C is selected as the higher experimental temperature in this experiment, and can be comprehensively analyzed with the results of the DM tests and modified DM tests. Although 35 °C and 50 °C are not the extreme temperatures for high-temperature failure of HMA, the decay of material properties among these two temperatures can still reflect the high-temperature performance of the materials.

An unconfined axial periodical load was applied on the HMA specimens in RLPD tests to further emphasize the viscoelastic and viscoplastic behavior of asphalt mixtures. Each loading cycle consisted of one section of haversine wave and one resting section, i.e., loading/rest process. In each loading period, the load amplitude of 500 N and 400 N were applied at 35 °C and 50 °C, respectively, to retain the strain amplitude in the range of 400~600$\mu\varepsilon$. The strain level exceeds the linear viscoelastic domain of the HMA [31,32] to further intensify the permanent deformation of materials compared to the modified DM tests. Due to the strain limit of the OFBG sensor, 100 and 50 loading cycles were selected for the temperature conditions of 35 °C and 50 °C, respectively. Three groups of loading/rest time combinations, ① 0.1 s/0.9 s, ② 1 s/1 s, and ③ 1 s/9 s, were adopted in this experiment, and the pairwise comparison was used to reflect the influence of loading time, rest time, and loading frequency on the deformation features of the material. The loading mode of 0.1 s/0.9 s was used for the specification AASHTO TP79 [33], and another two comparison groups were proposed on this basis. For different materials, two sets of parallel experiments have been carried out.

2.4.2. Data Analysis

The stiffness and permanent deformation rate of different HMA samples under repeated loading/unloading conditions were analyzed. Their differences in high-temperature stability and permanent deformation resistance were compared to investigate the synergistic effect of volcanic ash and the SBS modifier.

In each loading/rest time cycle, as shown in Figure 4, the loading curve and corresponding strain curve in the loading/rest time combination of 1 s/9 s were plotted as an example. As the curve of one periodic loading cycle plotted in Figure 4a, Point A_0 was the initial point of one loading period, B_0 was the peak point of the haversine loading, C_0 was the ending point of the haversine loading (i.e., the initial point of the rest time), and D_0 was the ending point of one periodic loading cycle. The loading curve $A_0B_0C_0D_0$ can be divided into three phases: the loading phase of A_0B_0, the unloading phase of B_0C_0, and the rest time phase of C_0D_0. As shown in Figure 4b, the corresponding strain curve ABCD was also divided into four sections: (1) total strain ε_{total} (phase AB); (2) instantly recoverable strain ε_e (phase BC); (3) delayed recoverable strain ε_{ve} (phase CD); and (4) residual strain ε_{vp}. Point A corresponded to the starting point of one loading cycle; Point B was the peak strain point, the time that was very close to that of the peak point B_0 on the haversine loading curve; Point C was the ending point of the applied haversine curve; and Point D was the ending point of this loading cycle. The four strain portions have the relationship shown in Equation (9).

$$\varepsilon_{total} = \varepsilon_e + \varepsilon_{ve} + \varepsilon_{vp} \tag{9}$$

where ε_{total} is the total strain, ε_e is the elastic strain, ε_{ve} is the viscoelastic strain, and ε_{vp} is the residual strain. It should be noted that the viscoplastic strain and plastic strain were not separated in the analysis and are represented by ε_{vp}.

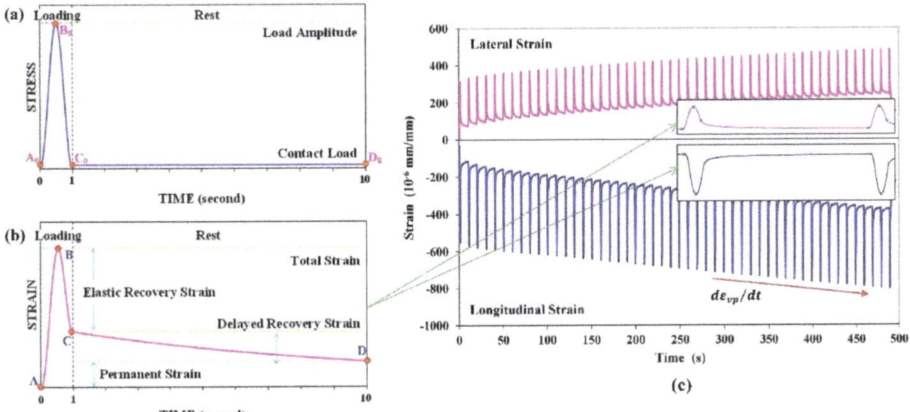

Figure 4. (**a**) Loading curve and (**b**) corresponding strain curve in the loading/rest combination of 1 s/9 s in RLPD tests, and (**c**) an illustration of the longitudinal and lateral permanent strain rate.

The total strain was divided into three parts of instantaneous recoverable strain, delayed recoverable strain, and residual strain to characterize the elastic, viscoelastic, and viscoplastic behaviors, respectively, of the HMA with different VA fillers and asphalt binders. The axial and lateral strain in 50 cycles of the loading/rest (1 s/9 s) process at 50 °C is shown in Figure 4c. With MATLAB code, for each loading/rest cycle, the starting point A_n, strain magnitude point B_n, starting point of rest C_n, and ending point of rest period D_n, which was also the starting point of next loading/rest cycle A_{n+1}, were obtained to analyze the different strains with Equations (10)–(14). The rate of permanent strain (RPS) was used to characterize the resistance of permanent deformation of the materials since the developing trend of the strain curve in each loading/rest cycle was determined by residual strain.

$$\varepsilon_{total,n} = \varepsilon_{B,n} - \varepsilon_{A,n} \tag{10}$$

$$\varepsilon_{e,n} = \varepsilon_{B,n} - \varepsilon_{C,n} \tag{11}$$

$$\varepsilon_{ve,n} = \varepsilon_{C,n} - \varepsilon_{A,n+1} \tag{12}$$

$$\varepsilon_{vp,n} = \varepsilon_{A,n+1} - \varepsilon_{A,n} \tag{13}$$

$$\left.\frac{d\varepsilon_{vp}}{dt}\right|_n \approx \frac{\varepsilon_{A,n+1} - \varepsilon_{A,n-1}}{t_{A,n+1} - t_{A,n-1}} \tag{14}$$

where t is the time period, n is the analyzed loading cycle, N is the total loading cycle, which is 50 in the example shown in Figure 4c, $\varepsilon_{total,n}$ is the total strain in the nth cycle, $\varepsilon_{e,n}$ is the elastic strain (or instantly recoverable) part of $\varepsilon_{total,n}$, $\varepsilon_{ve,n}$ and $\varepsilon_{vp,n}$ are the viscoelastic strain (or delayed recovered strain) and viscoplastic strain (or residual strain) part, respectively, and $\left.\frac{d\varepsilon_{vp}}{dt}\right|_n$ is the mean RPS in these n cycles.

The average value of each index in the last 10 cycles was calculated for analysis. In addition, the stiffness and RLSLS of each type of HMA can be further calculated with Equations (15) and (16), respectively.

$$S = \frac{\sigma_{amp}}{\varepsilon_{total}} \tag{15}$$

$$v_x = \frac{\varepsilon_{x,lat}}{\varepsilon_{x,long}} \tag{16}$$

where σ_{amp} is the amplitude of longitudinal stress, and the subscript x can be *total*, *e*, *ve*, or *vp*.

It should be noted that RLSLS is different from PR. PR is a concept in elastic theory, requiring the strain of materials to be within a linear elastic range. In this study, the strain magnitude of the HMA specimens in each loading cycle was approximately 500 $\mu\varepsilon$, which allowed materials to have more significant nonlinear behavior and a larger increasing rate of permanent strain.

In this study, the parameters of DM, PA, CPR, stiffness, and RPS for asphalt mixtures under the temperature conditions of 35 °C and 50 °C can be obtained by the DM test, modified DM test, and RLPD test, respectively, as shown in Table 5. Through the comprehensive analysis of multiple test parameters, the rationality of the test results can be verified with each other. Moreover, the performance of the materials can be evaluated from different perspectives. According to the experimental results, the change characteristics of the HMAs in the process of temperature increase were analyzed, and the performance difference of the material can be comprehensively evaluated. In addition, based on the strain analysis in the RLPD test, the elasto-visco-plastic behaviors of the mixtures during the deforming process can be further characterized to explore the relationship between the RLSLS index and the permanent deformation of the materials.

Table 5. The details of the analyzed indices of different testing methods.

Testing Method	Index	Temperature	Axial Strain Level	Sensor Direction
Standard DM test	DM, PA (mechanics parameter)	35 °C, 50 °C	<200 $\mu\varepsilon$ (viscoelastic behavior)	Axial
Modified DM test	CPR(DPR and phase lag) (mechanics parameter)	35 °C, 50 °C	<200 $\mu\varepsilon$ (viscoelastic behavior)	Axia Lateral
RLPD test	Stiffness, RPS (performance index) RLSLS (performance index)	35 °C, 50 °C	400~600 $\mu\varepsilon$ (elasto-visco-plastic behavior)	Axial Lateral

3. Results and Discussion

3.1. Standard DM Testing Results

The dynamic modulus (DM) and phase angle (PA) of the four types of HMA were measured in standard DM tests [29] at 35 °C and 50 °C under seven different testing frequencies ranging from 0.1 to 25 Hz. The value range and frequency curve of the four materials are arranged in Table 6 and Figure 5 respectively. As shown in Figure 5a, the DM values of the four mixes tested at both 35 °C and 50 °C all decreased with a smaller testing frequency, which was equivalent to the increase in temperature according to TTSP [30]. The moduli of each mixture at 35 °C were greater than that at 50 °C. This decrease of DM at high temperatures was induced by the more significant behavior of viscosity flow within the asphalt binder or mastic, which led to a softening of the HMA. Therefore, a high DM indicated a more stable mesostructure composition of the mix. As a result, unrecoverable relocation or restructuration was less prone to occur in the aggregate skeleton of HMA.

Table 6. The value range of DM and PA for the four mixtures at 35 °C and 50 °C.

Material	Index	Value Range (35 °C)	Value Range (50 °C)
LA-S		584~4482	317~1540
BA-S	Dynamic modulus	447~4254	207~1262
LA-P	(MPa)	381~4869	155~1298
BA-P		319~3735	142~913
LA-S		28.2~34.7	21.0~34.3
BA-S	Phase angle	29.5~36.1	21.2~35.8
LA-P	(degree)	31.2~39.2	21.9~39.1
BA-P		29.0~37.3	18.8~37.2

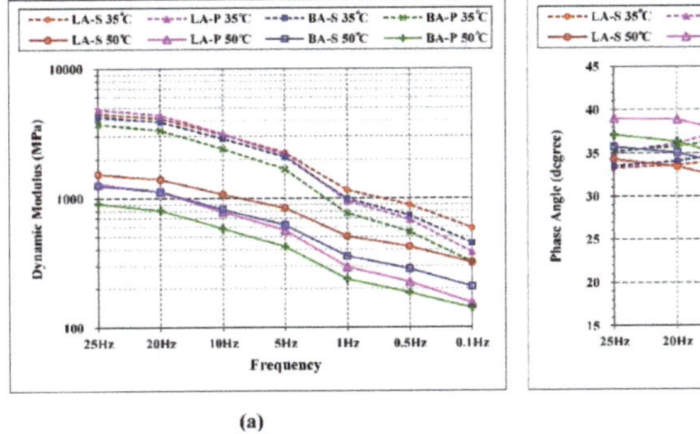
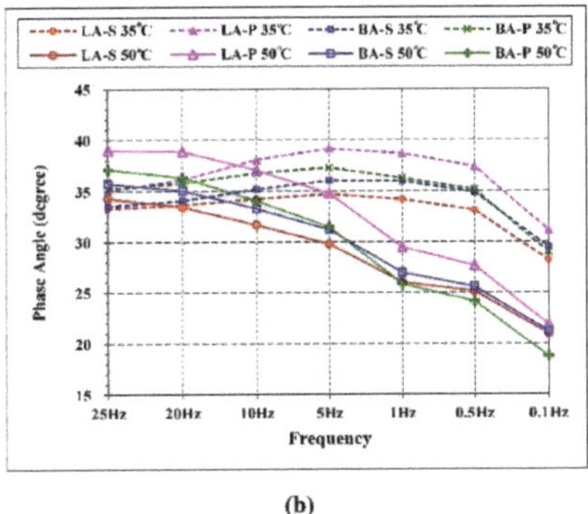

(a) (b)

Figure 5. The dynamic modulus of HMA at 35 °C and 50 °C (**a**), and phase angle at 35 °C and 50 °C (**b**).

On the other hand, the PA of all the mixtures at 35 °C increased from 25 Hz to 5 Hz and then decreased. In contrast, the PA of all the materials monotonically decreased with decreasing frequency at 50 °C, as shown in Figure 5b. This indicated that the temperature of approximately 35 °C was a transmission domain of the viscoelastic properties for these four types of materials. Therefore, the mechanical stability of the HMAs can be reflected by the 35 °C results. It should be noted that the tangent of PA represented the ratio between viscous loss energy and elastic storage energy. Low PA indicated more apparent elastic behavior, while high PA indicated a higher proportion of viscous behavior, which suggested more creep deformation occurrence. Furthermore, a smaller variation in PA under different frequencies at 50 °C also represented high stability, which can result in a higher rutting resistance.

The difference in testing results for the four types of HMAs were mainly induced by the different performances of asphalt mastic, which consists of VA filler and asphalt binder since the same aggregate gradation was applied. Based on the analysis of DM and PA, the performance of the four types of HMA at 35 °C can be ordered as LA-S > BA-S > BA-P > LA-P. This result indicated that the VA-SBS-modified HMA possessed better performance than the base asphalt mixture. Given that the temperature of 35 °C was not extremely high, the PA change at this temperature was more profound than DM change, which was more indicative of material stability. Since LA-P presented low DM and high change of PA, its performance was incomparable to the others. The performance of the four types of HMA at 50 °C presented a similar order to that at 35 °C, while LA-S had the most extraordinary high-temperature performance, followed by BA-S. Based on the DM and PA testing results, the two SBS-modified HMA presented better high-temperature performance than the other two that used base asphalt.

3.2. Modified DM Test

In the modified DM tests, the applicability of OFBG was explored in the measurement of the CPR of HMA cylindrical samples under harmonic loading. With the aid of OFBG sensors, the DPR ($|v^*|$) and the phase difference of the strain wave in the lateral and longitudinal directions of HMA, in terms of phase lag (δ_v), were obtained at two temperatures (35 °C and 50 °C) and three frequencies (1 Hz, 0.5 Hz, and 1 Hz), as summarized in Figure 6 and Table 7. Similar to the analyses of DM and PA in the standard DM test, the testing results at 35 °C also can be used to assess the mechanical stability of the mixes, while those at 50 °C were applied in the investigation of the permanent deformation properties of HMA at high temperatures.

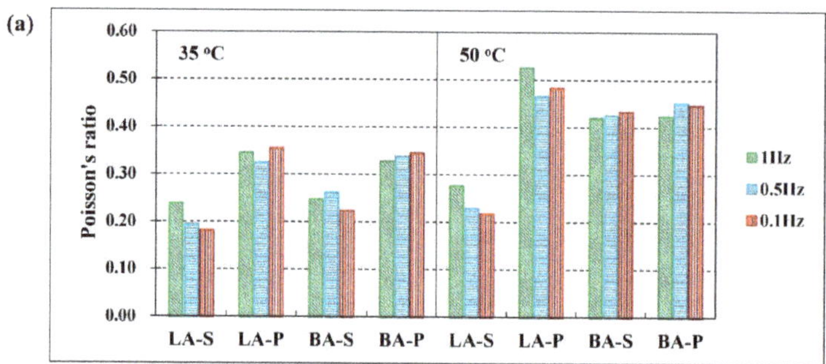

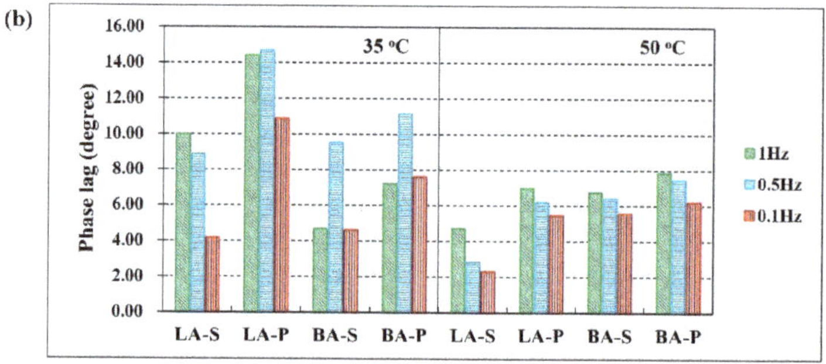

Figure 6. The DPR at 35 °C and 50 °C (**a**), and the phase lag of the mixes at 35 °C and 50 °C (**b**).

Table 7. The value range of DPR and phase lag of the four mixtures at 35 °C and 50 °C.

Material	Index	Value Range (35 °C)	Value Range (50 °C)
LA-S		0.18~0.24	0.22~0.28
BA-S	DPR	0.22~0.26	0.43~0.42
LA-P		0.32~0.35	0.46~0.53
BA-P		0.33~0.34	0.42~0.45
LA-S		4.17~9.98	2.30~4.71
BA-S	Phase lag	4.62~9.55	5.56~6.74
LA-P	(degree)	10.85~14.70	5.45~6.98
BA-P		7.21~11.16	6.20~7.87

Based on the DPR and phase lag of the four HMAs shown in Figure 6a,b, the $|v^*|$ values of the two VA-SBS HMAs at 35 °C were in the range of 0.2~0.25, while that of the VA-base HMAs were approximately 0.35. According to the results at the three testing frequencies, the four types of mixtures were in the order of LA-S < BA-S < BA-P ≈ LA-P with respect to DPR. On the other hand, the correlation among the phase lags was more complex. LA-P was more distinguished, with δ_V being approximately 13 degrees, while the δ_V of the other three materials was approximately 8 degrees. Considering the tests at low frequencies that had more adverse impacts on permanent deformation, the phase lags of the HMA followed the ascending order of LA-S < BA-S < BA-P < LA-P. Combined with the DM and PA results in Figure 5, the temperature stability of HMA at 35 °C was in the opposite order of LA-S > BA-S > BA-P > LA-P, which illustrated that CPR contains $|v^*|$ and δ_V can be used to reflect the performance difference of the asphalt mixtures. The lower $|v^*|$ and δ_V of the HMA were, the better temperature stability and mechanical properties.

The $|v^*|$ and δ_V at 50 °C showed different trends compared to those at 35 °C, as shown in Figure 6. LA-S presented the most significant difference compared with the other HMAs in both DPR and phase lags. First, the $|v^*|$ of LA-S at 50 °C was slightly higher than that at 35 °C and still remained in the range of 0.2~0.3, whereas the $|v^*|$ of the other three mixes apparently increased from 35 °C to 50 °C, which were all larger than 0.4. The smaller change in DPR indicated that LA-S has more stable mechanical properties with the varying temperatures. Furthermore, a lower $|v^*|$ suggested a less incompressible viscous flow deformation and more recoverable elastic deformation of LA-S during the loading process. Second, LA-S had the lowest δ_V at 50 °C, which was approximately 4 degrees, while the δ_V of the other materials were similar to each other, approximately 6~8 degrees. Since a larger δ_V would lead to extra squeezing and abrasing process due to incoordination between the internal lateral strain and longitudinal strain, more serious permanent deformation would result. The above analyses suggested that the extraordinary high-temperature performance was presented by LA-S, while BA-S was only slightly better than the two types of base asphalt mixtures. The high-temperature performance still had the order of LA-S < BA-S < BA-P < LA-P based on DPR values.

Based on any of the four test indices of DM, PA, DPR, and phase lag, the performance of LA-S was the best. Furthermore, the variation amplitude of its indices were also the smallest during the temperature increase. Further, the other three mixtures also approximately satisfy this rule. This indicates that the high temperature performance of the HMA was intrinsically related to the stability of mechanical properties under temperature changes. In addition, the CRP or DPR index, which was related to the lateral deformation, indicated that the smaller the value, the better performance of the material. This indicated that a parameter of Poisson's ratio also had the potential to be used as an evaluation index of material performance.

However, the strain level of HMA in the standard DM and modified DM tests was controlled under 200 $\mu\varepsilon$, which would not show significant permanent deformation. Therefore, the RLPD tests were conducted in this study to further analyze the mechanical properties of the HMA and their relevance to the RLSLS index, along with the development of permanent deformation.

3.3. RLPD Test

3.3.1. Stiffness

The stiffness of the HMA at specific temperatures, loads, and loading/rest time combinations was calculated using Equation (15) and compared to indicate the characteristics of mechanical responses of mixtures with different VA-binder mastics for multiple impact factors. The stiffness of the HMA samples at 35 °C and 50 °C are plotted in Figure 7.

First, the stiffness of all the mixtures at 35 °C under the condition of a short loading time per cycle (0.1 s/cycle) was approximately twice that of the stiffness under the other two loading conditions of 1 s loading time, while the stiffness under the long rest time per cycle (9 s/cycle) was slightly lower than that under the short rest time per cycle (1 s/cycle). The loading period had a significant influence on the experimental results. In addition, for the mixtures with the same asphalt binder, the LA filler HMA showed a slightly larger stiffness than the corresponding mix with BA.

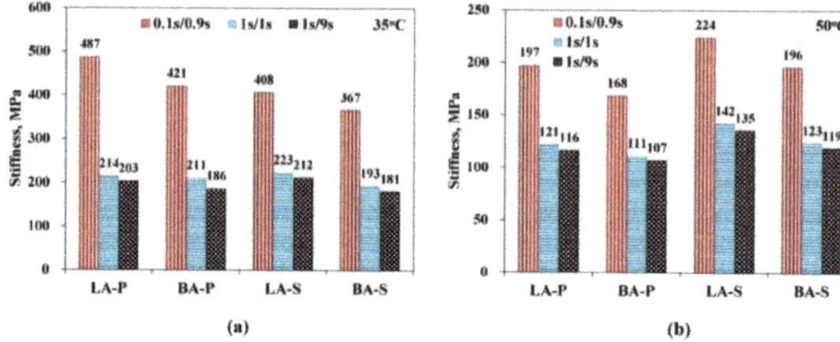

Figure 7. Stiffness of the four types of HMAs at (**a**) 35 °C and (**b**) 50 °C under the conditions of different loading/rest time combinations.

Under the 50 °C condition, the effects of the three loading/rest time combinations on stiffness were the same as those at 35 °C, which also presented the order of $S_{0.1/0.9} > S_{1/1} > S_{1/9}$. However, when the temperature rose to 50 °C, the stiffness of the VA-base asphalt mixture was lower than that of the corresponding VA-SBS mixture, and the four materials showed an order of LA-S > BA-S ≈ LA-P > BA-P.

Since the same gradation was used for all the investigated mixes, it indicated that the viscoelastic properties of the inner mastic can influence the modulus level of the mixture at high temperatures. However, the trend shown by the stiffness of the HMA was not as significant as those of DM, PA, and CPR in the DM and modified DM tests. As a result, it was less applicable to demonstrate more subtle differences of material performance by the only index of stiffness.

3.3.2. Rate of Permanent Strain (RPS)

The longitudinal and lateral RPS of the four types of HMA at 35 °C and 50 °C were calculated with their vertical and horizontal strain data using Equation (14), as summarized in Table 8 and Figure 8. According to its definition of $\frac{d\varepsilon_{vp}}{dt}$, RPS can directly indicate the rate of development of permanent deformation at specific temperatures, loading magnitude, and loading/rest period.

As illustrated in Figure 8, the lateral and longitudinal RPS of the same HMA at 35 °C developed similar trends, which were significantly affected by loading/rest time combinations. The RPS in the combination of 1 s/1 s was apparently higher than the counterparts in the other two combinations, while the RPS in the 0.1 s/0.9 s combination was larger than that in the 1 s/9 s one, indicating that (i) permanent deformation was highly dependent on the maximum loading period; and (ii) a relatively long rest period can decrease permanent deformation of the HMA with a longer time of strain recovery. In general, the two types of VA-SBS mixtures have lower RPS than the other two VA-base mixtures at 35 °C, which followed the order of LA-S ≈ BA-S < LA-P ≈ BA-P.

Table 8. The value range of RPS for the four mixtures at 35 °C and 50 °C.

Material	Sensor Direction	Value Range (35 °C)	Value Range (50 °C)
LA-S		0.24~1.30	0.22~1.59
BA-S	Vertical RPS	0.25~2.07	0.47~6.02
LA-P	(με/s)	0.58~3.24	0.89~9.72
BA-P		0.59~3.24	0.75~4.85
LA-S		0.11~0.67	0.11~1.33
BA-S	Horizontal RPS	0.14~1.31	0.24~4.17
LA-P	(με/s)	0.49~3.25	0.97~9.70
BA-P		0.53~3.00	0.89~7.54

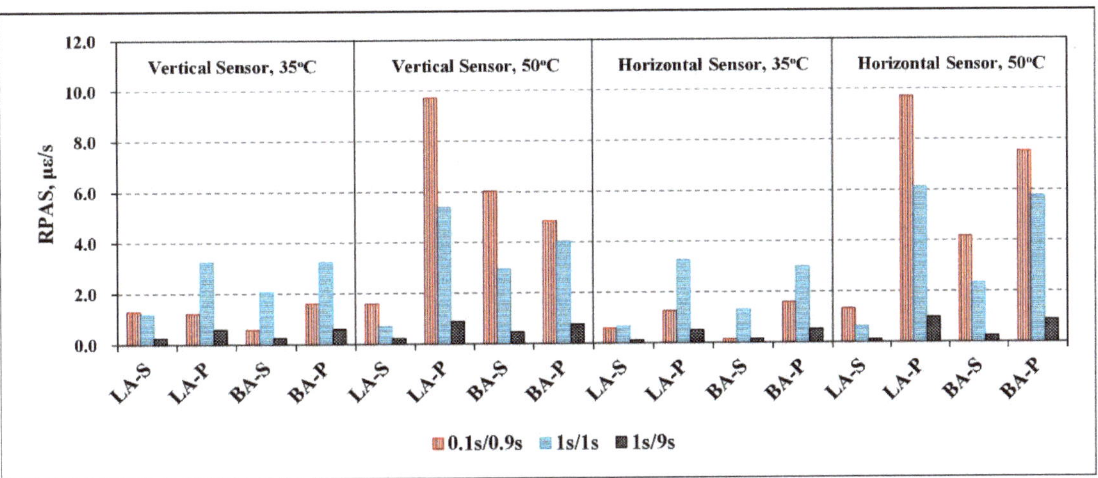

Figure 8. Longitudinal and lateral RPS of mixes at 35 °C and 50 °C under different loading/rest time combinations.

For the results at 50 °C, both the lateral and longitudinal RPS followed the ascending order of LA-S < BA-S < BA-P < LA-P, elucidating that LA-S had the best resistance of creep deformation in the high-temperature condition, as shown in Figure 8. For the same type of HMA tested at 50 °C, the RPS in the 1 s/9 s loading combination was still the lowest. However, the RPS in the 0.1 s/0.9 s combination became the largest at 50 °C, different from 35 °C, at which the largest RPS value occurred in the 1 s/1 s combination. This indicated that the viscoelastic property of the mixtures significantly changed at a high temperature of 50 °C, which was more apparent and raised the sensitivity to the loading frequency variation. In general, the RPS at 50 °C was apparently higher than that at 35 °C, suggesting the more significant creep deformation of HMA under high-temperature conditions.

In general, the high-temperature performance of the HMAs interpreted by stiffness and PRS was consistent with that analyzed based on the indices of DM, PA, and CPR at the same temperature. This consistency indicated that the three sides of HMA, high temperature, mechanical properties, and deformation behavior, were the different expressions of the constitutive relationship of the material. Therefore, these experimental results were essentially determined by the mesostructure constitutions of the material, which was the theoretical foundation of studies for establishing the related indices of permanent deformation using mechanical parameters of the HMA.

3.3.3. Ratio of Lateral Strain to Longitudinal Strain (RLSLS)

In the RLPD tests, the total strain of HMA in one loading/rest period was divided into instantaneous recovered strain, delayed recovered strain, and residual strain, according to Equation (9), to represent the elastic, viscoelastic, and viscoplastic behaviors, respectively. The longitudinal strains of all the HMA samples at 35 °C in different loading combinations are shown in Figure 9a,c,e, with their fractions in total strain marked. Correspondingly, the RLSLS values calculated with Equations (10)–(13), denoted by v_{tatal}, v_e, v_{ve}, and v_{vp}, were plotted in Figure 9b,d,f, respectively. The same types of results at 50 °C were plotted in Figure 10a–e.

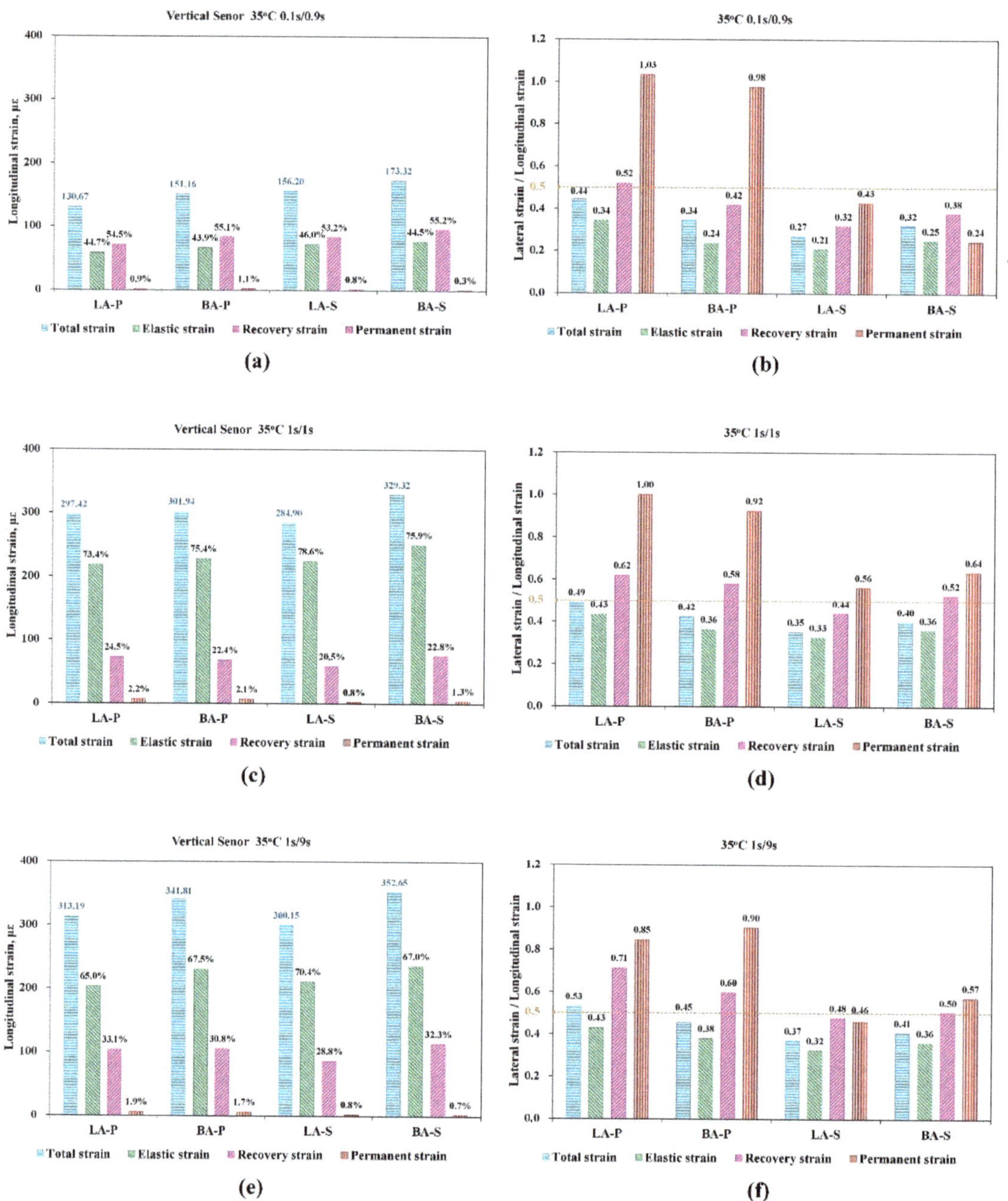

Figure 9. The longitudinal strain and the corresponding lateral/longitudinal strain ratios at the loading/rest time combinations of 0.1 s/0.9 s (**a,b**), 1 s/1 s (**c,d**), and 1 s/9 s (**e,f**) at 35 °C.

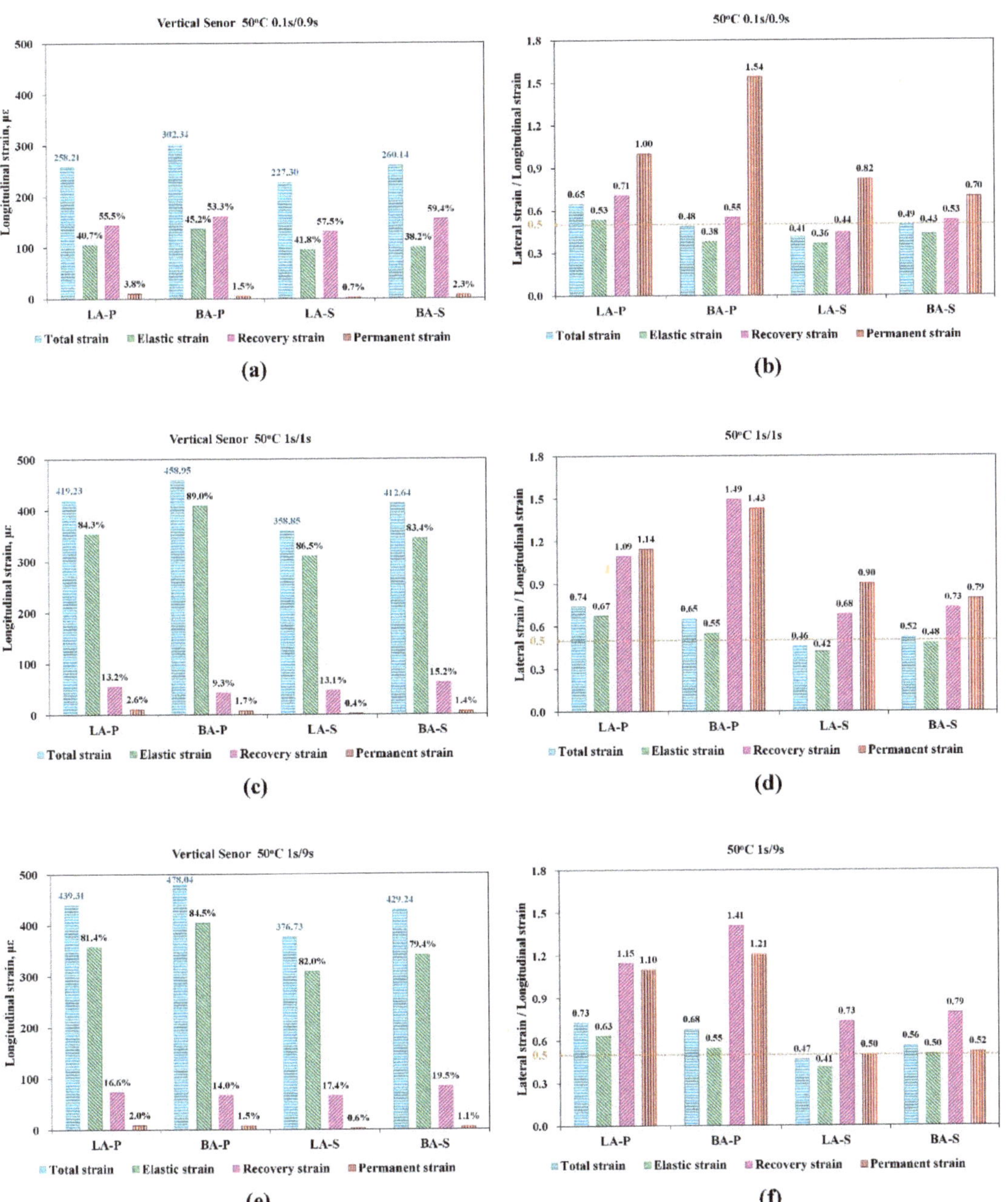

Figure 10. The longitudinal strain and the corresponding lateral/longitudinal strain ratios at the loading/rest time combinations of 0.1 s/0.9 s (**a,b**), 1 s/1 s (**c,d**), and 1 s/9 s (**e,f**) at 50 °C.

Strains and RLSLS at 35 °C

As shown in Figure 9a,c,e, the total strains of HMA at the 0.1 s/0.9 s loading combination at 35 °C were all approximately 150με, which was nearly one-half of the strains under the 1 s/1 s and 1 s/9 s loading combinations. The fraction of elastic recoverable strain in total strain for the 1 s/1 s and 1 s/9 s loading combinations (approximately 65~75%) were apparently higher than the proportion of delayed recovered strain (approximately 20~35%). At a loading combination of 0.1 s/0.9 s, the fraction of elastic strain (~45%) was slightly lower than that of the viscoelastic strain (~55%). The proportion of viscoelastic strain was increased from 20% to 30% when the loading combination changed from 1 s/1 s to 1 s/9 s, indicating the delayed recovery property of asphalt mixtures. In general, the proportion of viscoplastic strain in the total strain of the three types of loading conditions were all low, in the range of 1~3%. The total strain of the four HMAs showed the relationship of LA-S < BA-S and LA-P < BA-P, while the permanent strain followed the order of LA-S $\approx$ BA-S < BA-P $\approx$ LA-P.

On the other hand, the differences of the HMA samples in the performance at $35\,^\circ$C can be more distinctly demonstrated with RLSLS. As shown in Figure 9b,d,f, most of the RLSLS values corresponding to the four types of strains were lower than 0.5. In all of the three loading combinations, the v_{vp} of LA-P and BA-P exceeded 0.5 (approximately 1.0), and their v_{ve} also had a larger value (approximately 0.5~0.7). The v_{vp} and v_{ve} of the other two HMAs of LA-S and BA-S were relatively low, in the range of 0.25~0.65. The LA-S had the lowest RLSLS value among the four materials, and its corresponding VA-SBS mix of LA-S was largest in the same loading combinations. In general, the VA-base asphalt mixtures presented a trend of $v_e < v_{ve} < v_{vp}$ and the VA-SBS mixtures had a trend of $v_e < v_{ve} \approx v_{vp}$. As a result, the differences of the four types of HMA at 35 °C all resulted in the change of RLSLS, which were lower while the corresponding HMA had better performance.

Strains and RLSLS at 50 °C

As illustrated in Figure 10a,c,e, the total strain of all of the HMAs at 50 °C were apparently higher than those at 35 °C, which was approximately 300 $\mu\varepsilon$ under the 0.1 s/0.9 s loading condition, and exceeded 400 $\mu\varepsilon$ when the loading conditions of 1 s/1 s and 1 s/9 s were applied. This result was closely related to the modulus decrease induced by the temperature increase. The elastic strain, ε_e, under the loading conditions of 1 s/1 s and 1 s/9 s both had higher proportions, larger than 80%, while that in the loading condition of 0.1 s/0.9 s was lower than 50%, which was attributed to the increase in the viscous response of the HMA in high loading frequencies. Comparing the testing results in loading conditions of 1 s/1 s and 1 s/9 s, the proportion of viscoelastic strain, ε_{ve}, for 1 s/9 s was slightly higher than that for 1 s/1 s. This resulted from the more delayed recovered strain of the HMA under a longer rest time. Similar to the 35 °C conditions, the viscoplastic strain, ε_{vp}, had a low proportion within the total strain, which was approximately 1~3%.

In addition, the RLSLS values of the four HMAs at 50 °C were also analyzed and were all higher than their counterparts at 35 °C. Noting that the v_{ve} and v_{vp} of mixtures at 50 °C were all larger than 0.5. The v_{ve} and v_{vp} of VA-base mixes were close to 1.5, significantly higher than those of VA-SBS mixes (0.5~0.9). This result indicated that the change in performance and mechanical properties of the HMA can be characterized with the RLSLS parameter. Moreover, the v_e and v_{ve} of the HMA under loading conditions of 1 s/1 s and 1 s/9 s were both slightly higher than those at 0.1 s/0.9 s. All of the asphalt mixtures showed a trend of $v_e < v_{ve} \approx v_{vp}$, similar to that at 35 °C. It can be concluded that the v_{tatal}, v_e, v_{ve}, and v_{vp} of LA-S were the lowest, of which the high-temperature performance was also the best and the LA-P still obtained the worst result, which was the same as at 35 °C. As a result, the two types of VA-SBS HMA had lower RLSLS than the two types of VA-base mixtures.

As the analysis of RLSLS values of the four HMAs, when the more nonlinear behaviors introduced unrecoverable deformation of ε_{ev} and ε_{vp}, the lower material parameters of v_x would be obtained, which is consistent with the phenomenon shown by the CPR index in the modified DM tests.

Relationship between RLSLS and Permanent Strain

According to the RLPD testing results at 35 °C and 50 °C, the RLSLS indices corresponding to ε_e, ε_{ve}, and ε_{vp} were distributed in different value ranges, which followed the ascending order of $v_e<v_{ve}<v_{vp}$, as shown in Figure 11a. The v_e was the lowest since the corresponding elastic strain was mainly characterized by the elastic behavior of the HMA, which was consistent with Poisson's ratio of ideal elastic materials being smaller than 0.5. On the other hand, the v_{vp}, corresponding to the viscoplastic strain, represented unrecoverable viscoplastic and plastic defromation induced by the dislocation and reconstruction of the mastic and aggregate-skeleton, which were both uncompressible shear deformations. Although the theoretical upper limit of Poisson's ratio was 0.5, the value of the RLSLS index was allowed to exceed 0.5 for the complex elastic-visco-plastic behavior of materials. These results were attributed to the viscous behavior induced by strain rate-relevant shear flow in asphalt binder or mastics, which was also a theoretically uncompressible shear deformation. Therefore, the numerical distribution of RLSLS for v_e, v_{ve}, and v_{vp} corresponded to different mechanical behaviors and had apparent discrepancies during the deformation process of the HMA.

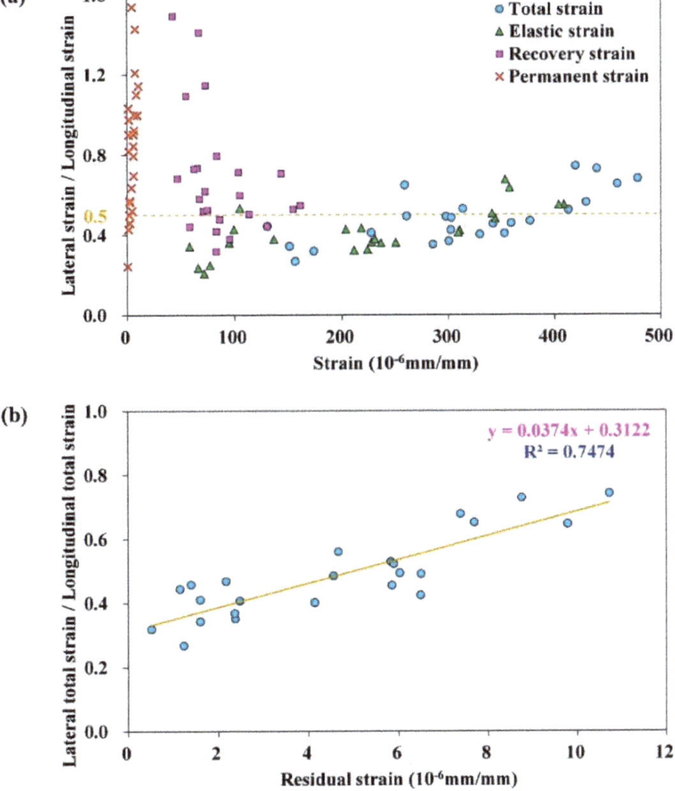

Figure 11. The (**a**) RLSLS value distribution of different strain types and (**b**) the linear fitting of v_{total} and ε_{vp}.

Therefore, this discrepancy can effectively characterize the permanent deforming property of the HMA. As illustrated in Figure 10b,d,f, the differences in v_{ve} and v_{vp} between VA-SBS mixtures and VA-base mixtures at 50 °C were more significant than the difference in v_e, indicating that (i) VA-SBS mixtures had better resistance to permanent deformation and a better deformation recovery property than VA-base mixtures; and (ii) this modification in mastic also decreased the proportion of viscous flow in the deformation process of HMA, presented by the lower v_{ve} and v_{vp} of LA-S and BA-S. Furthermore, LA-S had the lowest v_e, v_{ve}, v_{vp}, and v_{total} of the four types of HMA, consistent with its better performance characterized by DM, PA, CPR, stiffness, and RPS parameters (as shown in Table 9) than the others. This indicated that the proportion of viscous and plastic behaviors of LA-S was relatively low, while the recoverable elastic and viscoelastic behavior was more significant during the deformation process.

Table 9. The order of different indices according to high-temperature performance.

Index	Testing Method	High Temperature Performance
DM, PA	Standard DM test	LA-S > BA-S > BA-P > LA-P
CPR	Modified DM test	LA-S > BA-S > BA-P > LA-P
Stiffness	RLPD test	LA-S > BA-S ≈ LA-P > BA-P
RPS	RLPD test	LA-S ≈ BA-S > LA-P ≈ BA-P
RLSLS	RLPD test	LA-S > BA-S > BA-P > LA-P

Based on the above analyses, the v_{total} obtained in the RLPD tests was the integral expression of elastic, viscoelastic, and viscoplastic behaviors of the HMA, which determined the development of permanent deformation of the materials. The residual strain and the corresponding RLSLS index, v_{total}, of the four HMAs at 35 °C and 50 °C are plotted in Figure 11b, of which two of the coefficient of determination (R^2) values were approximately 0.75. This indicated that v_{total} can elucidate the permanent deformation property of the HMA. It should be noted that this correlation was calculated based on the four HMAs containing both base asphalt and SBS-modified asphalt at 35 °C and 50 °C, which was generally applicable and not limited to particular types of binders and specific temperatures.

Therefore, the RLSLS index of v_{total} obtained in the RLPD tests was a promising indicator for evaluating the flow-rutting investigated in this study. First, the external factors of temperature, loading, and confining pressure can be well controlled in RLPD tests, which significantly affects the development of permanent deformation. In this way, the performance of HMA can be evaluated based on the specific temperature and stress conditions of different pavement structures. Second, FR was generated by the fast accumulation of viscoplastic and plastic deformation in asphalt layers, which occurred in the complex conditions of adverse temperature and load. Therefore, the RLSLS index was highly accommodated to the generation mechanism of FR since it can characterize the mechanical behavior of the HMA during the deformation process. Finally, the total strain, corresponding to v_{total}, can directly indicate the performance of the HMA, rather than dividing total strain into three parts and calculating the corresponding RLSLS indices, which can further simplify the experimental assessment of asphalt mixtures.

In follow-up research, the rutting analysis of asphalt pavement with different thicknesses and material composition should be carried out based on the RLSLS index. The quantitative relationship between this index and the development of permanent deformation in a specific pavement structure under complex temperature field and stress field need to be explored, to establish a more effective evaluation method and prediction model for FR.

4. Conclusions

In this research, the axial and lateral deformation features of four HMAs, using two VA fillers combined with the binder of base asphalt and SBS-modified asphalt, were systematically analyzed at 35 °C and 50 °C through the DM test and RLPD test. The following conclusions can be drawn:

a. Based on the developed OFBG strain sensor, the high-frequency measurement of the axial and lateral strain of cylindrical HMA specimens under a dynamic loading mode was realized. This new sensor can be used to study the complex deformation behaviors of HMA;

b. In the standard DM tests and modified DM tests, LA-S had the best high temperature performance among the four asphalt mixtures, showing the largest DM and the smallest PA at 50 °C, of which the variation amplitude of mechanical parameters was the smallest. In addition, the DPR and phase lag of LA-S still showed the lowest value and change range;

c. The stiffness and PRS indices in the RLPD tests presented that the performance of the four HMAs can be ordered as LA-S > BA-S > BA-P > LA-P, which was consistent with the results evaluated by DM, PA, and CRP. From the perspective of DPR, the lateral deformation feature can reflect the high temperature stability and deformation resistance of the material. The smaller the value, the better the performance of the material;

d. In the RLPD tests, the total strain of the mixes was decomposed into three parts, being the elastic strain, viscoelastic strain, and viscoplastic strain. The RLSLS indices of different strain types presented the trend of $v_e < v_{ev} < v_{vp}$. The v_{tatal} and v_e values were generally less than 0.5, while the values of v_{ve} and v_{vp} could exceed 0.5;

e. The RLSLS index of v_{tatal} can be used to comprehensively evaluate the visco-elasto-plastic behavior of the HMA, and the R^2 value of the linear fitting with the permanent deformation ε_{vp} was approximately 0.75;

f. For different asphalt mixtures, the high temperature performance, mechanical properties; and deformation behavior were the different expressions of the constitutive relationship of the material. It made the various material parameters, DM, PA, CPR, stiffness, PRS; and RLSLS obtained based on different tests, finally show consistent results.

In order to establish an effective evaluation method and prediction model for flow-rutting, the relationship between the RLSLS index and the permanent deformation in a specific pavement structure under complex temperature and stress field should be further quantitated.

Author Contributions: Conceptualization, X.L. and W.L.; methodology, X.L. and W.L.; formal analysis, X.L.; investigation, X.L. and W.L.; resources, W.L.; data curation, X.L.; writing—original draft preparation, M.Z. and X.L.; writing—review and editing, M.Z. and X.L. All authors have read and agreed to the published version of the manuscript.

Funding: This research was funded by the Central Public-interest Scientific Institution Basal Research Fund, grant number 2020-9074.

Data Availability Statement: Not Applicable.

Conflicts of Interest: The researcher claims no conflict of interest.

References

1. Neifar, M.; Di Benedetto, H. Thermo-viscoplastic law for bituminous mixes. *Road Mater. Pavement Des.* **2001**, *2*, 71–95. [CrossRef]
2. Di Benedetto, H.; Mondher, N.; Sauzéat, C.; Olard, F. Three-dimensional thermo-viscoplastic behaviour of bituminous materials: The DBN model. *Road Mater. Pavement Des.* **2007**, *8*, 285–315. [CrossRef]
3. Zhang, Q.-S.; Chen, Y.-L.; Li, X.-L. Rutting in asphalt pavement under heavy load and high temperature. In Proceedings of the Asphalt Material Characterization, Accelerated Testing, and Highway Management: Selected Papers from the 2009 GeoHunan International Conference, Changsha, China, 3–6 August 2009; pp. 39–48.
4. İskender, E. Rutting evaluation of stone mastic asphalt for basalt and basalt–limestone aggregate combinations. *Compos. Part B* **2013**, *54*, 255–264. [CrossRef]
5. Pasquini, E.; Canestrari, F.; Cardone, F.; Santagata, F. Performance evaluation of gap graded asphalt rubber mixtures. *Constr. Build. Mater.* **2011**, *25*, 2014–2022. [CrossRef]
6. Zhao, S.; Huang, B.; Shu, X.; Ye, P. Laboratory investigation of biochar-modified asphalt mixture. *Transp. Res. Rec.* **2014**, *2445*, 56–63. [CrossRef]
7. Zhao, S.; Huang, B.; Ye, X.P.; Shu, X.; Jia, X. Utilizing bio-char as a bio-modifier for asphalt cement: A sustainable application of bio-fuel by-product. *Fuel* **2014**, *133*, 52–62. [CrossRef]

8. Movilla-Quesada, D.; Muñoz, O.; Raposeiras, A.C.; Castro-Fresno, D. Thermal suspectability analysis of the reuse of fly ash from cellulose industry as contribution filler in bituminous mixtures. *Constr. Build. Mater.* **2018**, *160*, 268–277. [CrossRef]
9. Lagos-Varas, M.; Movilla-Quesada, D.; Raposeiras, A.C.; Arenas, J.P.; Calzada-Perez, M.A.; Vega-Zamanillo, A.; Lastra-Gonzalez, P. Influence of limestone filler on the rheological properties of bituminous mastics through susceptibility master curves. *Constr. Build. Mater.* **2020**, *231*, 117126. [CrossRef]
10. Liu, X.; Liu, W.; Wang, S.; Wang, Z.; Shao, L. Performance evaluation of asphalt mixture with nanosized volcanic ash filler. *J. Transp. Eng. Part B Pavements* **2018**, *144*, 04018028. [CrossRef]
11. Liu, X.; Zhang, M.; Shao, L.; Chen, Z. Effect of volcanic ash filler on thermal viscoelastic property of SBS modified asphalt mastic. *Constr. Build. Mater.* **2018**, *190*, 495–507. [CrossRef]
12. Van Thanh, D.; Feng, C.P. Study on Marshall and Rutting test of SMA at abnormally high temperature. *Constr. Build. Mater.* **2013**, *47*, 1337–1341. [CrossRef]
13. Chaturabong, P.; Bahia, H.U. Mechanisms of asphalt mixture rutting in the dry Hamburg Wheel Tracking test and the potential to be alternative test in measuring rutting resistance. *Constr. Build. Mater.* **2017**, *146*, 175–182. [CrossRef]
14. Wen, H.; Wu, S.; Mohammad, L.N.; Zhang, W.; Shen, S.; Faheem, A. Long-term field rutting and moisture susceptibility performance of warm-mix asphalt pavement. *Transp. Res. Rec.* **2016**, *2575*, 103–112. [CrossRef]
15. Zhang, W.; Shen, S.; Wu, S.; Mohammad, L.N. Prediction model for field rut depth of asphalt pavement based on Hamburg wheel tracking test properties. *J. Mater. Civ. Eng.* **2017**, *29*, 04017098. [CrossRef]
16. Suh, Y.-C.; Cho, N.-H.; Mun, S. Development of mechanistic–empirical design method for an asphalt pavement rutting model using APT. *Constr. Build. Mater.* **2011**, *25*, 1685–1690. [CrossRef]
17. Ji, X.; Zheng, N.; Niu, S.; Meng, S.; Xu, Q. Development of a rutting prediction model for asphalt pavements with the use of an accelerated loading facility. *Road Mater. Pavement Des.* **2016**, *17*, 15–31. [CrossRef]
18. Tian, Y.; Lee, J.; Nantung, T.; Haddock, J.E. Development of a mid-depth profile monitoring system for accelerated pavement testing. *Constr. Build. Mater.* **2017**, *140*, 1–9. [CrossRef]
19. Lagos-Varas, M.; Movilla-Quesada, D.; Arenas, J.P.; Raposeiras, A.C.; Castro-Fresno, D.; Calzada-Pérez, M.A.; Maturana, J. Study of the mechanical behavior of asphalt mixtures using fractional rheology to model their viscoelasticity. *Constr. Build. Mater.* **2019**, *200*, 124–134. [CrossRef]
20. Lagos-Varas, M.; Raposeiras, A.C.; Movilla-Quesada, D.; Arenas, J.P.; Castro-Fresno, D.; Muñoz-Cáceres, O.; Andres-Valeri, V.C. Study of the permanent deformation of binders and asphalt mixtures using rheological models of fractional viscoelasticity. *Constr. Build. Mater.* **2020**, *260*, 120438. [CrossRef]
21. Ali, Y.; Irfan, M.; Ahmed, S.; Ahmed, S. Empirical correlation of permanent deformation tests for evaluating the rutting response of conventional asphaltic concrete mixtures. *J. Mater. Civ. Eng.* **2017**, *29*, 04017059. [CrossRef]
22. Witczak, M.W. *Simple Performance Tests: Summary of Recommended Methods and Database*; Transportation Research Board: Washington, DC, USA, 2005; Volume 46.
23. Witzcak, M.W. *Simple Performance Test for Superpave Mix Design*; Transportation Research Board: Washington, DC, USA, 2002; Volume 465.
24. Gandomi, A.H.; Alavi, A.H.; Mirzahosseini, M.R.; Nejad, F.M. Nonlinear Genetic-Based Models for Prediction of Flow Number of Asphalt Mixtures. *J. Mater. Civ. Eng.* **2011**, *23*, 248–263. [CrossRef]
25. Li, Q.; Yang, H.; Ni, F.; Ma, X.; Luo, L. Cause analysis on permanent deformation for asphalt pavements using field cores. *Constr. Build. Mater.* **2015**, *100*, 40–51. [CrossRef]
26. ASTM, D6927-06; Standard Test Method for Marshall Stability and Flow of Bituminous Mixtures; ASTM: West Conshohocken, PA, USA, 2006. [CrossRef]
27. Zhou, Z.; Liu, W.; Huang, Y.; Wang, H.; He, J.; Huang, M.; Ou, J. Optical fiber Bragg grating sensor assembly for 3D strain monitoring and its case study in highway pavement. *Mech. Syst. Signal Processing* **2012**, *28*, 36–49. [CrossRef]
28. Liu, W.; Wang, B.; Zhou, Z.; Cao, D.; Zhao, Y. Design and Testing of a Large-Scale Shape-Monitoring Sensor Based on Fiber-Bragg-Grating Sensing Technique for Pavement Structure. *J. Transp. Eng. Part A Syst.* **2017**, *143*, 04017009. [CrossRef]
29. AASHTO, TP62-07; Standard Method of Test for Determining Dynamic Modulus of Hot-Mix Asphalt (HMA); American Association of State Highway and Transportation Officials: Washington, DC, USA, 2007.
30. Zhao, Y.; Richard Kim, Y. Time–temperature superposition for asphalt mixtures with growing damage and permanent deformation in compression. *Transp. Res. Rec.* **2003**, *1832*, 161–172. [CrossRef]
31. Airey, G.D.; Rahimzadeh, B.; Collop, A.C. Viscoelastic linearity limits for bituminous materials. *Mater. Struct.* **2003**, *36*, 643–647. [CrossRef]
32. Di Benedetto, H.; Olard, F.; Sauzéat, C.; Delaporte, B. Linear viscoelastic behaviour of bituminous materials: From binders to mixes. *Road Mater. Pavement Des.* **2004**, *5*, 163–202. [CrossRef]
33. AASHTO, TP79-15; Standard Method of Test for Determining Dynamic Modulus and Flow Number for Asphalt Mixtures Using the Asphalt Mixture Performance Tester (AMPT); American Association of State Highway and Transportation Officials: Washington, DC, USA, 2015.

Article

Effects of Wax Molecular Weight Distribution and Branching on Moisture Sensitivity of Asphalt Binders

Wenqi Wang [1], Azuo Nili [2,3], Ali Rahman [2,3] and Xu Chen [2,3,*]

1   School of Architecture and Civil Engineering, Xihua University, Chengdu 610039, China; 0120030057@mail.xhu.edu.cn
2   School of Civil Engineering, Southwest Jiaotong University, Chengdu 610031, China; azuo.nili@my.swjtu.edu.cn (A.N.); arahman@swjtu.edu.cn (A.R.)
3   Highway Engineering Key Laboratory of Sichuan Province, Southwest Jiaotong University, Chengdu 610031, China
*   Correspondence: chenxuyouxiang@my.swjtu.edu.cn; Tel.: +86-186-8368-2989

Abstract: Wax is an important factor that affects the durability of asphalt binder. In order to understand the molecular weight distribution and branching of wax on the moisture sensitivity of asphalt binder, pure wax-doped asphalt binders are prepared and the performance of model asphalt binders are evaluated by surface free-energy (SFE) and binder bond strength (BBS) tests. In addition, asphaltene is regarded as an additive in this study. The results show that the addition of eicosane, triacontane, squalane and asphaltene can reduce the moisture sensitivity of asphalt, but not necessarily improve its moisture-induced damage resistance. The physical hardening effect of high-wax asphalt and its model asphalt is stronger than that of the corresponding low-wax asphalt and its model asphalt, and its moisture sensitivity is weaker than that of the low-wax asphalt. For all the model asphalts, there is a good correlation between the cohesion work, cohesion POTS (pull-off tensile strength), POTS ratio (the BBS moisture sensitivity index) and ER (the SFE moisture sensitivity index). When using the BBS test to characterize the moisture sensitivity of high-wax asphalt, it is recommended to leave the sample for some time until it is physically hardened and stable.

Keywords: molecular weight distribution; branching of wax; asphaltene; moisture sensitivity; surface free-energy; binder bond strength

Citation: Wang, W.; Nili, A.; Rahman, A.; Chen, X. Effects of Wax Molecular Weight Distribution and Branching on Moisture Sensitivity of Asphalt Binders. *Materials* **2022**, *15*, 4206. https://doi.org/10.3390/ma15124206

Academic Editors: Meng Ling, Yao Zhang, Haibo Ding and Yu Chen

Received: 29 May 2022
Accepted: 10 June 2022
Published: 14 June 2022

Publisher's Note: MDPI stays neutral with regard to jurisdictional claims in published maps and institutional affiliations.

1. Introduction

The strong moisture sensitivity of asphalt mixture is one of the main reasons for early distresses in asphalt pavement, such as grouting, spalling, looseness and potholes [1]. The process of moisture-induced damage to an asphalt mixture is complex, involving many reactions and theories, such as mechanics, physics, chemistry and thermodynamics [2]. It mainly includes two stages: (1) moisture transmission in asphalt mixture; (2) moisture damage on asphalt–asphalt (cohesive failure) and asphalt–aggregate (adhesion failure) interfaces; see Figure 1. The latter is regarded as the two main and direct mechanisms of moisture-induced damage on asphalt pavement [3]. Therefore, it is of great significance to study the moisture sensitivity of asphalt based on its cohesion and adhesion failure.

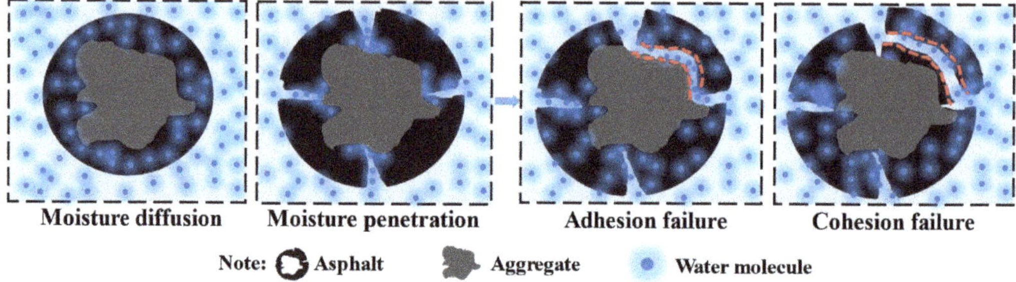

Figure 1. Moisture-induced damage process diagram of asphalt pavement. "Reprinted with permission from Ref. [4]".

The wax in asphalt is a mixture of saturated n-alkanes and a small amount of isoalkanes, with carbon atom numbers ranging from 20 to 40 [5,6]. The wax content has a significant effect on the thermal sensitivity of asphalt, which is in a melting state at high temperatures and is crystallized at low temperatures [7–9]. Kriz et al. [10] and Isaac et al. [11] believed that wax crystallization was the main cause of thermoreversible aging. Ding et al. [12,13], Qiu et al. [14] and Zhang et al. [15] studied the influence of a variety of wax-based warm-mix additives on thermoreversible aging in asphalt, indicating that the influence of wax-based warm-mix additives on thermoreversible aging was less than that of oxidation aging, and recommended the best combination of carbon atoms for wax-based warm-mix additives. On the other hand, due to the low mixing temperature of wax-based warm-mix asphalt mixture, moisture in the asphalt mixture cannot be completely removed, resulting in a wax-based warm-mix asphalt mixture being more vulnerable to moisture than a hot-mix asphalt mixture [16]. Nakhaei et al. [17] and Habal et al. [18] studied the influence of Sasobit (wax-based warm-mix additive) on the moisture sensitivity of asphalt binders based on surface free-energy, and the results showed that the addition of Sasobit increased the surface free-energy of asphalt, but also enhanced its moisture sensitivity.

However, the existing studies only analyzed the effect of wax on the thermoreversible aging and moisture sensitivity of asphalt from a mixture point of view (mainly wax-based warm-mix additive), and had not studied it from the molecular weight distribution and branching of wax perspectives [19,20]. In order to explore this meaningful research direction, Ding et al. [21] studied the effect of the branching of wax and asphaltene on the thermoreversible aging of asphalt binder. The results showed that $C_{20}H_{42}$ could significantly aggravate the thermoreversible aging performance of asphalt binder, while $C_{30}H_{66}$ and asphaltene have no similar effect.

To sum up, most of the existing studies on the wax of asphalt only focus on the wax-based mixture side, analyzing the effect of wax on the thermoreversible aging and moisture sensitivity of asphalt, but not the effect of the molecular weight distribution and branching of wax on the moisture sensitivity of asphalt binder. In view of this, this paper uses two mature moisture sensitivity testing methods, based on the surface free-energy (SFE) test and binder bond strength (BBS) test, to study the influence of the molecular weight distribution and the branching of wax on the moisture sensitivity of asphalt binder. In addition, because asphaltene has a strong polarity, it can explain the influence of wax on thermoreversible aging from the mechanism [22,23], and it has a strong influence on the adhesion between asphalt and aggregate. Therefore, this paper also studies the effects of asphaltene on the moisture sensitivity of the asphalt binder. It provides a theoretical basis and technical support for studying the effects of the molecular weight distribution and branching of wax on the moisture sensitivity of the asphalt binder.

2. Materials and Experiments

2.1. Materials

2.1.1. Asphalt

In this paper, asphalt samples A and B are selected as the base asphalt. Sample A is a low-wax asphalt and is produced in Venezuela. Sample B is a high-wax asphalt and is produced in China. The base asphalt, with a different wax content, is selected to analyze the influence of its own wax content on moisture sensitivity. The main technical indices of asphalt A and B are shown in Table 1.

Table 1. Main technical indexes of asphalt A and B.

Asphalt Sample	Penetration, 25 °C (0.1 mm)	Ductility (mm)	Soften Point (°C)	Continuous PG (°C)	Wax Content * (%)	Asphaltene Content ** (%)
A	53	141	37	32.8–57.2	0.53	23.4
B	42	26	52	29.5–61.7	2.52	24.1

* The wax content was determined by method of distillation (EN 12606-1: 2015) [24]. ** The asphaltene content was determined according to ASTM D4124 [25].

2.1.2. Additives

The number of carbon atoms of wax in asphalt is between 20 and 40 [26,27]. In this paper, the additives are eicosane (C20), triacontane (C30) and squalane (Sq), and Sq is the isomer of C30. All the alkane samples mentioned above are from Benzereagen Chemical Company, Inc. The model asphalts A/B + C20, A/B + C30 and A/B + Sq can be prepared by mixing the selected alkane with asphalt A and B at 165 °C and uniformly stirring. Asphaltene (As) comes from Karamay asphalt, extracted by the solvent deasphalting (SDA) process. After the asphaltene is ground to less than 100 mesh, it is added to the base asphalt binder, and then the model asphalt, A/B + As, can be prepared. The mixing method adopted in this paper is as follows. First, heat the base asphalt to 165 °C, then slowly add the additives to the base binder. A high-mixing shear device was adopted and the mixing speed was set to 4000 rpm. The mixing time for each additive is 1 h. The above-mentioned alkane and asphaltene content are all 3%, and the following model asphalt is expressed by abbreviations. The molecular formula structure diagram and the parameters of the additive are shown in Figure 2 and Table 2.

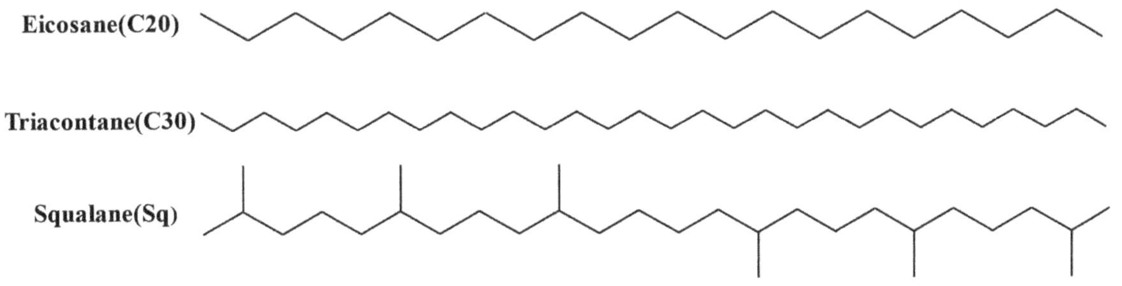

Figure 2. The molecular formula structure of additives.

Table 2. Main parameters of additives.

Parameters	Abbreviation	Formula	Purity	Density	Melting Point	Appearance
Eicosane	C20	$C_{20}H_{42}$	>99.0%	0.789 g/cm³	37 °C	Waxy crystals
Triacontane	C30	$C_{30}H_{62}$	>98.0%	0.810 g/cm³	66 °C	Waxy crystals
Squalane	Sq	$C_{30}H_{62}$	>99.0%	0.810 g/mL	−38 °C	Colorless liquid

2.2. Experimental Method

2.2.1. BBS Test Method

BBS tests started in the architectural coatings industry, and they can intuitively and conveniently measure the adhesive properties of materials in a short time. ASTM D4541 [28] and AASHTO TP-91 [29] have improved their testing methods to make them suitable for testing the bonding strength of the asphalt binder. In this study, the BBS tester—the pull-off adhesion tester, produced by the DeFelsko Company of the United States—is mainly composed of a test host, a drawing sleeve and a drawing head, as shown in Figure 3f. The test parameters refer to reference [4]; the pull-off tensile speed is 0.7 MPa/s and the asphalt-film thickness of the drawing head is 0.2 mm.

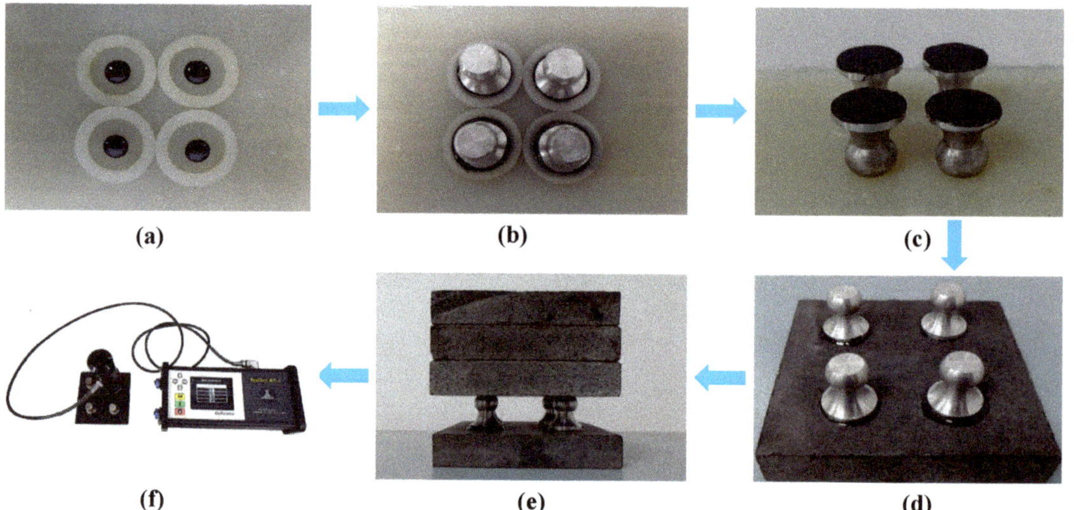

Figure 3. The BBS test method. (**a**) Pouring asphalt; (**b**) Sample preparation; (**c**) Attached sample (**d**) Sample on aggregate; (**e**) Loading sample (**f**) Detachment testing.

The forming steps of the BBS specimen are as follows: (1) cleaning slate; (2) place the asphalt, drawing head and granite plate into a 150 °C oven to heat for 30 min; (3) take the asphalt out of the oven, drop it into the silica gel pad, and immediately press the drawing head onto the top of the asphalt, as shown in Figure 3a,b, respectively; (4) after the asphalt under the drawing head is cooled, scrape off the extruded asphalt outside the drawing head, as shown in Figure 3c; (5) take out the granite slab from the oven, press the drawing head vertically onto the surface of the slab, place some heavy objects to further press the drawing head, and then place it back in the 25 °C constant temperature box for curing for some time, before testing, as shown in Figure 3d–f, respectively.

2.2.2. SFE Test Method

Surface free-energy γ is defined as the work that the outside world needs to do when a unit surface area is generated on the surface of an object [30]. It consists of the dispersion components γ^{LW} and polarity components γ^{AB}, as shown in Equations (1) and (2). Because asphalt is a viscous substance, it is difficult to ensure the same dripping quality of asphalt every time, and it is easy to draw wires during dripping. Therefore, in this paper, the DSA100 contact angle meter, produced by the KRUSS company, is used to test the contact angles of the distilled water and glycerin, and to form amide with the asphalt by the sessile-drop method. Then, the contact angles are brought into the Young–Dupre equation shown in Equation (3) to indirectly calculate the surface free-energy parameters of the asphalt [31]. The SFE parameters of the three chemical reagents are shown in Table 3.

Table 3. Surface free-energy parameters of chemical droplets and aggregate.

Liquids/Aggregate	γ^+	γ^-	γ^{AB}	γ^{LW}	γ
Distilled water (H_2O)	25.5	25.5	51	21.8	72.8
Formamide (CH_3NO)	2.28	39.6	19	39	58
Glycerol ($C_3H_8O_3$)	3.92	57.4	30	34	64
Granite	9.87	0.56	4.70	45.69	50.39

The test method of the contact angle is as follows: (1) Put the asphalt sample into a 160 °C oven and heat it to a flowing state; (2) soak the cleaned glass slide in hot asphalt for 4~5 s, then hang it vertically in the oven for 10 min, so that the excess asphalt drips freely; (3) after the glass slides are cured at room temperature for a certain period of time, the contact angle tester is used to test the contact angle θ between the chemical reagents and the asphalt. The larger the contact angle θ, the more distant the mutual combination level between them, and vice versa, the closer they are, as shown in Figure 4.

$$\gamma = \gamma^{LW} + \gamma^{AB} \tag{1}$$

$$\gamma^{AB} = 2\sqrt{\gamma^+\gamma^-} \tag{2}$$

$$\frac{(1 + COS\theta)\gamma_L}{2} = \sqrt{\gamma_S^+\gamma_L^-} + \sqrt{\gamma_S^-\gamma_L^+} + \sqrt{\gamma_S^{LW}\gamma_L^{LW}} \tag{3}$$

In Equations (1)–(3), γ, γ^{LW} and γ^{AB} represent surface free-energy, the dispersion component and polarity component, respectively; γ^+, γ^- and θ represents an acid component, basic component and contact angle, respectively. The lower corner marks S and L represent the asphalt and chemical reagents, respectively.

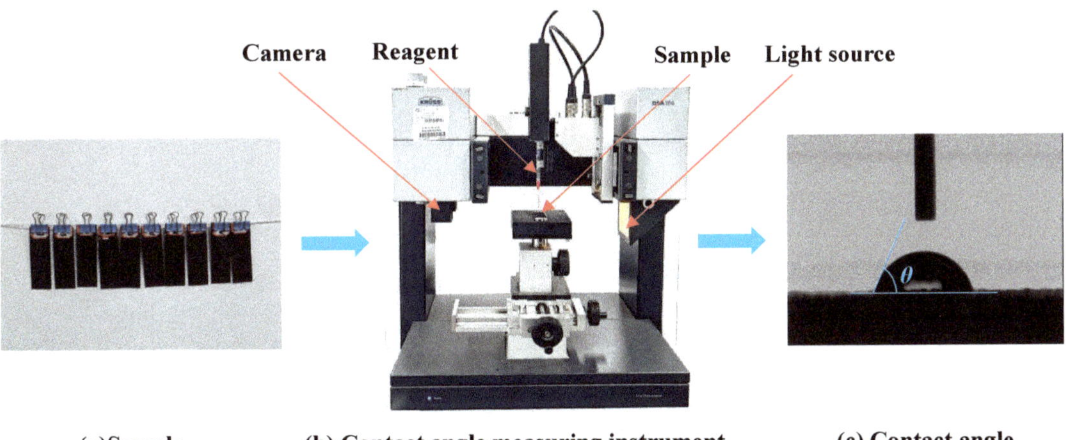

(a)Sample   **(b) Contact angle measuring instrument**   **(c) Contact angle**

Figure 4. Contact angle measurement method. (**a**) Sample preparation; (**b**) Contact angle measuring instrument; (**c**) Contact angle.

2.3. Experimental Scheme

In this paper, ten kinds of model asphalt (including base asphalt) are prepared by using two kinds of base asphalt and four kinds of additives, which are A/B, A/B + C20, A/B + C30, A/B + Sq and A/B + As, respectively. The BBS and SFE tests are used to test the moisture sensitivity parameters of the above ten kinds of model asphalt at room temperature and at the conditioning times of 1 h and 168 h, respectively. The parallel tests are all four groups. The specific test contents are shown in Figure 5.

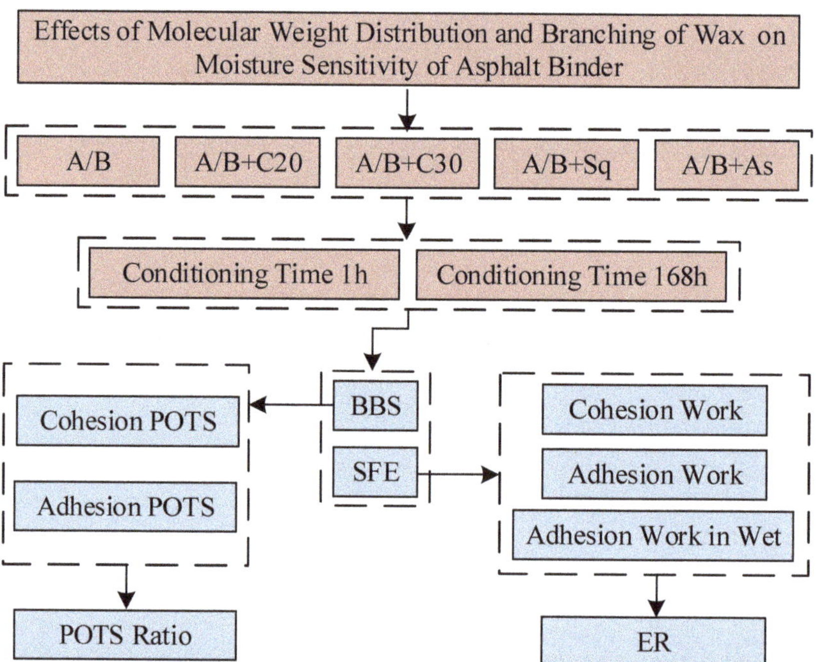

Figure 5. Flow chart of test scheme. Note: In this figure, cohesion POTS, adhesion POTS and the POTS ratio are the moisture sensitivity parameters and indices of the BBS test, respectively. Cohesion work, adhesion work in wet and ER are the moisture sensitivity parameters and indices of the SFE test, respectively.

3. Results and Discussion

The wax in asphalt has obvious thermal sensitivity. It dissolves in asphalt at high temperatures and precipitates in the form of crystals at normal and low temperatures, which makes asphalt hard. Struik [32] named this reversible reaction as physical hardening. In this paper, the degree of physical hardening (DPH) is used to describe the influence of the physical hardening of wax on the moisture sensitivity parameters of asphalt. DPH is obtained from the ratio of the moisture sensitivity parameters of the conditioning time of 1 h divided by the conditioning time of 168 h, as shown in Equation (4).

$$DPH = Value(1\,h)\,/\,Value(168\,h) \tag{4}$$

In Equation (4), DPH is the abbreviation for the degree of physical hardening, and the closer its value is to 1, the smaller the physical hardening effect is, otherwise, the larger it is; *Value (1 h)* and *Value (168 h)* represent the moisture sensitivity parameter values of the conditioning times of 1 h and 168 h, respectively.

3.1. BBS Test

The BBS test usually shows the cohesion failure of the asphalt–asphalt interface before moisture-induced damage occurs, and the adhesion failure of the asphalt–aggregate interface after 48 h of moisture-induced damage [33]. Therefore, the BBS test, in this paper, analyzes the effects of the branching of wax and asphaltene on the cohesion pull-off tensile strength (cohesion POTS) before moisture-induced damage occurs, and the adhesion pull-off tensile strength (adhesion POTS) after moisture-induced damage. The moisture sensitivity index is calculated from it.

3.1.1. Cohesion and Adhesion POTS Analysis Based on BBS

Figure 6 shows the cohesion POTS of model asphalt in a dry condition. The error bar in Figure 6 represents the variation range of four groups' parallel test results, and the same is true of the error bar in the figure below. It can be seen from Figure 6, that the cohesion POTS of asphalt A + C20 and asphalt B + C20 is obviously lower than that of A and B, respectively, while the cohesion POTS of asphalt A + C30/Sq and asphalt B + C30/Sq has no obvious difference with the corresponding base asphalts A and B, respectively, while the cohesion POTS of asphalt A + As and asphalt B+As is higher than that of the base asphalts A and B, respectively. The results show that the effects of adding C20, C30/Sq and As into asphalt A and B on the cohesion POTS are obviously reduced, without obvious influence, and increased, respectively, which may be due to the fact that the number of carbon atoms in wax in asphalt is between 20 and 40 [34], while the carbon atoms of C20 are at a relatively small level, and C30 and Sq are at an average level. The research results show that by adding C20 and C30, similar conclusions on the thermoreversible aging performance of base asphalt can be drawn [21]. In addition, adding asphaltene increases the cohesion POTS of base asphalts A and B because asphaltene can improve its modulus and viscosity.

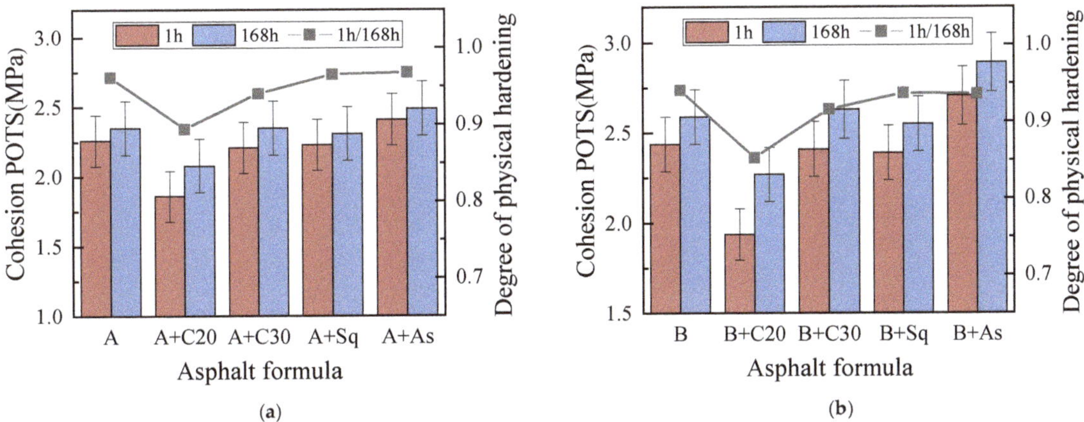

Figure 6. Cohesion POTS in dry conditions. (**a**) Asphalt A. (**b**) Asphalt B.

Figure 7 shows the adhesion POTS between the asphalt and aggregate after moisture-indued damage. It can be seen from Figure 7, that the adhesion POTS of asphalt A + C20 is not significantly different from that of asphalt A; the adhesion POTS of asphalt B + C20 is slightly less than that of asphalt B (within 3%); and the adhesion POTS of A + C30, B + C30, A + Sq and B + Sq are significantly higher than that of the corresponding base asphalts A and B. The results show that the influence of C20 and C30/Sq/As on the adhesion POTS of asphalt A and asphalt B has no obvious difference (asphalt B slightly decreases) and increases, and especially after asphalt B is mixed with asphaltene, it increases by about 17.5%. This is because asphaltene can increase the polarity of asphalt, thus improving the adhesion performance between the asphalt and aggregate [35].

From Figures 6 and 7, it can be seen that: (1) Asphalt B and its model asphalt are larger than the corresponding asphalt A and its model asphalt, both the cohesion and adhesion POTS, indicating that the higher the wax content of the base asphalt, the greater the adhesion and cohesion POTS for itself and its model asphalt. (2) For the cohesion and adhesion POTS, the degree of physical hardening (DPH) ranges of A and B, A+C20 and B+C20, A + C30 and B + C30, A + Sq and B + Sq, A + As and B + As are about 0.94~0.97, 0.85~0.95, 0.91~0.95, 0.94~0.98 and 0.94~0.98, respectively. The results show that when C20 is added, the DPH has the greatest influence on it, followed by C30, and there is no obvious difference in base asphalt for Sq and As. This is because C20 has the smallest molecular weight and good compatibility with base asphalt. After a conditioning time of 168 h hours, C20 can precipitate from base asphalt to form crystalline wax, which hardens the asphalt and leads to the greatest DPH. However, the molecular weight of C30 is larger than C20, and the number of carbon atoms is close to the average number of carbon atoms of wax in asphalt, which makes it difficult to precipitate crystals, unlike C20. Sq is an isomer of C30. Similar to C30, asphaltene does not belong to wax (Ding et al., 2021). (3) For the adhesion and cohesion POTS, the DPH of asphalt B and its model asphalt is less than that of corresponding asphalt A and its model asphalt, respectively. The results show that the higher the wax content of base asphalt, the greater the effect of physical hardening on the adhesion and cohesion POTS of itself and model asphalt. (4) The adhesion POTS of asphalt A, asphalt B and their model asphalt are both more DPH than the corresponding cohesive POTS, indicating that the influence of DPH on the adhesion POTS is lower than the cohesive POTS. This is because the specimen will be soaked in water for 48 h before the adhesion POTS test, during which time part of physical hardening can be completed. In addition, the adhesion POTS in a wet condition is jointly determined by asphalt, aggregate and water, while the cohesion POTS is only determined by asphalt itself, which weakens the influence of physical hardening on the test results.

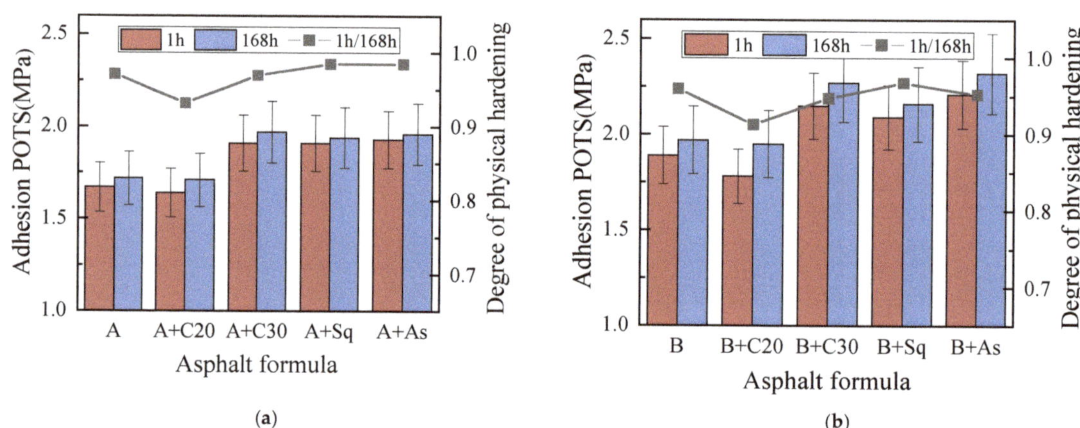

(a)

(b)

Figure 7. Adhesion POTS in wet conditions. (**a**) Asphalt A. (**b**) Asphalt B.

3.1.2. Moisture Sensitivity Analysis Based on BBS

In this paper, according to AASHTO T283 [36], the ratio of mechanical strength after moisture-induced damage to that before moisture-induced damage is taken as the moisture sensitivity index of the asphalt binder. A moisture sensitivity index of the BBS test, pull-off tensile strength ratio (POTS Ratio) is put forward. The larger the value, the smaller the moisture sensitivity of asphalt. On the contrary, the larger it is, as shown in Equation (5).

$$POTS\ Ratio = Adhesion\ POTS/Cohesion\ POTS \tag{5}$$

In Equation (5), Adhesion POTS represented POTS after moisture-induced damage, Cohesion POTS represented POTS before moisture-induced damage.

Figure 8 shows the POTS Ratio of asphalt A, B, and their model asphalt, from which we can see: (1) On the whole, adding C20, C30, Sq and asphaltene into asphalt A and B can increase their POTS Ratio and reduce moisture sensitivity, especially C20. It is worth noting that the asphalt samples with the most decrease in moisture sensitivity are not necessarily the best in moisture damage resistance. This is because the moisture sensitivity index is only the relative ratio of the POTS of the specimen before and after moisture-induced damage, and the moisture-induced damage resistance performance is also related to the absolute value of the adhesion and cohesion POTS of asphalt after moisture-induced damage. (2) For asphalt A, asphalt B and their model asphalt, the POTS Ratio after conditioning time of 168 h is less than that after 1 h, especially for the model asphalt added to C20. This is because the specimen will be soaked in water for 48 h before the adhesion POTS test, and the 48 h after conditioning time 1 h can complete partial physical hardening, which improves the adhesion POTS. However, after a conditioning time of 168 h, the physical hardening was almost completed, and the subsequent 48 h immersion in water could not improve its adhesion POTS. (3) The POTS Ratio of high-wax asphalt B and its model asphalt is higher than that of the corresponding low-wax asphalt A and its model asphalt, which indicates that the higher the wax content of base asphalt, the lower the moisture sensitivity of itself and its model asphalt.

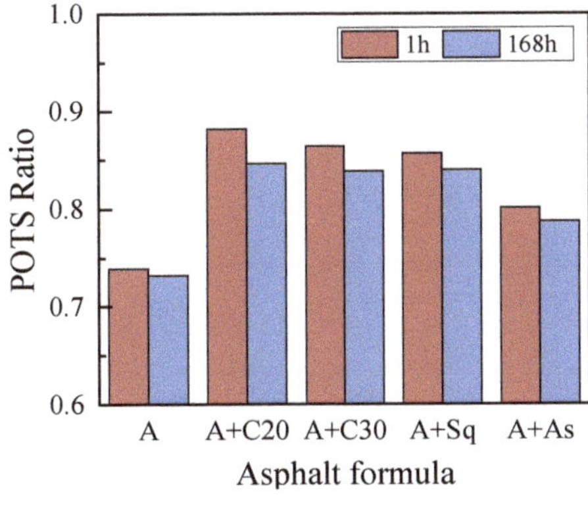
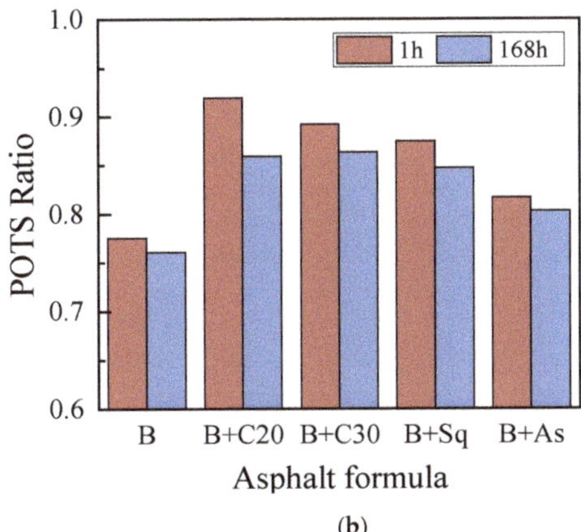

Figure 8. POTS ratio. (**a**) Asphalt A. (**b**) Asphalt B.

3.2. SFE Test

In order to verify the accuracy of the BBS test, this paper analyzes the effects of the molecular weight distribution and the branching of wax and asphaltene on the moisture sensitivity of asphalt, based on SFE. The solution process of the surface free-energy moisture sensitivity parameter ER of asphalt is as follows: Equations (8)–(10) can be derived from Equations (6) and (7). By placing the surface free-energy parameters of asphalt, aggregate and water into Equations (8)–(10), the cohesion work of the asphalt itself, the adhesion work of the asphalt–aggregate interface (dry condition) and the adhesion work of the asphalt–aggregate interface (wet condition) can be obtained. The ER can be obtained by substituting the adhesion work under dry and wet conditions into Equation (11). The larger the ER, the smaller the moisture sensitivity, and on the contrary, the greater the moisture sensitivity. The aggregate is granite and the surface free-energy parameters are shown in Table 3.

$$\Delta G_{ij} = \gamma_{ij} - \gamma - \gamma \tag{6}$$

$$\Delta G_{ikj} = \gamma_{ij} - \gamma_{ik} - \gamma_{jk} \tag{7}$$

In Equations (6) and (7), G_{ij} is the interfacial binding energy between two-phase materials, γ_{ij} is the interfacial energy of two-phase materials, γ_i is the surface energy of substance i, γ_j is the surface energy of j, and G_{ikj} is the interfacial binding energy of three-phase materials.

$$W_{AA} = 2\gamma_A = 2\sqrt{\gamma_A^{LW}} + 4\sqrt{\gamma_A^+ \gamma_A^-} \tag{8}$$

$$W_{AG} = 2\left(\sqrt{\gamma_A^+ \gamma_G^-} + \sqrt{\gamma_A^- \gamma_G^+} + \sqrt{\gamma_A^{LW} \gamma_G^{LW}} \right) \tag{9}$$

$$W_{AWG} = 2\left[\sqrt{\gamma_A^{LW}\gamma_W^{LW}} + \sqrt{\gamma_G^{LW}\gamma_W^{LW}} + \sqrt{\gamma_W^+}\left(\sqrt{\gamma_A^-} + \sqrt{\gamma_G^-}\right) + \sqrt{\gamma_W^-}\left(\sqrt{\gamma_A^+} + \sqrt{\gamma_G^+}\right) \\ -2\sqrt{\gamma_W^+\gamma_W^-} - \sqrt{\gamma_A^+\gamma_G^-} - \sqrt{\gamma_A^-\gamma_G^+} - \sqrt{\gamma_A^{LW}\gamma_G^{LW}} - \gamma_W^{LW}\right] \tag{10}$$

$$ER = \left|\frac{W_{AG}}{W_{AWG}}\right| \tag{11}$$

In Equations (8)–(11), $W_{AA'}$, $W_{AG'}$, W_{AWG} and ER are cohesion work, adhesion work in a dry condition, adhesion work in a wet condition and the moisture sensitivity index, respectively.

3.2.1. Cohesion and Adhesion Work Analysis Based on SFE

Figures 9 and 10 show the cohesion work and adhesion work in dry conditions, respectively, which shows that: (1) when C20, C30/Sq, and As are added to asphalts A and B, respectively, the cohesion work and adhesion work in dry conditions are decreased, have no obvious effect and increase, respectively, and the effect is similar to the effect on the cohesion POTS (BBS parameter). (2) In dry conditions, the cohesion work and adhesion work of high-wax asphalt B and its model asphalt are larger than that of the corresponding low-wax asphalt A and its model asphalt, and the effect of the wax content of the base asphalt on the cohesion POTS (BBS parameter) is similar. (3) After the addition of C20 to asphalts A and B, the DPH is the largest, followed by C30, and Sq and As are not significantly influenced. The DPH is similar to that of the cohesion POTS (BBS parameter). The above conclusions show that the cohesion POTS of the BBS test and cohesion work of SFE verified the accuracy of each other's test results. (4) In dry conditions, the adhesion work of all the model asphalts (including the base asphalt) is greater than that of the cohesion work, which explains, from the point of view of energy, that there is usually a cohesion failure rather than adhesion failure in the BBS test in dry conditions.

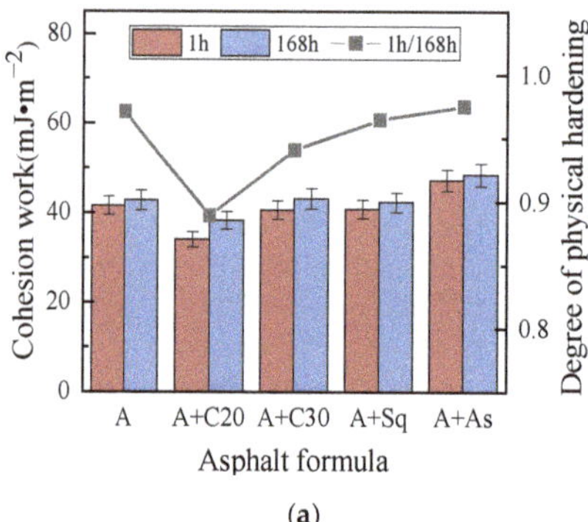

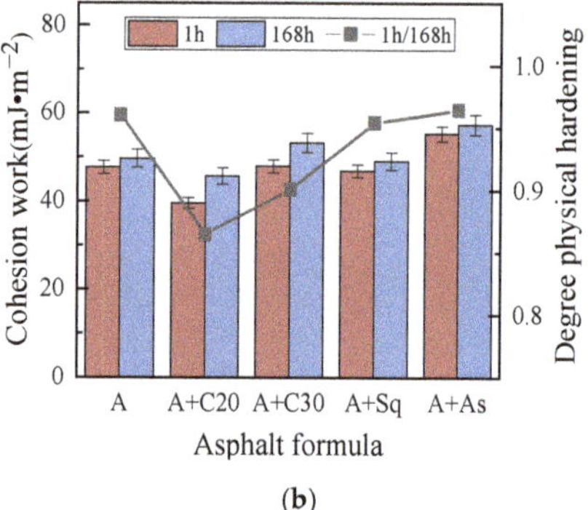

Figure 9. Cohesion work in dry conditions. (**a**) Asphalt A. (**b**) Asphalt B.

Figure 11 shows the adhesion work of asphalt samples in wet conditions, which shows that in wet conditions: (1) The adhesion work of all asphalt samples is a negative number, indicating that asphalt spalling from the aggregate surface occurs spontaneously in wet conditions, and the higher the value, the easier it is for asphalt to spontaneously peel off from the aggregate surface. (2) The addition of C20, C30 and Sq into A and B can reduce the absolute value of the adhesion work, however, the addition of As has no obvious effect on it. The results show that, in wet conditions, the addition of C20, C30 and Sq can alleviate the exfoliation of asphalt from the aggregate surface, while the addition of As has no obvious effect. (3) The addition of C20, C30, Sq and As to asphalts A and B, respectively, has no obvious regular effect on DPH, which is different from the effect on the cohesion work and adhesion work in dry conditions. A reasonable explanation is that the adhesion work in wet conditions is determined by the asphalt, aggregate and water— not only the asphalt and aggregate—which weakens the effect of the wax content on the physical hardening of asphalt. Existing studies believe that the effect of moisture on an asphalt–aggregate interface is extremely complex [4], and the specific reasons for the test results need to be further studied.

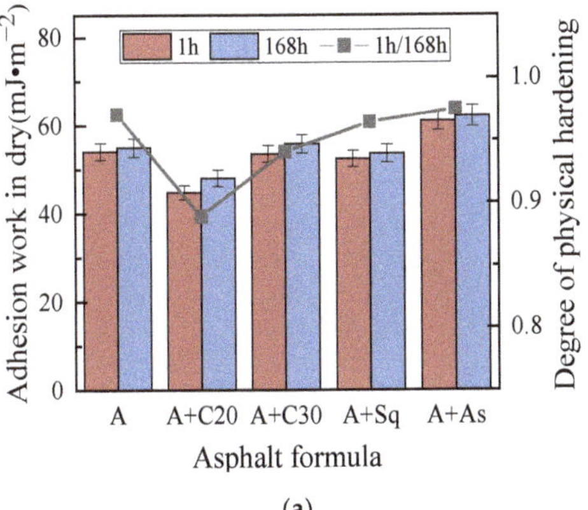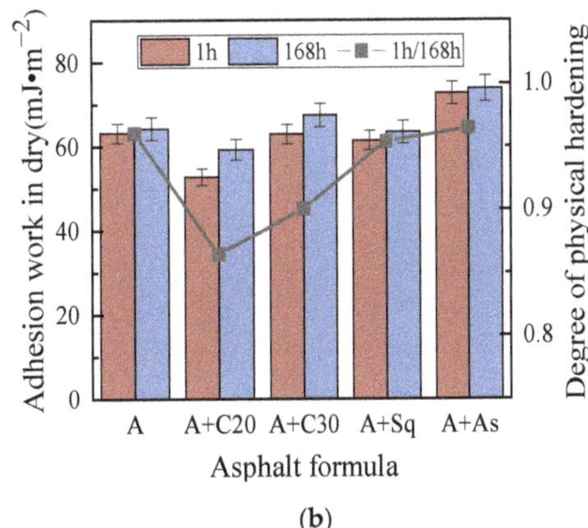

Figure 10. Adhesion work in dry conditions. (**a**) Asphalt A. (**b**) Asphalt B.

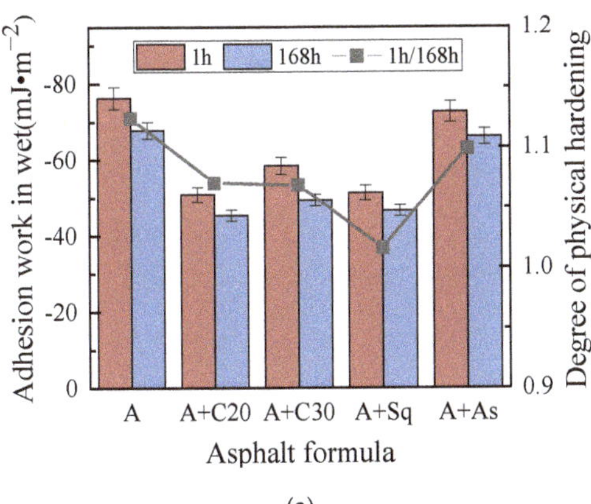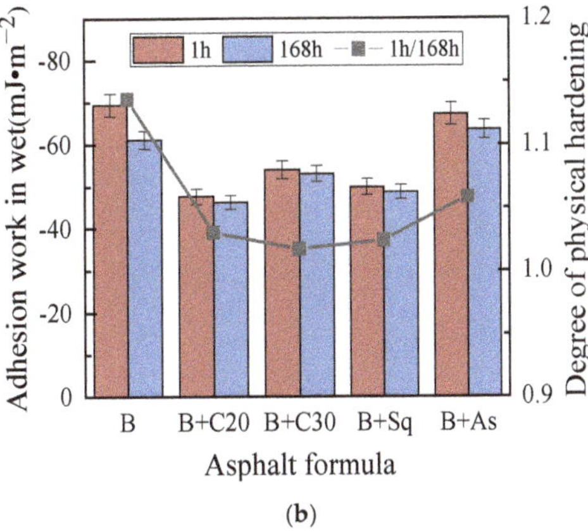

Figure 11. Adhesion work in wet conditions. (**a**) Asphalt A. (**b**) Asphalt B.

3.2.2. Moisture Sensitivity Analysis Based on SFE

Figure 12 shows the effects of the molecular weight distribution and the branching of wax and asphaltene on the moisture sensitivity index, ER. It can be seen from the figure that: (1) The ER of high-wax asphalt B and its model asphalt is higher than that of the corresponding low-wax asphalt A and its model asphalt. In addition, the moisture sensitivity of asphalts A and B can be improved by adding C20, C30, Sq and As. The SFE test conclusion above is consistent with the BBS test conclusion. It is worth mentioning, that when As is added, the increase in the adhesion and cohesion work of asphalts A and B in dry conditions is stronger than when C20, C30 and Sq are added, however, the moisture sensitivity was lower than that of C20, C30 and Sq. This is because the absolute value of the adhesion work of the A/B + As condition, shown in Figure 11, is much larger than that of the alkane (C20, C30 and Sq) model of asphalt in wet conditions, resulting in a lower ER than the alkane-model asphalt. (2) The ER of all asphalt samples at a conditioning time of 168 h was larger than that at 1 h, which was contrary to the BBS test result. This is because the BBS test required soaking in water for 48 h before testing the adhesion POTS of the conditioning time of 1 h, during which, a part of the wax could be precipitated from the base asphalt, which was equivalent to the conditioning time of 49 h (1 h + 48 h). However, this phenomenon did not exist in the SFE test method. Therefore, when using the BBS test to characterize the moisture sensitivity of high-wax asphalt, it is recommended that the sample be placed for a certain amount of time and then tested after the physical hardening is stable.

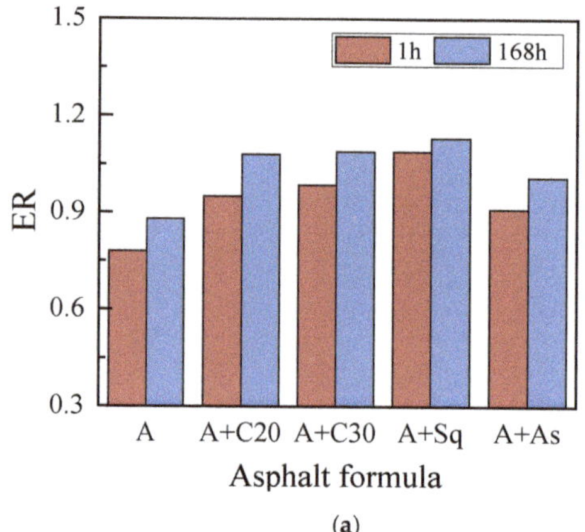

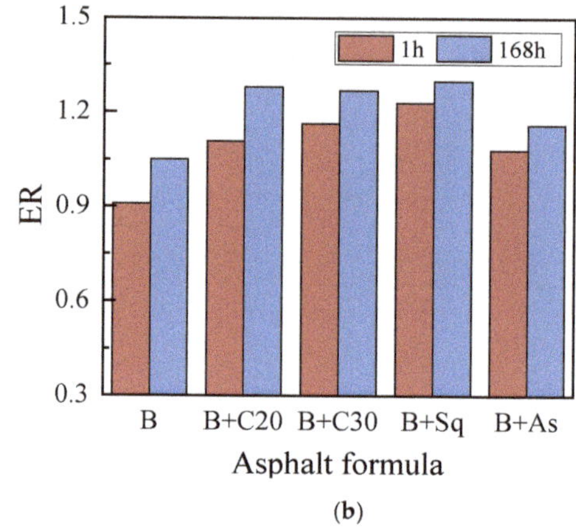

Figure 12. ER. (**a**) Asphalt A. (**b**) Asphalt B.

3.3. Correlation between SFE and BBS

In order to study the correlation between the SFE and BBS tests in evaluating the moisture sensitivity of the molecular weight distribution and the branching of wax and asphaltene of the asphalt binder, two groups of parameters or indices, representing the same meaning in the two test methods, were analyzed. Among them, the cohesion POTS (BBS) and cohesion work (SFE) both characterize the degree of failure of the asphalt–asphalt interface, while the POTS ratio (BBS) and ER (SFE) both characterize the moisture sensitivity of asphalt.

Figures 13 and 14 show the correlation between the above two groups of parameters or indices, respectively. It can be seen that: (1) The correlation between the cohesion POTS and cohesion work for the conditioning times of 1 h and 168 h are 0.96 and 0.93, and the correlation coefficient between the POTS ratio and ER are 0.71 and 0.78, respectively. The results show that the cohesion POTS and cohesion work, POTS ratio and ER both have a good correlation, especially the former. (2) For the correlation between the POTS ratio and ER, if A + C20 and B + C20 were removed from the samples at the conditioning times of 1 h and 168 h, the correlation coefficients would change from 0.71 to 0.81 and 0.78 to 0.77, respectively. The results showed that the removal of A + C20 and B + C20 significantly increased their correlation at a conditioning time of 1h, while no significant difference was observed at a conditioning time of 168 h, as shown in Figure 14. This is because the BBS test method can easily lead to a large POTS ratio, especially for the asphalt samples with high physical hardening, while the SFE test method does not affect ER. The reasons for a higher POTS ratio of the BBS test have been described many times above, and will not be repeated here.

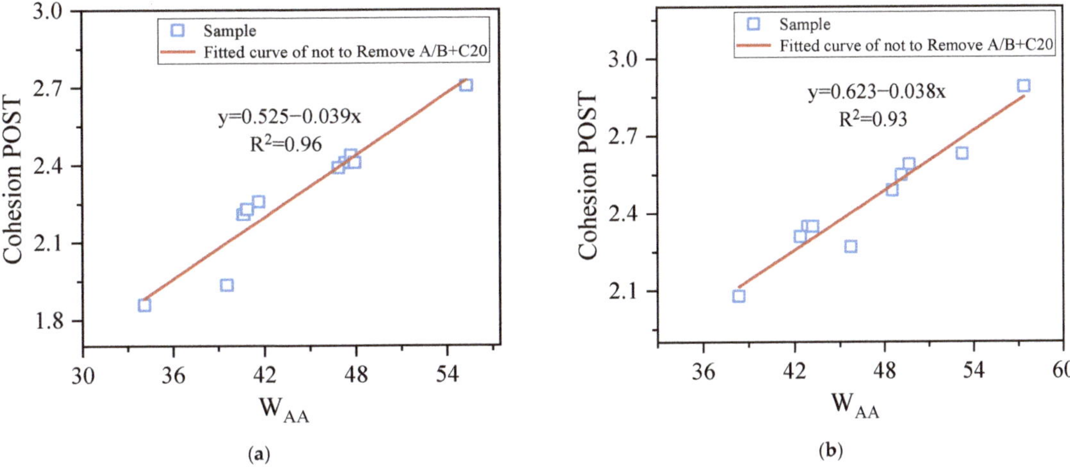

Figure 13. Correlation between cohesion POTS and cohesion work. (**a**) 1 h, (**b**) 168 h.

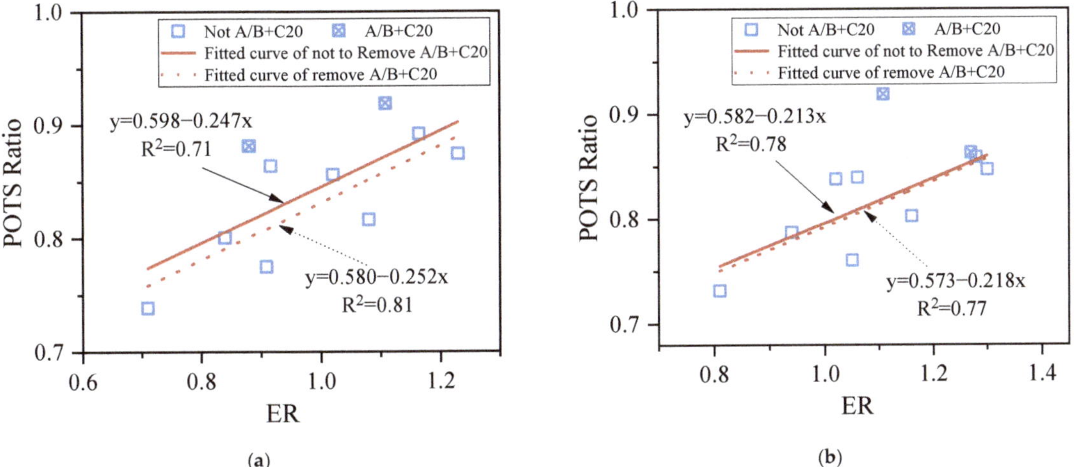

Figure 14. Correlation between POTS ratio and ER. (**a**) 1 h, (**b**) 168 h.

4. Conclusions

In this paper, the BBS and SFE tests are used to study the effects of the wax molecular weight distribution and branching on the moisture sensitivity of asphalt. The following conclusions can be drawn:

(1) The addition of n-eicosane, n-triacontane, squalane and asphaltene can reduce the moisture sensitivity of the base asphalt, but not necessarily improve its moisture-induced damage-resistance performance.

(2) For the cohesion POTS, regarding the cohesion and adhesion work in dry conditions, physical hardening has the greatest influence on n-eicosane, followed by n-triacontane. However, squalane and asphaltene have no obvious influence.

(3) The physical hardening effect of high-wax asphalt and its model asphalt is stronger than that of the corresponding low-wax asphalt and its model asphalt, and its moisture sensitivity is weaker than that of low-wax asphalt.

(4) Different molecular weight distributions and the branching of waxes will have different effects on the moisture sensibility of the asphalt binder.

Author Contributions: Conceptualization, X.C.; methodology, W.W.; validation, X.C. and A.R.; formal analysis, W.W.; investigation, W.W., A.N. and X.C.; writing—original draft preparation, W.W.; writing—review and editing, A.R. and X.C.; visualization, W.W.; supervision, X.C.; funding acquisition, W.W. All authors have read and agreed to the published version of the manuscript.

Funding: This research was funded by the Talent Introduction Project of Xihua University (School Key Project) (Z202111), the Scientific Research Project of the Sichuan Provincial Department of Education (16ZB0164), the Open Research Fund of Sichuan Provincial Key Laboratory of Road Engineering (15206569), and the Open project of Sichuan Key Laboratory of Green Building and Energy Saving (szjj2015-074).

Institutional Review Board Statement: Not applicable.

Informed Consent Statement: Not applicable.

Data Availability Statement: Not applicable.

Conflicts of Interest: The authors declare no conflict of interest.

List of Notations and Abbreviations

BBS	binder bond strength
POTS	pull-off tensile strength
Adhesion POTS	POTS after moisture-induced damage
Cohesion POTS	POTS before moisture-induced damage
POTS Ratio	BBS moisture sensitivity index
SFE	surface free-energy
ER	SFE moisture sensitivity index
γ	surface free-energy
γ^{LW}	dispersion component
γ^{AB}	polarity component
γ^{+}	acid component
γ^{-}	basic component
θ	contact angle
G_{ij}	interfacial binding energy between two-phase materials
γ_{ij}	interfacial energy of two-phase materials
γ_i	surface energy of substance i
γ_j	the surface energy of j
G_{ikj}	interfacial binding energy of three-phase materials
W_{AA}	cohesion work
W_{AG}	adhesion work
W_{AWG}	adhesion work in wet condition
DPH	abbreviation of degree of physical hardening
Value (1 h)	moisture sensitivity parameter values of conditioning time 1 h
Value (168 h)	moisture sensitivity parameter values of conditioning time 168 h

References

1. Ahmad, M.; Mannan, U.A.; Islam, M.R.; Tarefder, R.A. Chemical and mechanical changes in asphalt binder due to moisture conditioning. *Road Mater. Pavement Des.* **2018**, *19*, 1216–1229. [CrossRef]
2. Cheng, Z.; Kong, F.; Zhang, X. Application of the Langmuir-type diffusion model to study moisture diffusion into asphalt films. *Constr. Build. Mater.* **2021**, *268*, 121192. [CrossRef]
3. Cong, P.; Guo, X.; Ge, W. Effects of moisture on the bonding performance of asphalt-aggregate system. *Constr. Build. Mater.* **2021**, *295*, 123667. [CrossRef]
4. Chen, X.; Xiao, H.; Cao, C.; Yan, C.; Ren, D.; Ai, C. Study on Moisture Diffusion Behavior in Asphalt Binder Based on Static and Dynamic Pore Water Conditions. *J. Mater. Civ. Eng.* **2022**, *34*, 04022183. [CrossRef]
5. Lu, X.; Kalman, B.; Redelius, P. A new test method for determination of wax content in crude oils, residues and bitumens. *Fuel* **2008**, *87*, 1543–1551. [CrossRef]

6. Ding, H.; Zhang, H.; Zheng, X.; Zhang, C. Characterisation of crystalline wax in asphalt binder by X-ray diffraction. *Road Mater. Pavement Des.* **2022**, *23*, 1–17. [CrossRef]
7. Ding, H.; Hesp, S. Variable-temperature Fourier-transform infrared spectroscopy study of asphalt binders from the SHRP Materials Reference Library. *Fuel* **2021**, *298*, 120819. [CrossRef]
8. Ding, H.; Hesp, S. Balancing the use of wax-based warm mix additives for improved asphalt compaction with long-term pavement performance. *ACS Sustain. Chem. Eng.* **2021**, *9*, 7298–7305. [CrossRef]
9. Ding, H.; Hesp, S. Variable-temperature Fourier-transform infrared spectroscopy study of wax precipitation and melting in Canadian and Venezuelan asphalt binders. *Constr. Build. Mater.* **2020**, *264*, 120212. [CrossRef]
10. Kriz, P. *Glass Transition and Physical Hardening of Asphalts*; University of Calgary: Calgary, AB, Canada, 2009.
11. Omari, I.; Aggarwal, V.; Hesp, S. Investigation of two Warm Mix Asphalt additives. *Int. J. Pavement Res. Technol.* **2016**, *9*, 83–88. [CrossRef]
12. Ding, H.; Qiu, Y.; Rahman, A. Low-Temperature reversible aging properties of selected asphalt binders based on thermal analysis. *J. Mater. Civ. Eng.* **2019**, *31*, 04018402. [CrossRef]
13. Ding, H.; Qiu, Y.; Rahman, A. Influence of thermal history on the intermediate and low-temperature reversible aging properties of asphalt binders. *Road Mater. Pavement Des.* **2020**, *21*, 2126–2142. [CrossRef]
14. Qiu, Y.; Ding, H.; Su, T. Non-isothermal low-temperature reversible aging of commercial wax-based warm mix asphalts. *Int. J. Pavement Eng.* **2022**, *23*, 514–522. [CrossRef]
15. Zhang, H.; Zhang, H.; Ding, H.; Dai, J. Determining the Sustainable Component of Wax-Based Warm Mix Additives for Improving the Cracking Resistance of Asphalt Binders. *ACS Sustain. Chem. Eng.* **2021**, *9*, 15016–15026. [CrossRef]
16. Ji, J.; Yao, H.; Yuan, Z.; Suo, Z.; Xu, Y.; Li, P.; You, Z. Moisture Susceptibility of Warm Mix Asphalt (WMA) with an Organic Wax Additive Based on X-Ray Computed Tomography (CT) Technology. *Adv. Civ. Eng.* **2019**, *2019*, 7101982. [CrossRef]
17. Nakhaei, M.; Naderi, K.; Nasrekani, A.A.; Timm, D.H. Moisture resistance study on PE-wax and EBS-wax modified warm mix asphalt using chemical and mechanical procedures. *Constr. Build. Mater.* **2018**, *189*, 882–889. [CrossRef]
18. Habal, A.; Singh, D. Moisture Damage Resistance of GTR-Modified Asphalt Binders Containing WMA Additives Using the Surface Free Energy Approach. *J. Perform. Constr. Facil.* **2017**, *31*, 04017006. [CrossRef]
19. Edwards, Y.; Isacsson, U. Wax in bitumen: Part 1—Classifications and general aspects. *Road Mater. Pavement Des.* **2005**, *6*, 281–309. [CrossRef]
20. Edwards, Y.; Isaacson, U. Wax in bitumen: Part II characterization and effects. *Road Mater. Pavement Des.* **2005**, *6*, 434–468. [CrossRef]
21. Ding, H.; Liu, H.; Qiu, Y.; Rahman, A. Effects of crystalline wax and asphaltene on thermoreversible aging of asphalt binder. *Int. J. Pavement Eng.* **2021**, *22*, 1–10. [CrossRef]
22. Ding, H.; Zhang, H.; Liu, H.; Qiu, Y. Thermoreversible aging in model asphalt binders. *Constr. Build. Mater.* **2021**, *303*, 124355. [CrossRef]
23. Ding, H.; Hesp, S. Another look at the use of modulated differential scanning calorimetry to study thermoreversible aging phenomena in asphalt binders. *Constr. Build. Mater.* **2021**, *267*, 121787. [CrossRef]
24. *EN 12606-1*; In Bitumen and Bituminous Binders—Determination of the Paraffin Wax Content—Part 1: Method by Distillation. European Standard: Pilsen, Czech Republic, 2015.
25. *ASTM D4124*; In Standard Test Method for Separation of Asphalt into Four Fractions. ASTM: West Conshohocken, PA, USA, 2018.
26. Ding, H.; Zhang, H.; Zhang, H.; Liu, D.; Qiu, Y.; Hussain, A. Separation of wax fraction in asphalt binder by an improved method and determination of its molecular structure. *Fuel* **2022**, *322*, 124081. [CrossRef]
27. Ding, H.; Hesp, S. Quantification of crystalline wax in asphalt binders using variable-temperature Fourier-transform infrared spectroscopy. *Fuel* **2020**, *277*, 118220. [CrossRef]
28. *ASTM D4541*; In Standard Test Method for Pull-Off Strength of Coatings Using Portable Adhesion Testers. ASTM: West Conshohocken, PA, USA, 2017.
29. *AASHTO TP91*; In Standard Method of Test for Determining Asphalt Binder Bond Strength by Means of the Binder Bond Strength (BBS) Test. American Association of State and Highway Transport: Washington, DC, USA, 2013.
30. Hntsberger, J.R. Surface Energy, Wetting and Adhesion. *J. Adhes.* **1981**, *12*, 3–12. [CrossRef]
31. Tan, Y.; Guo, M. Using surface free energy method to study the cohesion and adhesion of asphalt mastic. *Constr. Build. Mater.* **2013**, *47*, 254–260. [CrossRef]
32. Struik, L.C.E. *Physical Aging in Amorphous Polymers and Other Materials*; Elsevier: Amsterdam, The Netherlands, 1978.
33. Huang, W.; Lv, Q.; Xiao, F. Investigation of using binder bond strength test to evaluate adhesion and self-healing properties of modified asphalt binders. *Constr. Build. Mater.* **2016**, *113*, 49–56. [CrossRef]
34. Lu, X.; Redelius, P. Compositional and structural characterization of waxes isolated from bitumens. *Energy Fuels* **2006**, *20*, 653–660. [CrossRef]
35. Kök, B.V.; Yilmaz, M.; Guler, M. Evaluation of high temperature performance of SBS+Gilsonite modified binder. *Fuel* **2011**, *90*, 3093–3099. [CrossRef]
36. *AASHTO T283*; In Standard Method of Test for Resistance of Compacted Asphalt Mixtures to Moisture-Induced Damage. American Association of State and Highway Transport: Washington, DC, USA, 2014.

materials

Article

Correlation Analysis between Mechanical Properties and Fractions Composition of Oil-Rejuvenated Asphalt

Rongyan Tian [1,2,3], Haoyuan Luo [1,2,*], Xiaoming Huang [1,2], Yangzezhi Zheng [1,2], Leyi Zhu [1,2] and Fengyang Liu [1,2]

[1] School of Transportation, Southeast University, Nanjing 211189, China; 230179751@seu.edu.cn (R.T.); huangxm@seu.edu.cn (X.H.); 230208809@seu.edu.cn (Y.Z.); 220213392@seu.edu.cn (L.Z.); 220213385@seu.edu.cn (F.L.)
[2] National Demonstration Center for Experimental Education of Road and Traffic Engineering, Southeast University, Nanjing 211189, China
[3] College of Engineering, Tibet University, Lhasa 850000, China
* Correspondence: luohaoyuan@seu.edu.cn; Tel.: +86-15-7-0841-2424

Abstract: To clarify the intrinsic relationship between the mechanical properties of asphalt and its fraction composition, the SARA fraction composition and six macroscopic mechanical properties (critical cracking temperature (T_{CR}), fatigue life (N_f), non-recoverable creep ($J_{nr3.2}$), penetration, ductility, and softening point) were investigated for 16 asphalt samples. Fraction contents of asphaltene and aromatic are strongly correlated with T_{CR} and ductility ($R^2 > 0.92$) that characterize the ability of asphalt to adapt to deformation at low and medium temperatures. Heavy fraction (asphaltene and resins) content is also strongly correlated with ($R^2 > 0.90$) penetration and $J_{nr3.2}$ that characterize the resistance of the asphalt to overall deformation at medium and high temperatures. To express the changes in the four fractions simultaneously with one indicator, a statistic, *average deviation of the fractions between the given asphalt and its original* (marked σ), is introduced in this study to characterize the degree of asphalt aging based on the fraction changes. It normalizes the four simultaneous change indicators (percentage of SARA fractions) during asphalt aging into one indicator. This new indicator has a strong correlation with several mechanical performance indicators of asphalt, where it is strongly correlated with T_{CR} ($R^2 > 0.90$), ductility, and penetration, which are also well correlated with $J_{nr3.2}$ ($R^2 > 0.85$), N_f ($R^2 > 0.75$), and softening point ($R^2 > 0.75$).

Keywords: asphalt binder; fractions composition; mechanical property; rheological properties; asphalt colloidal structure; asphaltene

Citation: Tian, R.; Luo, H.; Huang, X.; Zheng, Y.; Zhu, L.; Liu, F. Correlation Analysis between Mechanical Properties and Fractions Composition of Oil-Rejuvenated Asphalt. *Materials* **2022**, *15*, 1889. https://doi.org/10.3390/ma15051889

Academic Editor: Marek Iwański

Received: 30 December 2021
Accepted: 23 February 2022
Published: 3 March 2022

Publisher's Note: MDPI stays neutral with regard to jurisdictional claims in published maps and institutional affiliations.

1. Introduction

The asphalt binder is the residue from petroleum refining and has been used as a modern road material for 200 years owing to its suitable viscoelastic behavior. A quality asphalt binder is expected to have sufficient toughness, adaptability to deformation at low temperatures [1], sufficient modulus, and elastic recovery at high temperatures [2]. Moreover, it should be able to quickly absorb the energy from repetitive loads and release stresses during the temperature range in which it normally operates [3]. Subtle differences in crude oil origin, refining procedures, additives, and other factors can affect these mechanical properties. The basis for the differences in the external mechanical properties of asphalt is the difference in its internal chemical composition [4]. Scholars have sought to understand the fractions of the compounds that make up asphalt binders to redesign their entire structure [5] and to regulate their external mechanical properties [6] to produce asphalt binders that can be used in various complex environments [7].

The compounds within the asphalt include various n-/isomeric alkanes, aliphatic chain compounds, cyclic aromatic compounds, thick-ringed aromatic hydrocarbons, and various heteroatoms (such as nitrogen, sulfur, and iron) [8]. Many of these compounds are complex in structure and similar in molecular mass and chemical properties, making it difficult to separate and quantify them with the existing technical procedure [9]. A common and accepted practice is to divide the asphalt into four fractions according to molecular weight (SARA), namely Saturates (S), Aromatics (A), Resins (R), and Asphaltenes (As) [10]. Saturates have the lowest molecular masses among the four, their average molar mass ranges from approximately 470 to 800 g/mol, and their main components are waxy aliphatic chain compounds containing small amounts of linear n-alkanes and few aromatic rings [11]. Aromatic fractions are mainly composed of cyclic aromatic compounds with a molar mass between 600 and 980 g/mol [8]. Resins consist mainly of aromatic compounds with polar hybridized sulfur and nitrogen atoms, and their chemical composition is very close to that of asphaltenes, with only slightly lower molecular weights and polarities [12]. Asphaltenes have the highest polarity, with many thick-ringed aromatic compounds and various types of heteroatoms, and their molar mass usually ranges from approximately 1000 to 3500 g/mol [13]. Based on Pfeiffer's study [14], the molecular mass of these four fractions is continuous in asphalt. Furthermore, there is no clear boundary between them, and together, they form a stable colloidal structure. Considering this system, asphaltenes micelles are the dispersed phase stabilized by Resins, whereas Aromatics and Saturates serve as the continuous phase in the continuous matrix [15].

Controlling the macro-mechanical properties of the asphalt by regulating the fractions is currently being performed by many researchers, and one successful example is the rejuvenation of asphalt [16]. Because asphalt is exposed to the natural environment for a long time, the saturated and aromatic fractions in the asphalt are gradually lost under extreme temperature cycling, tire pressure, and rainfall. Furthermore, the deposition of resins may also transform it into asphaltene [17], resulting in hardening and deformation of the asphalt. This is reflected in the loss of mechanical properties such as crack resistance, fatigue resistance, and ductility on a macroscopic scale [18]. By artificially supplementing aging asphalt with maltenes, a portion of the asphaltene can be dissolved, and its proportion can be reduced, achieving a recovery of the external mechanical properties [19,20]. In addition, considering some cold regions such as Ontario and Canada, the addition of motor oil or petroleum-based oil to hard asphalt to configure soft asphalt for low-temperature applications is an economical and efficient solution [21]. However, the types and amount of fractions added in these methods are still based on experimental experience, and it is unclear whether the addition of new fractions changes properties other than those targeted [22]. For example, in the use of waste engine oil modified asphalt, there have been cases where additional magnesium and iron ions have promoted the oxidation of the asphalt, leading to increased aging [23,24].

In recent years, manufacturing asphalt rejuvenators from inexpensive light oils and recycled waste oils has become a hot research topic. A typical process for these rejuvenator productions is filtration, sorting, and blending. Owing to their low cost and wide availability, three laboratory products are expected to be commercialized: waste cooking oil (WCO) [25], waste bio-oil (WBO) [26], and waste engine oil (WEO) [19]. Rayhan et al. [27] have shown that WCO can rapidly recover the penetration, softening point, and ductility of aged asphalts. Hu [28] used WEO, a combination of WEO and furfural extraction oil (WEO+FEO), and a combination of WEO and epoxy resin (WEO+ER) to rejuvenate short-term aged asphalt. The results show that WEO-rejuvenated asphalt does not perform well in terms of moisture and fatigue resistance. Lekhaz et al. [29] tested the performance of a mixture of stone mastic asphalt (SMA) and WEO-rejuvenated asphalt, the results of which contradicted the previous results, i.e., WEO-rejuvenated asphalt concrete showed good moisture stability. Many studies of these three come to the almost uniform conclusion that they are all effective in restoring the performance of aged asphalt, especially in fatigue and low-temperature performance [30–32]. However, an obvious shortcoming of them is the lack of high temperature performance, i.e., rutting resistance [33,34]. Recently, Li et al. [35] comprehensively reviewed the research on using WCO and WEO as asphalt rejuvenators and concluded that both of them are effective in restoring the rutting resistance, fatigue, and low temperature properties of aged asphalt as long as they are dosed in appropriate amounts. They also point out that although both WCO and WEO have shortcomings, combining them to formulate new regenerants is expected to solve these shortcomings, and this is the direction of future research.

The characteristics of different types of waste-oil-rejuvenated asphalt differ significantly owing to the obvious differences in the source, composition, and refining technology of the recovered oil [36]. Haghshenas et al. [37] evaluated the effect of five regenerants (paraffinic oil, aromatic extracts, naphthenic oil, triglycerides/fatty acids, and tall oil) with different chemical compositions. Aromatic extracts had the most similar SARA structure to typical asphalt binders and had the best anti-aging performance. Triglyceride/fatty acid and tall oil did not perform well on low-temperature performance and cohesion after long-term aging due to excessive oxygen content and carbonyl and hydroxyl functions. Paraffinic and naphthenic that contain high saturate may create compatibility issues with asphalt binders. Ding et al. [38] concluded that $C_{20}H_{42}$ can significantly increase thermoreversible aging in the base binder; however, $C_{32}H_{66}$ and asphaltene additives did not produce a similar effect. Shariati et al. [39] proposed a hybrid bio-oil rejuvenator, which can revitalize the aged binder by simultaneously desorbing and peptizing aged binder molecules. Heterocyclic HY molecules (1-butyl-Piperidine and N-methyl-2-Pyrrolidone) in this hybrid bio-oil rejuvenator play an important role, which can effectively improve the resistance of revitalized binder to moisture-induced damages. Another study by Ding et al. [40] pointed out that residual crystalline waxes in WEO would reduce the strain rate of asphalt at low temperatures and increase the risk of cracking. A study by our team also showed that unfiltered metal residues in WEO will accelerate the secondary aging of recycled asphalt [41].

Preliminary studies [42–44] have been conducted to show that the asphaltene content affects the basic indices of penetration, softening point, Frass brittle point, etc. Among different structural fractions, asphaltenes display the lowest temperature susceptibility [45], and they significantly contribute to bitumen stiffness, rigidity, and elasticity [46,47]. Xin et al. [48] further investigated the effect of polycyclic aromatic compounds (PAC), a major fraction of asphaltene, on asphalt, and found that the elasticity and complex modulus of asphalt were reduced with increasing PAC. Lesueur [49] provides a detailed review of the effect of asphalt colloid structure on rheology and chemistry properties, concluding that, although the asphaltene content is small, it is the main cause of the high viscosity and non-Newtonian rheological properties of asphalt. Speight et al. [12] confirmed the role of resins as stabilizers for asphaltenes, which would precipitate from the oily bitumen components without the resins. The potential relationships between the fraction composition and mechanical properties of asphalt found in previous studies are collated in Table 1. Most of these studies just focused on the relationship between asphaltene and mechanical properties, and the selected indicators were usually simple indexes such as stiffness, ductility, and elasticity [46,50–52]. In addition, studies on the complex rheological properties of binders usually consider asphalt as a single-fraction material [53–55], and few studies have investigated the effect of SARA fraction on rheological properties. There are significant differences in the physicochemical properties between SARA fractions, which have a significant impact on the rheological properties of asphalt binders. Therefore, it is difficult to reveal the rheological nature of asphalt binders without a full understanding of the effects of each fraction on asphalt [56].

Table 1. Effect of fractions on mechanical properties of asphalt in previous studies.

Studies	Fractions	Influence on Mechanical Properties
Corbett, L.W. [42]	Saturates and aromatics	Positive correlation with hardness, temperature susceptibility, softening point
Sultana and Bhasin [57]	Saturates and aromatics	Negative correlation with tensile strength
Loeber, L. et al. [46]	Asphaltenes	Positively correlated with G* and stiffness
Ghasemirad, A. et al. [47]	Asphaltenes	Positively correlated with stiffness, elasticity, and high-performance grade (PG)
Hofko, B. et al. [50]	Asphaltenes	Positively correlated with stiffness and elasticity
Fernandez et al. [51]	Asphaltenes and resins	Positively correlated with penetration and negative correlation with ductility
Cooper et al. [52]	Asphaltenes	Negative correlation with fracture resistance
Xin et al. [48]	PAC in Asphaltenes	Positively correlated with elasticity and complex modulus
Speight, J.G. [12]	Resins	It is a stabilizer of asphaltene
Petersen, J.C. [58]	All the four fractions	Durability
Redeliusa, P. et al. [43]	Asphaltenes and aromatics	Respectively positive and negative correlation with viscosity
Haibo D. [38]	Asphaltenes	No significant correlation with low temperature performance

Note G* is the complex modulus of asphalt binders.

A lack of efficient fraction separation techniques is a major impediment to studying the impact of single fractions [4]. Conventional SARA separation methods (such as column chromatography (commonly known as Corbett method) [59], recommended by ASTM D2424 [60]), are time-consuming, have high solvent consumption, and only 1–2 g of asphalt can be separated in a single pass. Some new methods have also been developed by scholars to separate SARA fractions; nevertheless, they are limited by equipment and solvents, etc. It is difficult to obtain the scalability. Handel's new method, for instance, can separate 10 g in a single pass [61]; nonetheless, this amount is insufficient to be used as an additive to modify asphalt. Thin-Layer Chromatography with Flame-Ionization Detection (TLC-FID), originally a means of analyzing crude oil composition, is now also used for SARA analysis of asphalt. It can accurately and quickly obtain SARA distribution in one-fifth of the test time of the Corbett method [19]. Although this method is still unable to separate a large mass of independent fractions, it can be used to identify many oils with significant differences in composition and add them to the asphalt to observe their effects on the mechanical properties and fraction distribution of asphalt. This can be used to study a single change in asphalt properties.

It would help to advance the research work on asphalt regeneration and modification for specific application environments if the effect of each fraction on the macroscopic mechanical properties of asphalt could be identified. However, it seems that this work is currently limited by the lack of efficient methods for quantitative identification of fractions and uniform oil sources of asphalt, and no clear and uniform conclusions seem to have been reached in this work. Therefore, this study aims to investigate the correlation between the distribution of fractions of asphalt and its mechanical indicators. To achieve this objective, seven oil-rejuvenators with hugely different fraction compositions are selected to be added to the control of an aged asphalt 50/70 and are subjected to secondary aging, resulting in 16 asphalt samples. Basic property tests (such as penetration, ductility, and softening point), general rheological tests (such as critical cracking at low temperature (T_{CR}), fatigue life at medium temperature (N_f), non-recoverable creep compliance at high temperature ($J_{nr3.2}$), and SARA fraction distributions TLC-FID analyses are performed on these 16 samples derived from the same asphalt. Eventually, the correlations between the fraction distribution and these mechanical indices are investigated.

2. Materials and Methods

2.1. Raw Materials

2.1.1. Asphalt Binders

In this study, an aged asphalt 50/70 was used as the control and was analyzed for the mechanical properties and SARA fraction composition before and after adding oil-rejuvenators with different fractions composition. This control asphalt was recycled from the upper layer of the Chengdu-Chongqing expressway of China, which was completed and opened to traffic in 1995. After being in service for approximately 15 years, its performance grade was reduced from PG82-10 to PG70-22. The original asphalt (OA) had a needle penetration of 63 dmm(0.1 mm), a softening point of 47 °C, a 10 °C ductility of 26 cm, and a 135 °C viscosity of 0.44 Pa·s. After aging, it had a needle penetration of 31 dmm, a softening point of 65 °C, a 10 °C ductility of 5 cm, and a 135 °C viscosity of 1.32 Pa·s.

Its basic properties in the original and aged stated are summarized, where the penetration, softening point, ductility, viscosity, and performance grade were tested in accordance with the standard ASTM D5 [62], ASTM D36 [63], ASTM D113 [64], ASTM D4402 [65], and ASTM D6373 [66], respectively.

2.1.2. Oil-Rejuvenator

Table 2 shows the properties of all the seven oil-rejuvenators that can be divided into two categories according to base-oil types. One is the four bio-rejuvenators with raw materials such as waste edible oil, tung oil, biodiesel, and fish oil residue labeled Bio-1, Bio-2, Bio-3, and Bio-4, respectively. Many studies involving regeneration and aging have discussed the improvement of the rheological properties of aged asphalt using these four regenerants; nevertheless, few studies have analyzed their fraction compositions [23,67–69].

Table 2. Properties of the seven oil-rejuvenators.

Raw Materials	Base Oil	Label	Acid Value (mg KOH/g)	Iodine Value (g I/100 g)	Saponification Value (mg KOH/g)	Density @20 °C (cm)	Kinematic Viscosity @60 °C (mm^2/s)	Flash Point (°C)	Appearance
Waste edible oil	Bio-oil	Bio-1	≤7.1%	274	181	0.927	61.7	316	Brown, cloudy
Tung oil		Bio-2	≤0.4%	76	182	0.944	70.3	224	Yellow, transparent
Biodiesel		Bio-3	≤2.9%	143	192	0.965	68.6	251	Yellow, transparent
Fish oil residue		Bio-4	≤3.4%	173	198	0.994	100.5	210	Yellow, translucent
Light fraction oil	Petroleum extract	Pio-L	-	-	-	0.893	51.2	198	Yellow brown, translucent
Middle fraction oil		Pio-M	-	-	-	0.936	88.4	224	Colorless, transparent
Heavy fraction oil		Pio-H	-	-	-	1.016	152.4	240	Black, opaque

The other is the three petroleum-based regenerators separated at different temperatures during vacuum distillation. Light fraction oil is the product of the distillation temperature 200–220 °C, labeled Pio-L, and in its internal molecular weight composition, C_{12-18}, C_{6-12}, and C_{18-} account for approximately 14%, 85%, and 1%, respectively. Middle fraction oil is fractionated at 220–260 °C, labeled Pio-M, in which C_{12-18}, C_{6-12}, and C_{18-} account for 95%, 2%, and 3%, respectively. Heavy fraction oil is fractionated at 220–260 °C, labeled Pio-H, in which C_{12-18} and C_{18-} account for 33% and 65%, respectively. It has no C_{6-18}.

In Table 2, the acid value, iodine value, and saponification value of the four bio-rejuvenators were determined by the standard methods provided by ISO 660 [70], ISO 3961 [71], and ISO 3657 [72], respectively. For all the seven rejuvenators, their density at 20 °C, Kinematic viscosity at 60 °C, and flash point were determined by the standard methods provided by ASTM D4052 [73], ASTM D7279 [74], and ASTM D56 [75], respectively.

Fractions composition differences between these selected seven oil-rejuvenators are significant based on the TLC-FID test, which will be introduced in detail in Section 3.1.

2.2. Preparation of Oil-Rejuvenated Asphalt

Using high-speed shearing, oil-rejuvenated asphalts were made. The aged asphalt binder was heated to approximately 60 °C above the softening point (i.e., approximately 130 °C) and mixed with the rejuvenator, employing a shear mixer at a speed of 4000 r/min for 15 min [35]. The mixing condition was determined by the viscosity of aged asphalt 50/70 which is around 1.15~1.35 Pa·s at this temperature. In this specific range, the asphalt could be easily mixed with these rejuvenators. In previous mixing attempts, it was found that if the temperature was higher than this condition, some of the lighter rejuvenators (e.g., Pio-L) tend to volatilize (produce large amounts of white smoke) and thus affect the quality of the recycled asphalt. If the temperature is too low, the viscosity of the asphalt will be too high, and it will be difficult to miscible with heavy rejuvenators.

2.3. Design of Experiments

Figure 1 provides the experimental design of this study. First, the control aged asphalt 50/70 (RA) was rejuvenated using the seven oil-rejuvenators. Thereafter, these seven samples were subjected to secondary aging using the rolling film oven test (RTFOT) and 20-h pressurized aging vessel. Subsequently, all the asphalt samples were subjected to two parts of the test, i.e., mechanical property and fractions analysis tests. The mechanical property test includes multiple stress creep recovery (MSCR), linear amplitude scan (LAS), low-temperature critical cracking temperature calculation, and basic properties (penetration, softening point, ductility) tests. These were selected in many studies as a comprehensive set of indicators to evaluate the rutting, fatigue, and cracking resistance of the asphalt. Considering the fraction analysis tests, all the asphalt samples were separated and quantified for SARA (i.e., saturated, asphaltene, resinous, and aromatic) fractions using thin-layer chromatography with flame-ionization detection (TLC-FID, described in detail in Section 2.4.6). Finally, the correlation between the mechanical property indicators and fraction composition of these asphalts was analyzed separately.

2.4. Measurement and Characterization

2.4.1. Basic Properties Test

Basic properties, including penetration at 25 °C, softening point, and ductility at 15 °C, were evaluated in accordance with the standard ASTM D5 [62], ASTM D36 [63], and ASTM D113 [64], respectively. Their duplicate tests were performed three times.

2.4.2. Performance Grade (PG)

The temperature PG of all asphalt samples can be estimated using the dynamic shear rheological (DSR) and blending beam rheological (BBR) tests based on the AASHTO M320 [76] method. The DSR used in this study is the Discover HR-3 DSR manufactured by TA INSTRUMENT, and the BBR is the TE-BBR provided by CANNONTE.

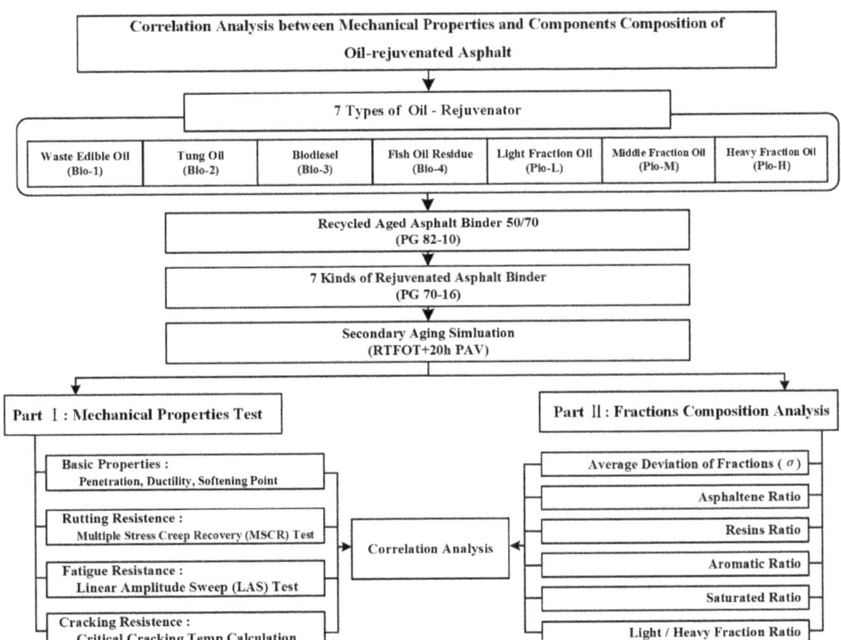

Figure 1. Schematic of the experimental test plan.

2.4.3. Low-Temperature Cracking Resistance Test

The critical cracking temperature (T_{CR}) calculation is employed to evaluate the anti-cracking performance of oil-rejuvenated asphalts at low temperatures. T_{CR} is a non-strength test index determined in the stiffness modulus data, while the stiffness modulus data are obtained from the extended bending beam rheological test [1]. This method can better reflect the ultimate performance of the asphalt at low temperatures than the PG test. It also has a high correlation with the actual pavement cracking [77].

The first step to calculate the T_{CR} is to obtain the low-temperature stress $\sigma(\xi)$ of the asphalt based on the continuous temperature decrease. This was calculated using the basic creep compliance $J(t)$ data obtained from the asphalt binder by the BBR test. The specific steps for calculating the low-temperature stress $\sigma(\xi)$ are based on AASHTO R49 [78] and the study by Roy and Hesp [77].

The second step is to find the T_{CR} in the curve of the $\sigma(\xi)$ based on the theory of single asymptote procedure (SAP), proposed by Shony et al. [79] Figure 2 shows the change in the $\sigma(\xi)$ as the temperature decreases. The temperature stress first increases gradually and then increases rapidly toward the end of the curve. The thermal stress limit is described using the asymptotes at the beginning and end of the thermal stress curve. The point where two asymptotic lines intersect is the T_{CR}, where the curvature changes the fastest and cracks are most likely to occur. In this study, T_{CR} is used as an indicator to assess the effect of oil-rejuvenators on the low-temperature crack resistance of asphalt. Smaller values of the T_{CR} indicate a better low-temperature cracking resistance, indicating that the asphalt can be used at low ambient temperatures without cracking due to thermal stress.

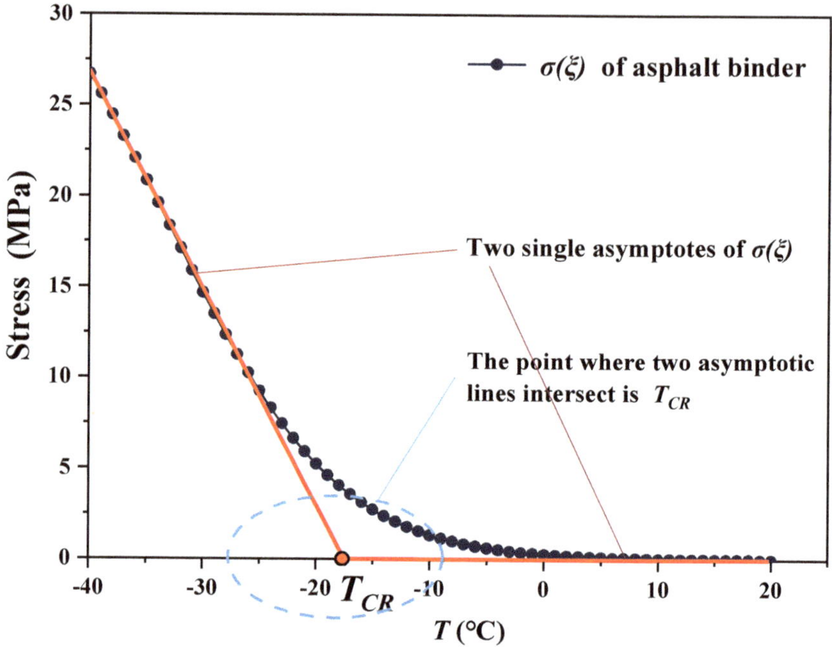

Figure 2. Temperature stress and TCR calculation according to the SAP theory [54].

2.4.4. Linear Amplitude Sweep (LAS) Test

The LAS test was employed to evaluate the anti-fatigue performance of all the asphalt samples, and for each sample, two duplicates are tested. The LAS test can better simulate the loss development of asphalt under repeated loads than the PG test [80]. The LAS test includes two steps. The first step is frequency scanning at 0.1% strain in the frequency range of 0.1–30 Hz to determine parameters α and B in Equation (1). The second step is linear amplitude scanning, where a round of oscillatory load cycles with linearly increasing amplitudes (from 0.1% to 30%) is conducted at a constant frequency (10 Hz) to generate the accelerated fatigue damage. The viscoelastic continuous damage theory VECD (viscoelastic continuous damage) is used to determine parameter A_{35} in Equation (1). The test method is based on AASHTO TP 101-12 [81], and a larger N_f indicates a better fatigue resistance. The asphalt fatigue failure life (N_f) is computed using Equation (1):

$$N_f = A_{35} \cdot \gamma_{max}^{-B} \tag{1}$$

where γ_{max} is the maximum expected asphalt strain for a given pavement structure, percent; B is equal to 2α, no unit; and N_f represents the number of loading cycles before failure.

2.4.5. Multiple Stress Creep Recovery (MSCR) Test

The MSCR test is employed to investigate the anti-rutting performance of all the asphalt5 samples, and the test method is based on ASTM D7405 [82]. Considering each sample, three duplicates are tested. The MSCR test has a better correlation with the anti-rutting performance of the asphalt compared to the rutting factor ($G^* / \sin \delta$). Therefore, it has gradually become a main method for identifying the high temperature performance of asphalts in the experiment system of Superpave [83]. For each asphalt sample, the MSCR test is performed at its performance grade (PG) high temperature. First, the specimen is loaded at constant creep stress for a 1-s length of time creep and followed with a zero-stress recovery of a 9-s length of time. Second, 20 creep and recovery cycles are performed at creep stress of 0.1 kPa. The first 10 cycles are for conditioning the specimen. The second ten cycles were designated as cycles from N = 1 to 10 and were employed for data collection and analysis. Thereafter, ten creep and recovery cycles are performed at creep stress of 3.2 kPa. The non-recoverable creep compliance measured at 3.2 kPa ($J_{nr3.2}$) is employed as an assessment of the endurance of the bitumen to permanent distortion under repeated loading state, and a smaller $J_{nr3.2}$ value represents a better rutting resistance [23].

2.4.6. Fraction Analysis of the Asphalt

Compositional harmonic and compatibility theories are the most recognized theories regarding the asphalt aging phenomena [84]. Reduction in the proportion of the light fraction or increase in the heavy fraction of asphalt is considered the basis for asphalt aging by both theories [85]. Many studies attempt to justify the change in the asphalt properties by investigating the fractions changes before and after aging; however, the separation and quantification of fractions have always been very difficult tasks. Common fractions analysis methods include Fourier transform infrared reflection (FTIR) [86] and solvent precipitation methods (introduced in standard ASTM D4242 [60]). The former has a good testing efficiency; nonetheless, it is difficult to quantify the composition of the fractions. Moreover, the latter can obtain the exact amount; nevertheless, the process is complex, time-consuming, and lacks reproducibility. In our previous study, an efficient quantitative analysis method of asphalt fractions, TLC-FID, was introduced. This was derived from a chromatographic method used in the petrochemical industry. The procedure and principle of the TLC-FID are illustrated in Figure 3. A constructed chromatographic column that leverages the different diffusion heights of the four fractions of the asphalt in a toluene solution is scorched, during which the intensity of the electrons emitted by each fraction at the point of aggregation is recorded and converted into the amount of this fraction [19].

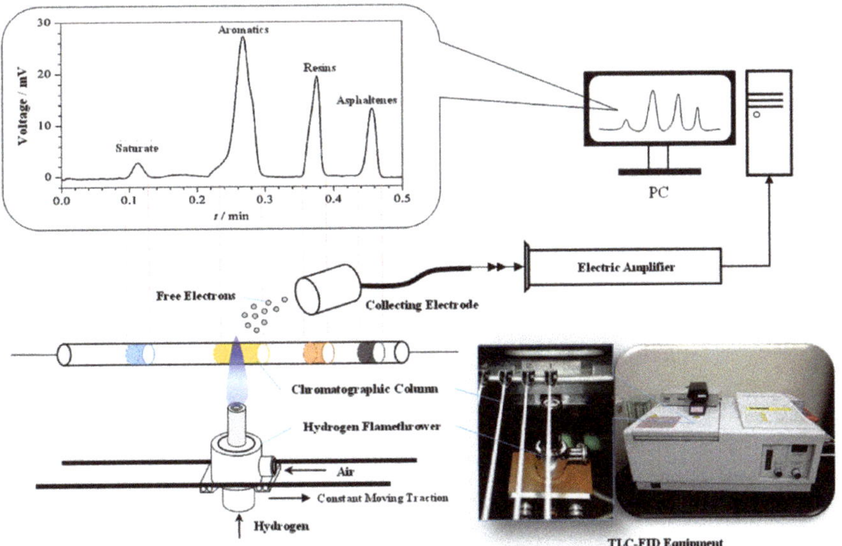

Figure 3. Test procedure and principle of the TLC-FID [19].

2.5. Dosage of Oil-Rejuvenators

To ensure the comparability between the asphalt samples, the dosage of each rejuvenator was determined following the rule: under the selected dosage, the rejuvenator should restore the performance grade (PG) of the asphalt closer to its original status (i.e., the PG of OA) as much as possible. Therefore, a series of pre-tests were designed to characterize the relationship between the dosage of each rejuvenator and the PG of the recycled asphalt. The dosages of 2%, 4%, 6%, 8%, and 10% were used. The results are shown in Figure 4.

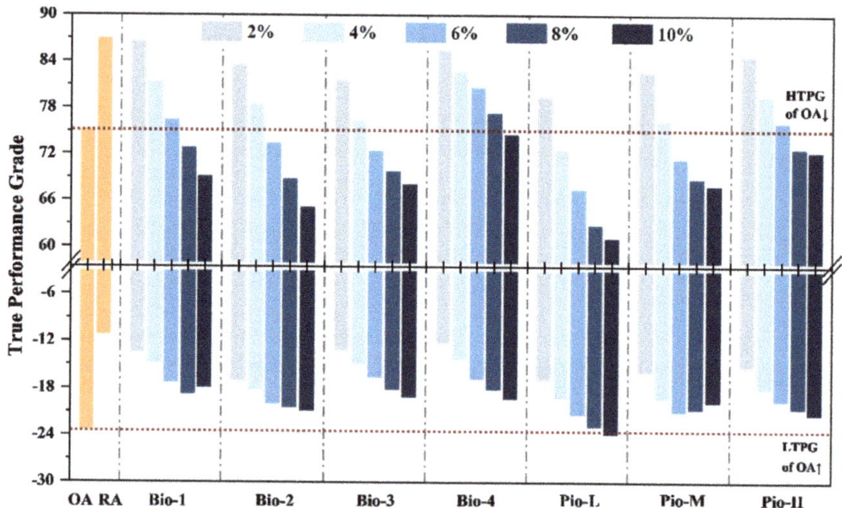

Figure 4. Relationship between the dosages and true performance grade of the seven oil-rejuvenators.

As shown in Figure 4, all the seven oil-rejuvenators are easy to recover the high-temperature PG (HTPG) of the RA to the level of OA (PG70-XX), where the critical dosages for the seven rejuvenators from Bio-1 to Pio-L are approximately 6%, 5%, 4%, 9%, 3%, 4%, and 6%, respectively. Once the dosages exceed the critical value, the HTPG of the rejuvenated asphalts will be worse than that of the OA. Considering the recovery effect of the low-temperature PG (LTPG), there are bottlenecks in these rejuvenators. When the dosage reached a certain value, the LTPG found it difficult to be further optimized or even deteriorate. No oil-rejuvenator could recover this index to the original level, except for 10% Pio-L. Obviously, there was no appropriate dosage for each oil-rejuvenator that could restore both the HTPG and LTPG of the RA to the original level simultaneously.

Considering most studies that focused on rejuvenated asphalts, a lack of high temperature performance has always been criticized [20,87]. Regarding this case, the consistency of the HTPG is prioritized, and the LTPG is in the same classification as much as possible. Therefore, the dosages of Bio-1, Bio-2, Bio-3, Bio-4, Pio-L, Pio-M, Pio-H were determined as 6%, 5%, 4%, 9%, 3%, 4%, and 6%, respectively, where the PG of all rejuvenated asphalt was PG 70-16.

3. Test Results

3.1. SARA Fractions Analysis

Figure 5 shows the SARA fractions results of all the samples. The top two bar graphs present the composition of the control asphalt OA and RA, where 24.7% (by mass ratio) of the aromatic and saturated (together called light fraction) were transferred to the asphaltene and resins (together known as heavy fraction) during the aging process. This is a typical aging process of asphalts [49].

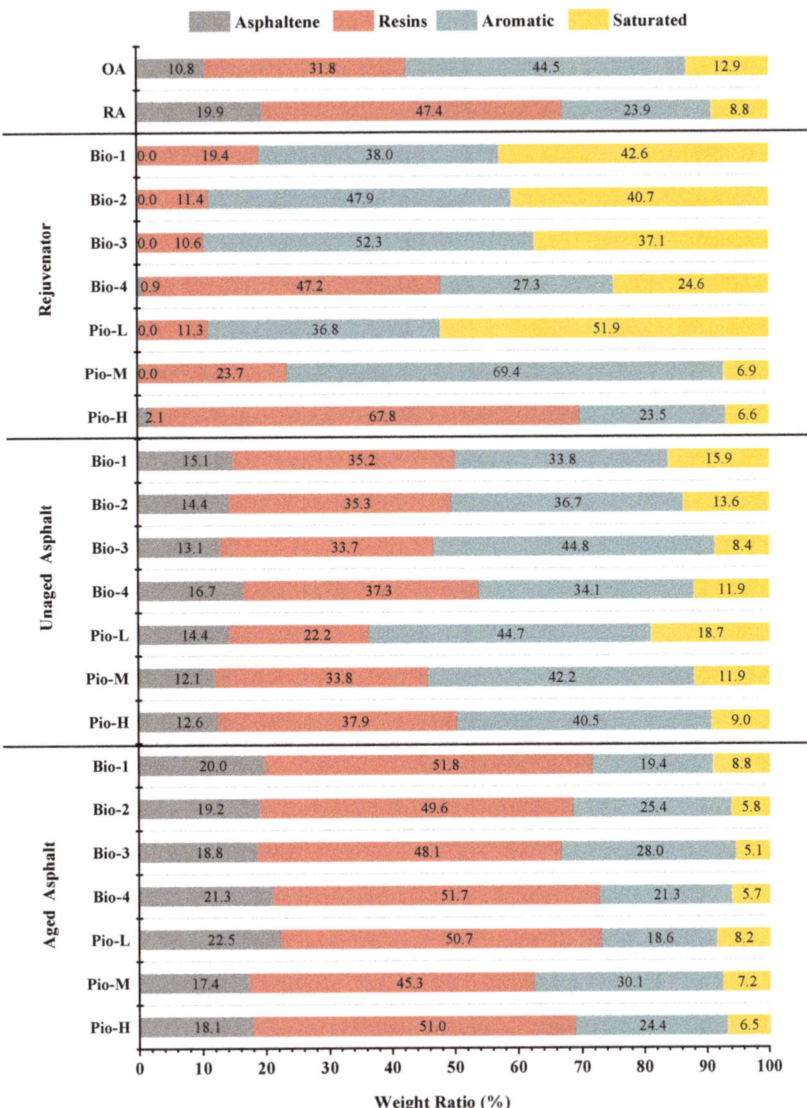

Figure 5. SARA fractions analysis of all the asphalt samples.

The subsequent seven bars present the composition of the seven oil-rejuvenators. Considering Bio-1, Bio-2, and Bio-3, the proportion of aromatic increases sequentially, whereas the saturated gradually decreases. Regarding the Bio-4, light and heavy fractions are almost 50/50, and there is even 0.9% asphaltene in them. For the three petroleum-based rejuvenators, the ratio of the heavy fraction increased sequentially from 11.3 for Pio-L to 23.7 for Pio-M and to 69.9% for Pio-H.

The fraction composition of rejuvenated asphalt (unaged) is affected by its corresponding rejuvenator. Considering the unaged samples (Bio-1 to Bio-3), the proportion of the aromatic increases, whereas the saturated decreases, similar to their corresponding rejuvenators. Heavy fractions in the Bio-4 asphalt are also significantly higher than the other three bio-oil-rejuvenated asphalts.

After aging, the trends of the fraction structure change were the same for all the samples, where the ratio of asphaltenes to resins increased and the aromatic and saturated decreased. Many fraction composition characteristics of the aged bio-oil-rejuvenated asphalts inherit the characteristics associated with them in the unaged stage. For example, the ratio of resins decreased from Bio-3 to Bio-1. This inherited relationship can also be found in Pio-M and Pio-H. Nevertheless, Pio-L is an exception. Unaged Pio-L has the highest light fraction content of 63.4%. However, after the second aging, its ratio of heavy fraction became the highest. This could be attributed to its unstable colloidal structure.

Koots and Speights [12] indicated that a resin acts as a surfactant, creating a so-called soluble layer and helping to maintain the suspension of the asphaltene in the aromatic fraction of the dispersion system. If a binder has resins amounting to 50%, approximately 75% of them are needed to stabilize the asphaltene dispersion. Obviously, the ratio of resin to asphaltene in the Pio-L is poor, resulting in inadequate dispersion and suspension of the asphaltene. Therefore, even if a large number of aromatic fractions are supplemented as the dispersion system, they may simply mix with other fractions, without forming a stable and homogeneous colloidal structure. In this situation, the free aromatics may collect and oxidize to resin and asphaltene [19,41].

In Figure 5, none of the rejuvenators could reduce the RA's asphaltene content to the level of the OA. Furthermore, the process for the resins content reduction appeared very difficult as well. This indicates that these rejuvenated asphalts produced by these rejuvenators are still essentially different from the OA, and aging has not been fully restored, although these rejuvenated asphalts have the same PG as the OA.

To characterize the degree of asphalt aging in its current state, a statistic, *average deviation of the fractions between the given asphalt and its original* (marked σ), was introduced in this study. Equation (2) gives the calculation of σ:

$$\sigma = \sqrt{\frac{\left(C_{Sat} - C_{Sat\ of\ OA}\right)^2 + \left(C_{Asp} - C_{Asp\ of\ OA}\right)^2 + \left(C_{Aro} - C_{Aro\ of\ OA}\right)^2 + \left(C_{Res} - C_{Res\ of\ OA}\right)^2}{df - 1}} \tag{2}$$

where C_{Sat}, C_{Asp}, C_{Aro}, and C_{Res} are the mass proportions of the four fractions (saturated, asphaltene, aromatic, and resins, respectively) of the asphalt samples in the current aging. $C_{Sat\ of\ OA}$, $C_{Asp\ of\ OA}$, $C_{Aro\ of\ OA}$, and $C_{Res\ of\ OA}$ are the mass proportions of the four fractions of this asphalt without any aging treatment (i.e., original asphalt). df is the degree of freedom of the variable and takes the value of four.

It is possible to quantify the degree of asphalt aging using σ, eliminating the need to describe the changes in the four variables (C_{Sat}, C_{Asp}, C_{Aro}, and C_{Res}) simultaneously. Particularly, σ reflects the average difference of the four fractions between the target asphalt and its original state, and it is used as an indicator to evaluate the degree of asphalt aging based on the composition. The smaller the σ value of asphalt, the less it differs from its original state, the less it deteriorates, and considering the rejuvenated asphalt, the better it recovers. The σ values of the seven oil-rejuvenated asphalt samples before and after the secondary aging are shown in Figure 6. Considering OA, σ is zero, and regarding the seven samples of the freshly regenerated asphalt, σ is ranked from smallest to largest as Pio-M, Bio-3, Pio-H, Bio-2, Pio-L, Bio-1, and Bio-4. This is also represented in the ranking of their regenerative effects (from best to worst). After the secondary aging, this ranking became Pio-M, Bio-3, Bio-2, Pio-H, Bio-4, Bio-1, and Pio-L, where σ values of Pio-H, Bio-4, Bio-1, Pio-L were greater than the RA. This indicates that the degree of these four rejuvenated asphalts after the secondary aging exceeded the ones before the regeneration. Characterizing the degree of asphalt aging, a detailed comparison between σ and the other mechanical indicators will be performed in Sections 3.2–3.5.

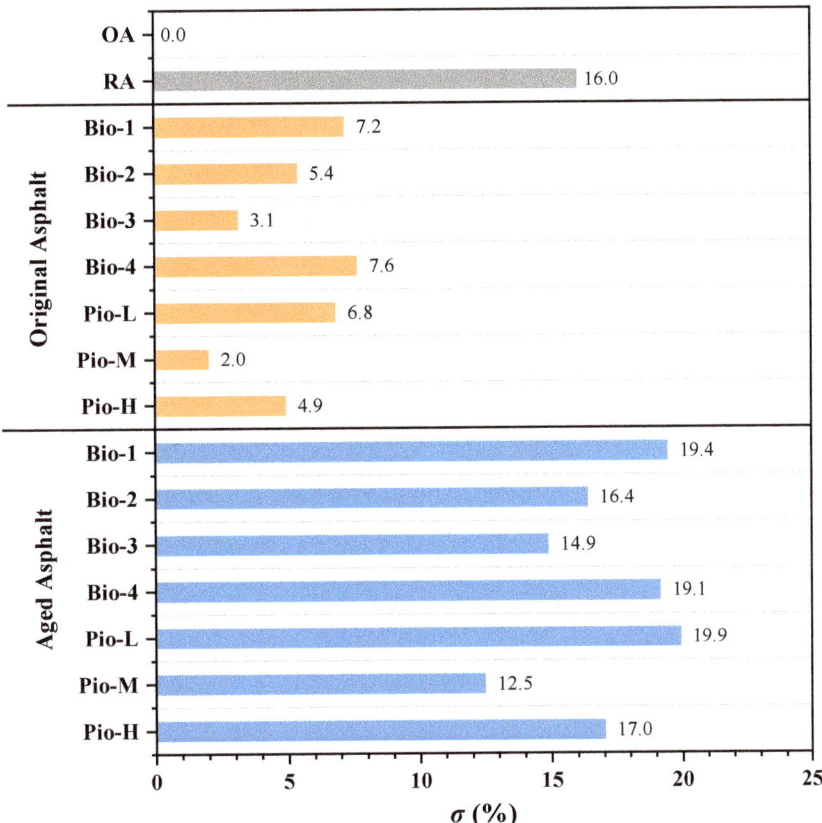

Figure 6. σ value of all the asphalt samples.

3.2. Cracking Resistance

Table 3 gives the results of the classical BBR tests, where $T_{s=300}$ and $T_{m=0.3}$ are the failure temperatures at stiffness equal to 300 MPa and m-value equal to 0.3, respectively. It difficult to accurately identify the low-temperature crack resistance of many samples referring the results, because multiple asphalts have the same LTPG. Therefore, the low temperature stress ($\sigma(\xi)$) and critical cracking temperature (T_{CR}) were further calculated based on the classical BBR results for further performance grading, as shown in Figures 7 and 8.

Table 3. Classical BBR test results of all the asphalt samples.

Asphalt	Aging State	Failure Temperature		LTPG
		$T_{s=300}$ (°C)	$T_{m=0.3}$ (°C)	
50/70	Unaged	−28.3	−24.9	PG XX-22
	Aged	−17.2	−14.5	PG XX-10
Bio-1	Unaged	−21.7	−17.8	PG XX-16
	Aged	−14.8	−10.4	PG XX-10
Bio-2	Unaged	−24.0	−21.1	PG XX-16
	Aged	−18.4	−13.6	PG XX-10
Bio-3	Unaged	−25.9	−22.2	PG XX-22
	Aged	−16.3	−14.6	PG XX-10

Table 3. *Cont.*

Asphalt	Aging State	Failure Temperature		LTPG
		$T_{s=300}$ (°C)	$T_{m=0.3}$ (°C)	
Bio-4	Unaged	−21.8	−16.9	PG XX-16
	Aged	−14.9	−11.2	PG XX-10
Pio-L	Unaged	−23.2	−17.7	PG XX-16
	Aged	−12.9	−9.7	PG XX-04
Pio-M	Unaged	−28.9	−24.0	PG XX-22
	Aged	−20.3	−16.9	PG XX-16
Pio-H	Unaged	−25.6	−21.5	PG XX-16
	Aged	−18.7	−13.6	PG XX-10

Figure 7 shows the curves of $\sigma(\xi)$ of all the samples. Before the secondary aging, the curves of these rejuvenated asphalts are all above those of the OA, indicating that their temperature stresses are greater than those of the OA (Figure 7a). Correspondingly, the T_{CR} values of the seven unaged rejuvenated asphalts are also higher than that of OA as shown in Figure 8. Considering the results of the T_{CR}, the cracking resistance of the seven unaged samples and control at low temperatures can be ranked from best to worst as OA, Pio-M, Bio-3, Pio-H, Bio-2, Pio-L, Bio-1, and Bio-4. This seems to be generally consistent with the previous ranking of the degree of regeneration given by the σ values in Section 3.1. After the secondary aging, the T_{CR} ranking changed to Pio-M, Bio-3, Bio-2, Pio-H, RA, Bio-4, Bio-1, and Pio-L, still consistent with the aged σ ranking. The $\sigma(\xi)$ and T_{CR} of the four samples, Pio-M, Bio-3, Bio-2, and Pio-L, are lower than the RA, indicating that their low-temperature crack resistance is better than that of the RA after secondary aging.

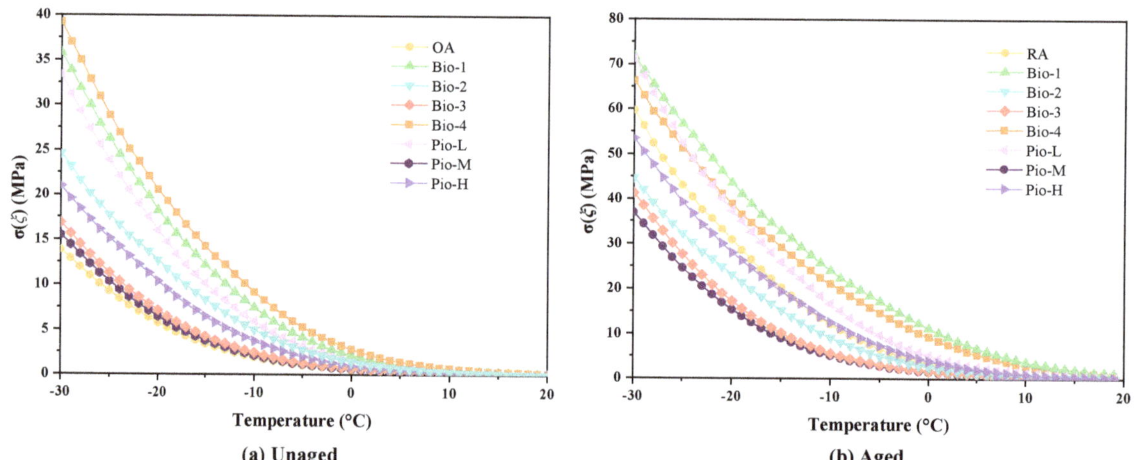

(a) Unaged

(b) Aged

Figure 7. Low-temperature stress curve of (a) unaged asphalts (b) aged asphalts.

There seems to be a correlation between the T_{CR} and σ, where the larger the value of σ (the more the fractions change), the larger the value of the T_{CR} (the worse the low-temperature performance of asphalt). To clarify the specific relationship between the fractions of these rejuvenated asphalts and their low temperature cracking performance, the correlation between the T_{CR} and σ was investigated. The results of the one-dimensional correlation analysis between the σ values of all the 16 asphalt samples and their T_{CR} are shown in Figure 9a, where the correlation coefficient (R^2) is 0.982, which can be considered a strong correlation ($R^2 > 0.900$). To further clarify which of the four fractions has a greater influence on the T_{CR}, the correlations between this mechanical indicator and the proportion of asphaltene, resins, aromatic, saturated, and heavy fraction are investigated as shown in Figure 9. The correlation between the T_{CR} and light fraction is the same as that with the heavy fraction because the light fraction is not an independent variable. The total content of the light and the heavy fraction is always 100%.

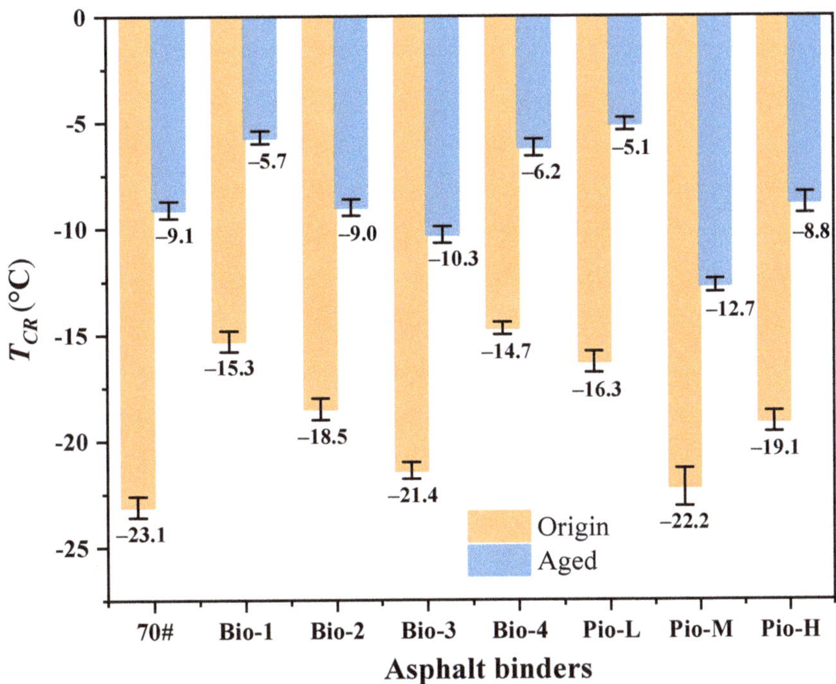

Figure 8. Critical cracking temperature (T_{CR}) of all the asphalt samples.

Considering the four fractions, T_{CR} is strongly linearly correlated ($R^2 > 0.900$) with asphaltene and aromatic ratio. The former is positively correlated (i.e., the higher the asphaltene content, the larger the T_{CR}, and the worse the asphalt cracking resistance), and the latter is negatively correlated (i.e., the higher the aromatic content, the smaller the T_{CR}, and the better the asphalt cracking resistance). Regarding the resins' ratio, the R^2 is 0.714, suggesting that it correlates (R^2 in 0.6–0.9) with T_{CR}. However, the R^2 is only 0.276 (no correlation, $R^2 < 0.4$) between the T_{CR} and saturated ratio. This does not seem to have a significant effect on the low temperature performance. Li et al. [35] indicated that the correlation between low temperature performances and saturated fraction content of asphalt was not high, but these performances were strongly influenced by the ratio of asphaltene to aromatic fraction. Based on these findings, the key reason for the superior performance of the four rejuvenated asphalts (Bio-2, Bio-3, Pio-M, and Pio-L) over the RA is because they all have less asphaltene and are more aromatic than the RA. Similarly, the T_{CR} of the seven unaged rejuvenated asphalts is worse than that of the OA because their asphaltene content was higher than that of the OA despite the dispersion and dissolution efforts.

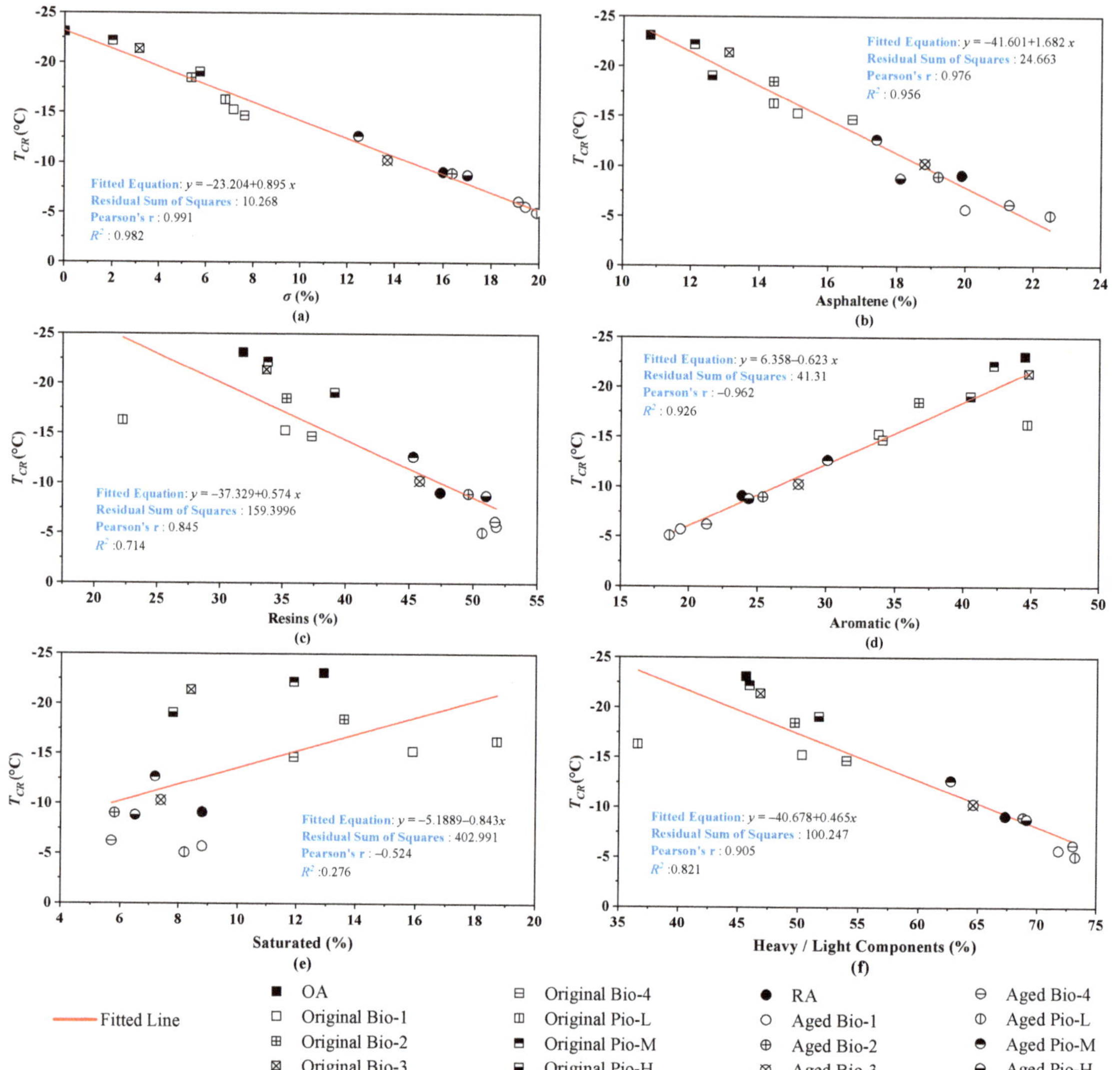

Figure 9. Correlation analysis between the T_{CR} and fraction of (**a**) σ; (**b**) asphaltene; (**c**) resins; (**d**) aromatic; (**e**) saturated; (**f**) heavy/light.

3.3. Fatigue Resistance

The LAS test at 25 °C under 5% strain was applied to all 16 asphalt samples. The curves in Figure 10 are the results of the amplitude scan in the LAS test. This illustrates the relationship between shear stress and strain. There is a shear stress peak in every curve, which can be referred to as the yield stress, and the strain of it is known as the yield strain. The asphalt that possesses smaller yield stress and much yield strain has a better performance to adapt to the transformation in the repeated load. The shear strain–stress curve of the eight asphalt samples before aging can be divided into two categories. The first includes OA, Bio-1-4, and Pio-L, whose stresses first increase with strain and gradually decrease after reaching the yield stress. A stress rebound occurs at approximately 19% of the strain. Another category, including Pio-M and Pio-H, did not have a rebound, and their stresses gradually decayed after reaching the yield stress. After aging, the stress rebound phenomenon in the first category of the curve disappeared and was replaced by a constant and moderate stress drop. Nonetheless, another type of curve maintained the same basic shape before and after aging, except for the stress increase.

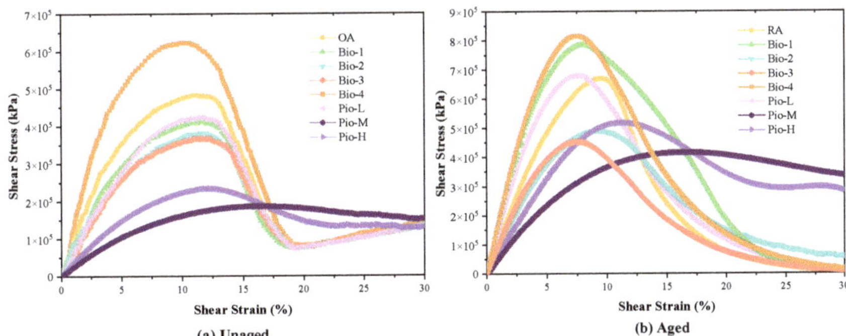

Figure 10. Shear strain versus shear stress curves of the LAS test of (**a**) unaged asphalts (**b**) aged asphalts.

The fatigue properties characterized by fatigue failure life (N_f) of the unaged samples are ranked from best to worst as Pio-M, Pio-H, OA, Bio-3, Bio-2, Bio-1, Pio-L, and Bio-4, whereas the ranking changes to Pio-M, Bio-3, Pio-H, Bio-2, RA, Pio-L, Bio-1, and Bio-4 after secondary aging as shown in Figure 11. Pio-M and Pio-H (which performed well in the low-temperature test) continued to perform well in the phase, whereas Pio-L, Bio-1, and Bio-4 (which performed poorly in the previous test) continued to perform poorly in this segment.

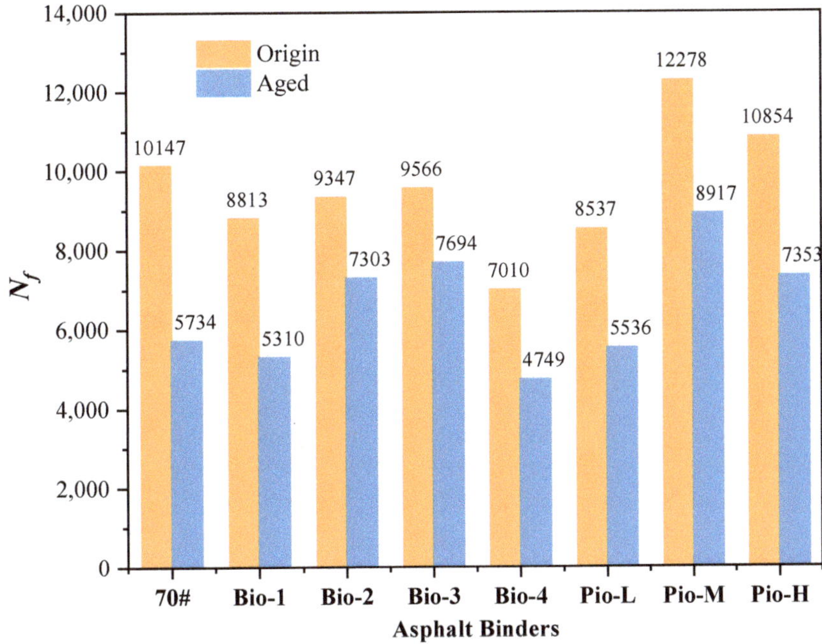

Figure 11. Fatigue failure life (N_f) of all the asphalt samples.

Similar to the T_{CR}, the correlations between N_f and the six fraction indicators are analyzed and shown in Figure 12. Considering the six fraction indicators, N_f had correlation coefficients greater than 0.6 with the σ, asphaltene, aromatic, and heavy/light fraction ratios. However, there was a weak correlation (R^2 in 0.4–0.6) between the N_f and resins' ratio, and almost no correlation ($R^2 < 0.4$) with the saturated ratio. The R^2 between the N_f and all the six fraction indicators is significantly smaller than that of the T_{CR}. This may be because the fatigue performance of the oil-rejuvenated asphalt is the result of the four fractions harmonizing with each other, rather than being determined by one or two key fractions as in the case of the T_{CR}.

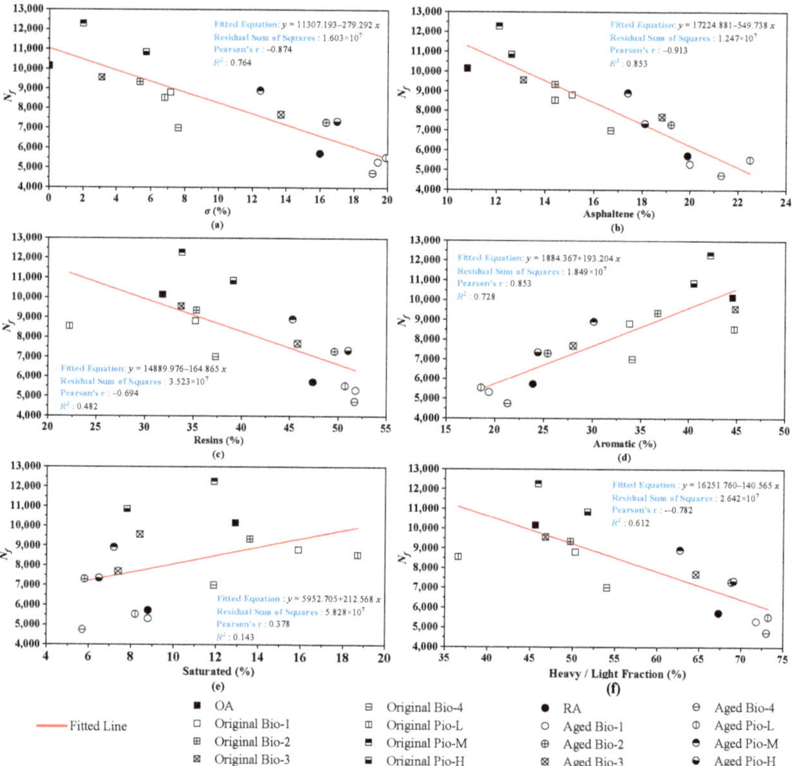

Figure 12. Correlation analysis between the N_f and fraction indicators. (**a**) σ; (**b**) asphaltene; (**c**) resins; (**d**) aromatic; (**e**) saturated; (**f**) heavy/light.

3.4. Rutting Resistance

All the asphalt samples were conducted using the MSCR test at the stress of 3.2 kPa. The test temperature for each sample is the high temperature with its performance grading without grade bumping. The two main indexes of this test, non-recoverable creep compliance ($J_{nr3.2}$) and strain recovery rate ($R_{3.2}$), are shown in Figure 13a,b. Asphalt with a good rutting resistance should have a small $J_{nr3.2}$ and large $R_{3.2}$.

The rutting resistance of these unaged asphalt samples is ranked (from best to worst) Bio-4, OA, Pio-H, Pio-M, Bio-3, Bio-2, Bio-1, and Pio-L based on the $R_{3.2}$ and $J_{nr3.2}$. After aging, this ranking changes to Bio-4, Pio-H, Bio-1, RA, Bio-2, Pio-L, Pio-M, and Bio-3. The ranking here is almost opposite to the previous ranking of the crack and fatigue resistances. Pio-M and Bio-3 have perfect N_f and T_{CR}; nonetheless, they exhibit the worst irrecoverable flexibility and strain recovery in the MSCR test. Moreover, Pio-L, Bio4, and Bio-1, which previously performed poorly, exhibit good high-temperature rutting resistance.

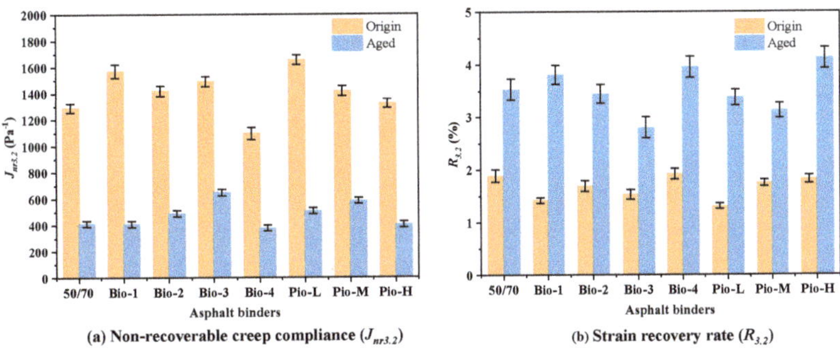

(a) Non-recoverable creep compliance ($J_{nr3.2}$)

(b) Strain recovery rate ($R_{3.2}$)

Figure 13. MSCR test results of all the asphalt samples.

The correlation analysis of $J_{nr3.2}$ with the six fraction indicators is shown in Figure 14, where all the six correlation coefficients are greater than 0.6. Particularly, Resins, which were not previously closely related to fatigue and low-temperature crack resistance, maintain a good correlation with $J_{nr3.2}$ ($R^2 = 0.891$). Furthermore, the heavy fraction strongly correlates with $J_{nr3.2}$ ($R^2 = 0.925$) because resins make up a large proportion of the heavy fraction. This indicates that the heavy fractions play a major role in the rutting resistance of these rejuvenated asphalts.

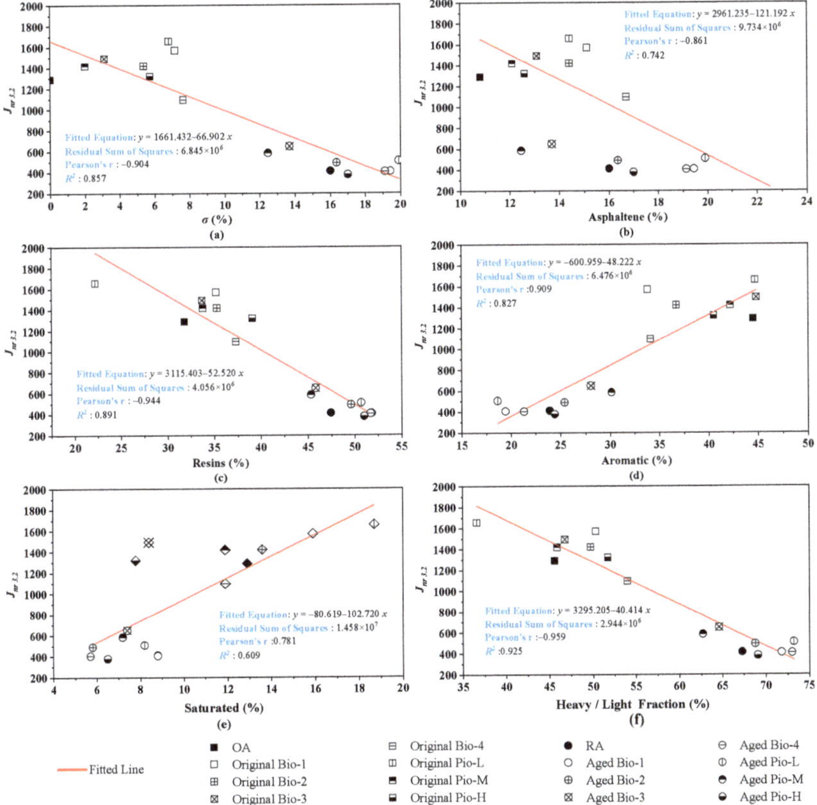

Figure 14. Correlation analysis between the $J_{nr3.2}$ and fraction indicators. (**a**) σ; (**b**) asphaltene; (**c**) resins; (**d**) aromatic; (**e**) saturated; (**f**) heavy/light.

3.5. Penetration, Ductility, and Softening Point

The test result of the three basic empirical properties: penetration, softening point, and ductility, are shown in Figure 15. Considering the unaged samples, their penetration and softening point were similar to those of the OA, and the differences were not more than 6%. However, regarding 15 °C ductility, none of the rejuvenated asphalts could be recovered to the level of the OA. In addition, there are significant differences in the ductility values of different rejuvenated asphalt. According to many asphalt material standards, ductility at 15 °C, greater than 100 cm, is a mandatory requirement for asphalt 50/70; however, only Pio-M, Pio-H, and Bio-3 meet this requirement. Based on the magnitude of the ductility, the recovery effects of the seven unaged samples can be ranked as Pio-M, Pio-H, Bio-3, Bio-2, Pio-L, Bio-1, and Pio-4 (from best to worst). It can be confirmed that oil regenerators easily completely recover the penetration and softening points; nevertheless, it is difficult for ductility.

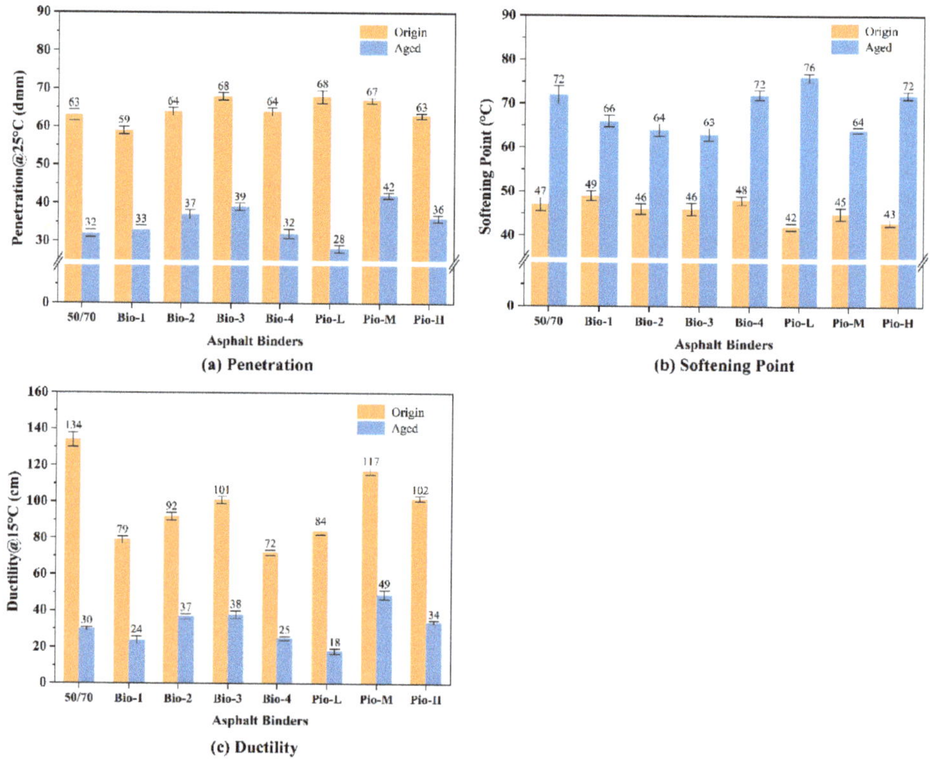

Figure 15. Basic empirical properties of all the asphalt samples: (**a**) penetration (**b**) softening point (**c**) ductility.

After secondary aging, the penetration and ductility of all the asphalt samples decreased, softening point increased, and there were significant differences among the samples. The ranking of the penetration from the largest to smallest is Pio-M, Bio-3, Bio-2, Pio-H, Bio-1, RA, Bio-4, and Pio-L, which reflects the degree of hardening of these binders. Considering the 15 °C ductility, the ranking from the best to worst is Pio-M, Bio-3, Bio-2, Pio-H, RA, Bio-4, Bio-1, and Pio-L, reflecting the deformability of these asphalts. These two rankings are almost identical, suggesting a correlation between the reduced deformability and hardening of these asphalt binders. This could be due to the changes in their internal fractions.

The correlations of the three basic empirical indicators with the six fraction indicators are analyzed in Figures 16–18. Considering these correlations, ductility and penetration generally have large R^2 with these fraction indicators, where the R^2 are greater than 0.9 of penetration with the σ and light/heavy fraction and an R^2 even greater than 0.95 in ductility of the σ and asphaltene. However, the correlations between the softening point and these fraction indicators are not as close as the two previous basic mechanical indicators, where the correlation coefficients only range from 0.6 to 0.8.

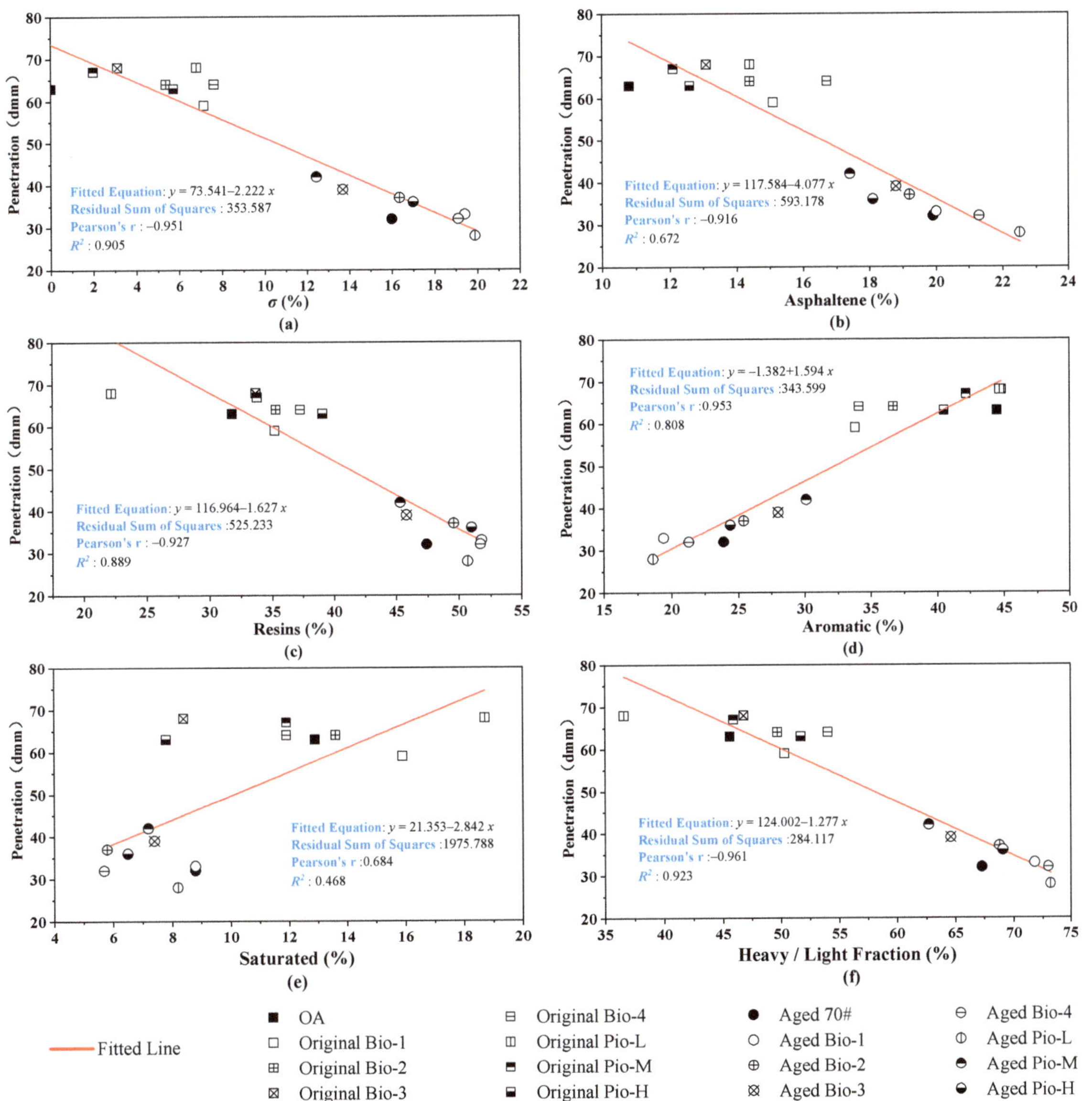

Figure 16. Correlation analysis between the penetration and fraction indicators. (**a**) σ; (**b**) asphaltene; (**c**)resins; (**d**) aromatic; (**e**) saturated; (**f**) heavy/light.

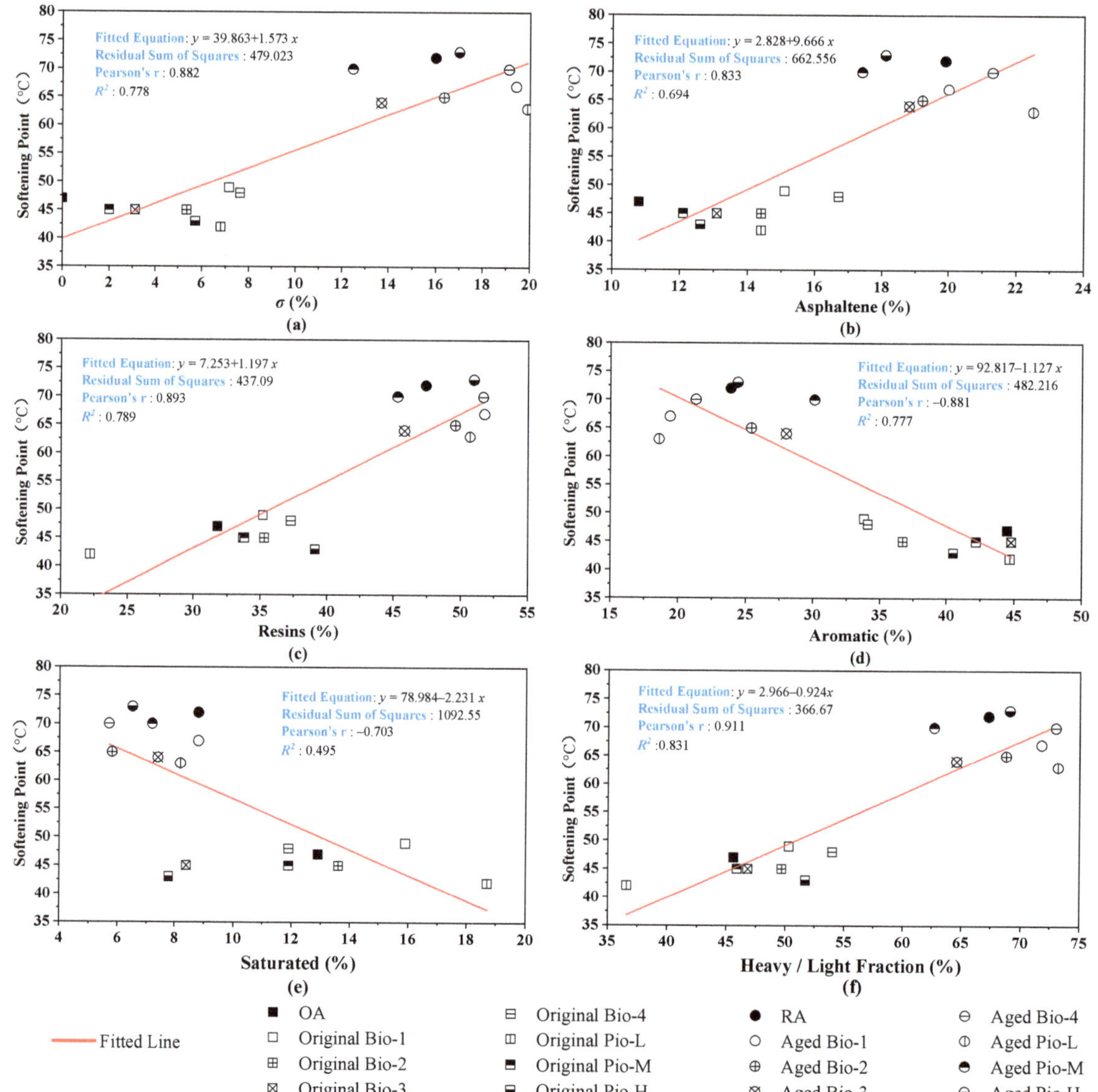

Figure 17. Correlation analysis between the softening point and fraction indicators. (**a**) σ; (**b**) asphaltene; (**c**)resins; (**d**) aromatic; (**e**) saturated; (**f**) heavy/light.

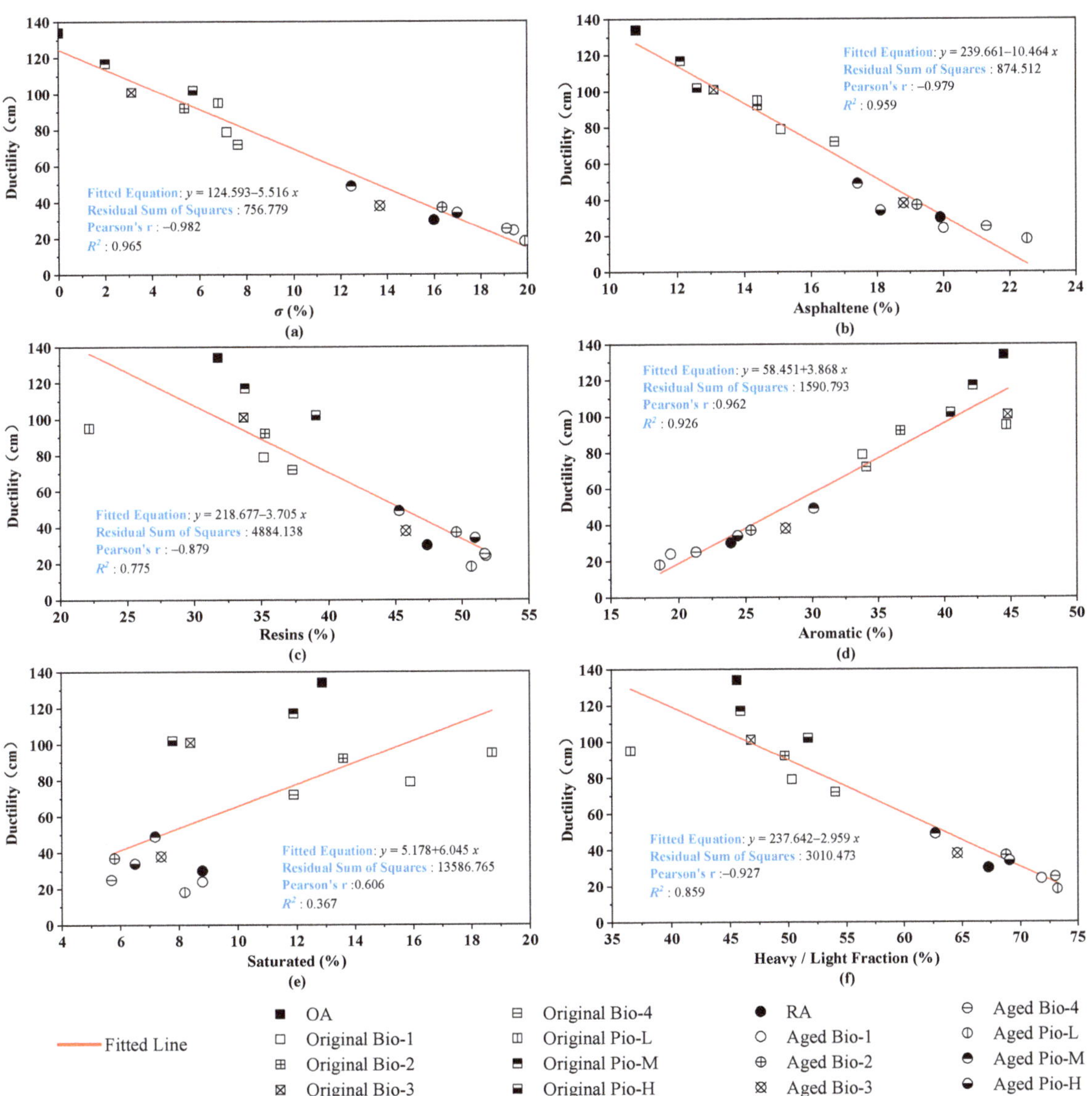

Figure 18. Correlation analysis between ductility and fraction indicators. (**a**) σ; (**b**) asphaltene; (**c**) resins; (**d**) aromatic; (**e**) saturated; (**f**) heavy/light.

4. Discussion

Table 4 summarizes all the rankings of the asphalt mechanical indicators for a systematic comparison, and some valuable conclusions can be drawn as follows.

Rankings of most asphalt samples before and after aging did not change significantly, generally within one to two places; however, the asphalt 50/70 was an exception. Before aging, it (OA) had the top performance of ductility, low-temperature crack, second rutting, and the third fatigue resistances; nevertheless, after aging (RA), they regressed to approximately ranking four or five. The OA has an absolute advantage, considering the low temperature cracking resistance and ductility; nonetheless, after aging, they could no longer be recovered regardless of the type of oil-rejuvenator used. After the second aging, if the benchmark of the aged asphalt 50/70 is still used for judgment, the comparison shows that only Pio-H and Bio-2 are better than the RA, considering the overall performance. Bio-1 and Bio-4 are excellent in rutting performance; nonetheless, they lag in cracking resistance, fatigue resistance, and ductility. By contrast, Pio-M and Bio-3 are superior to Bio-1 and Bio-4, considering the crack and fatigue resistances. They also performed poorly in the rutting resistance. Pio-L is almost always far behind the RA in all performance, considering only T_{CR}, N_f, $J_{nr3.2}$, and ductility because it is neither good nor bad for the penetration and softening point.

Table 4. Ranking of the mechanical indicators of all the asphalt samples.

Asphalt Samples	Raw Materials of Rejuvenator	Aging State	Ranking of Mechanical Indicators					
			Cracking Resistance (T_{CR})	Fatigue Resistance (N_f)	Rutting Resistance ($J_{nr3.2}$ & $R_{3.2}$)	Penetration	Ductility	Softening Point
50/70	-	Unaged	1	3	2	-	1	-
		Aged	5	5	4	3	5	3
Bio-1	Waste edible oil	Unaged	7	6	7	-	7	-
		Aged	7	7	3	4	7	5
Bio-2	Tung oil	Unaged	5	5	6		5	-
		Aged	3	4	5	6	3	6
Bio-3	Biodiesel	Unaged	3	4	5	-	4	-
		Aged	2	2	8	7	2	8
Bio-4	Fish oil residue	Unaged	8	8	1	-	8	-
		Aged	6	8	1	2	6	2
Pio-L	Light fraction oil	Unaged	6	7	8	-	6	-
		Aged	8	6	6	1	8	1
Pio-M	Middle fraction oil	Unaged	2	1	4		2	-
		Aged	1	1	7	8	1	7
Pio-H	Heavy fraction oil	Unaged	4	2	3	-	3	-
		Aged	4	3	2	5	4	4

Note: Only the aged asphalt samples have been ranked for penetration and softening point (according to their values from smallest to largest) because these two indicators have no significant difference in all the samples before aging. The indicators (T_{CR}, N_f, $J_{nr3.2}$, and ductility) are all ordered from best to worst, whereas the penetration and softening points follow the order of their values (from smallest to largest).

A detailed comparison revealed that the rejuvenated asphalt ranked almost the same for the low-temperature crack resistance, fatigue resistance, and ductility, regardless of the degree of aging. This is similar to the penetration, softening point, and high-temperature rutting resistance; however, it is opposite to the three previous indicators. This may be related to how these tests are performed, which can be broadly classified into two categories. The first category (T_{CR}, N_f, and ductility) examines the ability of the asphalt to adapt to deformation at medium or low temperatures [53,88,89], whereas the second category ($J_{nr3.2}$, $R_{3.2}$, penetration, and softening point) examines the ability of the asphalt to resist or recover deformation at high and medium temperatures [20]. Therefore, the ranking of the indicators obtained from the same category of the test may have some similarities.

Furthermore, there are large internal differences in the mechanical indicator among the three petroleum-based rejuvenated asphalts. Pio-M has an excellent deformation ability, Pio-L seems to harden severely and performs poorly in all aspects, whereas Pio-H has a more balanced ability (not particularly outstanding or particularly poor). All three of them were products of petroleum vacuum distillation; nonetheless, the different composition of the rejuvenator fractions leads to significant differences in the properties of the recovered binders [41]. Pio-L and Bio-1 are the two rejuvenators with the highest saturated fraction. They also brought in a large amount of saturated fraction into the aged asphalt, which seems to add a large amount of lighter fraction to the asphalt. However, many studies have reported that rejuvenators containing high amounts of saturates will cause compatibility issues with asphalt binders in long-term aging and may aggravate the imbalance of the asphalt colloidal structure [37,49]. More importantly, aging often occurs in functional groups with high oxygen content (e.g., carboxyl, aldehyde), which are enriched in saturated fractions [90]. Therefore, the mechanical properties of Pio-L and Bio-1 decay rapidly in secondary aging. However, since the aromatic fraction does not contain aging groups [91], Pio-M and Bio-3, which contain more than 50% aromatic fractions, performed well in long-term aging after the addition of aged asphalt.

Table 5 summarizes the correlation coefficients between the fraction composition and six indicators that can fully characterize the mechanical properties of the asphalt, from which a lot of valuable information can also be obtained.

Table 5. Correlation of the mechanical and fraction indicators.

Mechanical Indicators	Fractions Indicators					
	σ	Asphaltene Ratio	Resins Ratio	Aromatic Ratio	Saturated Ratio	Heavy/ Light Fraction
T_{CR}	0.982 ***	0.956 ***	0.714 **	0.926 ***	0.276	0.821 **
N_f	0.764 **	0.853 **	0.482 *	0.728 **	0.143	0.612 **
$J_{nr3.2}$	0.857 **	0.742 **	0.891 **	0.827 **	0.609 **	0.925 ***
Penetration	0.905 ***	0.773 **	0.889 **	0.871 **	0.468 *	0.923 ***
Softening Point	0.778 **	0.694 **	0.789 **	0.777 **	0.495 *	0.831 **
Ductility	0.965 ***	0.959 ***	0.775 **	0.926 ***	0.367	0.859 **

Note: $R^2 > 0.9$; strong correlation, marked ***. R^2 in 0.6~0.9; correlation, marked **. R^2 in 0.4~0.6; weak correlation, marked *. $R^2 < 0.4$; weak correlation, not marked.

The T_{CR} and ductility, which characterize the deformation performance of asphalts at low and medium temperatures, are strongly correlated with the asphaltene ratio, aromatic ratio, and σ ($R^2 > 0.900$). A comparative analysis of Figures 5, 8 and 15c shows that, considering any sample of the asphalt binder, the proportion of asphaltenes is usually low (unaged binder < 15%, aged binder < 22%) whereas aromatic is large (unaged binder in 30–45%, aged binder in 20–30%), both of which profoundly affect the ability of the asphalt to adapt to deformation [23]. Considering the asphalt samples with excellent T_{CR} and ductility, their ratio of aromatic to asphaltene is large, and because the ratio becomes smaller, these two mechanical indicators become progressively worse. Based on the explanation of the colloidal structure, because the saturated and aromatic contents in the dispersion medium decrease and the content of the protective substance resins and asphaltene in the dispersion phase increases, the asphalt changes from sol structure to sol-gel structure or even gel structure. Meanwhile, the asphalt becomes increasingly hard, and its deformation properties become worse [92,93]. The results of the correlation coefficients in Table 5 show that asphaltene and aromatic are the most important fractions of the dispersed phase and dispersion medium, respectively. Moreover, they play a decisive role in the deformation adaptability of asphalt. Based on the above reasons, considering the regeneration of the RA in this study, the T_{CR} and ductility of these regenerated asphalts are hardly comparable to the OA because none of the oil-rejuvenators can reduce asphaltene and raise the aromatic score to the level of OA. This also leads to asphalts that deviate more from the OA (i.e., the larger the σ possessed, the worse their T_{CR} and ductility), indicating the strong correlation between σ and these two mechanical indicators.

Moreover, $J_{nr3.2}$ and penetration are strongly correlated ($R^2 > 0.900$) with the heavy composition. Meanwhile, resins (the main component in the heavy fractions) also have a high correlation with these two mechanical indicators ($R^2 = 0.891$ for $J_{nr3.2}$ and $R^2 = 0.879$ for penetration). Its proportion in the unaged samples is approximately 30%, and this can increase to 50% after aging. In addition, considering the N_f and softening point, their correlations with these fraction indicators show a balanced relationship, with no significant strong correlation for one or two of them. This indicates the reconciliation effect between the four fractions that jointly affect these two mechanical properties.

In summary, although asphaltene is the smallest fraction of asphalt, it has a controlling effect on all mechanical performance of asphalt. Aromatic fraction content is closely related to the low temperature strain rate of asphalt. The larger its proportion, the stronger is asphalt's low-temperature cracking resistance and resistance to aging. Saturated fraction does not seem to be correlated with most mechanical performance, but its percentage should not be too high, otherwise, it will lead to accelerated aging of asphalt. In addition, the coordination between the four fractions determines the fatigue resistance and softening point of the asphalt. They do not seem to be influenced much by a single fraction as much as the other four mechanical indicators.

5. Conclusions

This study investigates the correlation between the distribution of asphalt fractions and mechanical indicators, especially the current commonly used rheological indicators, of 16 asphalt samples originating from the same asphalt; nonetheless, they have widely varying fraction compositions, and the following important conclusions are drawn:

- A new indicator, *average deviation of fractions between the given asphalt and its original* (marked σ) is proposed to characterize the degree of asphalt aging based on the variations of SARA fractional composition. It normalizes the four simultaneous change indicators (percentage of SARA fractions) during asphalt aging into one indicator. This new indicator has a strong correlation with several mechanical performance indicators of asphalt. Experiments and statistical analysis show that it is strongly correlated ($R^2 > 0.90$) with T_{CR}, ductility, and penetration, has a good correlation ($R^2 > 0.85$) with $J_{nr3.2}$, and is correlated with ($R^2 > 0.75$) fatigue life (N_f) and softening point. σ provides a potential approach to further investigate the inner relationship between SARA fraction composition and its mechanical properties of asphalts;
- The correlation between asphalt SARA fractions and their several mechanical indicators is clarified. Asphaltene fraction content is strongly correlated ($R^2 > 0.95$) with low temperature critical cracking temperature (T_{CR}) and ductility. Aromatic fraction content is also strongly correlated ($R^2 > 0.92$) with these two indicators, T_{CR} and ductility. Heavy fraction (asphaltene and resins) content is strongly correlated with ($R^2 > 0.90$) penetration and non-recoverable creep compliance ($J_{nr3.2}$ from MSCR test). Resins also have a good correlation ($R^2 > 0.88$) with penetration and non-recoverable creep compliance ($J_{nr3.2}$). Saturated fraction is not significantly correlated with these mechanical indicators (all $R^2 < 0.61$);
- Effects of some SARA fractions in asphalt regeneration are evaluated. Asphaltene has a controlling effect on all the mechanical performance of asphalt. The larger the aromatic proportion, the asphalt's low-temperature cracking resistance and resistance to aging ability are also stronger. The content of saturated fraction should not be too high, otherwise, it may accelerate aging of asphalt. In addition, all the oil-rejuvenators can barely restore the T_{CR} and ductility of the aged asphalt to their original levels because it is difficult to completely reduce the asphaltene content to the original level;
- More currently used mechanical properties indicators, especially those with complex rheology, can be examined for their correlation with fraction composition by the methods presented in this study. This type of work will provide a reference for the directional design of bitumen for applications in complex environments.

Author Contributions: Conceptualization, H.L. and R.T.; methodology, H.L.; validation, R.T., Y.Z. and L.Z.; formal analysis, R.T.; investigation, L.Z.; resources, R.T.; writing—original draft preparation, R.T.; writing—review and editing, H.L.; visualization, L.Z. and F.L.; supervision, X.H.; project administration, X.H.; funding acquisition, H.L., X.H. and R.T. All authors have read and agreed to the published version of the manuscript.

Funding: This research was funded by National Key R&D Project of China, grant number 2021YFB2600601 and 2021YFB2600600, Science Foundation of Tibet, China, grant number XZ201901-GB-14, and Postgraduate Research & Practice Innovation Of Jiangsu, China, grant number KYCX21_0139.

Institutional Review Board Statement: Not applicable.

Informed Consent Statement: Not applicable.

Data Availability Statement: Not applicable.

Acknowledgments: Constructive discussions Jinhui Huang in Sichuan University, China is gratefully acknowledged. All experiments were performed on the laboratory of Southeast University, China.

Conflicts of Interest: The authors declare no conflict of interest.

References

1. Xu, J.; Yang, E.; Luo, H.; Ding, H. Effects of warm mix additives on the thermal stress and ductile resistance of asphalt binders. *Constr. Build. Mater.* **2020**, *238*, 117746. [CrossRef]
2. Jamshidi, A.; Hamzah, M.O.; You, Z. Performance of Warm Mix Asphalt containing Sasobit®: State-of-the-art. *Constr. Build. Mater.* **2013**, *38*, 530–553. [CrossRef]
3. Angius, E.; Ding, H.; Hesp, S.A.M. Durability assessment of asphalt binder. *Constr. Build. Mater.* **2018**, *165*, 264–271. [CrossRef]
4. Wang, J.; Wang, T.; Hou, X.; Xiao, F. Modelling of rheological and chemical properties of asphalt binder considering SARA fraction. *Fuel* **2019**, *238*, 320–330. [CrossRef]
5. Martínez-Boza, F.; Partal, P.; Conde, B.; Gallegos, C. Influence of Temperature and Composition on the Linear Viscoelastic Properties of Synthetic Binders. *Energy Fuels* **2000**, *14*, 131–137. [CrossRef]
6. Lian, H.; Lin, J.-R.; Yen, T.F. Peptization studies of asphaltene and solubility parameter spectra. *Fuel* **1994**, *73*, 423–428. [CrossRef]
7. Ding, H.; Hesp, S.A.M. Quantification of crystalline wax in asphalt binders using variable-temperature Fourier-transform infrared spectroscopy. *Fuel* **2020**, *277*, 118220. [CrossRef]
8. Speight, J.G. Petroleum asphaltenes—Part 1: Asphaltenes, resins and the structure of petroleum. *Oil Gas Sci. Technol.* **2004**, *59*, 467–477. [CrossRef]
9. Speight, J.G. *The Chemistry and Technology of Petroleum*; CRC Press: Boca Raton, FL, USA, 2008. [CrossRef]
10. Yang, C.; Xie, J.; Wu, S.; Amirkhanian, S.; Zhou, X.; Ye, Q.; Yang, D.; Hu, R. Investigation of physicochemical and rheological properties of SARA components separated from bitumen. *Constr. Build. Mater.* **2020**, *235*, 117437. [CrossRef]
11. Branthaver, J.F.; Petersen, J.C.; Robertson, R.E.; Duvall, J.J.; Kim, S.S.; Harnsberger, P.M.; Mill, T.; Ensley, E.K.; Barbour, F.A.; Scharbron, J.F. Binder Characterization and Evaluation. Volume 2: Chemistry. *Chromatography* **1993**, *2*, 479. Available online: https://trid.trb.org/view/386723 (accessed on 12 January 2022).
12. Speight, J.G.; Koots, J.A. Relation of petroleum resins to asphaltenes. *Fuel* **1975**, *51*, 179–184. [CrossRef]
13. Di Primio, R.; Horsfield, B.; Guzman-Vega, M.A. Determining the temperature of petroleum formation from the kinetic properties of petroleum asphaltenes. *Nature* **2000**, *406*, 173–176. [CrossRef]
14. Pfeiffer, J.P.; Saal, R.N.J. Asphaltic Bitumen as Colloid System. *J. Phys. Chem.* **1940**, *44*, 139–149. [CrossRef]
15. Mangiafico, S.; Di Benedetto, H.; Sauzéat, C.; Olard, F.; Pouget, S.; Planque, L. Effect of colloidal structure of bituminous binder blends on linear viscoelastic behaviour of mixtures containing Reclaimed Asphalt Pavement. *Mater. Des.* **2016**, *111*, 126–139. [CrossRef]
16. Pahlavan, F.; Samieadel, A.; Deng, S.; Fini, E. Exploiting synergistic effects of intermolecular interactions to synthesize hybrid rejuvenators to revitalize aged asphalt. *ACS Sustain. Chem. Eng.* **2019**, *7*, 15514–15525. [CrossRef]
17. Aguiar, J.I.S.; Mansur, C.R.E. Study of the interaction between asphaltenes and resins by microcalorimetry and ultraviolet–visible spectroscopy. *Fuel* **2015**, *140*, 462–469. [CrossRef]
18. Janardhan, A.S.; Mansoori, G.A. Fractal nature of asphaltene aggregation. *Jounal Pet. Sci. Eng.* **1993**, *9*, 17–27. [CrossRef]
19. Luo, H.; Huang, X.; Tian, R.; Huang, J.; Zheng, B.; Wang, D.; Liu, B. Analysis of relationship between component changes and performance degradation of Waste-Oil-Rejuvenated asphalt. *Constr. Build. Mater.* **2021**, *297*, 123777. [CrossRef]
20. Zahoor, M.; Nizamuddin, S.; Madapusi, S.; Giustozzi, F. Sustainable asphalt rejuvenation using waste cooking oil: A comprehensive review. *J. Clean. Prod.* **2021**, *278*, 123304. [CrossRef]
21. Ding, H.; Tetteh, N.; Hesp, S.A.M. Preliminary experience with improved asphalt cement specifications in the City of Kingston, Ontario, Canada. *Constr. Build. Mater.* **2017**, *157*, 467–475. [CrossRef]

22. Al-Saffar, Z.H.; Yaacob, H.; Satar, M.K.I.M.; Kamarudin, S.N.N.; Mahmud, M.Z.H.; Ismail, C.R.; Hassan, S.A.; Mashros, N. A review on the usage of waste engine oil with aged asphalt as a rejuvenating agent. *Mater. Today Proc.* **2021**, *42*, 2374–2380. [CrossRef]
23. Qiu, Y.; Ding, H.; Rahman, A.; Wang, W. Damage characteristics of waste engine oil bottom rejuvenated asphalt binder in the non-linear range and its microstructure. *Constr. Build. Mater.* **2018**, *174*, 202–209. [CrossRef]
24. Paliukaite, M.; Assuras, M.; Hesp, S.A.M. Effect of recycled engine oil bottoms on the ductile failure properties of straight and polymer-modified asphalt cements. *Constr. Build. Mater.* **2016**, *126*, 190–196. [CrossRef]
25. Mamun, A.A.; Al-Abdul Wahhab, H.I. Comparative laboratory evaluation of waste cooking oil rejuvenated asphalt concrete mixtures for high contents of reclaimed asphalt pavement. *Int. J. Pavement Eng.* **2020**, *21*, 1297–1308. [CrossRef]
26. Cai, X.; Zhang, J.; Xu, G.; Gong, M.; Chen, X.; Yang, J. Internal aging indexes to characterize the aging behavior of two bio-rejuvenated asphalts. *J. Clean. Prod.* **2019**, *220*, 1231–1238. [CrossRef]
27. Ahmed, R.B.; Hossain, K. Waste cooking oil as an asphalt rejuvenator: A state-of-the-art review. *Constr. Build. Mater.* **2020**, *230*, 116985. [CrossRef]
28. Bai, T.; Hu, Z.A.; Hu, X.; Liu, Y.; Fuentes, L.; Walubita, L.F. Rejuvenation of short-term aged asphalt-binder using waste engine oil. *Can. J. Civ. Eng.* **2020**, *47*, 822–832. [CrossRef]
29. Lekhaz, D.; Saravanan, K.; Goutham, S. Effect of rejuvenating agents on stone matrix asphalt mixtures incorporating RAP. *Constr. Build Mater.* **2020**, *254*, 119298.
30. Dong, W.; Ma, F.; Li, C.; Fu, Z.; Huang, Y.; Liu, J. Evaluation of Anti-Aging Performance of Biochar Modified Asphalt Binder. *Coatings* **2020**, *10*, 1037. [CrossRef]
31. Elkashef, M.; Williams, R.C.; Cochran, E.W. Physical and chemical characterization of rejuvenated reclaimed asphalt pavement (RAP) binders using rheology testing and pyrolysis gas chromatography-mass spectrometry. *Mater. Struct.* **2018**, *51*, 1–9. [CrossRef]
32. Zhang, X.; Zhang, K.; Wu, C.; Liu, K.; Jiang, K. Preparation of bio-oil and its application in asphalt modification and rejuvenation: A review of the properties, practical application and life cycle assessment. *Constr. Build. Mater.* **2020**, *262*, 120528. [CrossRef]
33. Li, R.; Bahadori, A.; Xin, J.; Zhang, K.; Muhunthan, B.; Zhang, J. Characteristics of bioepoxy based on waste cooking oil and lignin and its effects on asphalt binder. *Constr. Build. Mater.* **2020**, *251*, 118926. [CrossRef]
34. Wang, H.; Ma, Z.; Chen, X.; Mohd Hasan, M.R. Preparation process of bio-oil and bio-asphalt, their performance, and the application of bio-asphalt: A comprehensive review. *J. Traffic Transp. Eng.* **2020**, *7*, 137–151. [CrossRef]
35. Li, H.; Feng, Z.; Ahmed, A.T.; Yombah, M.; Cui, C.; Zhao, G.; Guo, P.; Sheng, Y. Repurposing waste oils into cleaner aged asphalt pavement materials: A critical review. *J. Clean. Prod.* **2022**, *334*, 130230. [CrossRef]
36. Baghaee Moghaddam, T.; Baaj, H. The use of rejuvenating agents in production of recycled hot mix asphalt: A systematic review. *Constr. Build Mater.* **2016**, *114*, 805–816. [CrossRef]
37. Haghshenas, H.F.; Rea, R.; Reinke, G.; Yousefi, A.; Haghshenas, D.F.; Ayar, P. Effect of Recycling Agents on the Resistance of Asphalt Binders to Cracking and Moisture Damage. *J. Mater. Civ. Eng.* **2021**, *33*, 04021292. [CrossRef]
38. Ding, H.; Liu, H.; Qiu, Y.; Rahman, A. Effects of crystalline wax and asphaltene on thermoreversible aging of asphalt binder. *Int. J. Pavement Eng.* **2021**, *38*, 1–10. [CrossRef]
39. Shariati, S.; Rajib, A.I.; Fini, E.H. A multifunctional bio-agent for extraction of aged bitumen from siliceous surfaces. *J. Ind. Eng. Chem.* **2021**, *104*, 500–513. [CrossRef]
40. Ding, H.; Zhang, H.; Zheng, X.; Zhang, C. Characterisation of Crystalline Wax in Asphalt Binder by X-Ray Diffraction. *Road Mater. Pavement Des.* **2022**, *259*, 1–17. [CrossRef]
41. Luo, H.; Huang, X. Research on the Change of Performance and Component of Recycled Oil Regenerated Asphalt during Secondary Aging. *China J. Highw. Transp.* **2021**, *34*, 98. [CrossRef]
42. Corbett, L.W. Relationship between composition and physical properties of asphalt and discussion. In *Association of Asphalt Paving Technologists Proc*; Destech: St Paul, MN, USA, 1970; Volume 39, pp. 481–491.
43. Redelius, P.; Soenen, H. Relation between bitumen chemistry and performance. *Fuel* **2015**, *140*, 34–43. [CrossRef]
44. Calandra, P.; Caputo, P.; De Santo, M.P.; Todaro, L.; Turco Liveri, V.; Oliviero Rossi, C. Effect of additives on the structural organization of asphaltene aggregates in bitumen. *Constr. Build. Mater.* **2019**, *199*, 288–297. [CrossRef]
45. Thenoux, G.; Bell, C.A.; Wilson, J.E. Evaluation of physical and fractional properties of asphalt and their interrelationship. In *Asphalt Materials and Mixtures*; SAGE Publications: New York, NY, USA, 1988; pp. 82–97.
46. Loeber, L.; Muller, G.; Morel, J.; Sutton, O. Bitumen in colloid science: A chemical, structural and rheological approach. *Fuel* **1998**, *77*, 1443–1450. [CrossRef]
47. Ghasemirad, A.; Bala, N.; Hashemian, L. High-Temperature Performance Evaluation of Asphaltenes-Modified Asphalt Binders. *Molecules* **2020**, *25*, 3326. [CrossRef] [PubMed]
48. Qu, X.; Li, T.; Wei, J.; Wang, D.; Oeser, M.; Hao, P. Extraction of polycyclic aromatic compounds (PAC) and the influence on the mechanical and chemical properties of asphalt binder. *Constr. Build. Mater.* **2019**, *23*, 228. [CrossRef]
49. Lesueur, D. The colloidal structure of bitumen: Consequences on the rheology and on the mechanisms of bitumen modification. *Adv. Colloid Interface Sci.* **2009**, *145*, 42–82. [CrossRef]

50. Hofko, B.; Eberhardsteiner, L.; Füssl, J.; Grothe, H.; Handle, F.; Hospodka, M.; Grossegger, D.; Nahar, S.N.; Schmets, A.J.M.; Scarpas, A. Impact of maltene and asphaltene fraction on mechanical behavior and microstructure of bitumen. *Mater. Struct.* **2016**, *49*, 829–841. [CrossRef]

51. Fernandez-Gomez, W.D.; Quintana, H.; Rondon, A.; Daza, C.E.; Ca, L. The effects of environmental aging on Colombian asphalts. *Fuel* **2014**, *115*, 321–328. [CrossRef]

52. Cooper, S.B.; Negulescu, I.; Balamurugan, S.S.; Mohammad, L.; Daly, W.H.; Baumgardner, G.L. Asphalt mixtures containing RAS and/or RAP: Relationships amongst binder composition analysis and mixture intermediate temperature cracking performance. *Road Mater. Pavement Des.* **2017**, *18*, 209–234. [CrossRef]

53. Luo, H.; Leng, H.; Ding, H.; Xu, J.; Lin, H.; Ai, C.; Qiu, Y. Low-temperature cracking resistance, fatigue performance and emission reduction of a novel silica gel warm mix asphalt binder. *Constr. Build. Mater.* **2020**, *78*, 1–7. [CrossRef]

54. Qiu, Y.; Ding, H.; Rahman, A.; Luo, H. Application of dispersant to slow down physical hardening process in asphalt binder. *Mater. Struct.* **2019**, *52*, 9. [CrossRef]

55. Xiao, F.; Amirkhanian, S.N. Effects of liquid antistrip additives on rheology and moisture susceptibility of water bearing warm mixtures. *Constr. Build. Mater.* **2010**, *24*, 1649–1655. [CrossRef]

56. Shi, H.; Xu, T.; Zhou, P.; Jiang, R. Combustion properties of saturates, aromatics, resins, and asphaltenes in asphalt binder. *Constr. Build. Mater.* **2017**, *136*, 515–523. [CrossRef]

57. Sultana, S.; Bhasin, A. Effect of chemical composition on rheology and mechanical properties of asphalt binder. *Constr. Build. Mater.* **2014**, *72*, 293–300. [CrossRef]

58. Petersen, J.C. Chapter 14 Chemical Composition of Asphalt as Related to Asphalt Durability. *Dev. Pet. Sci.* **2000**, *40*, 363–399. [CrossRef]

59. Corbett, L.W. Composition of asphalt based on generic fractionation, using solvent deasphaltening, elution-adsorption chromatography, and densimetric characterization. *J Anal. Chem.* **1969**, *41*, 576–579. [CrossRef]

60. *ASTM-D2424*; Standard Test Method for Separation of Asphalt into Four Fractions. American Society for Testing Materials: West Conshohocken, PA, USA, 2009.

61. Handle, F.; Füssl, J.; Neudl, S.; Grossegger, D.; Eberhardsteiner, L.; Hofko, B.; Hospodka, M.; Blab, R.; Grothe, H. The bitumen microstructure: A fluorescent approach. *Mater. Struct.* **2016**, *49*, 167–180. [CrossRef]

62. *ASTM-D5*; Standard Test Method for Penetration of Bituminous Materials. American Society for Testing Materials: West Conshohocken, PA, USA, 2013.

63. *ASTM-D36*; Standard Test Method for Softening Point of Bitumen (Ring-and-Ball Apparatus). American Society for Testing Materials: West Conshohocken, PA, USA, 2014.

64. *ASTM-D113*; Standard Test Method for Ductility of Asphalt Materials. American Society for Testing Materials: West Conshohocken, PA, USA, 2017.

65. *ASTM-D4402*; Standard Test Method for Viscosity Determination of Asphalt at Elevated Temperatures Using a Rotational Viscometer. American Society for Testing Materials: West Conshohocken, PA, USA, 2015.

66. *ASTM-D6373*; Standard Specification for Performance-Graded Asphalt Binder. American Society for Testing Materials: West Conshohocken, PA, USA, 2021.

67. Yan, K.; Peng, Y.; You, L. Use of tung oil as a rejuvenating agent in aged asphalt: Laboratory evaluations. *Constr. Build. Mater.* **2020**, *239*, 117783. [CrossRef]

68. Chen, M.; Xiao, F.; Putman, B.; Leng, B.; Wu, S. High temperature properties of rejuvenating recovered binder with rejuvenator, waste cooking and cotton seed oils. *Constr. Build. Mater.* **2014**, *59*, 10–16. [CrossRef]

69. Ji-lu, Z. Bio-oil from fast pyrolysis of rice husk: Yields and related properties and improvement of the pyrolysis system. *J. Anal. Appl. Pyrolysis* **2007**, *80*, 30–35. [CrossRef]

70. *ISO-660*; Animal and Vegetable Fats and Oils—Determination of Acid Value and Acidity. International Organization for Standardization: Geneva, Switzerland, 2020.

71. *ISO-3961*; Animal and Vegetable Fats and Oils—Determination of Iodine Value. International Organization for Standardization: Geneva, Switzerland, 2018.

72. *ISO-3657*; Animal and Vegetable Fats and Oils—Determination of Saponification Value. International Organization for Standardization: Geneva, Switzerland, 2020.

73. *ASTM-D4052*; Standard Test Method for Density, Relative Density, and API Gravity of Liquids by Digital Density Meter. American Society for Testing Materials: West Conshohocken, PA, USA, 2019.

74. *ASTM-D7279*; Standard Test Method for Kinematic Viscosity of Transparent and Opaque Liquids by Automated Houillon Viscometer. American Society for Testing Materials: West Conshohocken, PA, USA, 2020.

75. *ASTM-D56*; Standard Test Method for Flash Point by Tag Closed Cup Tester. American Society for Testing Materials: West Conshohocken, PA, USA, 2021.

76. *AASHTO-M320*; Standard Specification for Performance-Graded Asphalt Binder. American Association of State Highway Transportation Officials: West Conshohocken, PA, USA, 2005.

77. Roy, S.D.; Hesp, S.A. Low-temperature binder specification development: Thermal stress restrained specimen testing of asphalt binders and mixtures. *Transp. Res. Rec.* **2001**, *1766*, 7–14. [CrossRef]

78. *AASHTO-R49*; Standard Practice for Determination of Low-Temperature Performance Grade. American Association of State Highway Transportation Officials: West Conshohocken, PA, USA, 2013.
79. Shenoy, A. Single-event cracking temperature of asphalt pavements directly from bending beam rheometer data. *J. Transp. Eng.* **2002**, *128*, 465–471. [CrossRef]
80. Bahia, H.U.; Hanson, D.; Zeng, M.; Zhai, H.; Khatri, M.; Anderson, R. *Characterization of Modified Asphalt Binders in Superpave Mix Design*; American Association of State Highway and Transportation Officials: Washington, DC, USA, 2001.
81. *AASHTO-TP101*; Standard Method of Test for Estimating Fatigue Resistance of Asphalt Binders using the Linear Amplitude Sweep. American Association of State Highway Transportation Officials: Washington, DC, USA, 2012.
82. *ASTM-D7405*; Standard Test Method for Multiple Stress Creep and Recovery (MSCR) of Asphalt Binder Using a Dynamic Shear Rheometer. American Society for Testing Materials: West Conshohocken, PA, USA, 2015.
83. Hossain, Z.; Ghosh, D.; Zaman, M.; Hobson, K. Use of the multiple stress creep recovery (MSCR) test method to characterize polymer-modified asphalt binders. *J. Test. Eval.* **2016**, *44*, 507–520. [CrossRef]
84. Chen, A.; Liu, G.; Zhao, Y.; Li, J.; Pan, Y.; Zhou, J. Research on the aging and rejuvenation mechanisms of asphalt using atomic force microscopy. *Constr. Build. Mater.* **2018**, *167*, 177–184. [CrossRef]
85. Kuang, D.; Liu, W.; Xiao, Y.; Wan, M.; Yang, L.; Chen, H. Study on the rejuvenating mechanism in aged asphalt binder with mono-component modified rejuvenators. *Constr. Build. Mater.* **2019**, *223*, 986–993. [CrossRef]
86. Chen, Z.; Zhang, H.; Zhu, C.; Zhao, B. Rheological examination of aging in bitumen with inorganic nanoparticles and organic expanded vermiculite. *Constr. Build. Mater.* **2015**, *101*, 884–891. [CrossRef]
87. Cong, P.; Guo, X.; Mei, L. Investigation on rejuvenation methods of aged SBS modified asphalt binder. *Fuel* **2020**, *279*, 118556. [CrossRef]
88. Luo, H.; Huang, X.; Rongyan, T.; Ding, H.; Huang, J.; Wang, D.; Liu, Y.; Hong, Z. Advanced method for measuring asphalt viscosity: Rotational plate viscosity method and its application to asphalt construction temperature prediction. *Constr. Build. Mater.* **2021**, *301*, 124129. [CrossRef]
89. Sharma, A.; Rongmei Naga, G.R.; Kumar, P.; Raha, S. Development of an empirical relationship between non-recoverable creep compliance & zero shear viscosity for wide-ranging stiffness of asphalt binders. *Constr. Build. Mater.* **2022**, *326*, 126764. [CrossRef]
90. Haghshenas, H.F.; Kim, Y.-R.; Kommidi, S.R.; Nguyen, D.; Haghshenas, D.F.; Morton, M.D. Evaluation of long-term effects of rejuvenation on reclaimed binder properties based on chemical-rheological tests and analyses. *Mater. Struct.* **2018**, *51*, 134. [CrossRef]
91. Singh, D.; Yenare, C.; Showkat, B. Rheological and Chemical Characteristics of Asphalt Binder Modified with Groundnut Shell Bio-oil. *Adv. Civ. Eng. Mater.* **2020**, *9*, 311–339. [CrossRef]
92. Read, J.; Whiteoak, D. *The Shell Bitumen Handbook*; Thomas Telford: London, UK, 2003.
93. Claudy, P.; Letoffe, J.; King, G.; Plancke, J. Characterization of asphalts cements by thermomicroscopy and differential scanning calorimetry: Correlation to classic physical properties. *Fuel Sci. Technol. Int.* **1992**, *10*, 735–765. [CrossRef]

Article

Research on the Combination of Firefly Intelligent Algorithm and Asphalt Material Modulus Back Calculation

Runmin Zhao [1], **Jinzhi Gong** [1,2], **Yangzezhi Zheng** [1] **and Xiaoming Huang** [1,3,*]

[1] School of Transportation, Southeast University, Nanjing 211189, China; 230228857@seu.edu.cn (R.Z.); gjz273664@126.com (J.G.); zyzz@seu.edu.cn (Y.Z.)
[2] Transportation Bureau of Jinhu, Huaian 211600, China
[3] National Demonstration Center for Experimental Education of Road and Traffic Engineering, Southeast University, Nanjing 211189, China
* Correspondence: huangxm@seu.edu.cn; Tel.: +86-13905174081

Abstract: The modulus of asphalt pavement material is a necessary parameter for the design, strength measuring and stability evaluation of asphalt pavement. To get more precise test data for asphalt pavement material modulus, a new modulus back calculation method is proposed in this article, named as the Firefly Asphalt Back Calculation Method (FABCM). This novel method uses the firefly optimization algorithm, which is a kind of particle swarm intelligence algorithm imitating the information transfer process among fireflies. To demonstrate the reliability and stability of FABCM, and to study the feasibility of multi-parameter modulus back calculation methods, this article used theoretical deflection curves calculated by BISAR3.0 and the actual measurement data of deflection curves and vertical pressures on the subgrade top surfaces on the full-scale test circular track in the Research Institute of Highway, Ministry of Transport (RIOHTrack) to conduct a modulus back calculation. The results show that FABCM only takes 0.5–1 s for each calculation, and the back calculation errors in the verification of FABCM are mostly smaller than 1%, which means that the firefly optimization algorithm was modified effectively in this article. Moreover, this article also indicates some key factors influencing the accuracy of modulus back calculation, and several reasonable suggestions to the application of modulus back calculation.

Keywords: back calculation of modulus; Falling Weight Deflectometer (FWD); firefly optimization algorithm; RIOHTrack

Citation: Zhao, R.; Gong, J.; Zheng, Y.; Huang, X. Research on the Combination of Firefly Intelligent Algorithm and Asphalt Material Modulus Back Calculation. *Materials* **2022**, *15*, 3361. https://doi.org/10.3390/ma15093361

Academic Editor: Francesco Canestrari

Received: 2 April 2022
Accepted: 5 May 2022
Published: 7 May 2022

Publisher's Note: MDPI stays neutral with regard to jurisdictional claims in published maps and institutional affiliations.

1. Introduction

The modulus of asphalt pavement material is an important parameter in asphalt pavement design, construction and maintenance, and both indoor and outdoor test methods, which can get the precise modulus of both bitumen and asphalt mixtures, are important in asphalt concrete research. For indoor tests, there are tensile creep tests [1], a dynamic mechanical analyzer (DMA) [2] and the M2F trapezoid beam loading facility [3] to determine the dynamic modulus of asphalt mixture, dynamic sheer rheometer (DSR) tests to determine the dynamic shear modulus of bitumen [4–6], and a simple performance tester (SPT) [7] to determine the dynamic shear modulus of asphalt mixture, etc. As for the outdoor methods, the back calculation of modulus based on the deflection curve obtained by a Falling Weight Deflectometer (FWD) is also widely used.

In 1968, Scrivner et al. [8] developed the first modulus back calculation method based on FWD deflection curves, and then designed and made the first Nomogram Abac. The modulus back calculation methods in the early stage are mostly based on regression analysis methods. For example, Sun et al. [9] determined the regression equations of subgrade modulus and surface layer modulus with 792 deflection curves of two-layer asphalt pavement structures. Yang et al. [10,11] developed a logarithmic regression equation with 1000 samples, and then conducted modulus back calculation on both the subgrade and surface layer. Although the modulus back calculation methods based on regression equations were developed earliest, it was proved that this kind of method could hardly give satisfying results in practical applications [12].

To pursue the smallest error between the theoretical deflection curve and measured deflection curve, nonlinear programming methods were used on modulus back calculations [13]. Typical modulus back calculation methods, based on nonlinear programming methods, are the "DEF" series (BISDEF and CHEVDEF) of programs by the U.S. Army Corps of Engineers, MODCOMP (Moduli Computation) by Irwin, WESDEF by Water-ways Experiment Station et al. [14]. To guarantee the convergence of nonlinear programming methods, it is usually necessary to give a strict modulus value range, which could restrict the application of these algorithms [15].

With the development of computer and data storage technology, the back calculation methods based on database search methods were also proposed, like the MODULUS developed by Uzah et al. [16,17]. In this kind of method, a direct search and interpolation algorithm are used to determine the modulus combination, which meets the deflection curve accuracy requirements [11]. In the research of Wang et al. [18], big errors appeared when the back calculated modulus combination was not included in the database, which meant the universality of back calculation methods based on the database search method was not satisfactory.

Furthermore, Cheng et al. [19] proved that the modulus back calculation methods based on the artificial neural network (ANN) are more stable and accurate compared with other methods. Due to the strong ability to conduct nonlinear dynamic processing, and the ability to adaptively build mapping relations between the input and output parameters, ANN has a strong fault tolerance for incomplete samples or uncertain samples containing fuzzy and random data, and is also suitable for big data processing [20]. Zaman et al. [21], Huang et al. [22], Wang et al. [23], Yang et al. [24–27], and Khazanovich et al. [28–30] proved that the rebound modulus of asphalt pavement materials could be calculated precisely with ANN. Moreover, another kind of intelligence-based algorithm used in the modulus back calculation is the swarm intelligence algorithm, which consists of population initializations, individual updates and group updates [31]. Fwa et al. [32], Himeno et al. [33], and Zha et al. [34] tried to apply the genetic algorithm in the modulus back calculation, which shows that the back calculation speed is usually slow due to the need for plenty of genetic calculations to obtain stable results; otherwise it is hard to reach a balance between calculation speed and the avoidance of being premature.

There are three factors which affect the accuracy of pavement modulus back calculation: the differences between a theoretical model and the actual pavement structure; the inherent defects of mathematical optimization algorithm; and the errors of input parameters [35,36]. To reduce the inherent defects of mathematical optimization algorithm, and to reduce the influence of single parameter errors on back calculation results, the Rosenbrock search method and the Gaussian perturbation strategy were introduced in this paper to modify the firefly optimization algorithm. Results of back calculation conducted both on theoretical deflection curves and measured deflection curves on RIOHTrack were analyzed by this new firefly asphalt back calculation method to verify whether the new algorithm has better practicality.

2. Materials and Methods

2.1. Material and Structure Parameters

2.1.1. Materials and Structures Used in the Theoretical Deflection Curve Calculation

To evaluate the influences of load levels and distribution of deflection curve points on back calculation results, 6 typical asphalt pavement structures including 2 flexible pavement structures, 2 semi-rigid pavement structures and 2 composite pavement structures were selected in this study. All the 6 asphalt pavement structures are simplified into the structures of 3 or 4 layers, which could be seen in Table 1:

Table 1. Material and structure parameters.

Structure Number	Material	Thickness (cm)	Modulus (MPa)	Poisson's Ratio
1	Asphalt concrete	18	2000	0.25
	Cement treated macadam	40	1700	0.25
	Soil base	/	40	0.4
2	Asphalt concrete	36	2000	0.25
	Graded macadam	40	300	0.35
	Soil base	/	40	0.4
3	Asphalt concrete	36	2000	0.25
	Graded macadam	40	300	0.35
	Soil base	/	60	0.4
4	Asphalt concrete	48	2000	0.25
	Cement treated macadam	40	1700	0.25
	Soil base	/	40	0.4
5	Asphalt concrete	24	2000	0.25
	Graded macadam	30	300	0.35
	Cement treated soil	30	600	0.25
	Soil base	/	40	0.4
6	Asphalt concrete	18	2000	0.25
	Cement treated macadam	20	1700	0.25
	Cement treated soil	20	600	0.25
	Soil base	/	40	0.4

The parameters were obtained from project design documentations.

2.1.2. Materials and Structures Used in the Actual Deflection Curve Measurement

As the first full-scale pavement testing loop in China, RIOHTrack was built in Tongzhou, Beijing. There are 19 kinds of asphalt pavement structures and 13 kinds of cement concrete pavement structures in RIOHTrack, as is shown in Figure 1:

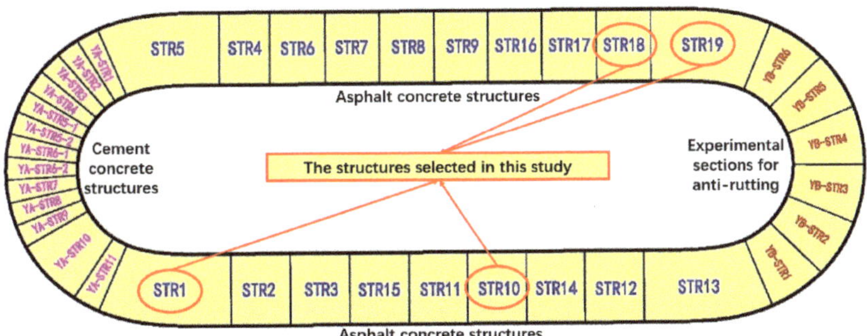

Figure 1. The structures in RIOHTrack.

There are a total of 4 kinds of pavement structures in RIOHTrack, which are namely STR1, STR10, STR18 and STR19, and were selected by investigating the most typical asphalt pavement structures around the world. All 4 structures were selected to conduct modulus back calculation by FABCM. STR1 consists of a thin asphalt concrete surface layer, a cement treated macadam base layer and cement treated soil subbase layer; STR10 consists of an asphalt concrete surface layer, a graded macadam base layer and a cement treated macadam subbase layer; STR18 consists of an asphalt concrete surface layer and a graded macadam base layer; STR19 consists of a thick asphalt concrete surface layer and a thin cement treated macadam base layer. To simplify the structures, the principle of equivalent modulus [37] (as is shown in Equations (1) and (2)) was used before the back calculation procedure:

$$h_i^* = h_{i1} + h_{i2} \tag{1}$$

$$E_i^* = \frac{E_{i1}h_{i1}^3 + E_{i2}h_{i2}^3}{(h_{i1} + h_{i2})^3} + \frac{3}{h_{i1} + h_{i2}}\left(\frac{1}{E_{i1}h_{i1}} + \frac{1}{E_{i2}h_{i2}}\right)^{-1} \tag{2}$$

where h_i^* refers to the equivalent thickness (mm); E_i^* refers to the equivalent modulus (MPa). The simplified structures of STR1, STR10, STR18 and STR19 are shown in Table 2:

Table 2. The simplified 4 structures on RIOHTrack.

Layer	STR1			STR10		
	Modulus (MPa)	Poisson's Ratio	Thickness (m)	Modulus (MPa)	Poisson's Ratio	Thickness (m)
1	9620	0.25	0.52	6300	0.25	0.28
2	2000	0.25	0.4	900	0.3	0.2
3	120	0.35		3250	0.25	0.4
4				120	0.35	

Layer	STR18			STR19		
	Modulus (MPa)	Poisson's Ratio	Thickness (m)	Modulus (MPa)	Poisson's Ratio	Thickness (m)
1	8830	0.25	0.48	13,200	0.25	0.48
2	900	0.3	0.48	6000	0.25	0.2
3	120	0.35		120	0.35	

The parameters were obtained from the documentations of RIOHTrack, and the elastic moduli of asphalt layers are under 20 °C, as is specified in Specifications for Design of Highway Asphalt Pavement [36].

2.2. The Firefly Asphalt Back Calculation Method (FABCM)

2.2.1. The Modified Firefly Optimization Algorithm

The firefly optimization algorithm is simulated according to the information transfer process among fireflies in nature [38–41]. The key factor of the firefly optimization algorithm is that fireflies with a low brightness are always attracted by fireflies with a high brightness and update their positions according to the position update equation. Furthermore, the firefly with the highest brightness will update its position immediately. After that, all fireflies will update their brightness and move continually according to the attraction rules.

The three basic hypotheses of the firefly optimization algorithm are as follows:

1. It is assumed that all fireflies are equally attracted to each other, and the less luminous fireflies are attracted to and move towards the brighter ones;
2. The attraction between two fireflies is only related to the distance between them and their luminous intensity. The attraction decreases with the increase in distance. Low intensity means less attraction to other fireflies;
3. The luminous intensity is determined by the objective equation.

The distance r_{ij} between firefly i and firefly j could be determined as:

$$r_{ij} = \|X_i - X_j\| = \sqrt{\sum_{k=1}^{d}(x_{i,k} - x_{j,k})^2} \tag{3}$$

where d refers to the dimension of the decision variable.

The attraction β of firefly could be expressed as follows:

$$\beta = \beta_0 e^{-\gamma r_{ij}^2} \tag{4}$$

where $\beta_0 \in [0,1]$, refers to the attraction when $r_{ij} = 0$; and $\gamma \in [0,10]$, refers to the fluorescence absorption coefficient.

The moving and updating of firefly positions could be described as: fireflies are attracted to brighter fireflies and shift their positions:

$$x_i^{t+1} = x_i^t + \beta_0 \cdot e^{-\gamma \cdot r_{ij}^2}(x_j^t - x_i^t) + \alpha \cdot (rand - 0.5) \tag{5}$$

where x_i^t refers to the position of firefly i at the time t; and $\alpha \in [0,1]$ refers to the random step, $rand \sim U(0,1)$ is a random number.

As for the brightest firefly, it will fly randomly:

$$x_{best}^{t+1} = x_{best}^t + \alpha \cdot (rand - 0.5) \tag{6}$$

where x_{best}^t is the optimal individual position in the firefly population of generation t.

Firefly optimization algorithm has plenty of advantages. For instance, due to the individual searching and the sensing ability of each firefly, it usually does not lead to local convergence, and it can be carried out concurrently. However, the traditional firefly optimization algorithm also faces some problems. As a result of the limited search range of each firefly, the firefly optimization algorithm could fall into local optimum or premature convergence. Finally, non-convergence might appear when all fireflies share the same brightness.

A modified firefly optimization algorithm was developed to improve the search ability of the firefly optimization algorithm in this study. In the modified algorithm, the reverse learning strategy was introduced to initialize firefly individuals, the Rosenbrock search method was introduced to improve the search abilities of firefly individuals, and the Gaussian perturbation strategy, which is a kind of random disturbance with Gaussian distribution, was added in the process of particle swarm optimization to prevent the rapid convergence of the intelligent particle algorithm in mathematical operation, thus to prevent incorrect results and to enhance the group evolution ability. The modified search procedure could be described as follows:

Step 1:

Initial population $x^{(1)} \in R^n$, which contains j decision individuals (number of firefly individuals); n refers to the dimension of the population variable. The unit orthogonal directions are $d^{(1)}, d^{(2)}, \ldots \ldots, d^{(n)}$; steps are $\delta_1^{(0)}, \delta_2^{(0)}, \ldots \ldots, \delta_n^{(0)}$; α refers to expansion factor where $\alpha > 1$, β refers to reduction factor where $\beta \in (-1,0)$; ε refers to the allowable error where $\varepsilon > 0$.

For $i = 1 : n$

$y^{(1)} = x^{(1)}$, $k = 1$, $j = 1$, $\delta_i = \delta_i^{(0)}$

End for

Step 2:

If $f(y^{(j)} + \delta_j d^{(j)}) < f(y^{(j)})$

$y^{(j+1)} = y^{(j)}$, $\delta_j = \alpha \cdot \delta_j$

Else

$y^{(j+1)} = y^{(j)}$, $\delta_j = \beta \cdot \delta_j$

End if

Step 3:

If $j < n$, then $j = j + 1$, GO TO step 2; else GO TO step 4

Step 4:

If $f(y^{(n+1)}) < f(y^{(1)})$, then $f(y^{(n+1)}) = f(y^{(1)})$, GO TO step 2; If $f(y^{(n+1)}) = f(y^{(1)})$, GO TO step 5.

Step 5:

If $f(y^{(n+1)}) < f(y^{(k)})$, then GO TO step 6; else if $|\delta_j| \leq \varepsilon$ for all j, then stop searching, regard $x^{(k)}$ as the approximate optimal solution; otherwise $y^{(1)} = y^{(n+1)}$, $j = 1$, GO TO step 2.

Step 6:

Make $x^{(k+1)} = y^{(n+1)}$, if $||x^{(k+1)} - x^{(k)}|| < \varepsilon$, stop searching, regard $x^{(k+1)}$ as the approximate optimal solution; else GO TO step 7.

Step 7:

Make $x^{(k+1)} = x^{(k)} + \sum_{i=1}^n \lambda_i d^{(i)}$, where $\lambda_i = \sum_1^j d^{(i)}$. Define a set of direction vectors: $p^{(1)}, p^{(2)}, \ldots \ldots, p^{(n)}$:

$$P^{(j)} = \begin{cases} d^{(j)}, & \lambda_j = 0 \\ \sum\limits_{i=j}^{n} \lambda_i d^{(i)} & \lambda_j \neq 0 \end{cases}$$

Normalize $\left\{ P^{(j)} \right\}$ with Gram–Schmidt orthogonalization equation, make:

$$q^{(j)} = \begin{cases} P^{(j)}, & j = 1 \\ P^{(j)} - \sum\limits_{i=1}^{j-1} \dfrac{q^{(i)T} P^{(j)}}{q^{(i)T} P^{(i)}} \cdot q^{(i)}, & j \geq 2 \end{cases}$$

Conduct vector unitization and make $\overline{d^{(j)}} = \dfrac{q^{(j)}}{||q^{(j)}||}$, then n new orthogonal search directions will be determined.

Step 8: Make $d^{(j)} = \overline{d^{(j)}}$, $\delta_j = \delta_j^{(0)}$, $j = 1, 2, \ldots\ldots, n$, $y^{(1)} = x^{(k+1)}$, $k = k+1$, $j = 1$, GO TO step 2.

Based on the Rosenbrock search method, the Gaussian perturbation strategy was also introduced to reduce the premature phenomenon and keep the population diverse. The Gaussian perturbation strategy was only used on the optimal individual x_{best}^t in the current population (the firefly population of the generation t) to increase the convergence rate of the modified firefly optimization algorithm in this study:

$$G_{best}^t = x_{best}^t \cdot (1 + Gaussian(\sigma)) \tag{7}$$

where: G_{best}^t refers to the optimal individual in the firefly population of the generation t after the implementation of the Gaussian perturbation strategy, $Gaussian(\sigma)$ is a random variable in Gaussian distribution.

The global optimal location is updated as:

$$x_{best}^{t+1} = \begin{cases} G_{best}^t & g(G_{best}^t) < g(x_{best}^t) \\ x_{best}^t \end{cases} \tag{8}$$

where: $g(\ )$ refers to the fitness value of individuals.

It could be known from the Equation (7) that if the current optimal solution is the local optimal solution, the Gaussian perturbation used on the current optimal individual x_{best}^t could effectively stop the algorithm from falling into the local optimum, improve the search efficiency of the Rosenbrock search method, and thus improve the traditional firefly optimization algorithm.

2.2.2. The Firefly Asphalt Back Calculation Method (FABCM)

Combined with the elastic layer system method (the program APBI) of road structure by FORTRAN, the modified firefly optimization algorithm introduced above is used to develop the asphalt pavement modulus back calculation program FABCM. The FABCM process is as shown in Figure 2:

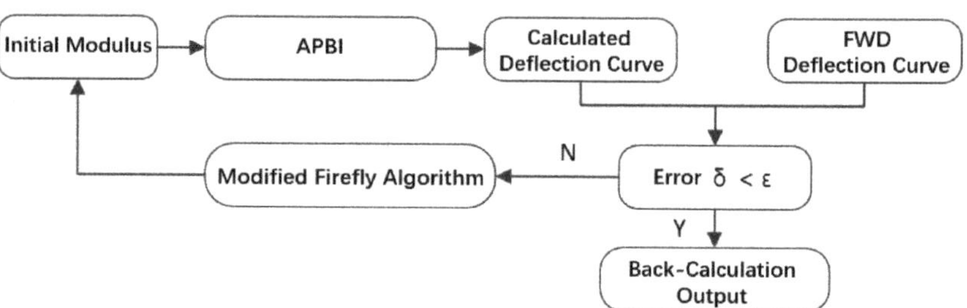

Figure 2. The Firefly Asphalt Back Calculation Method (FABCM) process.

2.2.3. The Multi-Parameter FABCM

Popular asphalt pavement modulus back calculation methods used now are based only on the FWD deflection curve. Considering the unavoidable system errors in FWD deflection curve measurement, which could lead to inaccuracy, based on the FABCM introduced above, a multi-parameter firefly back calculation method with the vertical strain at the bottom of one or several of the structure layers, or with subgrade compressive stress as input parameters, was designed in this study. The process is as shown in Figure 3:

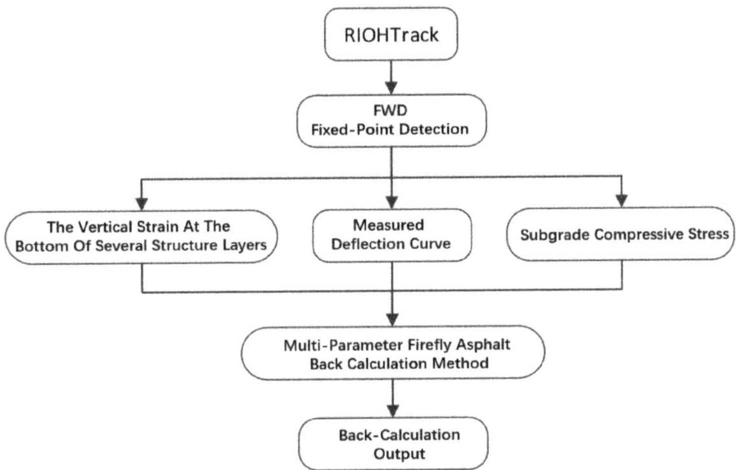

Figure 3. The multi-parameter FABCM process.

3. Results

3.1. The Verification of FABCM

Asphalt pavement structure systems of three and four layers were selected to verify the stability and accuracy of FABCM, and the theoretical deflection curves of them were calculated with BISAR 3.0. The structure of the three layers consists of asphalt concrete surface, cement treated macadam base and soil subgrade, and the structure of the four layers shares the same materials, but with a cement treated soil subbase between the base and subgrade. The thicknesses and moduli of those structure layers are as shown in Table 3. After that, those calculated deflection curves were brought into FABCM as input parameters to conduct modulus back calculation. The applied load is a circular uniform load with a concentration of 0.707 MPa (the load concentration of standard axle load BZZ-100 in China) and a radius of 15 cm (usually used bearing plate size for FWD [42]). The distances between the deflection measuring points and load geometric center are: P_0: 0 cm, P_1: 30 cm, P_2: 60 cm, P_3: 90 cm, P_4: 120 cm, P_5: 150 cm, P_6: 180 cm, P_7: 210 cm, P_8: 240 cm (as is shown in Figure 4). Poisson's ratios are determined by 0.25 for surface material, 0.25 for base material, 0.35 for subbase material and 0.4 for subgrade material. The calculated theoretical deflection curves are as shown in Table 4.

Table 3. The theoretical deflection curves calculated in BISAR3.0.

	Thickness (cm)	Modulus (MPa)	Theoretical Deflection Value (0.01mm)								
			0	30	60	90	120	150	180	210	240
			P_0	P_1	P_2	P_3	P_4	P_5	P_6	P_7	P_8
1	18/40	1800/1500/40	46.3	36.8	32.6	29.1	25.8	22.8	20.2	17.9	15.9
2	18/40/20	1800/1500/600/40	40.7	31.5	28.2	25.6	23.2	20.9	18.9	17.0	15.4

Table 4. The modulus back calculation accuracy.

	Back Calculated Modulus (MPa)	η (%)	δ (%)								
			δ_0	δ_1	δ_2	δ_3	δ_4	δ_5	δ_6	δ_7	δ_8
1	1802.0/1485.4/40.0	0.11/0.97/0.14	0.06	0.08	0.04	0.02	0.05	0.07	0.08	0.06	0.04
2	1809.7/1488.5/608.8/39.6	0.54/0.77/1.44/0.98	0.07	0.12	0.01	0.07	0.09	0.13	0.14	0.09	0.07

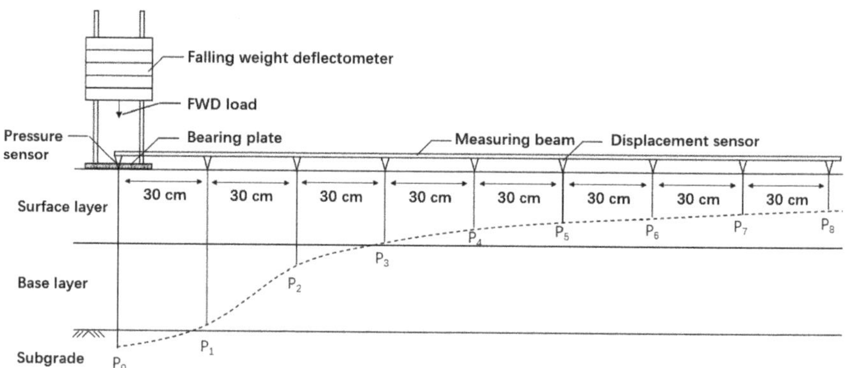

Figure 4. The distribution of deflection curve measuring points.

The accuracy of back calculation is calculated by the Equations (9) and (10):

$$\eta = \left| \frac{E - E_0}{E} \right| \times 100\% \tag{9}$$

$$\delta = \left| \frac{W_i - D_i}{D_i} \right| \times 100\% \tag{10}$$

where E_0 refers to the theoretical modulus; E refers to the back calculated modulus; D_i refers to the theoretical deflection curve; and W_i refers to the back calculated reflection basin. The back calculation accuracy is as shown in Table 4:

The errors of back calculated modulus are all smaller than 1%, except for the subbase of structure 2, and that of the back calculated deflection curves, which are all smaller than 0.2%. Furthermore, FABCM will take only 0.5–1 s for each calculation. It could be proved that both the accuracy and calculation speed of FABCM are satisfactory.

3.2. Modulus Back Calculation Based on Theoretical Deflection Curves Calculated with BISAR3.0

To evaluate the influence of different axle loads and deflection measuring types on the modulus back calculation accuracy, four kinds of circular uniform loads and eight kinds of deflection types are used in the theoretical deflection curve calculation. Compared with the relatively fixed size of bearing plates in FWD tests (the circular plates with the radius of 15 cm are usually used [42]), the change of impact loads is more available, to give reasonable guidance for filed FWD tests, the change of load concentration rather than that of load size was used to simulate different axle loads. The four circular uniform loads are: ① Concentration of 0.707 MPa and radius of 15 cm (equivalent to the axle load of 5 t); ② Concentration of 0.990 MPa and radius of 15 cm (equivalent to the axle load of 7 t); ③ Concentration of 1.273 MPa and radius of 15 cm (equivalent to the axle load of 9 t); ④ Concentration of 1.556 MPa and radius of 15 cm (equivalent to the axle load of 11 t). The theoretical deflection curves of 6 asphalt pavement structures under four circular uniform loads were calculated with BISAR3.0. With eight different types of theoretical deflection curves (as shown in Figure 4 and Table 5), the modulus back calculation was conducted in FABCM.

Table 5. The measuring points configurations of eight types of deflection curves.

Deflection Curve Type	Deflection Measuring Points
9	$P_0, P_1, P_2, P_3, P_4, P_5, P_6, P_7, P_8$
7−1	$P_0, P_1, P_2, P_3, P_4, P_5, P_6$
7−2	$P_0, P_1, P_2, P_3, P_4, P_6, P_8$
7−3	$P_0, P_1, P_2, P_3, P_4, P_7, P_8$
7−4	$P_0, P_1, P_2, P_4, P_6, P_7, P_8$
5−1	P_0, P_1, P_2, P_3, P_4
5−2	P_0, P_2, P_4, P_6, P_8
5−3	P_0, P_1, P_2, P_4, P_6

3.2.1. Back Calculation Errors of Surface Layer Modulus

The errors of back calculated surface modulus under different loads and pavement structures are shown in Figure 5.

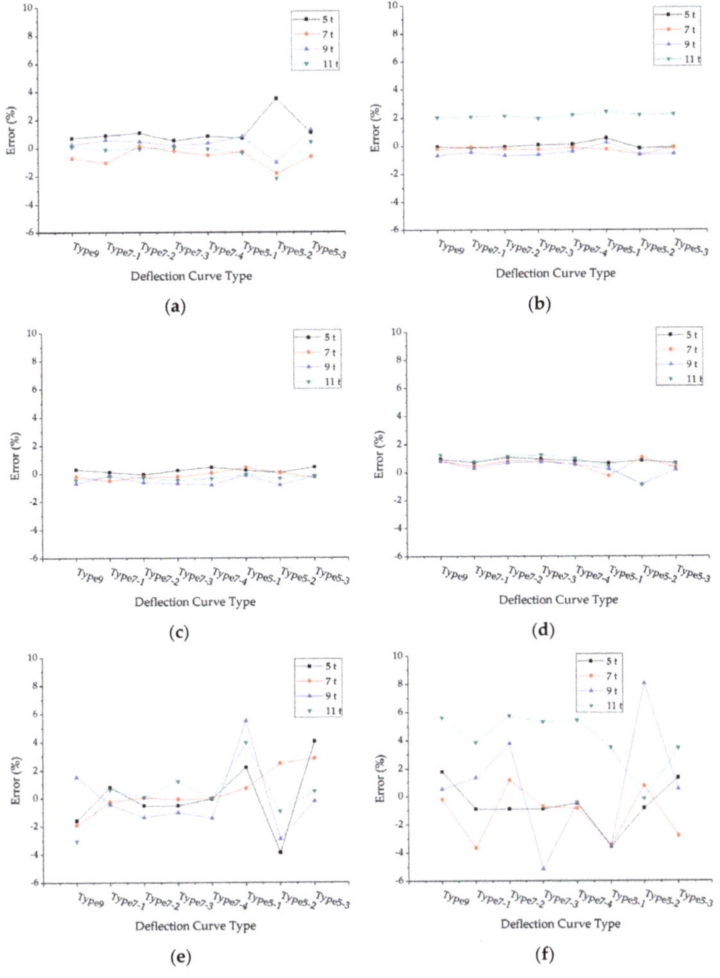

Figure 5. The errors of back calculated surface modulus under different deflection curve types: (**a**) Structure 1; (**b**) Structure 2; (**c**) Structure 3; (**d**) Structure 4; (**e**) Structure 5; (**f**) Structure 6.

The accuracies of this modulus back calculation method in each surface layer test are satisfying, and most of the errors are smaller than 2%. As for structure 5 and structure 6, the errors are relatively larger, but the maximum error is also smaller than 8%. Thus, the modulus back calculation results of asphalt pavement surface layers are reliable.

The average absolute values of surface modulus errors under different deflection curve types and axle loads were calculated and drawn in Figure 6. The average back calculation errors of the surface layer modulus fluctuate in a small amplitude under different forms of deflection curves, which means the accuracies of this new back calculation method are almost not affected by the forms of deflection curves. However, the errors decrease first and then rise with the increase in axle load.

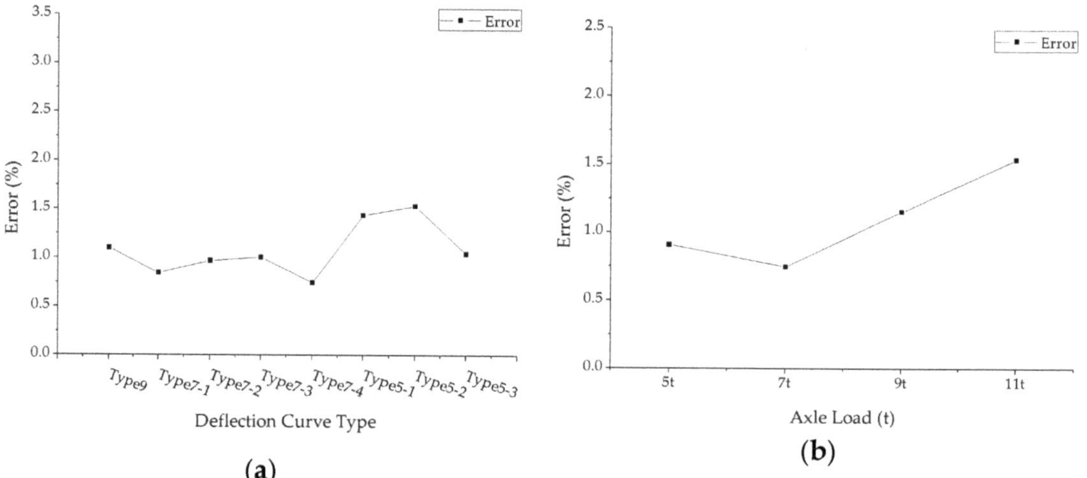

Figure 6. The average errors of back calculated surface modulus: (**a**) The average errors under different deflection curve types; (**b**) The average errors under different axle loads.

3.2.2. Back Calculation Errors of Base Modulus

The errors of back calculated base modulus under different loads and pavement structures are shown in Figure 7. The average absolute values of base modulus errors of each structure under different deflection curve types and axle loads were calculated and drawn in Figure 8.

It could be found from Figure 7 that the back calculation errors under each structure are basically within 2%, but the errors of structure 5 and structure 6 are larger than others. The error of structure 6 under the deflection curve type 9 and the axle load of 11 t even exceeds 15%. Compared with the surface layer, the modulus back calculation errors of the base layer are larger in general.

Figure 8a is the relation between the errors of back calculated base modulus and deflection curve types and this is still unclear. Whereas, it is clear that the errors are influenced by axle loads, as when the axle load increased into 11 t the error is as double what it would be in 5 t or 7 t (shown as Figure 8b) and this trend is similar with the one in Figure 6.

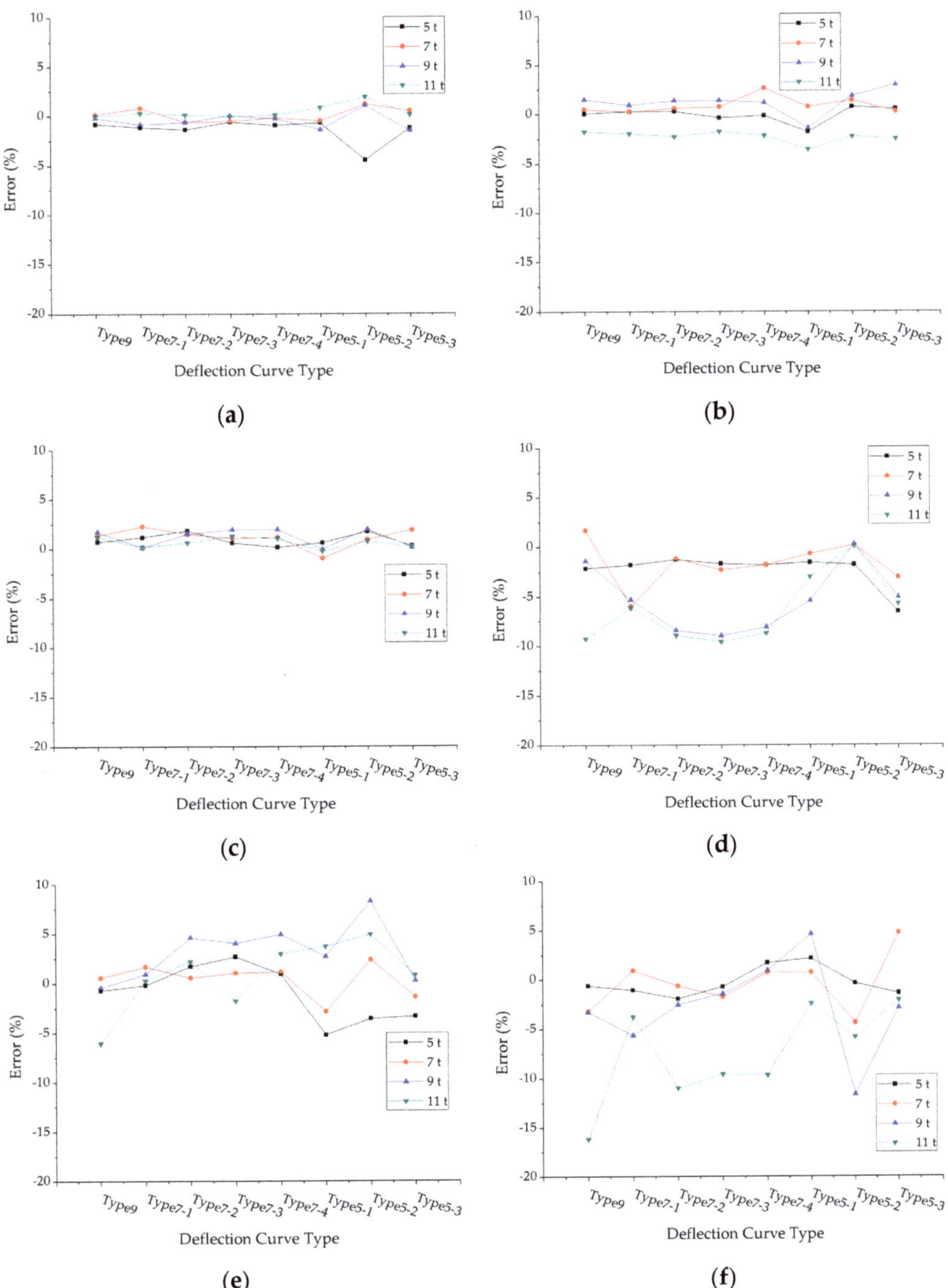

Figure 7. The errors of back calculated base modulus under different deflection curve types: (**a**) Structure 1; (**b**) Structure 2; (**c**) Structure 3; (**d**) Structure 4; (**e**) Structure 5; (**f**) Structure 6.

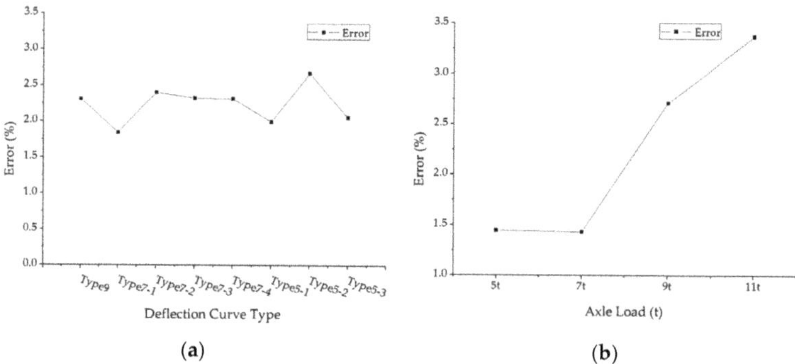

Figure 8. The average errors of back calculated base modulus: (**a**) The average errors under different deflection curve types; (**b**) The average errors under different axle loads.

3.2.3. Back Calculation Errors of Subbase Modulus

Only structure 5 and structure 6 possess subbase layers, the errors of back calculated subbase modulus under different loads and two pavement structures are shown in Figure 9. The average absolute values of the subbase modulus errors of each structure under different deflection curve types and axle loads were calculated and shown in Figure 10.

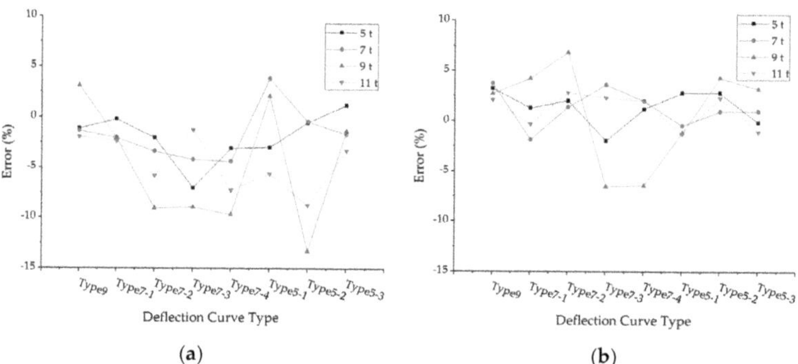

Figure 9. The errors of back calculated subbase modulus under different deflection curve types: (**a**) Structure 5; (**b**) Structure 6.

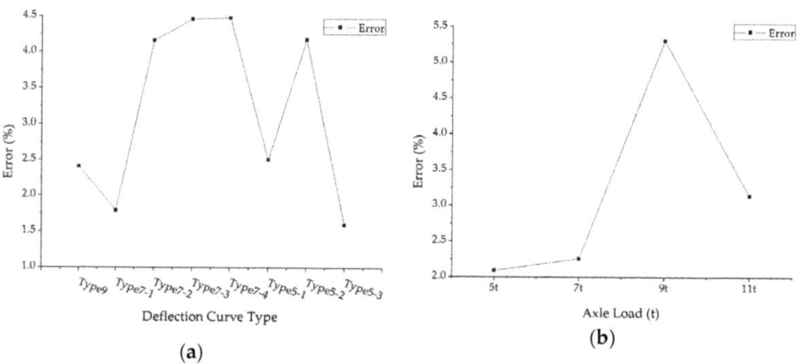

Figure 10. The average errors of back calculated subbase modulus: (**a**) The average errors under different deflection curve types; (**b**) The average errors under different axle loads.

The most errors in Figure 9 are within 10%, while the errors fluctuate with a big amplitude for both deflection curve types and axle loads, thus, Figure 10 shows no evident rules. A similar trend is also found in the subbase layer, which is where modulus errors increase sharply after a 7 t axle load.

3.2.4. Back Calculation Errors of Subgrade Modulus

The errors of back calculated subgrade modulus under different loads and pavement structures are shown in Figure 11. The average absolute values of subgrade modulus errors with each structure case under different deflection curve types and axle loads were calculated and shown in Figure 12.

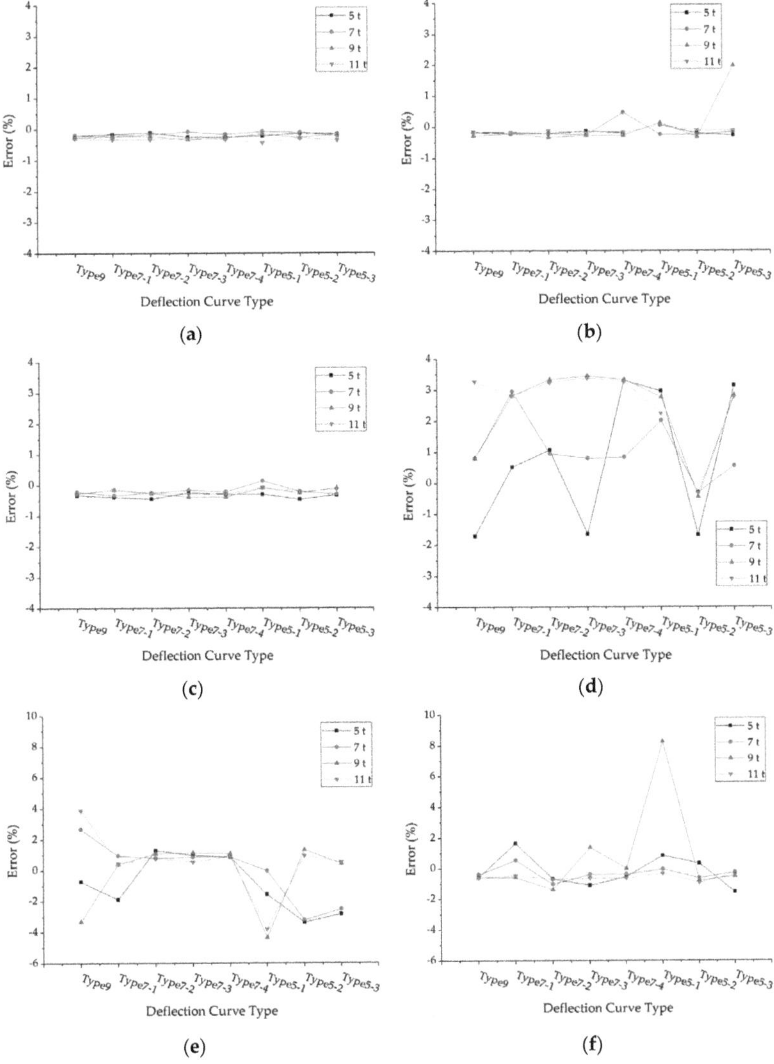

Figure 11. The errors of back calculated subgrade modulus under different deflection curve types: (**a**) Structure 1; (**b**) Structure 2; (**c**) Structure 3; (**d**) Structure 4; (**e**) Structure 5; (**f**) Structure 6.

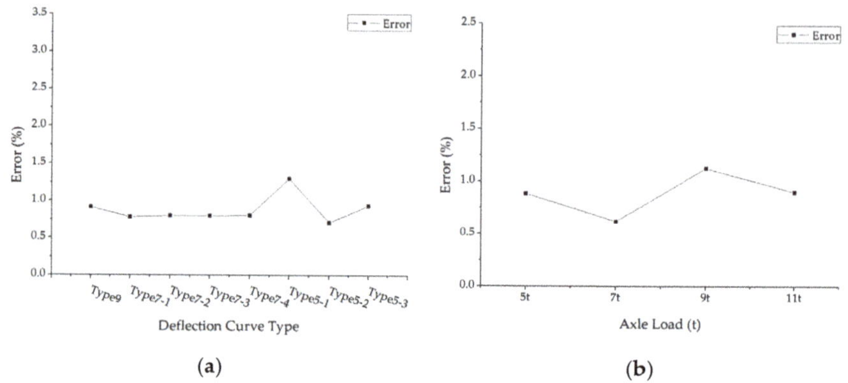

Figure 12. The average errors of back calculated subgrade modulus: (**a**) The average errors under different deflection curve types; (**b**) The average errors under different axle loads.

The back calculation errors of subgrade modulus are generally smaller, which means the accuracy of back calculated subgrade modulus is higher. It could be seen from Figure 12 that the errors of back calculated subgrade modulus have few effects by the deflection curve types and axle load levels.

3.3. Modulus Back Calculation Based on Actual Deflection Curves Measured on RIOHTrack

Considering the structure layer thickness as one of the most important parameters, which influences the modulus back calculation results greatly, it is important to measure the thickness of each layer in the four selected structures. The Mala Imaging Radar Array (MIRA) manufactured by Swedish MALA company was used to measure the thicknesses precisely, as is shown in Figure 13.

Figure 13. The MIRA.

According to the error analysis in Section 3.2, it could be known that there is no clear connection between modulus back calculation errors and deflection curve types, which indicates that the modulus back calculation accuracy is hardly influenced by the type of deflection curve. Based on this conclusion, considering that the deflection curve shape cannot be accurately described when the number of measuring points is too small, which could lead to the inaccuracy of modulus back calculation, the deflection curve type 9 was used in FWD tests due to its abundant measuring points. The FWD equipment used in this study is shown in Figure 14. The concentrated loads in FWD tests are 5 t, 7 t, 9 t and 11 t.

Figure 14. The FWD equipment.

The modulus back calculation results are shown in Tables 6–8.

Table 6. The modulus back calculation results of subgrade.

	5 t			7 t		
	Average Modulus (MPa)	Standard Deviation	Coefficient of Variation (%)	Average Modulus (MPa)	Standard Deviation	Coefficient of Variation (%)
STR1	275.03	29.3	11	294.63	37.3	13
STR10	359.78	137.85	38	405.67	193.46	48
STR18	265.26	191.81	72	281.56	77.77	28
STR19	257.51	29.39	11	320.84	177.39	55
Average	289	97	33	325	121	36
	9 t			11 t		
	Average Modulus (MPa)	Standard Deviation	Coefficient of Variation (%)	Average Modulus (MPa)	Standard Deviation	Coefficient of Variation (%)
STR1	273.1	20.47	7	276.35	24.2	9
STR10	320	60.81	19	375.86	90.38	24
STR18	308.65	88.32	29	356.73	204.07	57
STR19	243.3	9.44	4	280.17	116.8	42
Average	286.26	45	15	322	108	33

Table 7. The modulus back calculation results of middle layer.

	5 t			7 t		
	Average Modulus (MPa)	Standard Deviation	Coefficient of Variation (%)	Average Modulus (MPa)	Standard Deviation	Coefficient of Variation (%)
STR1	1702.55	1184.1	70	12,781	15,404.8	121
STR10 (1)	7455.04	3940.32	53	11,094	8477.23	76
STR10 (2)	13,726.8	14,815.3	108	4123.7	7077.58	172
STR18	1013.07	245.89	24	1020.9	347.57	34
STR19	4326.61	943.56	22	28,499	36,864.8	129
Average	5645	4225	55	11,504	13634	106
	9 t			11 t		
	Average Modulus (MPa)	Standard Deviation	Coefficient of Variation (%)	Average Modulus (MPa)	Standard Deviation	Coefficient of Variation (%)
STR1	3713.2	3047.01	82	3926.1	3262.34	83
STR10 (1)	8514.3	4090.67	48	9902.3	3996.46	40
STR10 (2)	10,365	5510.28	53	2732	1248.84	46
STR18	820.28	268.94	33	718.84	353.56	49
STR19	10,456	5109.43	49	3445.7	1464.08	42
Average	6774	3605	53	4145	2065	52

Table 8. The modulus back calculation results of asphalt concrete surface layer.

	5 t			7 t		
	Average Modulus (MPa)	Standard Deviation	Coefficient of Variation (%)	Average Modulus (MPa)	Standard Deviation	Coefficient of Variation (%)
STR1	6259	2748.5	44	10,625	12,632	119
STR10	19,564	14,164	72	20,200	14,622	72
STR18	15,688	10,266	65	17,937	12,517	70
STR19	19,978	15,288	77	30,391	21,156	70
Average	15,372	10,616	65	19,788	15,231	83
	9 t			11 t		
	Average Modulus (MPa)	Standard Deviation	Coefficient of Variation (%)	Average Modulus (MPa)	Standard Deviation	Coefficient of Variation (%)
STR1	10,959	6709	61	9697	6828	70
STR10	24,145	12,039	50	15,382	12,352	80
STR18	20,513	11,914	58	23,305	11,846	51
STR19	26,761	19,608	73	25,335	16,451	65
Average	20,594	12,567	60	18,430	11,869	66

The average back calculated moduli of subgrades are all around 300 MPa, the average standard deviations are around 100, and the average coefficients of variation are generally smaller than that of asphalt surface layers and middle layers. Both the average standard deviations and the average coefficients of variation in simplified middle layers are big, where the average standard deviations are all bigger than 3500, and the coefficients of variation are between 52% and 106%. This could be easily explained by the material diversity of middle layers. The back calculated moduli of asphalt concrete layers possess the biggest average standard deviation, but the average coefficients of variation are still between 60% and 83%, which means the dispersion of modulus back calculation of asphalt concrete surface is more stable than that of middle layers.

3.4. Modulus Back Calculation with Multi-Parameter FABCM

To test the reliability and accuracy of multi-parameter FABCM, and compare it with single-parameter FABCM, the vertical pressure on the top of the subgrade was used with a measured deflection curve together to conduct modulus back calculation.

When the RIOHTrack was constructed, different sensors including strain sensors, stress sensors, temperature monitors, hygrometers, etc., were buried in the structures. Hence, when different loads are applied on the road surface, the strain and stress states inside the pavement structures could be determined with data from those sensors. There are vertical stress sensors installed in the top-surface of the subgrade of each pavement structure, therefore, it is possible to know the vertical pressure on the top surface of the subgrade when loads are applied on the road, and the modulus back calculation based on the multi-parameter method is capable with those stress values.

The multi-parameter modulus back calculation was only performed on STR1 and STR19 for their relatively simple structures. The simplified structures are shown in Table 2, and the measured vertical pressure on the top surfaces of subgrades under different loads are shown in Table 9.

Table 9. The measured vertical pressures.

Structure	Vertical Pressure on the Top of Subgrade (kPa)			
	5 t	7 t	9 t	11 t
STR1	0.198	0.288	0.336	0.54
STR19	0.25	0.344	0.438	0.625

To compare the single-parameter FABCM and the multi-parameter FABCM, the back calculation results and the coefficients of variation in two methods are shown in Figures 15 and 16.

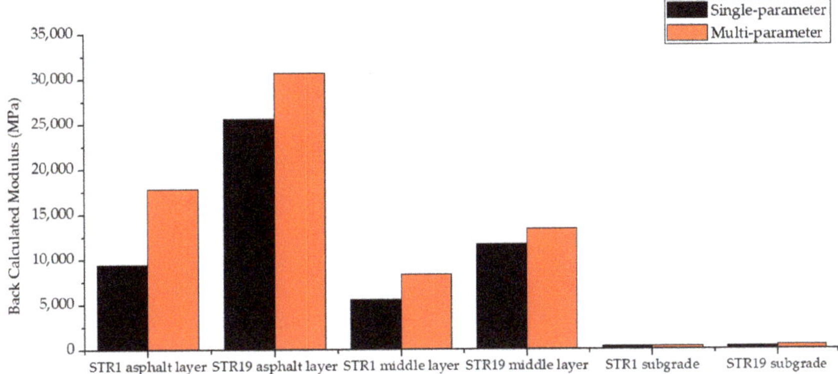

Figure 15. The back calculated modulus.

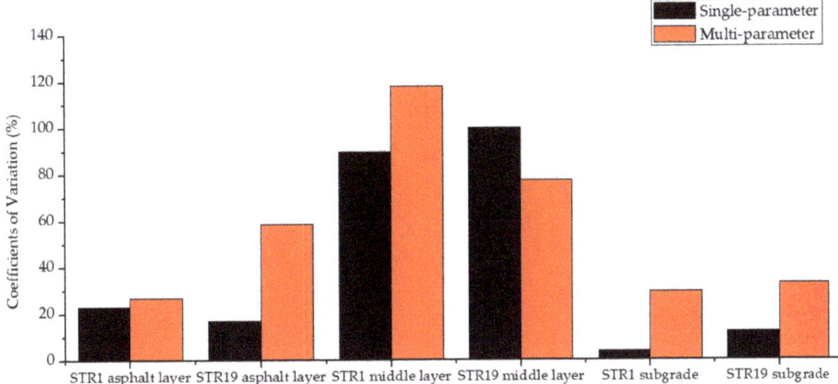

Figure 16. The coefficients of variation.

For the same structure layer, the back calculated modulus of multi-parameter FABCM differs greatly from that of single-parameter FABCM. The differences vary with different structure layers, and few relations could be found. When it comes to the coefficients of variation, it could be seen from Figure 16 that the variation coefficient of multi-parameter FABCM results is usually larger than that of single-parameter FABCM results for the same structure layer. Both of the two methods provide relatively stable modulus back calculation results for asphalt layers and the subgrade, while the results for the middle layers possess larger variation coefficients.

4. Discussion

The accuracy of this novel modulus back calculation is influenced greatly with pavement structures. When the pavement structure consists of three layers and the asphalt concrete surface layer is thin, like the structure 1, structure 2 and structure 3 in Section 3.2, the accuracy will be the highest. The iteration termination condition of back calculation is that $|\delta_j| \leq \varepsilon$, therefore, the final output modulus combination of the back calculation program is the optimal value of the mathematical optimization algorithm, rather than the modulus combination with the smallest error. When the number of pavement structure layers increase, more modulus and thickness parameters will be needed to be matched. The increase in the number of matching parameters increases the probability of the phenomenon mentioned above, which could lead to the inaccuracy of modulus back calculation. Thus, the smaller the number of structure layers, the higher the modulus back calculation accuracy.

For the subgrade material, the differences between the back calculated moduli of STR1, STR18 and STR19 gradually increase with the increase in the axle load. Under the same axle load, the back calculated modulus of STR10 is the largest. It shows that the subgrade material has typical nonlinear mechanical characteristics, and its back calculated modulus depends not only on the pavement structure, but also on the axle load applied on the road surface.

As for the middle layers, it could be seen that whatever the base type is, the back calculated modulus would be influenced by the level of axle load. Compared with that of flexible base and composite base, this influence on the modulus of semi-rigid base is not obvious.

When it comes to the asphalt layers, the back calculated modulus increases first and then decreases with the increase in axle load and finally tends to be stable, which is affected by the structure form. The thicker the asphalt layers are, the greater the load that the middle layer can bear, and the bigger the back calculated modulus of the asphalt layer would be.

The essence of the modulus back calculation is the match between the measured deflection curve and the theoretical deflection curve under the elastic layer system theory. Since FWD can be regarded as an instantaneous impact load (generally about 0.02 s), the response state of the pavement structure will also have nonlinear characteristics under the impact load of FWD due to the typical nonlinear characteristics of pavement materials. Therefore, no matter what kind of optimization algorithm is used in the modulus back calculation, the actual measured deflection curve is quite difficult to be consistent with the theoretical deflection curve under the current elastic mechanical model, which is also the reason that the back calculated modulus is not consistent with the design modulus of the pavement structure layer.

As is verified in Section 3.1, when the back calculation was conducted based on the theoretical deflection curves, which means there were no error caused by the differences between elastic mechanical model and non-linear characteristics of asphalt pavement materials, both the accuracy and calculation speed of FABCM are satisfactory. It could be told that the inherent defects of the firefly optimization algorithm could be avoided effectively with the modifying methods in this paper, while the calculation speed is not obviously affected.

The variation coefficient of multi-parameter FABCM results is larger than that of the single-parameter FABCM, which means in actual practice, the result of single-parameter FABCM is more reliable than that of multi-parameter FABCM due to the acquisition and accumulation errors of the input parameters in the multi-parameter FABCM program. Besides, compared with the deflection curve, the stress and strain inside the pavement structures are usually harder to be precisely measured, thus the single-parameter back calculation method is more recommended.

5. Conclusions

In this study, an asphalt pavement modulus back calculation method based on the firefly optimization algorithm was proposed to investigate the factors influencing back calculation accuracy and the way to select proper input parameters for modulus back calculation. The main findings are summarized as follows:

1. The firefly algorithm, a meta-heuristic intelligent optimization algorithm, is selected as the mathematical search method of the modulus back calculation algorithm, and the Rosenbrock search method and Gaussian perturbation strategy are used to improve the firefly algorithm in terms of search rules and intermediate optimal solution disturbance. Through a lot of calculation and comparative analysis, it is verified that FABCM has a satisfactory back calculation speed and accuracy;

2. There is no clear connection between the modulus back calculation errors and deflection curve types, which indicates that the modulus back calculation accuracy is not influenced by the type of deflection curve. Thus, considering that the deflection curve shape cannot be accurately described when the number of measuring points is too small, which could lead to the inaccuracy of the modulus back calculation, the deflection curve types of more measuring points are recommended in modulus back calculation;

3. It is found that when the axle load reaches 9 t or 11 t, the errors of back calculated modulus would increase a lot, and the back calculation on the 3-layer structure is more accurate than that on the 4-layer structure;

4. Due to the significant non-linear characteristics of pavement materials, no matter what kind of optimization algorithms are used, the back calculated modulus is hard to be consistent with the design modulus of the pavement structure layer under the current elastic mechanical model;

5. The variation coefficients of multi-parameter FABCM are generally larger than that of single-parameter FABCM in practice, which means the results of single-parameter FABCM are more stable. Considering the reasons discussed in Section 4, the single-parameter modulus back calculation method has its advantages in measurement of stress and strain state inside a pavement structure and shows potential application prospects.

The asphalt pavement mechanical model in FABCM is still an elastic model, which could not describe the non-linear characteristics of realistic asphalt pavement materials. Considering that the inherent defects, such as premature convergence and the local optimum of mathematical optimization algorithm, were effectively avoided with a Rosenbrock search method and Gaussian perturbation strategy in this study, the main reason why great errors occur when conducting modulus back calculation based on the measured deflection curves on RIOHTrack should be the difference between an elastic mechanical model and the non-linear asphalt pavement characteristics. To achieve the best possible match between the design modulus and the back calculated modulus in the future, we recommend the study and use of non-linear mechanical models, which could describe the asphalt pavement characteristics better in the back calculation methods.

Author Contributions: Conceptualization, R.Z. and J.G.; methodology, J.G.; validation, R.Z.; formal analysis, R.Z.; investigation, J.G.; resources, R.Z.; writing—original draft preparation, R.Z. and J.G.; writing—review and editing, Y.Z.; visualization, R.Z. and Y.Z.; supervision, X.H.; project administration, X.H. All authors have read and agreed to the published version of the manuscript.

Funding: This research was funded by National Nature Science Foundation of China (Grant Nos. 51908119 and 51890904); the National Key R&D Program of China, (Grant Nos. 21YFB2600600 and 21YFB2600601). Postgraduate Research and Practice Innovation of Jiangsu, China, (Grant Nos. KYCX21_0139).

Informed Consent Statement: Not applicable.

Data Availability Statement: Not applicable.

Conflicts of Interest: The authors declare that they have no known competing financial interest or personal relationship that could have appeared to influence the work reported in this paper.

References

1. Ling, M.; Luo, X.; Gu, F.; Lytton, R. Time-Temperature-Aging-Depth Shift Functions for Dynamic Modulus Master Curves of Asphalt Mixtures. *Constr. Build. Mater.* **2017**, *157*, 934–951. [CrossRef]
2. Dong, Z.; Li, L.P. Study on Dynamic Mechanical Properties and Microstructure of Epoxy Asphalt. In Proceedings of the 2015 International Conference on Applied Science And Engineering Innovation, Jinan, China, 30–31 August 2015; Volume 12, pp. 516–523.
3. Zhou, W. Strain-Dependent Model for Complex Modulus of Asphalt Mixture. *J. Chongqing Jiaotong Univ. Nat. Sci. Ed.* **2019**, *38*, 57–62.
4. Zhang, Y.; Ma, T.; Ling, M.; Zhang, D.; Huang, X. Predicting Dynamic Shear Modulus of Asphalt Mastics Using Discretized-Element Simulation and Reinforcement Mechanisms. *J. Mater. Civil Eng.* **2019**, *31*, 04019163. [CrossRef]

5.   Ding, H.; Qiu, Y.; Rahman, A. Influence of Thermal History on the Intermediate and Low-Temperature Reversible Aging Properties of Asphalt Binders. Road Mater. *Pavement* **2020**, *21*, 2126–2142. [CrossRef]

6.   Luo, H.; Huang, X.; Rongyan, T.; Ding, H.; Huang, J.; Wang, D.; Liu, Y.; Hong, Z. Advanced method for measuring asphalt viscosity: Rotational plate viscosity method and its application to asphalt construction temperature prediction. *Constr. Build. Mater.* **2021**, *301*, 124129. [CrossRef]

7.   Huang, Y.; Liu, Z.; Wang, X.; Li, S. Comparison of HMA Dynamic Modulus Between Trapezoid Beam Test and SPT. *J. Cent. South Univ. Sci. Technol.* **2017**, *48*, 3092–3099.

8.   Scrivner, F.H.; Moore, W.M.; Mcfarland, W.F.; Carey, G.R. A Systems Approach to the Flexible Pavement Design Problem. 1968. Available online: http://tti.tamu.edu/documents/32-11.pdf (accessed on 23 November 2021).

9.   Sun, R.; Tan, Z. Back Calculation Accuracy of Pavement Structural Layer Modulus. *Highway* **2005**, *4*, 92–95.

10.   Yang, G.; Wang, D.; Zhang, X. Inverse Calculation of Elastic Modulus of Asphalt Pavement Based on Curvature Index of Asphalt Pavement. *J. South China Univ. Technol. Nat. Sci. Ed.* **2007**, *35*, 5.

11.   Yang, G.; Wang, D.; Zhang, X. Back Calculation of Modulus of Resilience of Soil Foundation by Deflection Basin. *J. Harbin Inst. Technol.* **2009**, *41*, 137–141.

12.   Alkasawneh, W. *Backcalculation of Pavement Moduli Using Genetic Algorithms*; The University of Akron: Akron, OH, USA, 2007.

13.   Huang, X.; Deng, X. Evaluation of Pavement Structural Strength by Measured Deflection Basin. *J. Geotech. Eng.* **1996**, *18*, 4.

14.   Saltan, M.; Terzi, S.; Kuguksille, E.U. Backcalculation of Pavement Layer Moduli and Poisson's Ratio Using Data Mining. *Expert Syst. Appl.* **2011**, *38*, 2600–2608. [CrossRef]

15.   Zha, X. Summary of Back Calculation Methods of Pavement Structural Layer Modulus. *J. Transp. Eng.* **2002**, *2*, 1–6.

16.   Uzah, J.; Lytton, R.L.; Germann, F.P. General Procedure for Backcalculating Layer Moduli. In *Nondestructive Testing of Pavements and Backcalculation of Moduli*; STP 1026; ASTM: Conshohocken, PA, USA, 1989.

17.   Rohde, G.; Scullion, T. MODULUS 4.0: Expansion and Validation of the MODULUS Backcalculation System. *Res. Repart Tex. Transporation Inst.* **1990**, *10*, 1829–1833. Available online: https://trid.trb.org/view/1176898 (accessed on 23 November 2021).

18.   Ju, Z. Research on Pavement Modulus Backcalculation Based on Finite Element. Ph.D. Thesis, South China University of Technology, Guangzhou, China, 2019.

19.   Cheng, S.; Ni, F. Study on Back Calculation Method of Pavement Structure Modulus Based on Dynamic Deflection Analysis. *Mod. Transp. Technol.* **2011**, *8*, 4.

20.   Gao, S. Study on Dynamic Response and Modulus Back Calculation of Two-Layer System Subjected to LWD Load. Ph.D. Thesis, Hunan University, Hunan, China, 2018.

21.   Zaman, M.; Solanki, P.; Ebrahimi, A. Neural Network Modeling of Resilient Modulus Using Routine Subgrade Soil Properties. *Int. J. Geomech.* **2010**, *10*, 1–12. [CrossRef]

22.   Sun, X.; Huang, L. Application of Genetic Algorithm in Back Calculation of Pavement Structure Modulus. *J. Chang. Jiaotong Univ.* **2002**, *18*, 4.

23.   Zha, X.; Wang, B. Back Calculation of Pavement Modulus Based on Artificial Neural Network. *J. Transp. Eng.* **2002**, *2*, 4.

24.   Yang, G.; Zhang, X.; Wang, D. Study on Back Calculation of Resilient Modulus of Soil Foundation by BP Neural Network. *Highw. Eng.* **2007**, *32*, 44–46, 50.

25.   Yang, G.; Wu, K. Study on Back Calculation of Elastic Modulus of Asphalt Pavement Structure Layer by BP Neural Network. *J. Sun Yatsen Univ.* **2008**, *47*, 44–48.

26.   Yang, G.; Zhong, W.; Huang, X. Study on Inversion of Elastic Modulus of Asphalt Pavement by BP Artificial Neural Network Based on Road Surface Deflection Basin. *Sci. Technol. Innov. Guide* **2015**, *29*, 94–93, 95.

27.   Yang, G.; Zhong, W.; Huang, X. Study on Back Calculation of Elastic Modulus of Asphalt Pavement Base by BP Artificial Neural Network. *Subgrade Eng.* **2016**, *4*, 78–81.

28.   Khazanovich, L.; Roesler, J. Diploback: Neural-Network-Based Backcalculation Program for Composite Pavements. *Transp. Res. Rec. J. Transp. Res. Board* **1997**, *1570*, 143–150. [CrossRef]

29.   Khazanovich, L.E.V. *Lev Structural Analysis of Multi-Layered Concrete Pavement Systems*; University of Illinois at Urbana-Champaign: Champaign, IL, USA, 1994.

30.   Kothandram, S.; Ioannides, A.M. Diplodef: A Unified Backcalculation System for Asphalt and Concrete Pavements. *Transp. Res. Rec. J. Transp. Res. Board* **2001**, *1764*, 20–29. [CrossRef]

31.   Zhang, D. *MATLAB R2017a Artificial Intelligence Algorithm*; Electronic Industry Press: Beijing, China, 2018.

32.   Fwa, T.; Tan, C.; Chan, W. Backcalculation Analysis of Pavement-Layer Moduli Using Genetic Algorithms. *Transp. Res. Rec. J. Transp. Res. Board* **1997**, *1570*, 134–142. [CrossRef]

33.   Kameyama, S.; Himeno, K.; Kasahara, A. Backcalculation of Pavement Layer Moduli Using Genetic Algorithms. In Proceedings of the Eighth International Conference on Asphalt Pavements, Seattle, DC, USA, 1 January 1997.

34.   Zha, X.; Wang, B. Study on Back Calculation of Pavement Modulus Based on Homotopy Method. *Chin. J. Highw.* **2003**, *16*, 5.

35.   Liu, C. Back Calculation Method of Pavement Modulus Based on FWD and Analysis of Influencing Factors. *North. Commun.* **2010**, *9*, 3–5.

36.   Zou, H. *The Experimental Research of Asphalt Mixture Dynamic Modulus*; Chang'an University: Xi'an, China, 2013.

37.   Ministry of Transport of the People's Republic of China. JTG D50-2017. In *Specifications for Design of Highway Asphalt Pavement*; China Communication Press: Beijing, China, 2017.

38. Yang, X. *Nature-Inspired Metaheuristic Algorithms*, 2nd ed.; Luniver Press: Frome, UK, 2010.
39. Farahani, S.M.; Nasiri, B.; Meybodi, M.R. A Multiswarm Based Firefly Algorithm in Dynamic Environments. In Proceedings of the Third International Conference on Signal Processing Systems (ICSPS2011), Yantai, China, 27–28 August 2011.
40. Coelho, L.; Bernert, D.; Mariani, V.C. A Chaotic Firefly Algorithm Applied to Reliability-Redundancy Optimization. In Proceedings of the IEEE Congress on Evolutionary Computation, CEC 2011, New Orleans, LA, USA, 5–8 June 2011.
41. Coelho, L.; Mariani, V.C. Improved Firefly Algorithm Approach Applied to Chiller Loading for Energy Conservation. *Energy Build.* **2013**, *59*, 273–278. [CrossRef]
42. Ministry of Transport of the People's Republic of China. JTG E60-2008. In *Field Test Methods of Subgrade and Pavement for Highway Engineering*; China Communication Press: Beijing, China, 2008.

Article

Investigation of Adhesion Properties of Tire—Asphalt Pavement Interface Considering Hydrodynamic Lubrication Action of Water Film on Road Surface

Binshuang Zheng [1], Junyao Tang [2,*], Jiaying Chen [2], Runmin Zhao [2] and Xiaoming Huang [2,3]

1 School of Modern Posts, Nanjing University of Posts and Telecommunications, Nanjing 210023, China; zhengbs@njupt.edu.cn
2 School of Transportation, Southeast University, Nanjing 211189, China; cjiaying14@seu.edu.cn (J.C.); 230228857@seu.edu.cn (R.Z.); huangxm@seu.edu.cn (X.H.)
3 National Demonstration Center for Experimental Education of Road and Traffic Engineering, Southeast University, Nanjing 211189, China
* Correspondence: 230228855@seu.edu.cn; Tel.: +86-1879-588-8673

Citation: Zheng, B.; Tang, J.; Chen, J.; Zhao, R.; Huang, X. Investigation of Adhesion Properties of Tire—Asphalt Pavement Interface Considering Hydrodynamic Lubrication Action of Water Film on Road Surface. *Materials* 2022, 15, 4173. https://doi.org/10.3390/ma15124173

Academic Editor: Francesco Canestrari

Received: 14 May 2022
Accepted: 7 June 2022
Published: 12 June 2022

Publisher's Note: MDPI stays neutral with regard to jurisdictional claims in published maps and institutional affiliations.

Abstract: To obtain the tire–pavement peak adhesion coefficient under different road states, a field measurement and FE simulation were combined to analyze the tire–pavement adhesion characteristics in this study. According to the identified texture information, the power spectral distribution of the road surface was obtained using the MATLAB Program, and a novel tire hydroplaning FE model coupled with a textured pavement model was established in ABAQUS. Experimental results show that here exists an "anti-skid noncontribution area" for the insulation and lubrication of the water film. Driving at the limit speed of 120 km/h, the critical water film thickness for the three typical asphalt pavements during hydroplaning was as follows: AC pavement, 0.56 mm; SMA pavement, 0.76 mm; OGFC pavement, 1.5 mm. The road state could be divided into four parts dry state, wet sate, lubricated state, and ponding state. Under the dry road state, when the slip rate was around 15%, the adhesion coefficient reached the peak value, i.e., around 11.5% for the wet road state. The peak adhesion coefficient for the different asphalt pavements was in the order OGFC > SMA > AC. This study can provide a theoretical reference for explaining the tire–pavement interactions and improving vehicle brake system performance.

Keywords: tire–pavement contact; texture information identification; numerical modeling; fluid hydrodynamic lubrication theory; orthogonal experimental design; peak adhesion coefficient

1. Introduction

According to the tire–pavement contact mathematical model, the tire optimal slip rate is obtained when the road surface has the peak adhesion coefficient. In addition, the vehicle braking and tire–pavement contact mechanics indicate that the maximum braking deceleration can be achieved during the vehicle braking process under the condition of the peak adhesion coefficient of the road surface [1,2]. Therefore, it is necessary to study the variation characteristics of the asphalt pavement adhesion coefficient for predicting the peak adhesion coefficient of the road surface, which is regarded as the pressure threshold at which the braking system can realize the maximum braking force of the vehicle. Furthermore, the vehicle safety can be effectively improved during the braking process [3]. As one essential component of the tire–pavement friction force, the adhesion force is formed due to the rubber–pavement interaction and is highly dependent on the true area of contact [4,5]. Theoretically, the adhesion results from the shear force on the contact surface of the tire and road. Accordingly, many research institutes have studied the tire–pavement contact model and road safety problem in terms of the tire shear deformation effect, and many typical contact mechanic models between the tire and pavement were summarized in the work of

Huang et al. [6]. Recently, a formulation for a deformable solid was constructed using an arbitrary Lagrangian–Eulerian approach [7], and this formulation was applied to a new simplified tire model, represented by a circular ring including shear stresses and nonlinear effects due to the vehicle load. A numerical model of the flexible base asphalt pavement structure under nonuniform vertical tire contact pressure was established, and the effect of nonuniformity and uniformity on the pavement surface deflection was investigated [8]. Yet, no studies have been carried out to investigate the tire–pavement peak adhesion coefficient and its influencing factors.

Related studies have shown that the adhesion of the tire–pavement interface is determined by the macro texture of the road surface [9]. Meanwhile, a water film on the road surface can significantly reduce the tire–pavement adhesion [10–12]. Recently, many methods have been proposed to estimate the tire–pavement adhesion characteristics. These method can be divided into two types, cause-based and effect-based. Among them, cause-based methods detect the influencing factors of the adhesion coefficient, such as the road surface texture and fluid lubrication. The estimation method based on the slip rate μ–s curve is typically applied, which mainly considers vehicle acceleration, braking behavior, and lateral movement to estimate the road surface adhesion coefficient. For example, Bachmann [13] proposed the influencing factors of the tire–pavement adhesion coefficient; in particular, the road environmental parameters (road lubricant parameters, such as a water film) had a major impact, but the water film lubrication theory was ignored. German [14] proposed a typical polynomial model to approximate the relationship between the tire–pavement adhesion coefficient and the tire slip ratio using a simple polynomial function. However, this model is inapplicable when the slip ratio is high. The above studies indicated that the tire–pavement adhesion coefficient is closely related to the tire slip rate according to empirical estimation or tire dynamics; however, the adhesion coefficient variation rule has not been revealed from tire–pavement contact theory.

In the previous study, the surface water will generate a certain dynamic water pressure under the pressure of tire load when the vehicle drives on moist asphalt pavement, which can easily cause hydroplaning [15,16]. Most of the water in the tire–road contact area flows or splashes along the surface texture and tire pattern gap. According to Bernoulli theory, it can be known that the micro kinetic energy of the water flow is converted into the pressure of the tire against the road surface [17]. When the tire is running at a certain speed, tire hydroplaning occurs when the elastic hydrodynamic pressure of the water film in the vertical direction of the tire–fluid interface is equal to the tire load.

In theory, the rigidity of the road surface is infinitely flat. Assuming that the tire contact surface is flat, the angle between the tire and the road contact interface is actually very small. It can be considered that the component of the dynamic water pressure in the direction of the tire movement is close to 0. At this time, the vertical component of the hydrodynamic pressure at the stagnation point of the water flow is equal to the total micro kinetic energy, i.e.,

$$G = \int_{s} \frac{\rho_{\mathrm{w}} v_p^2}{2} ds, \tag{1}$$

where G is the tire load (kN), v_p is the critical hydroplaning speed of the tire (km/h), ρ_{w} is the fluid density (kg/m^3), and s is the action area of the water film (m^2).

Under the action of the vehicle load, the local elastic deformation of the contact between the tire and the road forms a water film between the lubricated surfaces. The hydrodynamic pressure generated by this water film can be regarded as the elastic deformation of the rubber tread. At this time, the dynamic lubrication state exhibits the phenomenon of elasto–hydrodynamic lubrication (EHL) [18].

Studies have shown that the factors mediating tire hydroplaning under wheel load are tire speed, road conditions (roughness, pavement type, water film thickness, etc.), and tire parameters (pattern structure, rubber material viscoelasticity, tire pressure, and load) [19]. Among them, the road conditions are the main objective influencing factors, and the water film thickness is related to the tire hydroplaning state. On a wet road, a vehicle at high speed can undergo dynamic hydroplaning, viscous hydroplaning, and micro EHL under the influence of the pavement texture and water film. When the water film thickness exceeds 1.0 mm, dynamic hydroplaning often occurs. On rainy days, as it just begins to rain and the road surface becomes wet, a thin water film of 0.01–0.1 mm is formed. According to EHL theory, the water film isolates the contact state between the tire and road surface, resulting in a decrease in the tire–pavement friction. In this case, micro EHL is more likely to occur.

In accordance with fluid hydrodynamic lubrication theory and tire–road contact theory, a tire hydroplaning finite element (FE) model was applied to determine the division of road states considering road surface water film thickness. Luo [20] proposed the limit standard of water film thickness on rainy days considering driving safety using the rainfall data of Hainan region in China, indicating that tire hydroplaning occurs easily when the water film thickness exceeds 4.0 mm. Moreover, Lufft Instrument and Equipment Corporation in Germany developed the advanced mobile road weather information sensor MARWIS, which can test road surface water film thicknesses up to 6 mm. Thus, the water film thickness ranging from 0 mm to 6 mm was taken as the research range to refine the asphalt pavement state.

In the literature, many studies have used the finite element (FE) method [21] and laboratory tests to study tire–pavement contact mechanics [22]. For example, Fwa [23,24] established an FEM for the tire–asphalt pavement interaction, revealing that the change rule of the friction coefficient varied with the thickness of the water film. However, these studies considered the pavement as a smooth flat surface and did not adequately consider the macro and micro texture of the asphalt pavement. On the basis of the tire–pavement contact mechanics, a tire hydroplaning FE model was established, and the tire–pavement adhesion coefficient characteristics were simulated [25,26]. A vehicle braking evaluation method based on the road adhesion coefficient was proposed to evaluate the road adhesion coefficient under the condition of tire sliding in an accident investigation [27]. According to the torque distribution of a single tire, a road adhesion coefficient prediction method was proposed by Ma [28]. Considering road surface fractal theory, Chen [29] tested the stress distribution of tires and found a positive correlation between the average effective stress and the friction coefficient. Guo [30] proposed a method for testing the tire adhesion coefficient and modified the steady–state semi–empirical sidetracking model of the tire. By experimentally examining the relationship between rubber–pavement adhesion and friction, a fair correlation was revealed between the adhesive bond energy and the measured coefficient of friction [5]. As reviewed above, most research results can explain the basic variation rule of the tire–pavement adhesion, but the adhesion characteristics are relative to the road surface fractal theory and rubber friction energy losses.

In view of the above research shortcomings, fluid hydrodynamic lubrication mechanics were applied in this study to demarcate the asphalt pavement states. Then, a field measurement and FE simulation were combined to reveal the tire–pavement adhesion characteristics under different road states. On the basis of the simulation results of the adhesion coefficient, an orthogonal experimental design method was applied to investigate the significant influencing factors for different road states. According to the variation of the simulated adhesion coefficient curves with tire slip rate, peak adhesion coefficient curves were obtained for different driving speeds.

2. Materials and Methods

2.1. Acquisition of Pavement Texture Information

Using the automatic close-range photogrammetry system (ACRP system) developed in [31], the surface texture of three typical asphalt pavements (Figure 1a) was collected. After the preprocessing of the collected images; reverse reconstruction technology was used to reconstruct the three–dimensional (3D) images of the asphalt pavement surface texture, and a 3D model of the asphalt pavement surface texture (Figure 1b) was established. Then,

GeoMagic and MeshLab 3D modeling software was adopted to preprocess the reverse reconstructed 3D model of the asphalt pavement surface texture, including hole filling correction, leveling, and definition of local coordinate axis attributes, and the road surface texture 3D elevation data were derived from the 3D model containing (x, y, z) 3D coordinate points (Figure 1c). According to the above 3D elevation data, the surface morphology of the asphalt pavement was reconstructed in MATLAB, as shown in Figure 2.

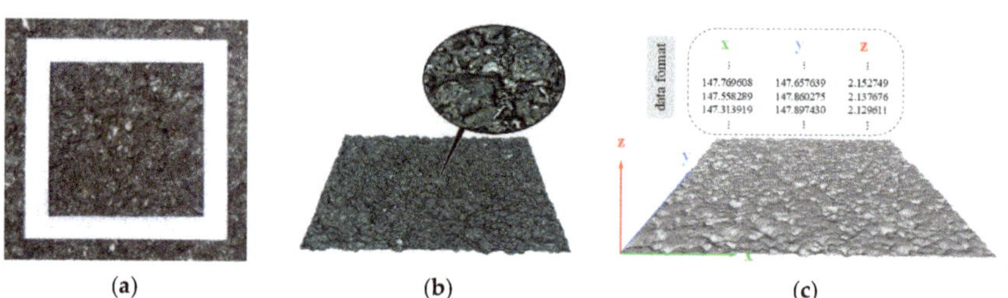

Figure 1. Acquisition of asphalt pavement texture information: (**a**) testing scope of pavement texture; (**b**) 3D digital pavement model; (**c**) 3D data of texture elevation.

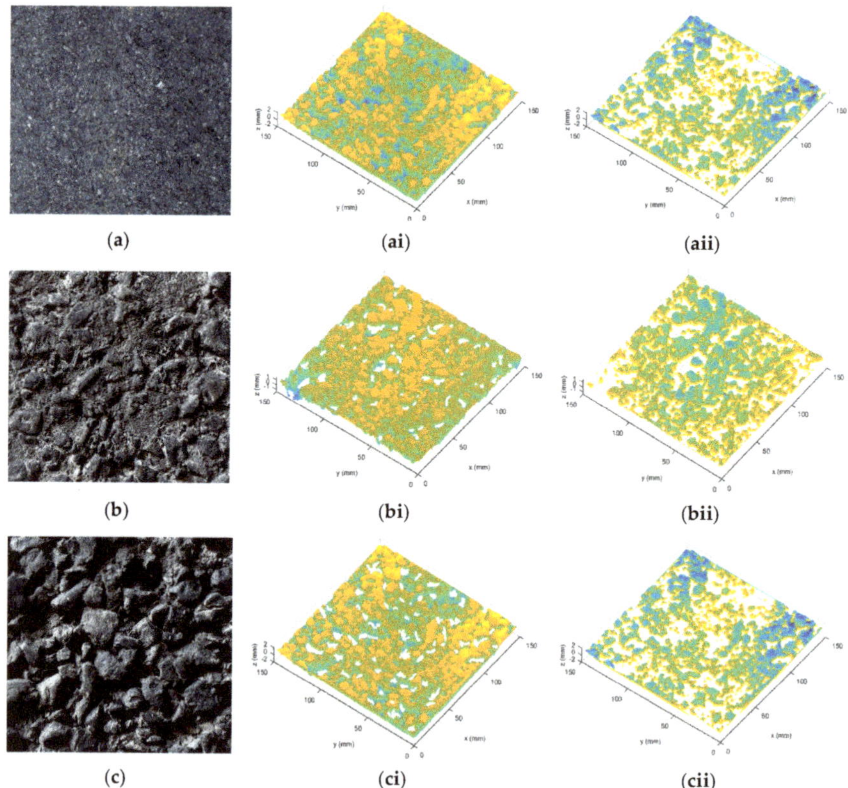

Figure 2. Pavement texture visualization under different conditions: (**a**) AC−13 pavement in (**ai**) dry conditions and (**aii**) wet conditions; (**b**) SMA−13 pavement in (**bi**) dry conditions and (**bii**) wet conditions; (**c**) OGFC−13 pavement in (**ci**) dry conditions and (**cii**) wet conditions. Notes: AC = dense-graded asphalt pavement; SMA = stone matrix asphalt; OGFC = open−grade friction course.

Three kinds of asphalt pavement were selected in this study: dense–graded asphalt concrete (AC), stone matrix asphalt (SMA), and open–grade friction course (OGFC). The gradation design of the three asphalt mixtures is shown in Table 1. To generate a wet pavement surface, water was uniformly sprayed on the dry surface until the concave asperities were sealed with water [32,33].

Table 1. Gradations for three kinds of asphalt pavement.

Sieve Size (mm)	Passing Rate of Each Sieve (%)									
Components	0.075	0.15	0.3	0.6	1.18	2.36	4.75	9.5	13.2	16
AC−13	6	10	13.5	19	26.5	37	53	76.5	95	100
SMA−13	10	13.2	16.3	19.5	22.7	25.8	29	63.5	97.9	100
OGFC−13	4.6	5.4	6.1	8.7	11.5	15.0	18.8	63.3	97.8	100

On the basis of the obtained 3D texture coordinate elevation data (x, y, z), the power spectral distribution (PSD) solver was written in MATLAB by applying the PSD calculation model for Persson friction theory [34,35]. Since the random variables of the fractal road surface were discrete points, further filtering, windowing, and sampling window compensation of the coordinate data were needed in the process of solving the PSD [19]. The PSD of the road surface under wet and dry conditions was calculated using Equation (2), and the PSD curves for AC−13, SMA−13, and OGFC−13 pavements under different road conditions were obtained as shown in Figure 3.

$$C(\mathbf{q}) = \frac{1}{(2\pi)^2} \int \langle h(x)h(0) \rangle e^{i\mathbf{q}x} dx, \tag{2}$$

where x is the wave vector direction, $h(0)$ is the surface elevation of the origin point, $h(x)$ is the surface elevation with the average elevation as the starting point, $\langle \dots \rangle$ represents the average on the plane, $\mathbf{q}$ is the wave vector, which can be obtained through the conversion of wavelength λ, and e is a natural constant.

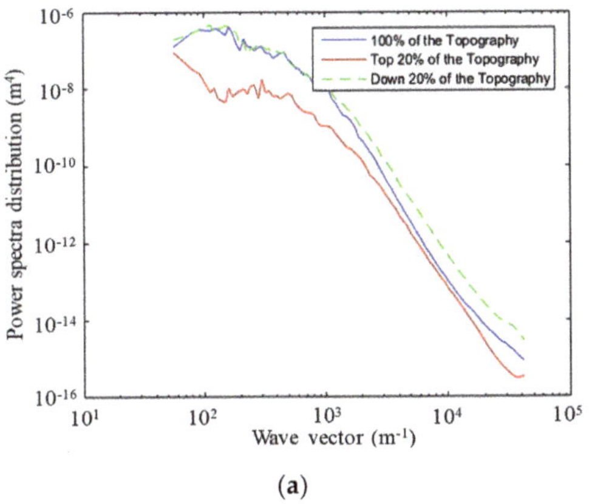

(**a**)

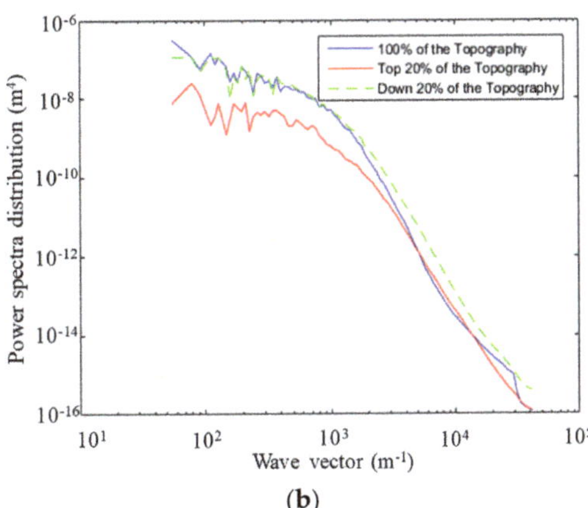

(**b**)

Figure 3. *Cont.*

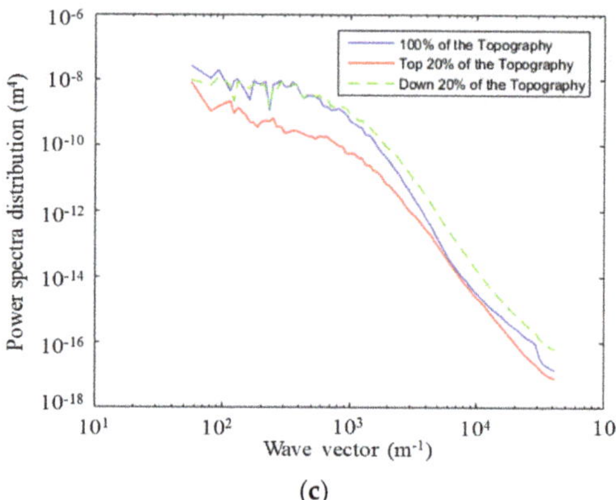

(**c**)

Figure 3. The 2D−PSD of asphalt pavement texture: (**a**) AC−13 asphalt pavement; (**b**) SMA−13 asphalt pavement; (**c**) OGFC−13 asphalt pavement.

As a function of the calculated PSD, the friction coefficient–velocity curves were obtained using the Persson friction coefficient formula (see Figure 4). The variation trends of the friction coefficient curves under different pavement conditions were similar, both of them decrease significantly with the increase in relative slipping speed.

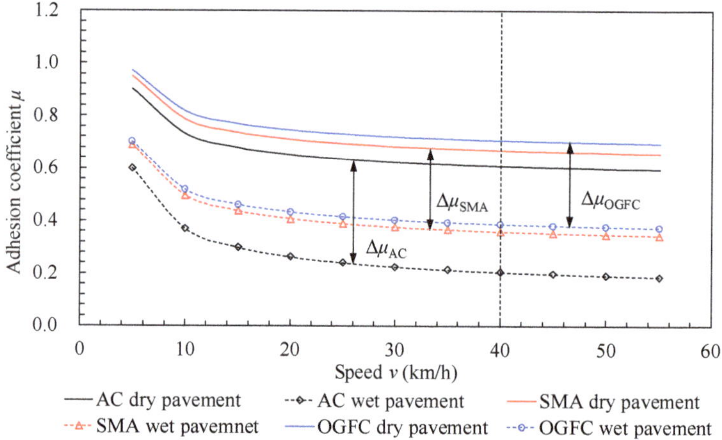

Figure 4. Variation of adhesion coefficient curves with tire speed.

When the speed exceeded 40 km/h, the curve tended to be gentle, indicating that the actual tire–road contact area was in a steady state at a relatively high speed. The friction coefficient of the wet road surface was lower than that of the dry pavement. In addition, a greater speed resulted in a greater difference in friction coefficient between the two states (dry and wet pavement). Regardless of the road state (dry or wet), the friction coefficient of the different pavements was in the order OGFC > SMA > AC, indicating that the surface roughness did not contribute to skid resistance caused by the water film barrier between the tire and the pavement.

2.2. Textured Pavement Modeling in ABAQUS

2.2.1. FE Model of Textured Pavement

For the 3D texture coordinate elevation data, the *.asc format file obtained by the ACRP system was used, containing a set of 3D coordinate points of the road surface, which allowed standardizing the network coordinate points. Furthermore, the .inp file for the 3D digital pavement model was processed in MATLAB and converted into .inp files suitable for the ABAQUS program. The ABAQUS program was adopted to realize the reconstruction of the 3D textured pavement. The specific processing steps for the FE model of the textured asphalt pavement are described below.

(1)   Generation of 3D digital pavement model

After image distortion removal and image adjustment, as well as point cloud reconstruction and registration, the 3D digital pavement model was built as shown in Figure 5a. The generated point cloud was matched, converting different sets of point clouds to the same coordinate system using the iterative closest point algorithm. Furthermore, the local axis properties, hole filling correction, and plane leveling were defined.

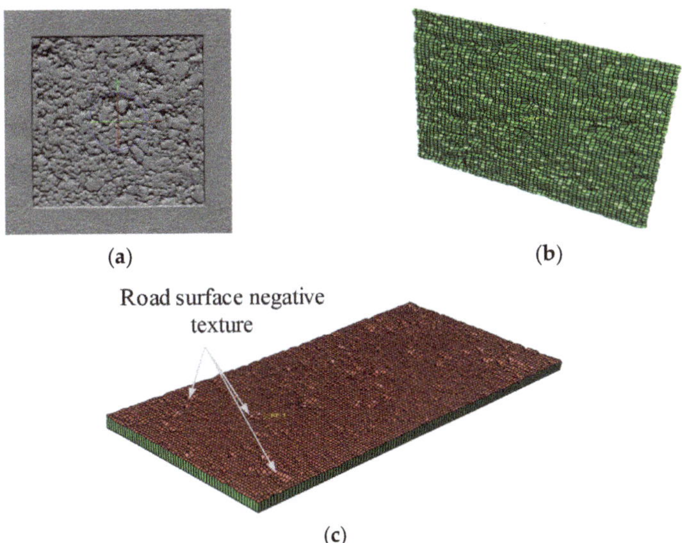

Figure 5. Generation process of textured pavement 3D FE model: (**a**) 3D model of texture morphology; (**b**) rigid shell element; (**c**) textured pavement 3D model.

(2)   Unit standardization of grid data

For the generated 3D digital pavement model, the Editplus software was used to extract the 3D texture coordinate elevation data of the .inp file before conversion into .txt format files, which was further realized by the MATLAB program. In MATLAB, the statement "$[m,n] = size$" was used to read the array size of the original data file (.txt data file) and generate the same size array storage space. Then, taking the number of data lines as the number of loops, the loop statement was adopted to read the original coordinate points (x, y, z) before unit processing using Equation (3).

$$\begin{cases} x(i,1) = \frac{data(i,1)}{100} \times 1000; \\ y(i,1) = \frac{data(i,2)}{100} \times 1000; \\ z(i,1) = \frac{data(i,3)}{100} \times 1000. \end{cases} \tag{3}$$

(3)   Sparsity of grid data

The minimum spacing between the point cloud data was 0.01 mm during the reconstruction process of the point cloud. A smaller spacing results in a denser point cloud, leading to a more realistic reconstructed surface topography. However, the density of the data grid directly exceeded the meshing capability of the ABAQUS preprocessing function, resulting in nonconvergence of the analysis calculation. Through repeated calculations, the minimum unit of the best grid element that allowed convergence was 5 mm. Since the scanned pavement texture data included a square pavement area of 30×30 cm^2, as shown in Section 3.1, the coordinate format of $-150{:}5{:}150$ was used.

After generating the datum coordinate points, the cubic interpolation method based on triangles was used to interpolate the elevation point *zk* through the "*zk = griddata (x, y, z, xi, yj, 'cubic')*" statement. After this interpolation, the sparsity of the road texture coordinate points was generated on the basis of 5 mm plane (*x, y*) coordinates. Considering the large difference between the stiffness of the road surface and the stiffness of the tire, the road surface could be regarded as a rigid body in the ABAQUS finite element model. The model of the road rut obtained using the above process was a rigid shell unit. Then, the grid of coordinate points was entered into the preprocessing .inp file in ABAQUS and visualized as shown in Figure 5b.

(4)   Extension processing of coordinate data

To provide enough rolling space for the tire in the FE model, it was necessary to expand the distribution characteristics of the pavement texture appropriately. The mirror image processing method was used to establish two mirror images for the coordinate points in step (3) to expand the simulated asphalt road to four times the original size along the tire rolling direction, i.e., 120 cm. This process was realized using the MATLAB "*mirror _ matrix*" statement.

(5)   Materialized processing of textured pavement

The rigid shell element cannot restrict the Euler fluid element; that is, the shell element cannot block the water film on the pavement surface. In explicit analysis, solid elements can play a role in blocking water. The digital model of the pavement surface texture generated in step (4) was extended along the *z*–direction by a certain distance to make it a 3D solid unit (in Figure 5c).

2.2.2. Tire Hydroplaning FE Model

The tire hydroplaning model building process in ABAQUS was based on the coupled Euler Lagrange algorithm, as described in [36,37], for which the inflated pattern tire Bridgestone 205–55–R16 was adopted [25]. Then, the dynamic friction coefficient curves (in Figure 6) between the tire and pavement under different road conditions were imported into the built hydroplaning model.

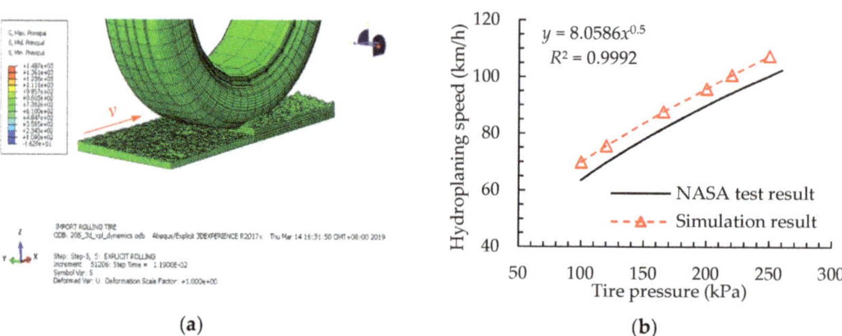

(**a**)   (**b**)

Figure 6. Validation of tire hydroplaning FE model: (**a**) simulation of tire hydroplaning process; (**b**) critical hydroplaning speed of the tire.

The critical hydroplaning speed of the tire calculated using the National Aeronautics and Space Administration (NASA) empirical formula was taken as the initial speed [38]. Then, the critical hydroplaning speed under different inflatable pressures was simulated by adjusting the rolling speed of the tires according to the tire hydroplaning FE model, as shown in Figure 6b. From the variation curve of the tire hydroplaning speed, it can be seen that, as the tire inflation pressure increased, the critical hydroplaning speed of the tire gradually increased. Moreover, it changed linearly with the square root of the tire pressure. According to the regression analysis, the predictive formula of the tire hydroplaning speed was obtained as $v_{crit} \approx 8.06 P^{0.5}$, which is consistent with the NASA test results.

Nevertheless, the fitting coefficient of the formula was slightly higher than that of the NASA empirical formula, which results from the difference in tire pattern parameters and road surface texture characteristics. However, the simulated critical hydroplaning speed was within the allowable range of error, verifying the accuracy of the hydroplaning FE model.

For the proposed hydroplaning model, Equation (4) was used to calculate the tire adhesion coefficient under wet conditions. In the model, the tire rolled in the z–direction; hence, the rolling resistance of the tire was calculated as a function of the joining force in the z–direction received by the reference point on the rim.

$$\mu_s = \frac{\mu(F_h - F_t) + F_d}{F_h} = \frac{F_z}{F_h}, \tag{4}$$

where μ_s is the tire adhesion coefficient under wet conditions, μ is the tire adhesion coefficient under dry conditions, F_d is the fluid drag force, F_h is the tire axle load, F_z is the tire rolling resistance, and F_t is the fluid lift force.

On a dry road, the adhesion coefficient is the tangential force divided by the normal load. The tire−pavement contact area contains the adhesion region and the slipping region. The tangential reaction force on the tire is the sum of the longitudinal force in the adhesion region and the sliding friction in the slipping region [39]. Hence, the adhesion coefficient can be calculated as follows:

$$\mu = \frac{F_{xn} + F_{xb}}{F_h} < \mu_f, \tag{5}$$

where μ is the adhesion coefficient under dry conditions, μ_f is the slipping friction coefficient, F_{xn} is the slipping friction in the adhesion region, and F_{xb} is the slipping friction in the slipping region.

3. Results

3.1. Pavement State Demarcation

According to the established tire hydroplaning FE model, the tire load of 3.922 kN and the inflation internal pressure of 250 kPa remained unchanged. The influence of the water film thickness on the critical hydroplaning speed for different asphalt pavements was analyzed. Considering the actual vehicle hydroplaning risk, the vehicle was deemed to already be in the hydroplaning state within the limit speed when the water film thickness was greater than 6 mm. Thus, the water film thickness within the range of 0–6 mm was considered. According to the definition of tire hydroplaning, the minimum driving speed is the critical hydroplaning speed when the tire–pavement contact force is equal to 0. The critical hydroplaning speed for different water film thicknesses was obtained as shown in Figure 7a.

Within a certain range of water film thickness, the trends of the critical hydroplaning speeds of the three typical asphalt pavements were generally consistent (Figure 7a). When the thickness of the water film was in the range of 0–2 mm, the critical hydroplaning speed of tire change rate was significant. When the water film thickness was greater than 2 mm, the critical hydroplaning speed curve tended to be stable. Driving at the specified speed limit of 120 km/h, the critical water film thickness h_{crit} for the three typical asphalt pavements for which tire hydroplaning occurred was as follows: AC, 0.56 mm; SMA, 0.76 mm; OGFC, 1.5 mm. This is mainly because the macro texture of the road surface provided a skid resistance force. In this case, the friction provided by the micro texture of the road surface was minimal. Obviously, when the water film thickness on the road surface was less than the critical value h_{crit} ($0 < h_{\text{w}} < h_{\text{crit}}$), the vehicle was not at risk of hydroplaning. Research results showed that the pavement conditions could be defined as the wet state when the water film thickness met the condition of $0 < h_{\text{w}} < h_{\text{crit}}$. Considering that the pattern depth limit of car tires is 1.6 mm, and that the highway speed limit is 120 km/h, the critical water film thickness h_{crit} was calculated to be 1.021 mm using Equation (6) [40,41]. Therefore, the wet state of the road surface could be defined as a water film thickness of $0 < h_{\text{w}} \leq 1$ mm.

$$v_{\text{p}} = v_{\text{crit}} + 12\frac{t}{h} + 60 \exp\left\{-3\left[h - \left(3 + \frac{t}{b}\right)\right]\right\}, \tag{6}$$

where v_{crit} is the critical hydroplaning speed of the tire ($v_{\text{crit}} = 6.35p^{1/2}$), p is the tire inflation pressure (kPa), t is the tread pattern depth (mm), h is the water film thickness on the road surface (mm), and b is the tread width of the tire (mm).

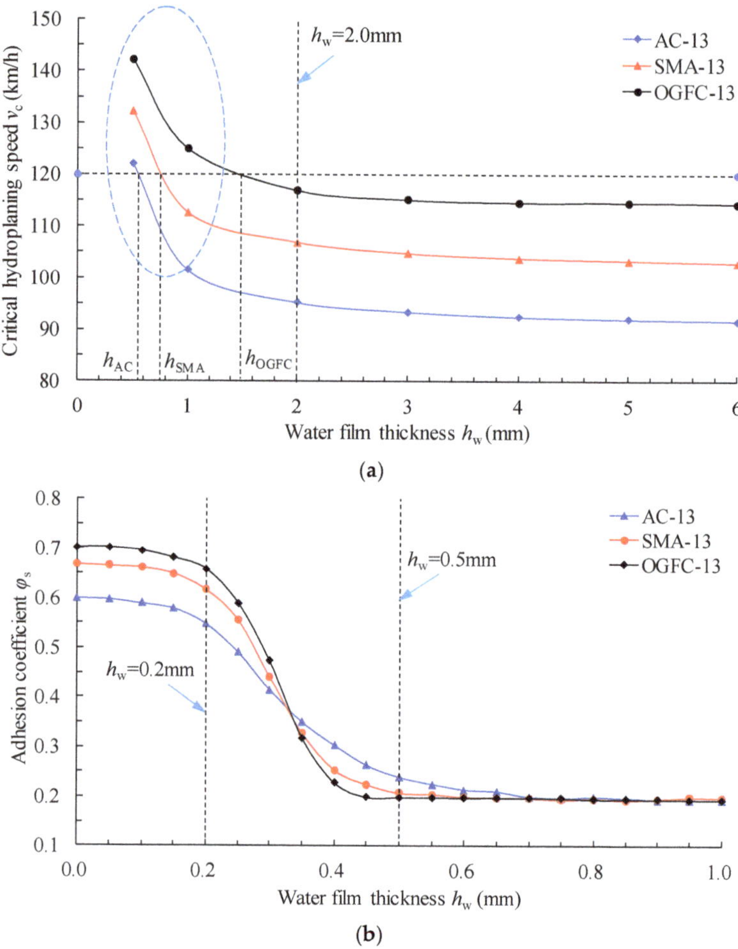

Figure 7. Influence of water film thickness on tire–pavement contact characteristics: (**a**) critical hydroplaning speed of the tire; (**b**) adhesion coefficient between tire and pavement.

According to the tire–pavement contact mechanism, there is a minimum skid resistance of the road surface before the critical water film thickness appears. In this case, the micro convex body on the road surface is completely wrapped by the water film, and the frictional resistance suffered by the tire is almost entirely derived from the viscous force of the lubricating medium (water). This critical state is called the dividing line between the wet state and ponding state, at which the skid resistance of the road surface is the worst. In view of this, it is necessary to refine the adhesion characteristics of the tire–fluid–road surface with a water film thickness of less than 1 mm from the micro contact theory.

Setting the initial speed of the tire to 60 km/h and keeping the other parameters unchanged, the water film thickness in the established hydroplaning FE model was adjusted. The influence of the water film thickness within the range of 0–1 mm on the adhesion coefficient was discretized. The variation of the adhesion coefficient curves with the water film thickness is shown in Figure 7b. It can be seen that the adhesion coefficient curves could be divided into three phases:

- Boundary lubrication stage (the water film thickness ranges from 0 to 0.2 mm). Here, the fluid lubrication effect was very small, but the road surface micro convex body had a large contribution rate. The friction between the tire and pavement depended on the asperity of the road surface.
- Mixed lubrication stage (the water film thickness ranges from 0.2 mm to 0.5 mm). Part of the road surface micro convex body was blocked by the water film and became an "anti−skid noncontributing area". At this time, the friction characteristics of the road surface were determined by the fluid viscosity and surface roughness.
- Elastic fluid lubrication stage (the water film thickness ranges from 0.5 mm to 1.0 mm). At this stage, the road surface micro convex body was completely submerged by the water film. However, the road surface water film thickness was low, and no dynamic water pressure was generated.

According to the theory of elastic hydrodynamic lubrication, the pavement could be divided into four states considering the water film thickness and the corresponding sources of the adhesion coefficient:

- In the dry state, i.e., $h_w = 0$ mm, the adhesion coefficient between the tire and pavement mainly depended on the texture characteristics of the road surface contact surface.
- In the wet state or moist state, i.e., $0 < h_w = 0.5$ mm, the adhesion coefficient was determined by the roughness of the road surface and the fluid viscosity.
- In the lubrication state, i.e., 0.5 mm $< h_w \leq 1$ mm, the adhesion coefficient was negligible, the adhesion force depended on the fluid viscosity resistance, and the tire slip rate increased rapidly.
- In the ponding state, i.e., $h_w > 1$ mm, the adhesion force of the tire depended entirely on the fluid viscosity resistance. At this time, there was a risk of tire hydroplaning.

According to the above analysis, different road states had different sources of tire–pavement adhesion coefficients, and the corresponding adhesion coefficient could be determined by the specific road surface type, water film thickness, vehicle speed, and tire inflation pressure. In a dry road state, the adhesion coefficient is mainly determined by the roughness of the road surface. On a wet road, the "anti−skid noncontributing area" of the road surface should be considered.

3.2. Influencing Factors of Adhesion Coefficient

3.2.1. In a Dry Road State

Through the ACRP system, three kinds of asphalt pavement texture information were acquired. The MATLAB program was applied to write a code for realizing 3D coordinate point data import and visualization of the asphalt pavement surface texture. Then, the mean texture depth (MTD) values were calculated; the visual interface is shown in Figure 8. Compared with the sand patch method, the results show that the errors were all within 5%, further verifying the accuracy of the MTD value obtained through the ACRP system, as shown in Table 2.

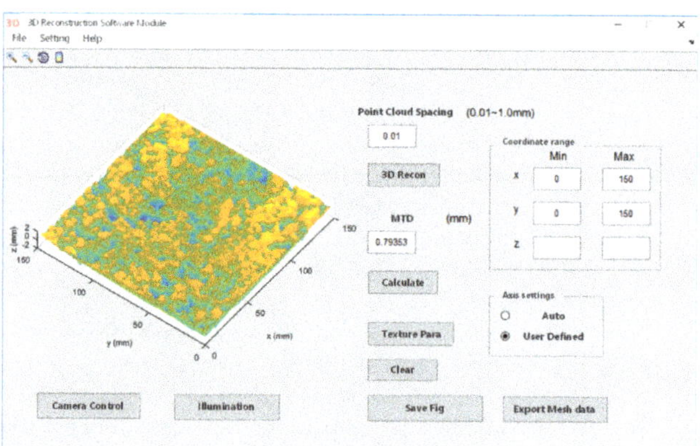

Figure 8. The 3D reconstruction software module.

Table 2. The MTD values for three kinds of asphalt pavement.

Asphalt Pavement Type	Specimen	Sand Patch Method (mm)	ACRP System (mm)	Relative Error (%)
AC−13	1	0.45	0.44	0.16
	2	0.48	0.48	0.37
	3	0.49	0.49	0.47
SMA−13	1	0.61	0.61	1.01
	2	0.75	0.74	0.71
	3	0.84	0.84	1.20
OGFC−13	1	0.86	0.85	0.81
	2	0.92	0.91	0.80
	3	0.93	0.93	0.48

The tire internal pressure was set to 240 kPa, and the load was set to 3.922 kN. According to the test data in Table 2, the MPD values of the road surface macro texture of 0.32 mm, 0.47 mm, 0.63 mm, 0.83 mm, 1.01 mm, and 1.21 mm were chosen to analyze the variation of the road surface adhesion coefficient under different driving speeds, as shown in Figure 9a. As the MPD value gradually increased, the adhesion coefficient between the tire and the road increased, and the increasing trend was more significant at higher driving speeds. For example, when the driving speed was 40 km/h, the adhesion coefficient increased by 33.7%, whereas it increased by 47.1% at a speed of 100 km/h. Apparently, the macro texture of the pavement helped to improve the adhesion of the pavement.

The tire speed was set to 60 km/h, and the load was set to 3.922 kN; then, the tire inflation pressure was adjusted in the established hydroplaning FE model. The influence of tire inflation pressure on the adhesion coefficient was investigated, and the results are shown in Figure 9b. It can be seen that the adhesion coefficient between the tire and the road surface increased with the increase in tire inflation pressure; the adhesion coefficient curves were distributed as a parabola, and the rate of increase decreased with the increase in the inflation pressure. Comparing the change trend of the adhesion coefficient of three typical asphalt pavements, it can be found that the OGFC asphalt pavement had the largest adhesion coefficient under the same tire inflation pressure, followed by SMA and AC asphalt pavement. A greater inflation pressure resulted in a greater change rate of the OGFC asphalt pavement adhesion coefficient.

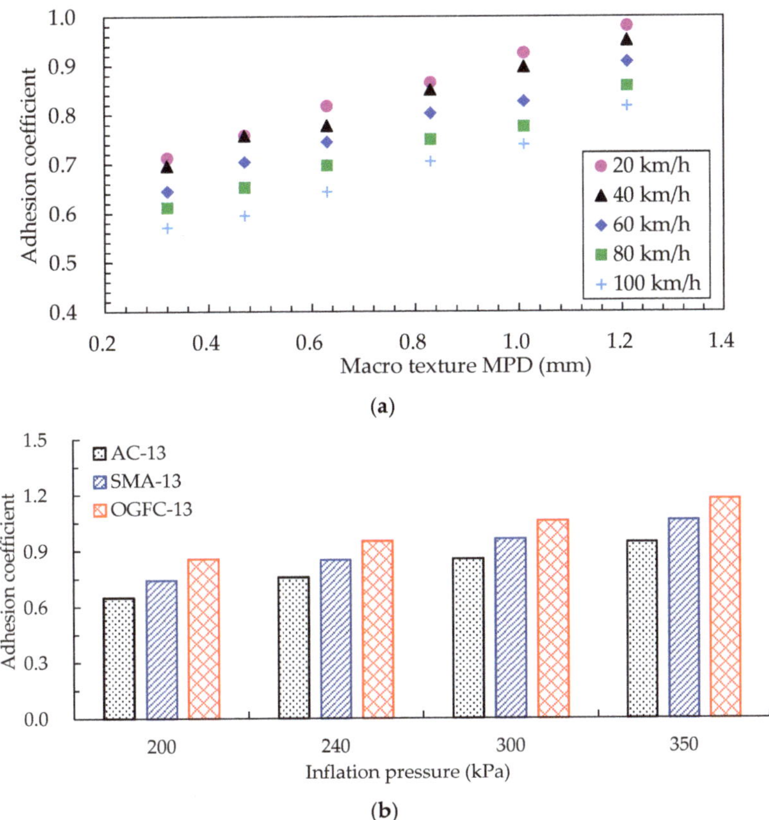

Figure 9. Influencing factors of adhesion coefficient in a dry road state: (**a**) influence of macro texture on adhesion coefficient; (**b**) influence of inflation pressure on adhesion coefficient.

3.2.2. Under Wet Road Conditions

The calculation of the road surface adhesion coefficient under different water film thicknesses was simulated using Equation (4). Furthermore, the influence of the speed and the macro texture of the road surface on the adhesion coefficient between the tire and the road surface was analyzed. The adhesion coefficient curves under different water film thicknesses were obtained as shown in Figure 10. In addition, the variable values of water film thickness covered all road states.

For certain macro texture parameters and rolling speeds, the road surface adhesion coefficient gradually decreased as the water film thickness increased. When the water film thickness was $h_w \leq 1.0$ mm, the pavement adhesion coefficient was larger and the change rate of the adhesion coefficient was higher as the macro texture increased. When the water film thickness was $h_w > 1.0$ mm, the adhesion coefficient gradually decreased. Moreover, the influence degree of the road surface macro texture decreased rapidly, indicating that the road surface adhesion characteristics mainly depended on the viscosity of the water flow when the water film thickness was greater than 1 mm (in a moist road state).

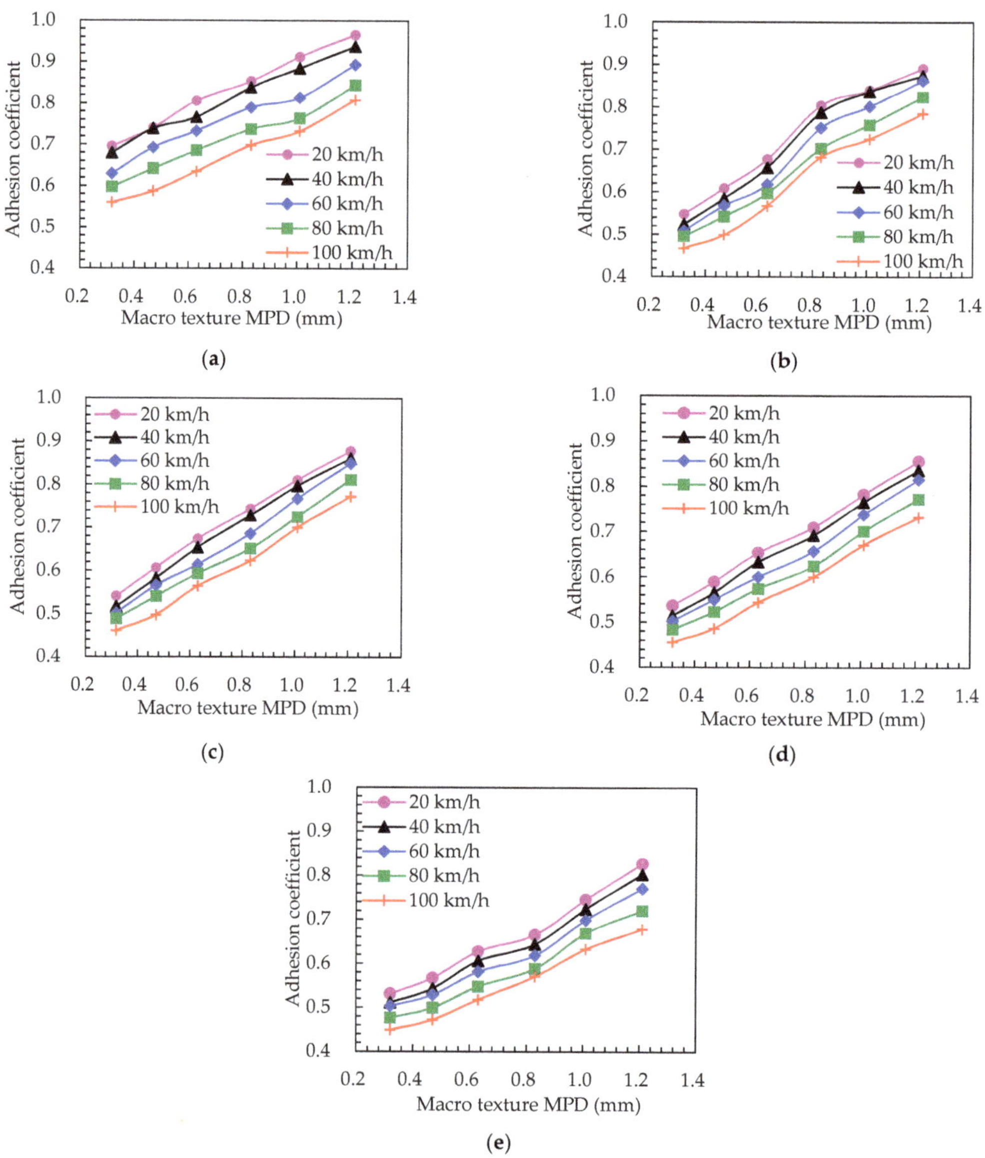

Figure 10. (**a**) h_w = 0.2 mm; (**b**) h_w = 1.0 mm; (**c**) h_w = 2.0 mm; (**d**) h_w = 3.5 mm; (**e**) h_w = 5.0 mm. Influence of water film thickness on adhesion coefficient in a wet road state.

4. Discussion

4.1. Sensitivity Analysis Based on Orthogonal Experimental Design

4.1.1. In a Dry Road State

Based on the simulation result of the adhesion coefficient in Section 3.2.1, the adhesion coefficient between the tire and pavement was taken as the objective function, the macro texture MPD, speed, and tire inflation pressure were considered as test factors in a dry road state (Table 3). An orthogonal table $L_{16}(4^5)$ is used to represent orthogonal experiment with five factors and four levels was selected to arrange the orthogonal test. Here, k_1, k_2, k_3, k_4, and k_5 represented the MPD value, tire speed, inflation pressure, and two blank columns, respectively. The analysis results of the orthogonal experimental design are shown in Table 4.

Table 3. Levels of influence factors in a dry road state.

No.	MPD Value (mm) k_1	Tire Speed (km/h) k_2	Tire Inflation Pressure (MPa) k_3
1	0.47	40	200
2	0.63	60	240
3	0.83	80	300
4	1.01	100	350

Table 4. Range and variance analysis of factors in a dry road state.

Analysis Item	Test Factors				
	k_1	k_2	k_3	k_4	k_5
x_1	0.670	0.813	0.742	0.740	0.745
x_2	0.716	0.778	0.741	0.749	0.748
x_3	0.785	0.719	0.744	0.751	0.751
x_4	0.810	0.671	0.756	0.742	0.738
R	0.140	0.142	0.015	0.011	0.013
df	3	3	3	3	3
SS	0.049	0.047	0.001	0.0014	0.0016
F	2.526	2.423	0.052	—	—
F_{crit} 0.10	2.490	2.490	2.490	—	—
Significance level	Significant	Insignificant	Insignificant	—	—

As shown in Table 3, the factors were arranged from significant to insignificant as follows: MPD value, speed, and tire inflation pressure. Moreover, the optimal combination of factors was k_1x_1, k_2x_4, and k_3x_2, and the most disadvantageous combination was k_1x_4, k_2x_1, and k_3x_4. According to variance analysis results, the significance level of each factor was consistent with the range analysis results. As shown in Table 4, the macro texture MPD value of the road surface was the most significant factor, whereas the influence of speed and tire inflation pressure could be neglected.

4.1.2. In a Wet Road State

Similarly, the adhesion coefficient between the tire and pavement was taken as the objective function in a wet road state according to the simulation results in Section 3.2.2. Furthermore, the water film thickness, macro texture MPD, and tire speed were considered as the test factors (Table 5). An orthogonal table $L_{25}(5^6)$ representing an orthogonal experiment with six factors and five levels was selected to arrange the orthogonal test. Here, k_1, k_2, k_3, k_4, k_5, and k_6 represented the water film thickness, macro texture MPD, tire speed, and three blank columns, respectively. The analysis results of the orthogonal experimental design are shown in Table 6.

Table 5. Levels of influence factors in a wet road state.

No.	Water Film Thickness (mm) k_1	MPD Value (mm) k_2	Tire Speed (km/h) k_3
1	0	0.32	20
2	0.2	0.47	40
3	1.0	0.63	60
4	2.0	0.83	80
5	5.0	1.01	100

Table 6. Range and variance analysis of factors in a wet road state.

Analysis Item	Test Factors					
	k_1	k_2	k_3	k_4	k_5	k_6
x_1	0.741	0.566	0.722	0.665	0.662	0.673
x_2	0.733	0.611	0.714	0.668	0.660	0.659
x_3	0.630	0.655	0.665	0.667	0.675	0.677
x_4	0.631	0.708	0.627	0.673	0.666	0.653
x_5	0.582	0.777	0.590	0.645	0.654	0.650
R	0.159	0.211	0.132	0.028	0.021	0.027
df	4	4	4	4	4	4
SS	0.099	0.136	0.063	0.00074	0.00055	0.00071
F	1.954	2.684	1.243	1	1	1
F_{crit} 0.10	2.190	2.190	2.190	—	—	—
Significance level	Insignificant	Significant	Insignificant	—	—	—

According to the results of orthogonal experimental, it can be seen that the macro texture of road surface was the most significant factor, followed by water film thickness and tire speed, which is consistent with the results in a dry road state. Whether in a dry state or wet state, the macro texture parameter of the road surface was the most significant factor. Thus, the three typical kinds of asphalt pavement were selected as the macro texture parameter variables to analyze the change rule of the peak adhesion coefficient.

4.2. Peak Adhesion Coefficient for Different Road States

4.2.1. In a Dry Road State

Considering the significant influencing factors of the adhesion coefficient, only the adhesion coefficient variation of different asphalt pavement types with tire slip rates was considered in this study. The test speed of the tire was defined as 60 km/h, and the simulation results are shown in Figure 11a. When the slip rate was around 15%, the adhesion coefficient reached the maximum value, indicating the peak adhesion coefficient.

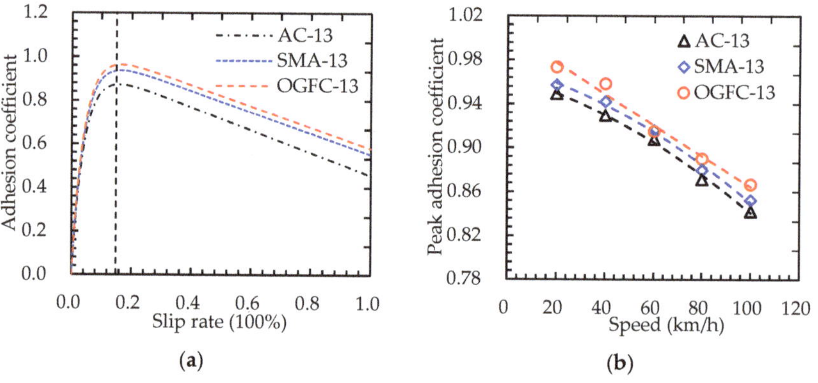

Figure 11. Adhesion characteristics of asphalt pavement in a dry road state: (**a**) variation of adhesion coefficient with slip rate; (**b**) peak adhesion coefficient.

The tire slip rate was maintained in the optimal range (around 15%), and the rolling speed of the tire model was adjusted while keeping other parameters constant. Then, the peak adhesion coefficient curves for different asphalt pavements were obtained as shown in Figure 11b. The peak adhesion coefficient of the road surface gradually decreased with the increase of vehicle speed, and the peak adhesion coefficient of the road surface was distributed as a convex parabola. This is mainly because the tire rolling radius became large at high speed, resulting in a greater contact area between tire and road surface. Thus, the adhesion force provided by the road surface was reduced.

4.2.2. In a Wet Road State

Considering the influencing factors of adhesion characteristics under wet road conditions analyzed in Section 3.2.2, it can be seen that the adhesion coefficient in a wet state ($0 < h_\mathrm{w} \leq 0.5$ mm) was the largest. With the accumulation of rainfall and time, the thickness of the surface water film increases gradually. Under the effect of water lubrication, the road surface texture is "sealed" and cannot provide frictional resistance. The adhesion force between the tire and the road surface mainly results from the fluid viscous force. Thus, the adhesion coefficient of the road surface in a typical wet road state ($h_\mathrm{w} = 0.2$ mm) was taken as research object. The influence of macroscopic texture parameters on the peak adhesion coefficient of the road surface was analyzed. Three typical kinds of asphalt pavement characterizing different macro texture intervals were selected to obtain the variation of adhesion coefficient curves with the tire slip rate, as shown in Figure 12a. With an increase in the tire slip rate, the adhesion coefficient between the tire and pavement quickly increased and then gradually decreased. When the slip rate was around 11.5%, the adhesion coefficient curves attained their peak.

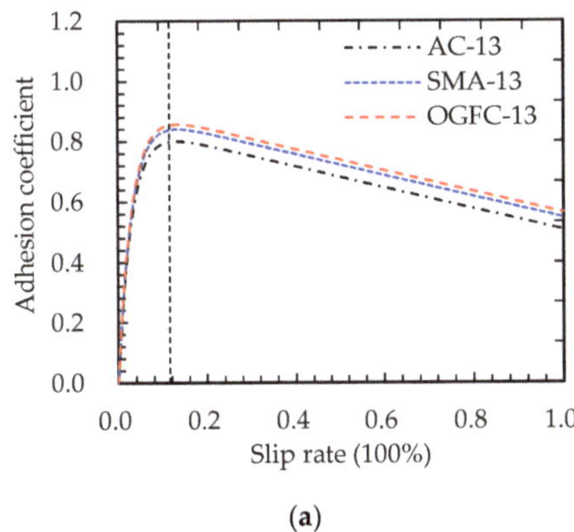

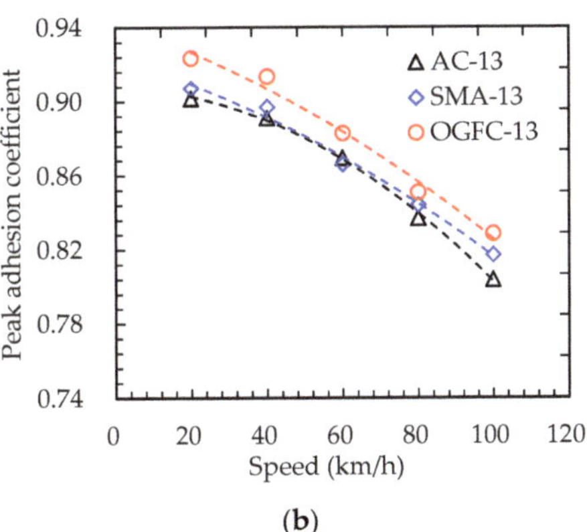

Figure 12. Peak adhesion coefficient curves of asphalt pavement: (**a**) adhesion coefficient varied with slip rate; (**b**) peak adhesion coefficient.

Similarly, the tire slip rate was maintained in the optimal range (around 11.5%), and the rolling speed of the tire model was adjusted while keeping other parameters constant. According to the simulation results, the peak adhesion coefficient curves for three kinds of asphalt pavement under different vehicle speeds were obtained, as shown in Figure 12b.

It can be seen that the peak adhesion coefficients for all three kinds of asphalt pavement at the same vehicle speed had the following relationship in a wet road state: OGFC pavement > SMA pavement > AC pavement. Moreover, the peak adhesion coefficient on OGFC pavement in a dry road state was larger than that on AC pavement in a wet road state. Apparently, the peak adhesion coefficient in a wet road state was slightly lower than that in a dry road state, which was mainly determined by the contribution rate of the pavement texture.

5. Conclusions

In this study, an automatic close-range photogrammetry system was adopted to acquire asphalt pavement texture information. Then, the road power spectrum density curve was calculated using MATLAB according to Persson friction theory. Using the established tire hydroplaning model, the influence of water film thickness on the tire–pavement contact characteristics was analyzed, and the road states were demarcated according to fluid hydrodynamic lubrication theory. For different road states, the significant influencing factors of the tire–pavement adhesion characteristics were investigated using an orthogonal experimental design, and the peak adhesion coefficient was calculated. The main results are presented below.

1. For both dry and wet road conditions, the adhesion coefficient of different types of pavements could be ordered as OGFC pavement > SMA pavement > AC pavement. The change rules of the adhesion coefficient curves under different pavement conditions were basically similar, both decreasing significantly with the increase in relative slipping speed. When the speed exceeded 40 km/h, the curves tended to be gentle. For the insulation and lubrication of the water film, there existed an "anti-skid noncontribution area".

2. When the thickness of the water film was in the range of 0–2 mm, the critical hydroplaning speed of tire change rate was significant. When the water film thickness was greater than 2 mm, the critical hydroplaning speed curve tended to be stable. When driving at the specified speed limit of 120 km/h, the critical water film thickness for the three typical asphalt pavements that resulted in hydroplaning was as follows: AC, 0.56 mm; SMA, 0.76 mm; OGFC, 1.5 mm. According to the theory of elastic hydrodynamic lubrication, the pavement could be divided into four states: dry state, wet sate, lubricated state, and ponding state.

3. In a dry road state, as the MPD value gradually increased, the adhesion coefficient between the tire and the road increased, and the increasing trend was more significant at higher driving speeds. Upon increasing the tire inflation pressure, the adhesion coefficient increased and was distributed as a parabola. The OGFC asphalt pavement had the largest adhesion coefficient under the same tire inflation pressure, followed by SMA and AC asphalt pavement. A greater inflation pressure led to a greater change rate of the adhesion coefficient for the OGFC asphalt pavement.

4. In a wet road state, for certain macro texture parameters and rolling speed, the road surface adhesion coefficient gradually decreased as the water film thickness increased. When the water film thickness was $h_w \leq 1.0$ mm, the pavement adhesion coefficient was larger and the change rate of the adhesion coefficient was higher as the macro texture increased. When the water film thickness was $h_w > 1.0$ mm, the adhesion coefficient gradually decreased.

5. According to the sensitivity analysis results, whether in a dry state or wet state, the macro texture parameter of the road surface was the significant factor. In a dry road state, when the slip rate was around 15%, the adhesion coefficient reached the peak value, i.e., around 11.5% for the wet road state. At the same speed, the peak adhesion coefficient for different asphalt pavements could be ordered as OGFC pavement > SMA pavement > AC pavement for both the dry state and the wet state. Moreover, the peak adhesion coefficient on OGFC pavement in a dry road state was larger than that on AC pavement in a wet road state.

In this paper, a field measurement and FE simulation were combined to reveal the tire–pavement adhesion characteristics under different road states, which can provide a comprehensive information for explaining the tire–pavement interaction. As a function of the different road surface peak adhesion coefficients μ_{max} obtained, the desired braking deceleration a_{des} of autonomous vehicles can be calculated using the formula $a_{des} = \mu_{max} \, g$. Thus, the research results in this paper can provide key parameters for further research on autonomous braking performance.

Author Contributions: Conceptualization, B.Z. and J.T.; methodology, J.C.; software, R.Z.; validation, B.Z.; formal analysis, J.T.; investigation, J.C.; resources, B.Z.; writing—original draft preparation, B.Z.; writing—review and editing, J.T.; visualization, J.C.; supervision, R.Z. and X.H.; project administration, X.H. and B.Z. All authors have read and agreed to the published version of the manuscript.

Funding: The research was financially supported by the Natural Science Research Start-up Foundation of Recruiting Talents of Nanjing University of Posts and Telecommunications (Grant No. NY221150) and the National Natural Science Foundation of China (Grant No. 51778139).

Institutional Review Board Statement: Not applicable.

Informed Consent Statement: Not applicable.

Data Availability Statement: The data used to support the findings of this study are available from the first author upon request. In this paper, some models and codes used during the study are proprietary or confidential in nature and may only be provided with restrictions, such as the PSD calculation code foe MATLAB software and the subroutine of the tire–pavement interface in ABAQUS.

Conflicts of Interest: The authors declare no conflict of interest.

References

1. Zhang, L. Research on Key Technology of Active Collision Avoidance Braking System Based on Pavement Identification. Master's Thesis, Jiangsu University, Zhenjiang, China, 2017.
2. Barber, J.R. *Contact Mechanics, Soild Mechanics and Its Applications*; Springer International Publishing: Berlin, Germany, 2018.
3. Zhang, B. Tire Tread Pattern Optimization with Regard to Adhesion on Wet Surface. Master's Thesis, South China University of Technology, Guangzhou, China, 2013.
4. Hall, J.W.; Smith, K.L.; Titus-Glover, L.; Wambold, J.C.; Yager, T.J.; Rado, Z. Guide for Pavement Friction. In *Contractors Final Report for NCHRP Project 01-43*; Transportation Research Board: Washington, DC, USA, 2009; p. 108.
5. Al-Assi, M.; Kassem, E. Evaluation of Adhesion and Hysteresis Friction of Rubber—Pavement System. *Appl. Sci.* **2017**, *7*, 1029. [CrossRef]
6. Huang, X.; Zheng, B. Research Status and Progress for Skid Resistance Performance of Asphalt Pavements. *China J. Highw. Transp.* **2019**, *32*, 32–49.
7. Vu, T.D.; Duhamel, D.; Abbadi, Z.; Yin, H.P.; Gaudin, A. A nonlinear Circular Ring Model with Rotating Effects for Tire Vibrations. *J. Sound Vib.* **2017**, *388*, 245–271. [CrossRef]
8. Jiang, X.; Zeng, C.; Gao, X.; Liu, Z.; Qiu, Y. 3D Fem Analysis of Flexible Base Asphalt Pavement Structure under Non-uniform Tyre Contact Pressure. *Int. J. Pavement Eng.* **2019**, *20*, 999–1011. [CrossRef]
9. Xiao, S.; Zhao, F.; Zhou, X.; Tan, Y.; Li, J. Spatiotemporal Evolution Analysis of Pavement Texture Depth on RIOH Track Using Statistical and Rescaled Range Approaches. *Constr. Build. Mater.* **2022**, *338*, 127560. [CrossRef]
10. Persson, B.N.J.; Scaraggi, M. Lubricated Sliding Dynamics: Flow Factors and Stribeck Curve. *Eur. Phys. J. E* **2011**, *34*, 113–119. [CrossRef]
11. Priyantha, W.J.; Gary, L.G. Use of Skid Performance History as Basis for Aggregate Qualification for Seal Coats and Hot-Mix Asphalt Concrete Surface Courses. *Transp. Res. Rec.* **1995**, *1501*, 31–38.
12. Singh, D.; Patel, H.; Habal, A.; Das, A.K.; Kapgate, B.P.; Rajkumar, K. Evolution of Coefficient of Friction between Tire and Pavement under Wet Conditions Using Surface Free Energy Technique. *Constr. Build. Mater.* **2019**, *204*, 105–112. [CrossRef]
13. Bachmann, T. Wechselwirkungen Im Prozeß der Reibung Zwischen Reifen und Fahrbahn. Ph.D. Thesis, TU Darmstadt, Düsseldorf, Germany, 1998.
14. Germann, S.; Wurtenberger, M.; Daiss, A. Monitoring of the Friction Coefficient between Tire and Road Surface. In Proceedings of the Third IEEE Conference on Control Applications, Glasgow, UK, 24–26 August 1994.
15. Do, M.; Cerezo, V.; Beautru, Y.; Kane, M. Influence of Thin Water Film on Skid Resistance. *J. Traffic. Transp. Eng.* **2014**, *2*, 36–44.
16. Wang, K.C.P.; Luo, W.; Li, J.Q. Hydroplaning Risk Evaluation of Highway Pavements Based on IMU and 1 mm 3D Texture Data. In Proceedings of the Second Transportation & Development Congress, Orlando, FL, USA, 8–11 June 2014; pp. 511–522.
17. Christensen, H. Stochastic Models for Hydrodynamic Lubrication of Rough Surfaces. *Proc. Inst. Mech. Eng.* **1969**, *184*, 1013–1026. [CrossRef]
18. Lugt, P.M.; Morales-Espejel, G.E. A Review of Elasto-Hydrodynamic Lubrication Theory. *Tribol. Trans.* **2011**, *54*, 470–496. [CrossRef]
19. Johannesson, P.; Rychlik, I. Laplace Processes for Describing Road Profiles. *Procedia Eng.* **2013**, *66*, 464–473. [CrossRef]
20. Li, R. Study on Unsaturated Hydraulic Asphalt Concrete Pavement Internal Hydrodynamic Pressure Distribution. Master's Thesis, Chongqing Jiaotong University, Chongqing, China, 2013.
21. Yu, L.; Hu, J.; Li, R.; Yang, Q.; Guo, F.; Pei, J. Tire-Pavement Contact Pressure Distribution Analysis Based on ABAQUS Simulation. *Arab. J. Sci. Eng.* **2022**, *47*, 4119–4132. [CrossRef]
22. Yu, M.; You, Z.; Wu, G.; Kong, L.; Gao, J. Measurement and Modeling of Skid Resistance of Asphalt Pavement: A Review. *Constr. Build. Mater.* **2020**, *260*, 119878. [CrossRef]
23. Ong, G.P.; Fwa, T.F. A Mechanistic Interpretation of Braking Distance Specifications and Pavement Friction Requirements. *Transp. Res. Rec.* **2010**, *2155*, 145–157. [CrossRef]
24. Fwa, T.F.; Pasindu, H.R.; Ong, G.P. Critical Rut Depth for Pavement Maintenance Based on Vehicle skidding and hydroplaning consideration. *J. Transp. Eng.* **2012**, *138*, 423–429. [CrossRef]
25. Zheng, B.; Chen, J.; Zhao, R.; Tang, J.; Tian, R.; Zhu, S.; Huang, X. Analysis of Contact Behaviour on Patterned Tire-Asphalt Pavement with 3-D FEM Contact Model. *Int. J. Pavement Eng.* **2022**, *23*, 171–186. [CrossRef]

26. Zheng, B.; Huang, X.; Zhang, W.; Zhao, R.; Zhu, S. Adhesion Characteristics of Tire-Asphalt Pavement Interface Based on A Proposed Tire Hydroplaning Model. *Adv. Mater. Sci. Eng.* **2018**, *2018*, 5916180. [CrossRef]
27. Novikov, I.; Lazarev, D. Experimental Installation for Calculation of Road Adhesion Coefficient of Locked Car Wheel. *Transp. Res. Procedia* **2017**, *20*, 463–467. [CrossRef]
28. Ma, B.; Lv, C.; Liu, Y.; Zheng, M.; Yang, Y.; Ji, X. Estimation of Road Adhesion Coefficient Based on Tire Aligning Torque Distribution. *J. Dyn. Syst. Meas. Control* **2018**, *140*, 051010. [CrossRef]
29. Chen, B.; Zhang, X.; Yu, J.; Wang, Y. Impact of Contact Stress Distribution on Skid Resistance of Asphalt Pavements. *Constr. Build. Mater.* **2017**, *133*, 330–339. [CrossRef]
30. Guo, K.H. Uni-Tire: Unified Tire Model. *Chin. J. Mech. Eng.* **2016**, *52*, 90–99. [CrossRef]
31. Chen, J.; Huang, X.; Zheng, B.; Zhao, R.; Liu, X.; Cao, Q.; Zhu, S. Real-Time Identification System of Asphalt Pavement Texture Based on the Close-Range Photogrammetry. *Constr. Build. Mater.* **2019**, *226*, 910–919. [CrossRef]
32. Granshaw, S.I. Close Range Photogrammetry: Principles, Methods and Applications. *Photogramm. Rec.* **2010**, *25*, 203–204. [CrossRef]
33. Tanaka, H.; Yoshimura, K.; Sekoguchi, R.; Aramaki, J.; Hatano, A.; Izumi, S.; Sakai, S.; Kadowaki, H. Prediction of the Friction Coefficient of Filled Rubber Sliding on Dry and Wet Surfaces with Self-Affine Large Roughness. *Mech. Eng. J.* **2016**, *3*, 15-00084. [CrossRef]
34. Persson, B.N.J. On the Fractal Dimension of Rough Surfaces. *Tribol. Lett.* **2014**, *54*, 99–106. [CrossRef]
35. Michele, C. A Simplified Version of Persson's Multiscale Theory for Rubber Friction Due to Viscoelastic Losses. *J. Tribol.* **2018**, *140*, 011403.
36. Zhu, S.; Liu, X.; Cao, Q.; Huang, X. Numerical Study of Tire Hydroplaning Based on Power Spectrum of Asphalt Pavement and Kinetic Friction Coefficient. *Adv. Mater. Sci. Eng.* **2017**, *2017*, 5843061. [CrossRef]
37. Zheng, B.S.; Zhu, S.Z.; Chen, Y.Z.; Huang, X.M. Influence Factors Analysis of Adhesion Characteristic on Tire-Asphalt Pavement Based on Tire Hydroplaning Model. *J. Southeast Univ. Nat. Sci. Ed.* **2018**, *48*, 719–725.
38. Yager, T.J. *Comparative Braking Performance of Various Aircraft on Grooved and Ungrooved Pavements at the Landing Research Runway*; Pavement Grooving and Traction Studies; NASA Wallops Station: Wallops Island, VA, USA, 1969.
39. Zhuang, J. *Vehicle Tire Mechanics*; Beijing Institute of Technology Press: Beijing, China, 2009.
40. Seta, E.; Nakajima, Y.; Kamegawa, T.; Ogawa, H. Hydroplaning Analysis by FEM and FVM: Effect of Tire Rolling and Tire Pattern on Hydroplaning. *Tire Sci. Technol.* **2000**, *28*, 140–156. [CrossRef]
41. Zhang, H. Study on Braking Behavior of Typical Vehicle under Moist or Water Asphalt Pavement Condition. Master's Thesis, Southeast University, Nanjing, China, 2016.

MDPI

St. Alban-Anlage 66

4052 Basel

Switzerland

Tel. +41 61 683 77 34

Fax +41 61 302 89 18

www.mdpi.com

Materials Editorial Office

E-mail: materials@mdpi.com

www.mdpi.com/journal/materials